Resilience and Risk

NATO Science for Peace and Security Series

This Series presents the results of scientific meetings supported under the NATO Programme: Science for Peace and Security (SPS).

The NATO SPS Programme supports meetings in the following Key Priority areas: (1) Defence Against Terrorism; (2) Countering other Threats to Security and (3) NATO, Partner and Mediterranean Dialogue Country Priorities. The types of meetings supported are generally "Advanced Study Institutes" and "Advanced Research Workshops". The NATO SPS Series collects together the results of these meetings. The meetings are co-organized by scientists from NATO countries and scientists from NATO's "Partner" or "Mediterranean Dialogue" countries. The observations and recommendations made at the meetings, as well as the contents of the volumes in the Series, reflect those of participants and contributors only; they should not necessarily be regarded as reflecting NATO views or policy.

Advanced Study Institutes (ASI) are high-level tutorial courses to convey the latest developments in a subject to an advanced-level audience.

Advanced Research Workshops (ARW) are expert meetings where an intense but informal exchange of views at the frontiers of a subject aims at identifying directions for future action.

Following a transformation of the programme in 2006, the Series has been re-named and re-organised. Recent volumes on topics not related to security, which result from meetings supported under the programme earlier, may be found in the NATO Science Series.

The Series is published by IOS Press, Amsterdam, and Springer, Dordrecht, in conjunction with the NATO Emerging Security Challenges Division.

Sub-Series

A.	Chemistry and Biology	Springer
B.	Physics and Biophysics	Springer
C.	Environmental Security	Springer
D.	Information and Communication Security	IOS Press
E.	Human and Societal Dynamics	IOS Press

http://www.nato.int/science
http://www.springer.com
http://www.iospress.nl

Series C: Environmental Security

Resilience and Risk

Methods and Application in Environment, Cyber and Social Domains

edited by

Igor Linkov
US Army Corps of Engineers, Concord, Massachusetts, USA

and

José Manuel Palma-Oliveira
University of Lisbon, Lisbon, Portugal

Published in Cooperation with NATO Emerging Security Challenges Division

Proceedings of the NATO Advanced Research Workshop on
Resilience-Based Approaches to Critical Infrastructure
Safeguarding
Azores, Portugal
26–29 June 2016

Library of Congress Control Number: 2017946847

ISBN 978-94-024-1126-3 (PB)
ISBN 978-94-024-1122-5 (HB)
ISBN 978-94-024-1123-2 (e-book)
DOI 10.1007/978-94-024-1123-2

Published by Springer,
P.O. Box 17, 3300 AA Dordrecht, The Netherlands.

www.springer.com

Printed on acid-free paper

All Rights Reserved
© Springer Science+Business Media B.V. 2017
This work is subject to copyright. All rights are reserved by the Publisher, whether the whole or part of the material is concerned, specifically the rights of translation, reprinting, reuse of illustrations, recitation, broadcasting, reproduction on microfilms or in any other physical way, and transmission or information storage and retrieval, electronic adaptation, computer software, or by similar or dissimilar methodology now known or hereafter developed.
The use of general descriptive names, registered names, trademarks, service marks, etc. in this publication does not imply, even in the absence of a specific statement, that such names are exempt from the relevant protective laws and regulations and therefore free for general use.
The publisher, the authors and the editors are safe to assume that the advice and information in this book are believed to be true and accurate at the date of publication. Neither the publisher nor the authors or the editors give a warranty, express or implied, with respect to the material contained herein or for any errors or omissions that may have been made. The publisher remains neutral with regard to jurisdictional claims in published maps and institutional affiliations.

Preface

As society relies more upon integrated cyber-physical systems, the potential for security systems breaches increases. As new safeguards are developed and implemented, adversaries continue to develop new ways to breach and disrupt critical infrastructure. While significant advances in the field of risk assessment have been achieved, risk-based solutions tend to focus on assessing and strengthening individual components of complex systems under specific threat scenarios. Realization of the inability to predict threats resulted in significant interest in resilience-based management. The US National Academy of Sciences (NAS) defines resilience as "the ability to prepare and plan for, absorb, recover from, and more successfully adapt to adverse events." This definition calls for a system view of resilience and provides the basis for several interagency efforts in the USA and EU on developing metrics for resilience management and for integrating temporal capacity of a system to absorb and recover from heterogeneous adverse events and then to adapt; resilience provides an entity with the ability to repair, replace, patch, or otherwise reconstitute lost capability or performance in physical, cyber, social, and cognitive domains. Resilience thus uses strategies of adaptation and mitigation to augment traditional risk management.

The idea for this book was conceived at the NATO Advanced Research Workshop (ARW) on "Resilience-Based Approaches to Critical Infrastructure Safeguarding." This meeting—held in Azores, Portugal, in June 2016—focused on ways in which military commanders and civilian decision-makers could utilize resilience management in their operations. Military and civilian applications of resilience concepts are concerned with similar threats and need to be harmonized. Ongoing attacks require immediate response and thus real-time decision-making; resilience, as a property of a system, must transition from concepts and definitions to an operational paradigm for system management, especially under emergent and future threats. Methods and tools that are able to reconcile conflicting information, as well as the complex context of the decision-making environment, will be discussed.

The organization of the book reflects major topic sessions and discussions during the workshop. Workshop participants were organized into four working groups, which addressed (1) the methodologies of risk- and resilience-based management for critical infrastructure, as well as (2) infrastructure, (3) cybersecurity, and (4) social domain aspects of critical infrastructure risk- and resilience-based management. The papers in Parts I, II, III, and IV review and summarize the risk- and resiliency-based management approaches discussed for each of the four working group domains from the NATO workshop. Part V of the book provides a series of studies that illustrate applications of resiliency-based approaches for critical infrastructure use and needs across the world. The first chapter of Parts I, II, III, and IV starts with a group report summarizing the consensus principles and initiatives for each of the working groups at the NATO workshop. Each part of the book reviews achievements, identifies gaps in current knowledge, and suggests priorities for future research in topical areas. The wide variety of content in the book reflects the workshop participants' diverse views as well as their regional concerns.

The workshop discussions and papers in the book clearly illustrate that, while existing risk assessment and risk management frameworks provide a starting point for addressing risks, emerging risks such as terrorism, new technologies, and climate change add a significant level of complexity to this process. Resilience is emerging as a complementary tool to risk assessment that can be used to address these challenges. The goals of the workshop included the identification of strategies and tools that could be implemented to reduce technical uncertainty and prioritize research to address the immediate needs of the regulatory and risk assessment communities. The papers in the book illustrate the various applications of advanced risk assessment, resilience assessment and management, policy and applications ranging from environmental management all the way to cybersecurity, and other approaches to assist researchers and policymakers with benefiting the world at large.

Concord, MA, USA — Igor Linkov
Lisbon, Portugal — José Manuel Palma-Oliveira
July 2017

Acknowledgments

We would like to acknowledge Dr. Bojan Srdjevic (NATO workshop codirector) for his help in the organization of the event that resulted in this book. We also wish to thank the workshop participants and invited authors for their contributions to the book. We are deeply grateful to Kelsey Poinsatte-Jones and Valerie Zemba for their editorial and technical assistance. Additional technical assistance in the workshop organization and facilitation was provided by Decision Partners (Ms. Sarah Thorne). The workshop agenda was prepared in collaboration with the Society of Risk Analysis; we would like to express our special thanks to Prof. James Lambert, SRA president at the time of the meeting. Financial support for the workshop was provided by NATO. Additional support was provided by the US Army Engineer Research and Development Center, Future Resilient Systems at the Singapore-ETH Centre (SEC), which is funded by the National Research Foundation of Singapore (NRF) under its Campus for Research Excellence and Technological Enterprise (CREATE) programme, and Factor Social. We are forever in debt to our spouses, Elena Belinkaia and Margarida Goulão, for all of their patience in dealing with the demanding nature of the conference organization and book preparation.

Contents

Part I
Introduction

Chapter 1
An Introduction to Resilience for Critical Infrastructures

Igor Linkov and José Manuel Palma-Oliveira

Abstract Wide ranging and uncertain threats to public health, energy networks, cybersecurity, and many other interconnected facets of infrastructure and human activity, are driving governments, including those of the United States, European Union and elsewhere to further efforts to bolster national resilience and security. Resilience offers the capability to better review how systems may continually adjust to changing information, relationships, goals, threats, and other factors in order to adapt in the face of change and uncertainty – particularly those potential changes that could yield negative outcomes. Specific to this need, fifty scholars and practitioners of risk and resilience analysis from some twenty countries met in Ponta Delgada in the Azores Islands from June 26 – 29, 2016 to discuss the challenges associated with the emerging science of resilience theory and applications. Sponsored and funded in part by the North Atlantic Treaty Organization (NATO) Science for Peace and Security Programme, the overall topic of this meeting was "Resilience-Based Approaches to Critical Infrastructure Safeguarding." The workshop focused on ways in which military commanders and civilian decision makers alike could utilize resilience analysis in operations. More specifically, workshop discussion centered on both general resilience theory and analysis as well as various applications of resilience in topics ranging from cybersecurity to infrastructure resilience to ecosystem health. This chapter serves as a general introduction to the perspectives of various participants, as well as a reflection of discussion regarding how resilience thinking and analysis may be applied to critical infrastructure in various applications.

Keywords Resilience • Risk • Critical infrastructure • Network science • Policy • Communication

---

I. Linkov (✉)
US Army Corps of Engineers Research and Development Center, Concord, MA, USA
e-mail: Igor.Linkov@usace.army.mil

J.M. Palma-Oliveira
University of Lisbon, Lisbon, Portugal

© Springer Science+Business Media B.V. 2017

I. Linkov, J.M. Palma-Oliveira (eds.), *Resilience and Risk*, NATO Science for Peace and Security Series C: Environmental Security,
DOI 10.1007/978-94-024-1123-2_1

Wide ranging and uncertain threats to public health, energy networks, cybersecurity, and many other interconnected facets of infrastructure and human activity, are driving governments, including those of the United States, European Union and elsewhere to further efforts to bolster national resilience and security. Concerns arise from an increasingly interconnected world, where infrastructure systems rely on novel technologies that, while expanding services and promoting system maturation and growth, expose such systems to new and cascading risks that could devastate the normal functioning of important systems. Such risks – ranging from cybersecurity to loss of biodiversity to important ecosystem services – represent growing challenges for risk managers in the twenty-first century. They require developing conventional risk management strategies, but also resilience-driven strategies to adequately protect against undesirable consequences of uncertain, unexpected and often dramatic events.

The National Academy of Sciences (NAS) defines disaster resilience as "the ability to plan and prepare for, absorb, recover from, and adapt to adverse events" (NAS 2012). The NAS definition highlights a societal need to address highly uncertain and consequential risk events that are not easily addressed through traditional approaches of risk management. With this in mind, the paragraph above defines a scientific challenge about complexity, interdependencies, forms of adaptation, scale that requires a new synthesis across complexity, biology, computers, social and cognitive sciences. Connecting the science challenge to the societal need will require engineering advances — and critically those advances will necessarily bridge the traditional divide between engineering disciplines and social sciences.

With this in mind, decision makers and policymakers have utilized the concept of resilience to evaluate the capability of various complex systems to maintain safety, security and flexibility, and recover from a range of potential adverse events. Further, resilience offers the capability to better review how systems may continually adjust to changing information, relationships, goals, threats, and other factors in order to adapt in the face of change – particularly those potential changes that could yield negative outcomes. Preparation for reducing the negative consequences of such events when they occur is generally thought to include enhancing resilience of systems in desirable states, and have been described as including considerations of risk assessment as well as necessary resilience actions before, during, and after a hazardous event takes place. As such, resilience efforts inherently consider the passage of time and shifting capabilities and risks that may accrue due to changes in system performance and capacity to absorb shocks. Resilience strategies have the potential to radically change how a nation prepares itself for the potential disruptions of key services such as its energy, water, transportation, healthcare, communication and financial services. When nations prepare for recovery from external shocks of a significant magnitude, resilience strategies must be considered.

Despite the promise of resilience analysis to improve the safety and security of the variety of industries mentioned, and others, the field remains relatively new to the risk management community. Some risk managers oppose risk and resilience, some articulate the two concepts for their complementarity, some say that risk is part of resilience, others say that resilience is part of risk. One recurring complication is the lack of standardization in the field. Practitioners employ a variety of definitions, metrics, and tools to assess and manage resilience in differing applications. Another complication includes the sheer breadth of what resilience analysis implies, both from the standpoint of methodology as well as case applications. These issues motivate the need to provide an overview of various perspectives on the definitions, interpretations, and methodological underpinnings of resilience analysis and thinking as it relates to more traditional risk management. Such an exercise is necessary for, and vital to, the future of the field, where further structuration will be needed to facilitate a more common set of definitions and working tools that practitioners can use to deploy resilience into various fields in the future.

Specific to this need, 50 scholars and practitioners of risk and resilience analysis from some 20 countries met in Ponta Delgada in the Azores Islands from June 26–29, 2016 to discuss the challenges associated with the emerging science of resilience theory and applications. Sponsored and funded in part by the North Atlantic Treaty Organization (NATO) Science for Peace and Security Programme, the overall topic of this meeting was "Resilience-Based Approaches to Critical Infrastructure Safeguarding." The meeting utilized the collective experience and insight of scholars and experts across industry, government, academia, and other organizations to explore various interpretations and understandings of resilience in order to provide a comprehensive and universal understanding of how the methodology might be applied to critical infrastructure systems in various disciplines and applications.

The workshop focused on ways in which military commanders and civilian decision makers alike could utilize resilience analysis in operations. More specifically, workshop discussion centered on both general resilience theory and analysis as well as various applications of resilience in topics ranging from cybersecurity to infrastructural resilience to ecosystem health. In this vein, uncertain yet consequential shocks and stresses that challenge a system's resilience require immediate response and thus real time decision making. As such, this workshop sought to discuss how methods and tools are able to address such concerns by reconciling conflicting inputs, overcoming high uncertainty, and facilitating context-driven decision making within various resilience applications. These topics were addressed via a collection of panel discussions and presentations as well as within smaller working groups. For the former, seven panels were organized across the three-day workshop to review various considerations and applications of resilience.

Within such discussion, topics included areas such as with Resilience Needs in Partner Countries, the Integration of Risk and Resilience into Policy, Cyber Risk and Resilience, and others. For each panel discussion, invited participants were asked to organize a presentation in order to stimulate discussion regarding various panel topics and applications of resilience theory and practice. For the latter, workshop participants were organized into four working groups that addressed risk and resilience based management in (1) infrastructure, (2) cyber, (3) social domains, and (4) methodology and tools for cross-domain integration. This chapter serves as a general introduction to the perspectives of various participants, as well as a reflection of discussion regarding how resilience thinking and analysis may be applied to critical infrastructure in various applications. While specific topics of resilience will be addressed within each of the individual chapters, remaining sections of this chapter will discuss the participants' perspectives on (i) a comparison of risk-based and resilience-based strategies, (ii) features of resilience, and (iii) a layout of topics covered in throughout the book.

1.1 Comparison of Risk and Resilience Management Strategies

Resilience analysis fundamentally maintains much of the same philosophical background as traditional risk assessment. However, resilience analysis additionally delves into the unknown, uncertain and unexpected at the scale of systems rather than individual components. Resilience thinking requires practitioners to ponder potential future threats to system stability and develop countermeasures or safeguards to prevent longstanding losses. Resilience analysis maintains one primary difference in the sense that it primarily focuses on outcomes: practitioners are directly concerned by the ability of the impacted organization, infrastructure, or environment to rebound from external shocks, recover and adapt to new conditions. In other words, where traditional risk assessment methods seek to harden a vulnerable component of the system based upon a snapshot in time, resilience analysis instead seeks to offer a 'soft landing' for the system at hand. Resilience management is the systematic process to ensure that a significant external shock – i.e. climate change to the environment, hackers to cybersecurity, or a virulent disease to population health – does not exhibit lasting damage to the functionality and efficiency of a given system. This philosophical difference is complex yet necessary in the face of the growing challenges and uncertainties of an increasingly global and interconnected world.

In reviewing the similarities and differences in the fields of risk and resilience (approaches and methodologies), it is necessary to consider the philosophical, analytical, and temporal factors involved in each field's deployment (Aven 2011). Philosophical factors include the general attitude and outlook that a risk or resilience analyst expresses when understanding and preparing for risks in a given model. Analytical factors include those quantitative models and qualitative practices deployed to formally assess risk in a particular model. Lastly, temporal factors include the timeframe over which risk is traditionally considered using the analytical models available. Overall, consideration of these and other factors will demonstrate that, while resilience analysis does differ somewhat from more conventionally utilized risk assessment, resilience thinking is highly compatible with existing methods and are synergistic with traditional risk analysis approaches.

Philosophically, risk and resilience analysis are grounded in a similar mindset of (a) avoiding negative consequences of bad things happening and (b) reviewing systems for weaknesses and identifying policies or actions that could best mitigate or resolve such weaknesses. Risk is the operative term for both methodologies, and the overall goal is to lessen as much as possible the damages that could accrue from a hazardous external shock or other undesirable event. As such, practitioners of both mindsets are explicitly required to identify and categorize those events that could generate hazardous outcomes to humans, the environment, or society in general (i.e. commerce, infrastructure, health services, etc.), and subsequently identify countermeasures to meet such hazards.

However, the two methodologies contrast on two key aspects: how to assess and understand uncertainty, and how to judge outcomes of hazardous events (Scholz et al. 2012; Fekete et al. 2014; Aven and Krohn 2014). For the former, a traditional risk analysis approach would seek to identify the range of possible scenarios in an ad hoc or formalized manner, and protect against negative consequences of an event based upon the event's likelihood, consequences and availability of funding, to cover an array of issues for a given piece of infrastructure or construct. In this way, conventional risk assessors generally construct a conservative framework centered upon system hardness, such as with system protections, failsafe mechanisms, and/or response measures to protect against and respond to adverse events. Such a framework has its benefits, but as we discuss in the next section, if the risk philosophy that supports the analysis is too rigid and inflexible, this can hinder event response efforts to rebound from a severe or catastrophic event.

For judging outcomes of hazardous events, resilience analysis fundamentally seeks to provide the groundwork for a 'soft landing', or the ability to reduce harms while helping the targeted system rebound to full functionality as quickly and efficiently as possible, which may imply adaption to new conditions. This is consistent with The National Academy of Sciences (NAS) definition of resilience, which denotes the field as "the ability to plan and prepare for, absorb, recover from, and adapt to adverse events." While this difference may appear subtle, it carries a significantly different operating statement that causes resilience analysts to focus more on 'flexibility' and 'adaptation' within their targeted systems. This differs from the conventional approach commonly deployed by traditional risk analysis, which instead seeks to identify a system that is fail-safe in nature yet inherently conservative. However, the intrinsic uncertainty of the world, the various actors and forces at work, and the systemic nature of many risks, make it significantly unlikely that inflexible systems would prevent all risks in the long run, or would adequately protect against severe events that could cause lasting and sweeping damage to society and the environment. This is particularly true for low-probability events, which have a significant chance of being written off in a traditional risk assessment report as being excessively unlikely enough to not warrant the proper resources to hedge against (Park et al. 2013; Merz et al. 2009). Even high-consequences events are often written off of many decision-makers' agendas, when they have a low probability of occurrence.

Analytical differences between traditional risk analysis and resilience analysis are less understood and developed due to the relatively recent attention to resilience. However, it is possible to derive some understanding based upon the philosophical frameworks that underlie the risk management process. Both risk analysis and resilience analysis permit the use of both quantitative data and qualitative assessment, which allows for greater overall flexibility in applications ranging from well-known hazards to highly uncertain and futuristic hazards through the utilization of subject expert insight where quantitative data is limited. Such information is generally integrated into a specific index or model in order to translate the findings into a meaningful result for the risk analyst, who is then able to offer either an improved understanding of the real risk that certain hazards pose against targeted infrastructure and/or an improved review of which alternative actions or policy options may be taken to mitigate the harms presented by such risks.

Quantitative data may be derived from engineering tests in the field, climate models, design specifications, historical data, or experiments in a laboratory, among others, where policymakers and stakeholders are able to view and assess the likelihood and consequence of certain risks against identified anthropologic or natural infrastructure. Likewise, qualitative assessment is generally derived from meetings with subject experts, community leaders, or the lay public, and can be can be used for narrative streamlined assessment such as with content analysis. In most cases, it is optimal to include both sources of information due to the ability of quantitative field data to indicate more accurate consequences and likelihoods of hazard alongside qualitative assessment's ability yield greater context to an existing understanding of risk data. However, it is often not possible for both sets of information to be generated with full confidence, either because of a lack of reliability within qualitative sources of assessment or because of lack or insufficience of quantitative data (due to the rarity of the situation that is studied, or concerns of ethical experimentation, and/or cost and time issues), leaving policymakers and stakeholders to make the best decisions with what is available to them. This is universally true for both traditional risk analysis and its fledgling partner in resilience analysis, and is likely to be the case for any risk assessment methodology to be developed in the future.

However, conceptualizations of risk and resilience are different. Resilience quantification is less mature than its peer methodology in traditional risk assessment, which otherwise has decades of practical use. This is because resilience is particularly relevant for dealing with uncertain threats, which are always difficult, if not impossible, to quantify. Nonetheless, several quantitative, semi-quantitative, and qualitative approaches have been proposed and deployed to measure systemic resilience at local, national, and international levels for a variety of catastrophic events (generally those with low-probability, high-consequences). Some of these approaches could be relatively simplistic, for example with a qualitative classification system. Others are more complex, for example with resilience matrices or highly complex network analysis, where the availability of information and user preferences determines the level of sophistication deployed for a given resilience case. Despite these differences, however, resilience thinking and analysis will be similarly dogged by the potential for 'garbage-in, garbage-out' analysis, where resilience practitioners must be vigilant, rigorous and robust in their use of relevant and valid quantitative data or qualitative information for whichever risk classification they to employ (Hulett et al. 2000).

Temporally, risk analysis and resilience analysis are required to consider the near-term risks that have the potential to arise and wreak havoc upon complex systems (Hughes et al. 2005). Both engage in exercises that identify and chart out those potential dangers that threaten to damage the infrastructure in question. This exercise can range from being unstructured and ad hoc to organized and iterative, yet ultimately analysts consider a series of threats or hazards that can have some measurable impact upon natural or man-made structures. These hazards are then reviewed based on their likelihood of occurrence and consequences on outcome, which is another iterative process. Lastly, risk analysts are required to assess the immediate aftermath of the various adverse events that were initially identified, and gain a greater understanding into how different components of infrastructure may be damaged and what the consequences of this may be.

Resilience analysis differs temporally from traditional risk analysis by considering recovery of the system once damage is done. Thus, in addition to considering system decline immediately after an event (i.e. risk), resilience adds consideration of longer term horizons that include system recovery and adaptation. Traditional risk analysis *can* integrate recovery and adaptation (for example, by considering probability of system to recover by specific time after event or likelihood that it will be able to adapt), yet this is not necessarily the prime focus of the overall risk analytic effort. Instead, a traditional risk analysis project constructs the ideal set of policies that, given available money and resources, would offer the best path forward for risk prevention and management. Attention to longer term and lower probability threats is often neglected in favor of more intermediate and likely dangers, with only limited emphasis or focus on the need for infrastructural and organizational resilience building, in the face of uncertain and unexpected harms. In this way, traditional risk assessment may not accurately or adequately prepare for those low-probability yet high-consequence events that could dramatically impact human and environmental health or various social, ecological, and/or economic systems that have become ubiquitous within modern life.

1.2 Features of Resilience

Globalization is increasing and strengthening the connectivity and interdependencies between social, ecological, and technical systems. At the same time, increasing system complexity has led to new uncertainties, surprising combinations of events, and more extreme stressors. Confronted by new challenges, the concept of resilience, as an emergent outcome of complex systems, has become the touchstone for system managers and decision-makers as they attempt to ensure the sustained functioning of key societal systems subject to new kinds of internal and external threats. Ecological, social, psychological, organizational, and engineering perspectives all contribute to resilience as a challenge for society. However, there are weak linkages between concepts and methods across these diverse lines of inquiry. Useful ideas and results accumulate and partially overlap but it is often difficult to find the common areas. Further, the different technical languages hamper communication of ideas about resilience across of the different contributing disciplines and application problems.

Connelly et al. (2016) identified features of resilience that are common across conceptualizations of resilience in various fields including (i) critical functions (services), (ii) thresholds, (iii) recovery through cross-scale (both space and time) interactions, and (iv) memory and adaptive management. These features are related to the National Academy of Science definition of resilience through the temporal phases of resilience (Table 1.1). The concept of *critical functionality* is important to understanding and planning for resilience to some shock or disturbance. *Thresholds* play a role in whether a system is able to absorb a shock, and whether recovery time or alternative stable states are most salient. *Recovery time* is essential in assessing system resilience after a disturbance where a threshold is not exceeded. Finally, the concepts of *memory* describe the degree of self-organization in the system, and adaptive management provides an approach to managing and learning about a system's resilience opportunities and limits, in a safe-to-fail manner.

Critical Functions (Services) Understanding the resilience of systems focuses on assessing how a system responds to sustained functioning or performance of critical services while under stress from an adverse event. In assessing resilience, it is necessary to define the critical functions of the system. Stakeholders play a key role in defining critical functions (Palma-Oliveira et al. 2017). Operationalizing resilience concepts depends on identifying the resilience of what, to what, and for whom. In addition, system resilience depends on how the boundaries of the system are drawn (i.e., the chosen scale of interest) and the temporal span of interest. Scale is often dictated by the social organizations responsible for managing the system based on temporal and spatial dimension (Cumming et al. 2006). Thus, stakeholders influence how resilience is assessed both in terms of defining critical functions and system scale. For example, the Resilience Alliance workbooks for practitioners assessing resilience in socio-ecological systems asks stakeholder groups to envision the system and scale of interest, possible disturbances, and to identify vulnerabilities (Resilience Alliance 2010). Further, with respect to psychological resilience, individuals are responsible for assessing resilience through self-reported inventories of protective factors (e.g., adaptable personality, supportive environment, fewer stressors, and compensating experiences) (Baruth and Caroll 2002). It is common practice to use questionnaire responses of stakeholders to assess resilience in psychological and organizational systems.

Thresholds The concept of resilience involves the idea of stable states or regimes in which a system exists prior to a disruptive event. Systems are able to absorb changes in conditions to a certain extent. Further, resilient systems have higher ability to anticipate and use other forms of information and have different ways to synchronize over multiple players (Woods 2003). However, if a shock perpetuates changes in conditions that exceed some intrinsic threshold, the system changes regimes such that the structure or function of the system is fundamentally different. It is the balance of positive and negative feedbacks that can cause a system trajectory to exceed a threshold and degrade system performance (leading to the "collapse" phase of the adaptive cycle) (Fath et al. 2015). The nested nature of systems contributes to the possibility of cascading effects when a threshold at one

Table 1.1 Resilience features common to socio-ecology, psychology, organizations, and engineering and infrastructure, which are related to the temporal phases from the National Academy of Science definition of resilience (discussed in Connelly et al. 2017)

Description by application domain					
NAS phase of resilience	Resilience feature	Socio-ecological	Psychological	Organizational	Engineering & Infrastructure
Plan	Critical function	A system function identified by stakeholders as an important dimension by which to assess system performance			
		Ecosystem services provided to society	Human psychological well-being	Goods and services provided to society	Services provided by physical and technical engineered systems
Absorb	Threshold	Intrinsic tolerance to stress or changes in conditions where exceeding a threshold perpetuates a regime shift			
		Used to identify natural breaks in scale	Based on sense of community and personal attributes	Linked to organizational adaptive capacity and to brittleness when close to threshold	Based on sensitivity of system functioning to changes in input variables
Recover	Time	Duration of degraded system performance			
		Emphasis on dynamics over time	Emphasis on time of disruption (i.e., developmental stage: childhood vs adulthood)	Emphasis on time until recovery	Emphasis on time until recovery
Adapt	Memory/ adaptive management	Change in management approach or other responses in anticipation of or enabled by learning from previous disruptions, events, or experiences			
		Ecological memory guides how ecosystem reorganizes after a disruption, which is maintained if the system has high modularity	Human and social memory, can enhance (through learning) or diminish (e.g., post-traumatic stress) psychological resilience	Corporate memory of challenges posed to the organization and management that enable modification and building of responsiveness to events	Re-designing of engineering systems designs based on past and potential future stressors

scale is crossed and causes disruptions at other scales (Kinzig et al. 2006). The sensitivity of system and sub-system performance to changes in inputs can be used to determine resilience thresholds. Resilience thresholds within organizations are linked to the adaptive capacity of the organization and of the management scheme utilized. Identifying thresholds prior to exceeding them is difficult and an area of intense research (Angeler and Allen 2016). When a threshold is crossed, return is difficult, especially where hysteresis is present. Where or when a threshold is not exceeded, resilience is still relevant, but measures of return time are more appropriate. These concepts are interlinked, and return time may slow as the resilience limits of a system are approached (i.e., critical slowing) (Dakos et al. 2008; Gao et al. 2016).

Scale Resilience is often considered with respect to the duration of time from a disruptive event until recovery (or until the system has stabilized in an alternate regime), and the spatial extent of the system of interest. We consider space and time scales as inextricably linked. Changes in critical functionalities are highly correlated in time and space. It is a flawed approach when one aspect of scale is considered without co-varying the other. There is frequently an emphasis on minimizing time to recovery where full or critical levels of services or functions are regained. *Engineering resilience*, in particular, has a focus on the speed of return to equilibrium, but this measure of resilience does not adequately consider the possibility of multiple stable states, nor account for non-stationarity (Walker et al. 2004). However, return to equilibrium provides important information about the resilience of a system to perturbations that don't cause the system to exceed a threshold and enter into an alternative regime. In the psychological domain, there is also a consideration for the timing of disruptive events within an individual's lifetime. For example, children might be more susceptible than adults to negative psychological impacts from an event, though this is not always the case. Further, resilience requires an appreciation for system dynamics over time. It is thought that resilience is linked to the dynamics of certain key variables, some of which are considered "slow" changing and constitute the underlying structure of the system while others are "fast" changing representing present-day dynamics. Panarchy theory captures this cross-scale structure in complex systems (Allen et al. 2014).

Memory Memory of previous disruptions and the subsequent system response to a shock can facilitate adaptation and make systems more resilient. For example, Allen et al. (2016) observe that ecological memory aids in reorganization after a disruptive event. It has also been noted that socio-ecological resilience is enhanced by a diversity of memories related to the knowledge, experience, and practice of how to manage a local ecosystem (Barthel et al. 2010). Institutional memory can extend beyond individuals. For example, institutional memory is responsible for maintaining lessons learned from previous challenges to the organization or to similar organizations (Crichton et al. 2009). In each case system-wide sensing or monitoring is essential to capture changes in salient driving conditions and critical functions.

Memory of an event in the short term often results in increased safety or resilience through anticipation of a shock or disruptive event through enhanced resistance or adaptive capacity, though in the long-term the memory of the event fades (Woods 2003). Memory tends to be maintained if the system has high modularity or diversity.

In human physiology, responding to repeated stressors produces long run changes in the physiological systems affected by the series of events that evoke stress responses. Although memory of a past experience can have a negative impact on an individual, in some cases, memory can enable positive adaptation whereby these individuals are better able to cope with future stressors. Social memories tend to influence individuals' interpretations of reality, and thus maladaptive social memories can decrease individual and societal resilience.

Adaptive Management Under changing conditions, however, memory of past disturbances and responses may not be sufficient for maintaining system performance or critical functionality. The concept of adaptive management acknowledges uncertainty in knowledge about the system, whereby no single management policy can be selected with certainty in the impact. Instead, alternative management policies should be considered and dynamically tracked as new information and conditions arise over time. Accordingly, management is able to adapt to emergent conditions, reduce uncertainty, and enhance learning in a safe-to-fail manner. By adjusting response strategies in advance to disruptive events, management is able to build a readiness to respond to future challenges. Anticipation and foresight lead organizations to invest in capabilities to deal with future disruptions and prepare for multi-jurisdictional coordination and synchronization of efforts such that the system adapts prior to disturbances. Thus, system-wide sensing (and monitoring), anticipating disruptions, adapting and learning (from both success and failure) occur proactively and in a perpetual cycle, or until key uncertainties are reduced (Park et al. 2013).

There are a number of common features of resilience linked to the planning, absorbing, recovering and adapting phases identified in the NAS definition. Preparing or planning for resilience involves stakeholder identification of critical functions of the system and the strategic monitoring of those functions. Intrinsic thresholds or boundaries determine the amount of disturbance a system can absorb before the system enters an alternate regime, whereby the structure and/or critical functions of the system are different. Whether the system transitions to a new regime or remains the same, the time until the system (performance and critical functionality) recovers from a disturbance is used to assess resilience. Finally, memory and adaptive management facilitate system coping to changing conditions and stressors, even in an anticipatory sense. These features, along with stakeholders and scale, are important across domains in understanding and communicating resilience concepts.

1.3 Benefits of Resilience Thinking Over Traditional Risk Analysis

Traditional risk analysis and resilience analysis differ, yet overall they must be considered complementary approaches to dealing with risk (Fig. 1.1). One way to assess how they are complementary is to consider Risk Assessment as bottom-up approach starting from data and resilience as Top-Down approach starting with mission and decision maker needs with obvious need for integration. Risk assessment process starts with data collection and progresses through modelling to characterization and visualization of risks for management while resilience starts with assessing values of stakeholders and critical function and through decision models progresses towards generation of metrics and data that ultimately can inform risk assessments.

Resilience analysis focuses on both everyday dangers and hazards to organizational and infrastructural condition along with longer term or lower probability threats that have significantly negative outcomes. The purpose of such focus is to improve the target's ability to 'bounce back' from an adverse event, or reduce the time and resources necessary to return the impacted infrastructure back to normal operating procedures. In this way, resilience analysts are by default required to consider risk over the extended or long term and review those events which could prevent a system or infrastructure from returning to full functionality for an extended period. Though not universally true, resilience management *may* afford policymakers

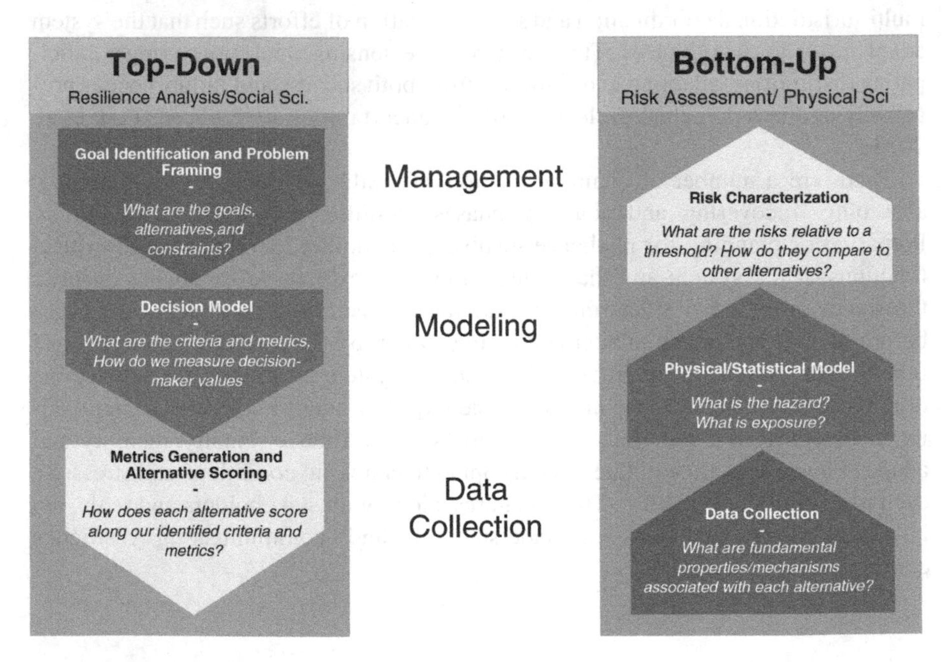

Fig. 1.1 Risk and resilience integration (After Linkov et al. 2014)

and stakeholders a greater upfront defense against system endangering hazards such as those that occurred in the case of Hurricane Katrina or Superstorm Sandy.

A conventional way to determine how risk and resilience are complementary is to consider that risk assessment is the preliminary phase to resilience analysis. It provides the first elements needed to trigger, or not, the need for resilience assessment. This is particularly true in the case of low-probability, high consequence risks of the distance future, such as those associated with climate change, large-scale cybersecurity threats, or severe weather events on the coasts. In this way, resilience analysis adds a different perspective that traditional risk analysts may otherwise miss – the ability to understand the capacity of an organization or infrastructural system to rebound from a massive external shock. While it is impossible to fully predict a highly uncertain and infinitely diverse future, a robust resilience analysis can offer system level preparation across physical, information and social domains thus improving the functionality of the system in the midst of a crisis. While low-probability high-severity events are rare, several have been experienced in recent memory (ranging from the September 11th terrorist attacks to the Fukushima Daiichi nuclear disaster), making resilience assessment both a realistic and highly useful tool to minimize unnecessary losses to infrastructure, capital, and most importantly, human wellbeing.

These benefits of resilience analysis do not immediately mean that resilience analysis is an all-around improvement over conventional risk analytic methods. For traditional risk analysis, risk planning is a multistage effort that requires advanced threat identification for hazardous events prior to their occurrence with follow-up risk mitigation focused on hardening vulnerable system components. Resilience analysis starts with identifying critical functions of the system and stakeholder values with subsequent assessment of system improvement alternatives. Resilience analysis centers on the integration of risk perception (the active identification of risk and hazard in the midst of uncertainty), risk mitigation (steps taken to reduce harms before they occur), risk communication (the need for a clear and meaningful discourse on the seriousness of risk to the general population), and risk management (post hoc measures to address a realized hazard) collectively guide any risk or resilience effort. In this way, resilience analysis *is* far more than a focus on rebounding from a serious risk event, but rather a series of similar steps as with conventional risk analysis that has its own angle on how to best prepare for such hazards.

Resilience analysis cannot, however, replace risk assessment. Its systems approach is characterized by a higher complexity of conceptualization and disconnect from specific system components that needs to be engineered individually. Moreover, less severe and better characterized hazards are better served by existing conventional methods that adequately assess perceived cost and benefits for a given action.

1.3.1 Resilience Discussion from the NATO Workshop on Safeguarding Critical Infrastructure

Overall, the Azores workshop focused on ways in which military commanders and civilian decision makers could utilize resilience management in operations. Military and Civilian applications of resilience concepts are concerned with similar threats and need to be harmonized. Ongoing attacks require immediate response and thus real time decision making; resilience, as a property of a system, must transition from concepts and definitions to an operational paradigm for system management, especially under emergent and future threats.

This chapter serves as a general introduction to the concept and application of resilience, specifically as it relates to traditional risk management, and in particular about suggestions for metrics or indicators that can be developed to assess resilience in a system, and the performance of resilience strategies. Further to this point, the following chapters describe applications of resilience to critical infrastructure from various methodological and analytical perspectives.

Workshop participants were organized into four working groups, which address risk and resilience based management in (1) infrastructure, (2) cyber, and (3) social domains and (4) methodology and tools for cross-domain integration. "State of the Science and Practice". Each of these groups produced reports on their particular domain of resilience thinking and management, which are further amplified by individual submissions by various international authors.

Further Suggested Readings

Allen CR, Angeler DG, Garmestani AS, Gunderson LH, Holling CS (2014) Panarchy: theory and application. Ecosystems 17:578–589

Allen CR et al (2016) Quantifying spatial resilience. J Appl Ecol 53:625–635

Angeler DG, Allen CR (2016) Quantifying resilience. J Appl Ecol 53:617–624

Aven T (2011) On some recent definitions and analysis frameworks for risk, vulnerability, and resilience. Risk Anal 31(4):515–522

Aven T, Krohn BS (2014) A new perspective on how to understand, assess and manage risk and the unforeseen. Reliab Eng Syst Saf 121:1–10

Barthel S, Sörlin S, Ljungkvist J (2010) The urban mind. In: Sinclair PJJ, Nordquist G , Herschend F, Isendahl C (eds) (Uppsala Universitet, Uppsala, Sweden, http://www.diva-portal.org/smash/record.jsf?pid=diva2:395721), pp 391–405

Baruth KE, Caroll JJ (2002) A formal assessment of resilience: the Baruth protective factors inventory. J Individ Psychol 58:235–244

Connelly EB, Allen CR, Hatfield K, Palma-Oliveira JM, Woods DD, Linkov I (2017) Features of resilience. Environ Syst Decis 37(1):46–50

Crichton MT, Ramsay CG, Kelly T (2009) Enhancing organizational resilience through emergency planning: learnings from cross-sectoral lessons. J Conting Crisis Manag 17:24–37

Cumming GS, Cumming DHM, Redman CL (2006) Scale mismatches in social-ecological systems: causes, consequences, and solutions. Ecol Soc 11(20):1–20

Dakos V, Scheffer M, van Nes EH, Brovkin V, Petoukhov V, Held H (2008) Slowing down as an early warning signal for abrupt climate change. Proc Natl Acad Sci 105(38):14308–14312
Fath BD, Dean CA, Katzmair H (2015) Navigating the adaptive cycle: an approach to managing the resilience of social systems. Ecol Soc 20. doi:10.5751/ES-07467-200224
Fekete A, Hufschmidt G, Kruse S (2014) Benefits and challenges of resilience and vulnerability for disaster risk management. Int J Disaster Risk Sci 5(1):3–20
Gao J, Barzel B, Barabási AL (2016) Universal resilience patterns in complex networks. Nature 530(7590):307–312
Hughes TP, Bellwood DR, Folke C, Steneck RS, Wilson J (2005) New paradigms for supporting the resilience of marine ecosystems. Trends Ecol Evol 20(7):380–386
Hulett DT, Preston JY, CPA PMP (2000) Garbage in, garbage out? Collect better data for your risk assessment. In Proceedings of the Project Management Institute Annual Seminars & Symposium, pp 983–989
Kinzig AP, Ryan PA, Etienne M, Allison HE, Elmqvist T, Walker BH (2006) Resilience and regime shifts: assessing cascading effects. Ecol Soc 11(1):20
Linkov, I., Anklam, E., Collier, Z. A., DiMase, D., & Renn, O. (2014). Risk-based standards: integrating top–down and bottom–up approaches. Environment Systems and Decisions, 34(1), 134-137.
Merz B, Elmer F, Thieken AH (2009) Significance of high probability/low damage versus low probability/high damage flood events. Nat Hazards Earth Syst Sci 9(3):1033–1046
Palma-Oliveira J, Trump B, Wood M, Linkov I (2017) The tragedy of the anticommons: a solutions for a "NIMBY" post-industrial world. Risk analysis
Park J, Seager TP, Rao PSC, Convertino M, Linkov I (2013) Integrating risk and resilience approaches to catastrophe management in engineering systems. Risk Anal 33(3):356–367
Resilience Alliance (2010) Assessing resilience in social-ecological systems: A practitioner's workbook Version 2.0 (available at http://www.resalliance.org/3871.php)
Scholz RW, Blumer YB, Brand FS (2012) Risk, vulnerability, robustness, and resilience from a decision-theoretic perspective. J Risk Res 15(3):313–330
Walker B, Holling CS, Carpenter SR, Kinzig A (2004) Resilience, adaptability and transformability in social – ecological systems. Ecol Soc 9:5
Woods D (2003) Creating foresight: how resilience engineering can transform NASA's approach to risky decision making

Part II
Methods

Chapter 2
Towards a Generic Resilience Management, Quantification and Development Process: General Definitions, Requirements, Methods, Techniques and Measures, and Case Studies

Ivo Häring, Giovanni Sansavini, Emanuele Bellini, Nick Martyn, Tatyana Kovalenko, Maksim Kitsak, Georg Vogelbacher, Katharina Ross, Ulrich Bergerhausen, Kash Barker, and Igor Linkov

Abstract Generic standards on risk management and functional safety (e.g. ISO 31000 and IEC 61508) and similar frameworks proved to be surprisingly efficient to trigger and consolidate a widely accepted and ever more effective best practice frontier for risk control. In particular, this includes fundamental and applied research activities to improve processes and to provide more advanced, interlinked and effective methods for risk control. However, this also included the identification of yet unresolved challenges and lacks of completeness. The present work goes beyond these frameworks to address the need for a joint approach to frame resilience management and quantification for system development and improvement. It is

---

I. Häring (✉) • G. Vogelbacher • K. Ross
Fraunhofer Institute for High-Speed Dynamics, Ernst-Mach-Institut, EMI,
Am Klingelberg 1, 79588 Efringen-Kirchen, Germany
e-mail: haering@emi.fraunhofer.de

G. Sansavini
Reliability and Risk Engineering Laboratory, Institute of Energy Technology, Department of Mechanical and Process Engineering, ETH Zürich, Zürich, Switzerland

E. Bellini
Distributed System and Internet Technology, Information Engineering Department, University of Florence, Florence, Italy

N. Martyn
RiskLogik, a Division of Deep Logic Solutions Inc.,
14 Bridge Street, Box 1060, Almonte, ON K0A 1A0, Canada

T. Kovalenko
Department of Management, Technology and Economics, ETH Zurich,
Scheuchzerstrasse 7, 8092 Zurich, Switzerland

M. Kitsak
Department of Physics, Northeastern University,
110 Forsyth Street, 111 Dana Research Center, Boston, MA 02115, USA

© Springer Science+Business Media B.V. 2017

I. Linkov, J.M. Palma-Oliveira (eds.), *Resilience and Risk*, NATO Science for Peace and Security Series C: Environmental Security,
DOI 10.1007/978-94-024-1123-2_2

understood as extending classical risk control to creeping or sudden disruptive, unexpected (unexampled) events, as strongly focusing on technical systems and organizational capabilities to bounce back (better) and as providing generic (technical) resilience capabilities for such resilience response performance. To this end, the article presents general resilience requirements, a resilience management process, which systematically refers to a resilience method taxonomy, resilience levels as well as an applicability table of methods to different resilience management steps for each resilience level. Three case studies elucidate the approach: (i) disruption effect simulation for the Swiss energy grid, (ii) data-driven resilience of the urban transport system of Florence, and (iii) Ontario provincial resilience model in Canada. The approach comprises representative existing resilience concepts, definitions, quantifications as well as resilience generation and development processes. It supports the development of further refined resilience management and quantification processes and related improved methods in particular to cover jointly safety and security needs as well as their practical application to a wide range of socio-technical cyber-physical hybrid systems. This will foster credible certification of the resilience of critical infrastructure, of safety and security critical systems and devices.

Keywords Resilience management • Resilience quantification • General requirements • Process • Method taxonomy • Resilience levels • Resilience method rigor • Case study • Resilience concept • Resilience definition • Safety • Security • Technical safety • Safety II • Cyber resilience • Resilience engineering • Technical science-driven resilience improvement

2.1 Introduction and Motivation

In recent years an increasing number of resilience concepts (e.g. Rose 2004; Thoma 2011, 2014; Kovalenko and Sornette 2013; Righi et al. 2015; Häring et al. 2016a), assessment (e.g Bruneau et al. 2003; Tierney and Bruneau 2007; ISO 31010 2009; Baumann et al. 2014; Larkin et al. 2015; Schoppe et al. 2015; Thoma et al. 2016), quantification (e.g., Havran et al. 2000; Bruneau et al. 2003; Pant et al. 2014), generation (e.g. AIRMIC et al. 2002; Steenbergen 2013; Häring et al. 2016a), enhancement (e.g., Chang and Shinozuka 2004; Baird 2010) and development (e.g. Boyd

U. Bergerhausen
Department Bridges and Structural Technology, Federal Highway Research Institute (BASt), Berlin, Germany

K. Barker
School of Industrial and Systems Engineering, University of Oklahoma,
202 W. Boyd St, Norman, OK 73069, USA

I. Linkov
US Army Corps of Engineers Research and Development Center,
Concord, MA, USA

1995; Osinga 2007; Cavallo and Ireland 2014; Linkov et al. 2014; Righi et al. 2015; Häring et al. 2016a) processes have been proposed for technical and socio technical systems.

In particular the following areas have been covered: business and organizational safety (La Porte 1996; Friedenthal et al. 2011; Steenbergen 2013; Sahebjamnia et al. 2015), socio-technical (e.g. Bruneau et al. 2003; Chang and Shinozuka 2004; MCEER 2006; O'Rourke 2007; Tierney and Bruneau 2007; Rose 2009; Cimellaro et al. 2010; Renschler et al. 2011; Tamvakis and Xenidis 2013; Häring et al. 2016a) and social (organizational) systems (e.g. Boyd 1995; Walker et al. 2002; Dekker et al. 2008; Edwards 2009; Baird 2010; Larkin et al. 2015).

Main application domains and industry sectors include: aviation air traffic control systems (e.g. MIT 2006; Seidenstat and Splane 2009; Mattsson and Jenelius 2015; Renger et al. 2015), health care (e.g. 2000, 2006, Johansson et al. 2006; Cooper and Chiaradia 2015), hospitals (Gertsbakh and Shpungin 2011; Cooper and Chiaradia 2015), electric energy generation (e.g. McDaniels et al. 2008; Gopalakrishnan and Peeta 2010; Gertsbakh and Shpungin 2011; Ouyang and Wang 2015; Nan and Sansavini 2017) and distribution (e.g. Mansfield 2010), gas grids (e.g. Antenucci and Sansavini 2016), oil pipeline grids (e.g. Vugrin et al. 2011) and refineries (e.g. Vugrin et al. 2011), river dams and levees (e.g. Naim Kapucu et al. 2013), hydro plants (e.g. Khakzad and Reniers 2015), inland waterways (e.g. Baroud et al. 2014a, b; Hosseini and Barker 2016), fresh water (e.g. Rose and Liao 2005; Hosseini and Barker 2016) and sewage systems (Holling 1973; Djordjević et al. 2011; Kerner 2014), telecommunication grids (e.g. Sterbenz et al. 2013), rail networks (e.g. Khaled et al. 2015), main road networks (e.g. Ip and Wang 2011; Reggiani 2013; Faturechi and Miller-Hooks 2014; Jenelius and Mattsson 2015; Khademi et al. 2015; Khaled et al. 2015; Koulakezian et al. 2015; Oliveira et al. 2016), supply grids (e.g. Linkov et al. 2013b) and financial sector core services (e.g. 2006, Linkov et al. 2013b; Sahebjamnia et al. 2015).

Only a rather small number of studies concentrates on the (technical) resilience of smaller systems, for example for (autonomous) cars (Sivaharan et al. 2004) or mobile phones (Ramirez-Marquez et al. 2016). Possible further smaller sample systems further include: smart homes (Mock et al. 2016), again autonomous cars (Fenwick et al. 2016; Pearl 2016), or mobile phones (Nnorom and Osibanjo 2009; Jing et al. 2014).

However, a generic and tailorable method-supported framework and process is missing that allows understanding most of the existing and published work done as part of an emerging and general resilience management, quantification and development process, including system improvements. A further challenge is that such an approach should be reproducible, certifiable and auditable, in particular to make it practically applicable, scientifically acceptable and accepted in practice.

Such a framework should be sufficiently general to cover a wide range of approaches as well as be sufficiently specific and novel to be distinguishable from existing approaches, e.g. in terms of feasibility of implementation, in particular from classical risk management. It should allow for both state of the art of science and best practice in industry as well as open the door wide for much needed further innovation and improvement.

In a wider context, UN concepts like the 17 sustainable development goals ask for more resilience for sound development (UN General Assembly 2015). In a similar way, also in the Sendai framework for disaster risk reduction, resilience is a key concept (Aitsi-Selmi et al. 2015). It strongly advocates for resilience to counter natural and anthropogenic threats. However, it is expected that the key drivers for resilience are also more daily needs for a future-proof risk control in an ever more networked system environment, which asks for the full spectrum of possible risk control before, during and after events, e.g. by looking at the phases preparation, prevention, detection, protection, response, recovery and adaption.

Such a resilience management, analysis, development and improvement (generation) process should be such that further standardization is supported, in particular for take up in auditing and certification processes. It should be aware of existing generic frameworks for risk control and be open to learning and exchange.

In the field of (cyber-physical) threats for critical infrastructure, like telecommunication systems or energy supply systems, security and resilience concepts should take into account threats from outside and inside the infrastructure systems (Aitsi-Selmi et al. 2015) as well as caused by natural, anthropogenic, man-made and man-made terroristic (malicious) events. Selected sample events include major subsystem and system failures caused by systematic or statistic system failures, cyber attacks, internal sabotage, attacks caused by aircrafts or drones as well as natural and natural-technical (natech) hazards like earthquakes, flooding or strong wind events (Seidenstat and Splane 2009; Sterbenz et al. 2013).

There is a strong and expected further increasing need for controlling an ever wider range of threats and their combinations to systems: of known, non-expected, unexampled, and (locally) unknown (zero-exploit) natural, anthropogenic (natech), accidental, malicious and terroristic events.

Last but not least, the ever more connected world provides a wealth of data that enables up to real time analytics thus requiring novel concepts of risk control of technical and socio technical systems, in particular for controlling undesired emergent systems states and evolving undesired events. In particular, applied resilience concepts in this domain will strengthen end users and decision makers as well as enable management and policy levels to cope more successfully with undesired events.

Taking the background and the selected main needs into account, the article addresses the challenges as follows. Section 2.2 gives ranked arguments and rationales for the present approach, further elucidation of the context and existing research gaps and industry needs. Section 2.3 lists general requirements for resilience management and quantification, including for resilience improvement and development. Section 2.4 describes in detail the proposed generic and tailorable resilience management process covering resilience quantification and resilience development and improvement. Section 2.5 introduces the minimum set of resilience assessment process quantities that are deemed necessary within the framework. Section 2.6 provides a preliminary classification and taxonomy of methods recommended for implementing the process and for fulfilling its requirements. Section 2.7 presents sample resilience level quantifications for selected system functions and related sample methods selection within three case studies. Section

2.8 discusses how the presented approach relates to representative existing resilience concepts, definitions and quantifications. Sections 2.9 and 2.10 conclude and give a broad outlook, respectively.

2.2 Arguments for Generic and Tailorable Resilience Management, Quantification and Development Approach for Socio Technical Systems

While the introduction already envisioned and detailed some of the main arguments for a generic resilience quantification, management and development process for existing and future socio technical systems, the present section gives a ranked list of the main arguments and rationales for such an approach.

The following list can also be understood as a short top-level objectives list of the presented approach. The main objectives include:

(2.1) Advancement and extension of risk management approaches for emerging needs.
(2.2) Provision of a seamless and orthogonal extension of existing frameworks for risk control.
(2.3) Need for advanced decision support (leave system as is, insure or improve) and improvement options for legacy, emerging and future systems.
(2.4) Efficient coping with an ever increasing variety, unpredictability, scale and number of events: natech (anthropogenic), accidental, malicious, terroristic.
(2.5) Coping with the convergent needs of safety, security, IT-security, reliability, availability, maintainability and transformability of systems.
(2.6) Take-up of needs of (big) data-driven (real-time) risk control, e.g. when utilizing anomaly detection, forensics and counter-action evidence mining and analytics.
(2.7) Allowance of novel business models due to advanced resilience-informed risk control.
(2.8) Meeting multi-dimensional system requirements including efficiency, user acceptance, sustainability, low carbon footprint, acceptable control of risks, trustworthiness, dependability (e.g. see discussion in Häring et al. 2016a).
(2.9) Providing an efficient way to cope with the multitude of possible damage events/disruptions, up to unexampled (unknown-unknown, black swan) events due to the increasing complexity of systems.
(2.10) Provision of a minimum set of common terms and definitions.
(2.11) Along the timeline of evolving and coping with disruptive events: even stronger focus on preparation, response and recovery.
(2.12) Along the functional capability side of systems: provision of a process to identify relevant (technical) resilience capabilities for sensing, modeling, inferring, acting and adapting for sufficient overall risk control/resiliency.
(2.13) Along the layer build-up of systems: stronger taking up the interfaces e.g. between physical, technical, cyber, organizational-social, and environmental-economic layers.

(2.14) Support of envisioned standardization efforts, in particular for technical system resilience, to make resilience a certifiable system design and behavior (technical) property.
(2.15) Support to meet top level UN development goals or the UN Sendai framework for disaster risk reduction, which both strongly advocate for resilience for building sustainable systems.
(2.16) Take-up of strands of discussions under the headlines of Normal Accident Theory (Perrow 2011) and High Reliability Organizations (HRO) (La Porte 1996) in a framed process to meet some of the identified challenges, e.g. in building post event capabilities and permanent self-assessment, in particular from a technical perspective.

2.3 General Requirements for Resilience Management, Quantification and Development

In a formal way, in this section standard requirements as collected in generic standards could be repeated, respectively tailored to the resilience management, quantification and development context. This necessary exercise will be much shortened by concentrating on key top-level requirements, most of which will be fulfilled in more detail in Sects. 2.4, 2.5 and 2.6:

(3.1) Seamless and orthogonal extension and uptake of existing risk control definition terminologies wherever possible.
(3.2) Tailorable, reproducible, certifiable processes.
(3.3) Guidance on methods, techniques and measures to be used along the process.
(3.4) Quantitative guidance for the overall process, in particular for the resilience needs quantification and the development rigor level.
(3.5) Sufficient intra and inter organizational independence of personnel conducting the approach.
(3.6) Adequate professional level and background of involved personnel.
(3.7) Strong take up of societal context and of societal, individual and ethical needs.
(3.8) Traceable and structured documentation of approach.
(3.9) Clear requirements regarding authoritative knowledge on system domain, system context, system interdependency and system interfaces.
(3.10) Well-defined iterative and updatable resilience quantification, management and improvement process.
(3.11) Clearly defined process step main objectives and sub-objectives, expected inputs and outputs.
(3.12) Description of expected key approaches (concepts) used within all steps, e.g. system function definitions deemed relevant for resilience assessments.

(3.13) Strong guidance on method type and rigor selection within all process steps as well as for fulfilling generic requirements.
(3.14) References to existing approaches and methods wherever possible for clarification.

2.4 Resilience Management and Quantification Process

Resilience management is defined in the following as an iterative process that can be decomposed into sequential steps. Figure 2.1 represents the resilience management cycle. As listed in Sect. 2.3, the process of resilience management should be governed by approved principles and corresponding frameworks, general requirements to resilience quantification and development, as well as specific requirements for the process and steps. The resilience management should be supported by a wide range of quantification methods, methods, techniques and measures, in particular decision making techniques.

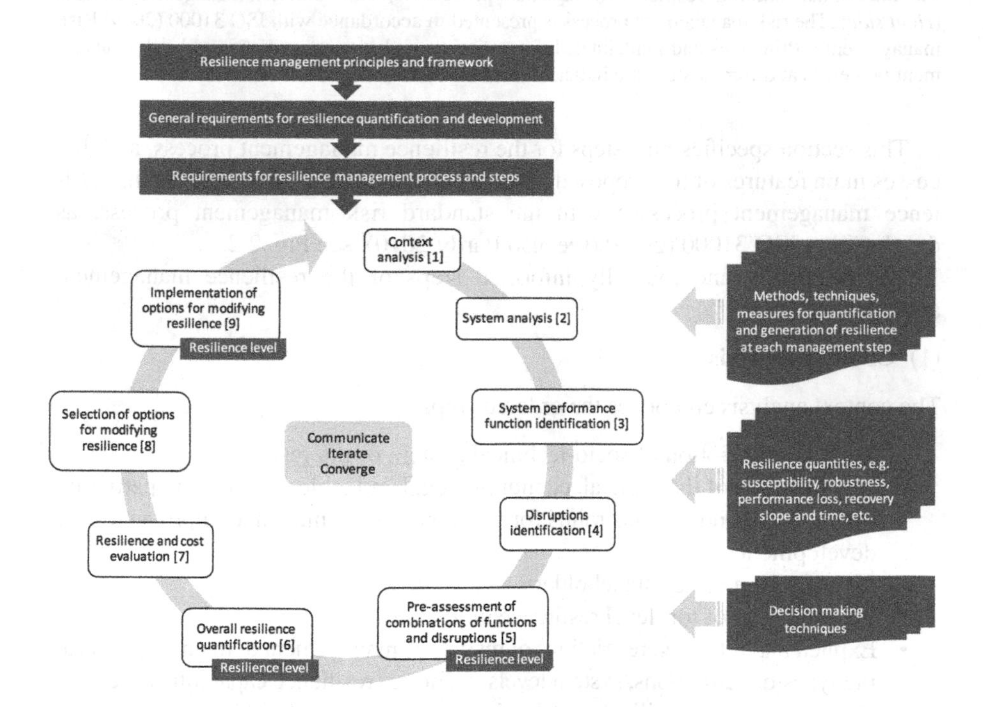

Fig. 2.1 Generic resilience management process that consists of nine steps and covers resilience quantification and development. The iterative process is governed by approved principles and framework, general requirements, specific process and steps requirements. Methods are used to support the approach in all steps (*right side*). Selected resilience quantities are used mainly in steps 5–9

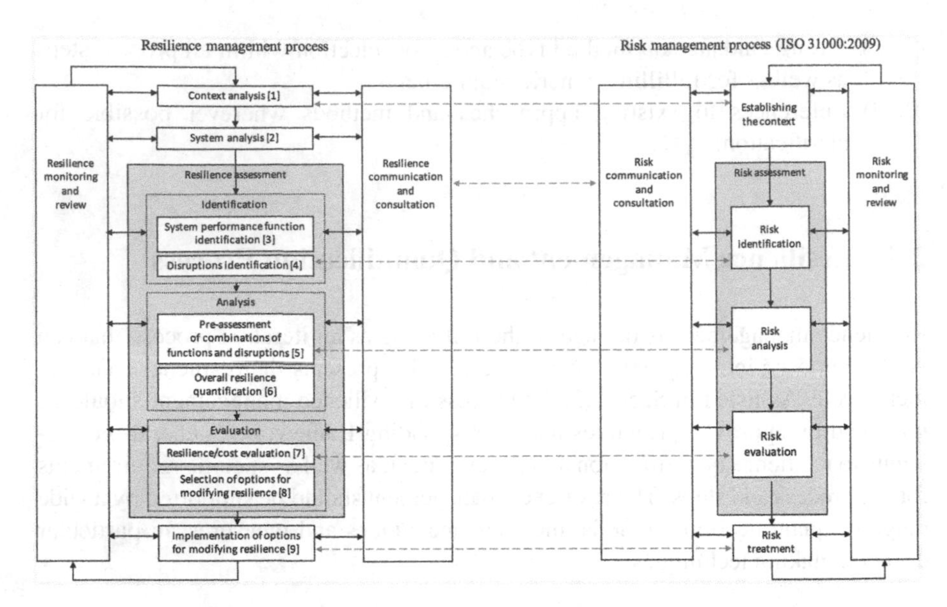

Fig. 2.2 Juxtaposition of resilience management process (*left side*) and risk management process (*right side*). The risk management process is presented in accordance with ISO 31000 (2009) Risk management – Principles and guidelines. Information flows between resilience and risk management processes at different steps are indicated by *dashed arrows*

This section specifies nine steps for the resilience management process, and discusses main features of the proposed approach. It concludes by comparing the resilience management processes with the standard risk management process, as developed in ISO 31000 (2009) (see also Purdy 2010), see Fig. 2.2.

The iteratively and mutually informed steps of the resilience management approach read:

(1) **Context analysis**

The context analysis comprises the ordered steps:

- Verbose description of socio-technical system of interest
- Identification of the societal, economic, legal and ethical context, in particular timeline and budget constraints for resilience assessment and improvement or development
- Identification of key stakeholders
- Identification of top-level resilience objectives
- Explicit and informed restrictions of resilience management domain, e.g. regarding types of disruptions, system levels, technical resilience capabilities, etc.
- Determination of resilience evaluation criteria, e.g. individual and collective, local and non-local (profile-wise), respectively

(2) **System analysis**, comprising the ordered steps

- System (technical) environment and interface analysis
- System boundary definition (spatial, with respect to time, resolution, etc.)

- System interface identification, inter and intra system boundary definitions
- System dynamic behavior assessment
- (Top-level) System static and dynamic (graphical) modelling/representation

(3) **System performance function identification**, comprising the ordered steps

- Identification of system functions, services, properties expected to be relevant for resilient behavior of system
- Definition of system performance functions and generation of qualitative (verbose description of function) and quantitative (relating to availability, reliability, etc.) descriptions
- Equivalently, identification and description of non-performance functions
- Summary/Inventory of system performance and non-performance function space relevant for resilience

(4) **Disruptions identification**, comprising the ordered steps

- Threat/Hazards/Disruptions identification (possible root causes), classical risk events
- Identification of service function disruptions
- Elicitation of means to cover (as far as possible) unexampled (unknown unknown, black swan) events, e.g. in terms of their effects on system (service) functions
- Identification of loss of (technical) resilience capabilities
- Consideration of potentially affected system layers, e.g. physical, technical, cyber, organizational, etc.
- Summary/Inventory of disruptions space relevant for resilience
- Assessment of uncertainty of disruptions identification

(5) **Pre-assessment of the criticality of combinations of system functions and disruptions**, comprising the ordered steps

- Completion of system (non-)performance space and disruption space by considering all possible (multiple) pairings, including ordered along the timeline
- Method selection and application of fast/resource effective (e.g. qualitative up to semi-quantitative) pre-assessment methods for all combinations of system functions (as identified in resilience management step 3) and potential disruptions (as identified in step 4) considering at least the following resilience dimensions:–
  - resilience cycle phases (timeline assessment)
  - (technical) resilience capabilities
  - management domains, e.g. physical, technical, cyber, organizational, social
- Determination of resilience levels (resilience level 1–3) of all system (non-) performance functions (where feasible) taking account of all identified disruptions

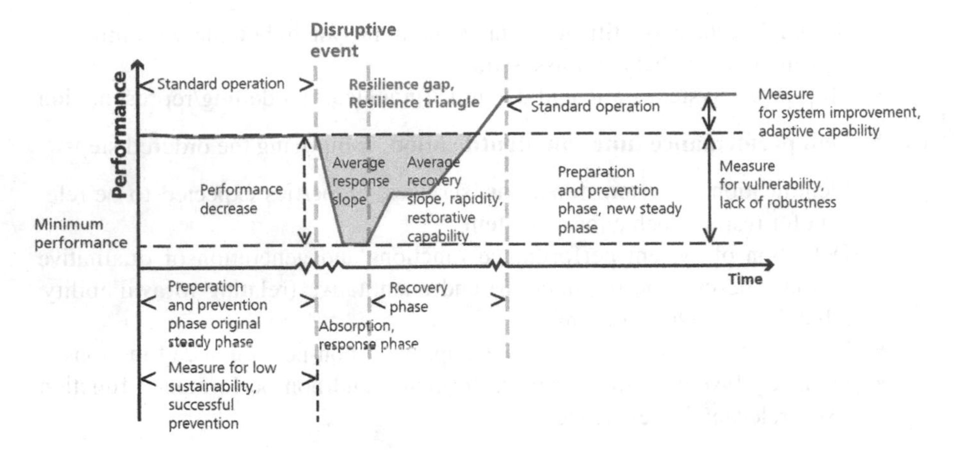

Fig. 2.3 Resilience quantities derived from time-dependent system function performance curve in case of bounce back with improvement (bounce back better)

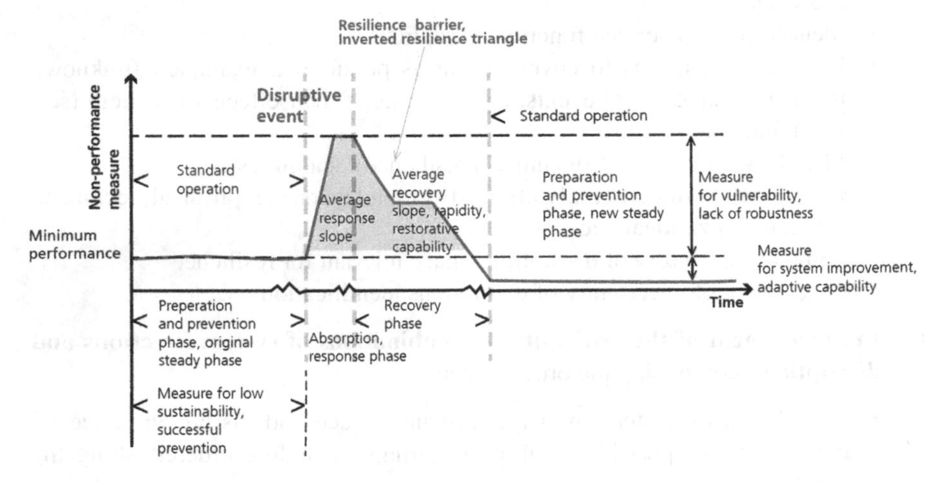

Fig. 2.4 Resilience assessment quantities based on time-dependent system function or system service non-performance curve in case of bounce back with improvement (bounce back better)

- Identification of critical pairings regarding criteria per pair and overall resilience criteria, e.g. using a semi-quantitative chance/risk assessment approach for system (non-)performance function objectives
- Identification of pairings that cannot be assessed with limited effort

(6) **Overall resilience quantification,** comprising the ordered steps

- Selection of resilience quantities of interest, e.g. based on an assessment of system performance or non-performance functions, see Figs. 2.3 and 2.4
- Resilience quantification methods selection
- System modelling sufficient for methods selected

- Application of system resilience quantification methods
- Overall resilience quantification (taking account of all critical combinations and beyond if necessary)
- Determination of resilience level of system (non-)performance functions taking account of all identified disruptions
- Determination of other resilience assessment quantities needed for assessment, e.g.
  - Mean time till disruption
  - Vulnerability/What-if-damage in case of disruptions
  - Time to bounce back (better)
  - Performance loss (area of resilience triangle)
  - Relative performance increase after recovery
- Aggregation and Visualization of resilience quantities

(7) **Resilience evaluation,** comprising the ordered steps

- Resilience performance comparison (e.g. with historic quantities of system performance functions)
- Illustration of effects of system performance loss
- Selection and application of decision making methods
- Evaluation of the acceptance of the obtained system resilience performance level and system resilience quantities for all identified threats: e.g. in terms of
  - acceptable,
  - improvement as high as reasonably practicable (AHRAP principle of resilience management),
  - not acceptable (must be modified)

(8) **Selection of options for improving resilience,** comprising the ordered steps

- Generation of overview up to inventory of resilience improvement options
- Selection and application of decision making methods for the selection of improvement measures
- Iterative re-execution of the resilience management steps for assessing the resilience gain
- Selection of improvement options

(9) **Development and implementation of options for improving resilience,** comprising the ordered steps

- Selection and application of domain-specific standards as far as possible
- Transformation of qualitative and quantitative resilience system performance function descriptions in (multi-) domain-specific traceable technical requirements
- Determination of the resilience levels for subsystems taking account of the system design

- Design, development, integration and testing of system or system improvements using appropriate and efficient methods that correspond to the resilience level identified

The overall characteristic of the described resilience management approach is its strong focus on system functions that are expected to be available before and acceptably available during and after crisis events, in the recovery and response phase or when transforming to an even better system. The approach puts an emphasis on potential disruptions of such (technical) system functionalities, ways to assess and to control them. Several further main features are discussed below.

The first feature is a distinction between context definition and system understanding. This implies sufficient understanding of external connections to other systems, internal links within the system and its subsystems, as well as understanding of system management layers: from physical, technical, cyber, organizational up to the social and policy level, if necessary.

The second feature is concentration on performance functions of a system (service) for comprehensive identification of potential disruptions, independently of (and beyond) known well-describable risk events. This allows to broaden the scope of considered threats, and to cover at least some of unexampled events. Furthermore, the performance functions of systems deemed relevant for assessing system resilience can be understood as a dynamic or behavioral description of the system, thus extending the classically more static system understanding and assessment.

The third feature is the required pre-assessment of all combinations of system functions and disruptions for a pre-identification of potentially critical combinations. In a conservative approach, only the latter is subject to more sophisticated resilience quantification.

However, and also as the fourth feature, the approach asks for an overall resilience assessment of critical combinations of system performance functions and disruptions. This includes to potentially assess cascading and snowball-like effects of such disruptions. As well as to consider combinations of disruptions. It also incudes to consider the overall resilience of the system rather than single isolated resilience issues, e.g. similar as in risk assessment.

The fifth feature is the strong distinction between overall resilience quantification, resilience evaluation as well as resilience improvement measure selection.

As sixth feature, the step of implementation of options for improving resilience is a summarizing step that comprises up to a whole development processes, e.g. of hardware and software. This step is not in the focus of the present work. A well-known example for an established development process is for instance the V-model.

However, the present approach provides verbosely (in terms of their functionality) and quantitatively (in terms of the resilience level) specified system (service) functions that can be implemented. This allows to link with existing standards and frameworks (see e.g. La Porte 1996; Kaufman and Häring 2011; Steenbergen 2013). Hence as the seventh feature, the present approach lists how many steps are necessary to define sufficient system requirements for resilient system design identification while trying to resort as far as possible to existing standards.

The approach uses (semi-)quantitative expressions at least in the following steps: 5, 6, 7 and 9, which is the eighth feature.

The ninth feature of the approach is that it aims at the complete assessment of combinations of system functions and disruptions in a stepwise approach progressing from fast and efficient methods to more resource-intensive and hence often expensive methods. This is indicated by the resilience management steps 4–6 and 6–7, respectively. Of course, an even more layered approach could be chosen.

Figure 2.2 allows to compare the 8-step resilience management process as described above with the risk management process in accordance with ISO 31000. If applied to a whole organization, the implementation of the resilience management process should be driven by the needs of the organization and integrated into its corporate structure.

The resilience management is especially valuable for critical system functions or services. Thus, it can also be applied selectively to areas of high importance, rather than to systems or organizations as a whole. A tailored resilience management process can be designed to be an extension of risk management, or an independent process. When risk and resilience are managed separately, it is important to insure transparency and continuous information exchange between these processes. In the following, resilience management is considered as an independent process.

In summary of the above discussion, the following main extensions and modifications can be identified when comparing the proposed resilience management process with the ISO 31000 risk management process: (i) static and dynamic system function understanding; (ii) focus on system disruptions rather than risk events; (iii) conservative semi-quantitative resilience and overall resilience quantification for combinations of system functions and disruptions; (iv) identification of qualitative and quantitative system resilience performance function definitions.

2.5 Resilience Assessment Quantities for Resilience Management and Development

As indicated in Fig. 2.1, at least the resilience management and development steps 5, 6 and 9 use the resilience level quantity, which is introduced in the present section, see Table 2.1.

For each combination of system performance function and disruption, in step 5 the risk on or alternatively chances for acceptable system behavior are semi-quantitatively assessed and labeled using the resilience level. In step 6, the resilience level determines the rigor of the resilience quantification effort. In step 9, it determines the effort conducted for the resilience system function development.

The different types of uses of the resilience level and hence also its characterization are further discussed below Table 2.1. Especially in the outlook in Sect. 2.10, it will be discussed how further refinements of the resilience management and generation process might lead to more step-specific resilience level definitions. Using a

Table 2.1 Resilience levels 1–3

Characterization/attribute	Resilience level		
	Resilience level 1	Resilience level 2	Resilience level 3
Semi-quantitatively determined risk of critical loss of system performance function in case of disruption	Low risk	Medium risk	High risk
Semi-quantitatively determined chance of no disruption of performance function in case of disruption	High chance	Medium chance	Low chance
Risk in terms of expected individual casualties per year	3E-06/a	2E-05/a	1E-04/a
Risk in terms of collective or group risk of events with one or more casualties per year	3E-03/a	2E-02/a	1E-01/a
Risk in terms of total monetary damage per year in Million Euro or Dollar	2E + 00/a	1E + 01/a	7E + 01/a
Probability of failure (including non-availability) of resilience system function on demand	30%	5%	1%
Availability of resilience functionality on demand	70%	95%	99%
Continuous failure rate of system function per hour	3E-05/h	6E-06/h	1E-06/h
Model resolution used for quantification	Low	Medium	High
Level of rigor of methods employed	Low	Medium	High
Level of (deep) uncertainty of method	Low	Medium	High
Level of completeness	Low	Medium	High
Time effort needed for conducting method	Low	Medium	High
Degree of acceptance of methods	Accepted; common practice	Best practice	State of the art; emerging
Input data needed	Small quantity	Medium quantity	High quantity
Level of quantification	Qualitative, semi-quantitative	Semi-quantitative, quantitative	quantitative
Level of expertise (domain and method)	Low	Medium	High
Method level categorization according to (Linkov et al. 2013a, b): tiered approach	Tier 1	Tier 2	Tier 3

(continued)

Table 2.1 (continued)

Characterization/attribute	Resilience level		
	Resilience level 1	Resilience level 2	Resilience level 3
Type of modeling	Symbolic, graphical, emulative, animative	Simplified network approaches, input-output models	Multi-Domain-specific coupled, resorting to domain specific knowledge; complex network approaches

Characterization of criticality of combinations of system performance functions and disruptions on system level (as output of resilience management step 5). Characterization of rigor of resilience quantification method (as input for step 6). Characterization of reliability of system resilience performance functions (on demand or continuous), of reliability of systems and subsystems contributing to resilience performance functions and of development effort used during the development of resilient system performance functions (respectively for step 9)

single resilience level for all phases has the advantage of simplicity. The disadvantage is a kind of overloading or multiple definition of the resilience level definition.

When considering the use of the resilience levels in more detail (see also Sect. 2.3 on the resilience quantification and development steps), at least five types of use can be listed:

- In steps 5 and 6, the resilience levels are used to quantify the level of risk or chance for each combination of system performance function identified in step 3 and disruption as identified in step 4. In this case, the resilience levels label the level of criticality of combinations of system performance functions and disruptions.
- In steps 5 and 6, the resilience level is also used to quantify the level of confidence, rigor and absence of statistical, systematic and deep uncertainty regarding the resilience quantification results.
- At least in step 6, the resilience level is in addition used to label the level of availability of the system performance function or at least selected features of the system performance function. Examples for the latter are sufficient robustness of the system performance function in case of disruptions, fast response and recovery, steep resilience recovery slope, small resilience loss triangle area, etc. For resilience quantification also additive and multiplicative measures can be used. See Figs. 2.3 and 2.4 for further potential resilience features of interest for which resilience levels need to be found.
- In step 9, the resilience levels are used to allocate, design and assess reliability properties of (sub)systems that are used to generate resilience. This is similar to the so-called allocation of safety integrity levels (SIL) in case of safety-critical systems (SIL allocation, SIL decomposition, see the generic standard IEC 61508 Ed. 2). Therefore, it is recommended to resort to existing standards wherever possible in step 9. This also implies that a system resilience performance function can be realized using an independent combination of other functions.
- In step 9, the resilience levels are also used to determine the development effort of the system performance functions. This is again very similar to the use of SILs in IEC 61508 Ed. 2 and similar derived functional safety standards.

For consistency and simplification, these listed various uses of the introduced resilience levels are not distinguished. Hence Table 2.1 contains characterizing attributes for all of this usages. In the following an example is given. If there is only a single system performance function and single potential disruption believed to be of interest, this combination could be assessed in step 5 to have resilience level 1. In step 6 it is analyzed in detail which features or combination of features of the system performance function have to be provided in case of the disruption. This could be a fast response and recovery within a time interval, for which a resilience level 2 is found in the detailed resilience assessment of step 6. This means that on demand the response and recovery will be in time in 95% of all cases, or alternatively, only in 5% of all cases the response and recovery fails. If the step 7 cvaluates that this risk on resilience has to be mitigated, in step 8 it could be decided that the system function is developed with resilience level 3. This means that in step 9 corresponding resilience level 3 processes, techniques and measures need to be used for the development and implementation of the system function to be sufficiently resilient.

The reliability numbers in Table 2.1 are computed as follows. The failure rate on demand for resilience level 1 of features of system functions relevant for resilience or the overall function is defined as

$$F = 0.3 = 30\%. \tag{2.1}$$

The survivability rate on demand reads

$$S = 1 - F = 0.7 = 70\%. \tag{2.2}$$

Assuming one event per year as rare event, the continuous failure rate reads

$$F_{cont.} = \frac{F}{\Delta t} = \frac{0.3}{365 \cdot 24\,h} \approx 3 \cdot 10^{-5}\,h^{-1}. \tag{2.3}$$

Similar equations are obtained for resilience level 2 and 3 as in Eqs. 2.1, 2.2, 2.3, by reducing the failure rate on demand by a spreading scaling factor of

$$s = 6, \tag{2.4}$$

respectively, see Table 2.1.

Correspondingly, a resilience level 4 can be defined with $F = 0.001 = 0.1\%$, $S = 0.999 = 99.9\%$ and $F_{cont.} = 2 \cdot 10^{-7} h^{-1}$.

These definitions allow referring to existing standards, e.g. ISO 31508, for the development and implementation of sufficiently resilient system performance functions, which can be identified to be safety functions. For instance, it is straightforward to demand, e.g., safety integrity levels (SILs), for system performance functions, identified to be critical. In this way resilience level 1 is certainly fulfilled by SIL 1 functions, resilience level 2 by SIL1 to SIL2, and resilience level 2 is at least SIL2.

In general, within the present context, and as argued in Sect. 2.4, it should be referred to dependable system standards, safety critical system standards or just domain specific system development standards that contain such system performance levels, as far as possible.

Further resilience quantities as mentioned in resilience management step 6 could be for instance defined as in Figs. 2.3 and 2.4. Such quantities include for instance:

- Initial average performance (or non-performance),
- Susceptibility (mean time till disruption),
- Vulnerability (lack of robustness, initial performance decrease or non-performance increase),
- Average absorption slope,
- Mean time of absorbing the disruption,
- Average response slope,
- Average recovery slope (resourcefulness),
- Performance loss (or non-performance increase),
- Time to recover (rapidity) and
- Final performance (or non-performance).

In Table 2.1, the last six lines list whole model classes, in particular an approach using tiers or layers of analysis of resilience. In Sect. 2.6, along with the method taxonomy for each method or method class, a more detailed assessment of the suitability of methods and method classes for each resilience management phase and each resilience level is given. In this sense, the method classes used within Table 2.1 can be seen as a link to Sect. 2.6, which resolves to a higher degree which methods are deemed suitable for each resilience level. Therefore Table 2.1 gives only an indication which method classes are deemed appropriate for each resilience level. Furthermore, as for instance also in the tier approach, higher resilience levels require the use of more methods rather than the selection of only a single method or method class.

With respect to resilience categorization, Linkov et al. (2013b) discuss a tiered approach to resilience assessment that can be iteratively followed by stakeholders to meet resilience needs of a given organization. As noted in Table 2.1, this includes three tiers for assessment. Tier 1 includes screening models and indices that are utilized to identify improvements and investigation needs for further analysis – effectively reviewing the system at hand alongside its critical components whose resilience is important to bolster and maintain. Building from such assessment, Tier 2 includes more detailed models and formal decision analytic operations that prioritize the various components and critical functions of a given system, where such prioritization will help identify performance needs/capabilities while also providing a comparative approach to review investment needs for system resilience at various time intervals. Lastly, Tier 3 engages with a complex modeling of interactions between systems and sub-systems in order to review potential cascading interactional effects, a phenomenon referred to in literature as panarchy. For Tier 3 analysis, a robust scenario analysis is needed that reviews interaction effects within and between systems under various conditions of shock and stress.

As three discrete steps of a resilience management approach, each Tier produces different outcomes and may individually meet the needs of stakeholders. Tier 1 will likely yield relative rather than absolute results that comparatively addresses the performance of a system to existing, well-known examples. Subsequently, Tier 2 seeks to unveil the structure of a system and its various interconnected parts. Lastly, Tier 3 reviews mathematically how these interconnected parts (described as critical functions) interact with each other normally, and how a disruption in service for one sub-system can generate harmful effects to others. As such, a tiered approach promotes resilience thinking and assessment by reviewing system interactions, performance, and recovery from shock to a universe of threats – something essential to combat low probability high risk events that have come to describe many applications of resilience in literature (Linkov et al. 2013b).

2.6 Resilience Generation Method Taxonomy

Table 2.2 lists a taxonomy of methods and method classes. The taxonomy contains a bag of categories that are potentially overlapping. Their suitability for the three resilience levels according to Table 2.1 and each resilience quantification and development phase (resilience management phase) according to Fig. 2.1 is assessed in Table 2.3.

Inspecting and adding up the recommendations for each column of Table 2.3, i.e. for each resilience management phase and for each resilience level, results in a list of methods and method classes recommended for each resilience management phase and resilience level, respectively. The presented categories and methods are representative and suffice for the sample cases presented in Sect. 2.7. Table 2.3 shows that in most cases, as the resilience level increases, more methods are required to be applied rather than a restriction to a set of best or cutting edge methods or method classes. The reason behind this is that resilience assessment and development is a process that needs rather more and different perspectives when higher resilience levels are required than only more advanced methods, in particular to avoid systematic errors in resilience assessment and development.

Table 2.3 shows a tabular assessment approach of the suitability of a method or method class for each resilience quantification and development process step. This assessment is exemplarily also visualized in Fig. 2.5 for a single method using a spider diagram.

In Table 2.3, for each resilience level 1, 2 and 3 as defined in Table 2.1 and for each resilience management step as shown in Fig. 2.1, the suitability of the method or method class is assessed using the following equivalent scales, which are color coded:

$$\begin{aligned}
&\left\{\begin{matrix}\text{strongly } \textit{not} \text{ recommended},,,,\textit{not} \text{ recommended},,,,\textit{no} \text{ recommendation},,,, \\ \text{recommended},,,,\text{strongly recommended}\end{matrix}\right\}, \\
&\{--,,,,-,,,,0,,,,+,,,,++\}, \\
&\{-2,,,,-1,,,,0,,,,1,,,,2\}, \\
&\{1,,,2,,,,3,,,,4,,,,5\}, \\
&\{\textit{red},,,,\text{orange},,,,\text{yellow},,,,\text{green},,,,\text{blue}\}.
\end{aligned} \tag{2.5}$$

Table 2.2 Methods up to method classes suitable for semi-quantitative risk assessment of combinations of system performance functions and disruptions, for resilience quantification, and for resilient system function development: short name and description; related references

Title of method; short description	References
1. Qualitative/(semi-)quantitative analytical resilience assessment approaches	Edwards (2009), Baird (2010), Thoma (2011), Linkov et al. (2014), Thoma (2014), Finger et al. (2016), and Häring et al. (2016a)
Such assessments incorporate several resilience dimensions in nested approaches, e.g. taking within a chance/risk management approach for top level resilience objectives into account all resilience cycle phases (e.g. prepare, prevent, protect, respond, recover), all (technical) resilience capabilities (sense, model, simulate and infer, act, adopt) and all system layers (physical, technical, cyber, organizational, social). It also include to use for instance two resilience dimensions and defining semi-quantitative scales for each combination of attributes, e.g. using resilience cycle phases and system layers. Typically, such approaches are used for expert or crowd input. However, they can also be used for quantitative input. This method class also comprises index-based approaches, e.g. weighted combinations of resilience indices.	
2. Resilience order expansion approaches and their quantification using statistical and probabilistic approaches	Bruneau et al. (2003), Cimellaro et al. (2010), Tamvakis and Xenidis (2013), Fischer et al. (2015), Finger et al. (2016), and Häring et al. (2016a)
Feasible are time independent and time-dependent approaches. Implementations may use statistical-historical, empirical and data mining approaches for empirically based resilience assessments, e.g. based on historical event data and records. A typical example is to ask for the probability of events characterized by the combinations of resilience dimensions assuming that the resilience dimensions determine overlapping sets. Such approaches allow for assessing upper and lower bounds of resilience quantities.	
3. Resilience trajectory/propagation/transition matrix/dynamic approaches	Bruneau et al. (2003), Chang and Shinozuka (2004), Tierney and Bruneau (2007), Linkov et al. (2014), Khademi et al. (2015), Bellini et al. (2016a), and Häring et al. (2016a)
Examples include technical-engineering approaches for cases where the transition between resilience assessment layers is available, e.g. from verbose threat description to hazard source characterization, from hazard source characterization to mechanical local loading, from mechanical loading to object physical response, from object physical response to damage effects, from damage effects to management decisions or computations and simulations that take advantage of combined domain knowledge along with established human and societal behavior modelling approaches.	

(continued)

Table 2.2 (continued)

Title of method; short description	References
4. Socio technical cyber physical- engineering system modelling, simulation and analysis, including agent-based	Bruneau et al. (2003), Chang and Shinozuka (2004), Rose (2004), O'Rourke (2007), Tierney and Bruneau (2007), Rose (2009), Cimellaro et al. (2010), Renschler et al. (2011), Tamvakis and Xenidis (2013), Linkov et al. (2014), Podofillini et al. (2015), Häring et al. (2016a), and Nan et al. (2016)
Key idea is that domain-specific modelling and simulation approaches are combined, e.g. electricity grid, water and telecommunication simulations. In addition, the interfaces between the systems, operators and users are modelled. A modelling level is chosen that suffices for the generation of (time-dependent) resilience curves, indicators and resilience density distributions. Examples include: coupled simulations of multi-technology and multi-domain small and large socio technical systems at various scales, complexity and levels of abstraction, allowing as well for complex human and societal models, like for instance agent-based approaches, using graph modeling or coupled engineering simulations. Such models and simulations may take into account the physical-technical layer, the cyber and the human layer (e.g. operators, users, decision makers). Examples include coupled network simulations using engineering models for node simulations, agents for coupling and agents for human behavior modelling. Such approaches can be extended up to societal modelling, if third party and policies are considered.	
5. Network/grid models and simulation	Boccaletti et al. (2006), Taylor et al. (2006), Fortunato (2010), Newman (2010), Barthélemy (2011), Gertsbakh and Shpungin (2011), Holme and Saramäki (2012), Burgholzer et al. (2013), Sterbenz et al. (2013), Vlacheas et al. (2013), Bellini et al. (2014b), Boccaletti et al. (2014), Vugrin et al. (2014), Zhang et al. (2015), Ganin et al. (2016), Gao et al. (2016), and Hosseini and Barker (2016)
Network or grid models are based on graph theory, e.g. directed graphs. Nodes typically resemble infrastructure or system entities with certain properties and internal states. Edges represent dependency relationships both upstream and downstream. Graph models are suitable for abstract representation of key features of systems. Based on graph models, including initial conditions and state transition rules, the time-development can be simulated. Well known graph models include Markov models and (colored) Petri nets.	

(continued)

Table 2.2 (continued)

Title of method; short description	References
6. Physical-engineering (multi-domain, combinational, coupled) simulations based on 2D, 3D, CAD, GIS data models	Gröger and Plümer (2012), Fischer et al. (2014), Riedel et al. (2014), Abdul-Rahman (2016), Fischer et al. (2016), Lu et al. (2016), and Vogelbacher et al. (2016)
This category covers a wide range of physical-numerical simulation approaches based e.g. on finite element methods or other numerical discretizations. Examples include: propagation of explosive hazards/loadings, chemicals dispersion, hydraulic and geophysical modelling as well as water flow modeling in grids. Often the effect of the loadings on buildings/infrastructure is simulated as well, e.g. the structural response due to wind loading. This is the case for multi-domain combined coupled simulations.	
Typical input data include 2D/3D/4D GIS, e.g. elevation data, hydrological maps, water distributions, semantic city data models, e.g. CityGML, BIM and CAD models.	
The simulations can be analyzed and visualized using e.g. 3D geospatial analysis layers, e.g. by computing the effect of explosive loadings on infrastructure components using constructive simulations based on 3D modeled components of plants.	
7. Cyber logic/layer modelling and simulation	Linkov et al. (2013b), Schoppe et al. (2013), and DiMase et al. (2015)
These methods comprise network/graph models and simulations applied to the cyber/digital layer of systems. Examples include the simulation of industry control systems (ICS)/Supervisory Control and Data Acquisition (SCADA) systems, internet connected to subsystems/components, command-and-control lines, etc. This includes the modelling of the effect of logic commands and interfaces on physical components. In particular, one may resort to modelling of control systems, e.g. according to ISO/IEC 27002 (Jendrian 2014; Stouffer et al. 2015), which can also be used to model internet port (IP) masking.	
8. Procedure and process modelling and analysis	van Someren et al. (1994), Schoppe et al. (2013), Shirali et al. (2013), Christmann (2014), and Khakzad and Reniers (2015)
Procedure and process modelling and analysis comprise heuristics and top-level models of business processes, structures and processes within systems. Examples include organizational hierarchy models, decision making models, iterative systematic improvement models, monitoring and maintenance models, etc. They can be used to elucidate the structure and behavior of a system or organization.	
9. Human factor approaches, human-machine-modelling, and mental modeling technologies	Siebold, van Someren et al. (1994), Augustinus (2003), Tochtermann and Scharl (2006), Linkov et al. (2013a), Schäfer et al. (2014), Bellini et al. (2016b), and Grasso et al. (2016)
Mental modeling technologies comprise mental representations, models and simulations for human behavior. Mental models can be developed for operators, decision makers, responsible persons and third party. In particular, mental models can be used for modelling agents in agent-based simulations. More established approaches include human factor approaches and (simple) man-machine interface modelling approaches. In each case, the level of sophistication has to be selected.	

(continued)

Table 2.2 (continued)

Title of method; short description	References
10. State machine modelling and simulation, including Boolean failure state evaluation, forward and backward simulation	Ouyang, Satumtira and Dueñas-Osorio (2010), Esmiller et al. (2013), Schoppe et al. (2013), Siebold (2013), Ouyang (2014), and Renger et al. (2015)
State machine modeling and simulation is understood as finite discrete state modelling of overall systems, including transitions between states. This allows the propagation of states through the system model. Furthermore failure states may be identified in terms of sets of sates (failure vectors). Also Boolean logic (e.g. fault tree analysis) may be used to identify sets of failure states. This allows to assess operational effects, time behavior and costs. The models have to avoid/be aware of possible cyclic relationships resulting potentially in endless control loops without effect.	
The approach can be used for forward (pathways resulting from a failure) and backward propagation (searching for events that lead to a failure).	
11. Domain specific models and simulations for specific infrastructure types	Ouyang, Australian Government (2010), Satumtira and Dueñas-Osorio (2010), Suter (2011), Kaufmann and Häring (2013), Francis and Bekera (2014), Ouyang (2014), and Stergiopoulos et al. (2015)
This refers to domain specific models accepted by the respective communities, e.g. high voltage grid models, etc. In the USA 18 infrastructures have been defined, in Canada 8. However, such models typically rather exist for standard operation than for the modelling of disruptions or major damage events.	
12. Resilience and risk visualization	Law et al. (2006), Cimellaro et al. (2010), Zobel (2011), Keybl et al. (2012), Kaufmann and Häring (2013), Bellini et al. (2015), and Ramirez-Marquez et al. (2016)
Examples include versions of risk/chance matrix/map (e.g. frequency and consequences of lack of resilience capabilities), local resilience/risk heat maps, relevance clusters of risk, risk flow maps, etc. Any combination of resilience dimensions (e.g. resilience cycle phases and resilience capabilities) can be used also for resilience and risk visualization of corresponding indicators. See also semi-quantitative/analytical approaches.	
13. Interoperability models, Input-Output models	Rose and Liao (2005), Cimellaro et al. (2010), and Renschler et al. (2011)
Network models where nodes are modeled at hoc, e.g. with linear algebra models, for determining their state in dependence of other nodes. For instance a water pumping station needs water and electricity for functioning. If the water input fails it can supply water for 2 days.	

(continued)

Table 2.2 (continued)

Title of method; short description	References
14. Probabilistic and stochastic approaches, Markov processes, probabilistic network approaches	Barker et al. (2013) and Podofillini et al. (2015)
Key elements of these approaches use probabilities, e.g. resilience behavior as conditional probability. In a similar way, resilience behavior can be assessed based on empirical data using statistic approaches, for instance to determine the time to recovery from historical data. Markov processes, e.g. extensions of random walk approaches, and probabilistic networks, e.g. Bayesian belief networks, can be understood as further extensions of such approaches. In particular, it is possible to interpret functional quantities as time-dependent probability quantities.	
15. Empirical and field studies	Norros (2004), Schäfer et al. (2014), Bellini et al. (2016c, b), and Vogelbacher et al. (2016)
Empirical and field studies can be applied to obtain data for user and expert assessment. An example is to ask for actual performance in different resilience dimensions or combinations thereof. In particular, qualitative interviews are effective in identifying system behavior in case of disruptions.	
16. Engineering approaches	Dekker et al. (2008), Hollnagel (2009, 2011), Hollnagel et al. (2010), Voss et al. (2012), Esmiller et al. (2013), Riedel et al. (2014, 2015), Schäfer et al. (2014), DiMase et al. (2015), Siebold et al. (2015), and Fischer et al. (2016)
Fast computational methods using analytical-empirical domain-specific expressions, in particular for civil engineering, mechanical engineering, chemical engineering, process engineering, safety and security engineering	
17. Modified inductive system analysis methods: inductive analytical resilience assessment	Alberts and Hayes (2003), Vugrin et al. (2011), Shirali et al. (2013), Fox-Lent et al. (2015), Ouyang and Wang (2015), and Häring et al. (2016b)
Modifications of classical inductive methods that determine the resilience behavior of systems by propagating the effect of a single subsystem or component failure behavior to determine the effect on resilience for the overall system. Examples include variants of failure mode effects analyses, namely failure mode and effects criticality and/or diagnostic analyses (FMEA, FMEDA, FMECA), as well as variants of event tree analysis (ETA). In each case, it is key to tailor the method to the analysis goal at hand. For instance, the effects on the response and recovery capabilities of disruptive events can be investigated using a variation of an FMEA, an FMEDA will in addition determine whether the system is capable to assess its own capabilities.	

(continued)

Table 2.2 (continued)

Title of method; short description	References
18. Modified classical hazard analysis methods: analytical disruptions analysis Modifications of the classical hazard list (HL), preliminary hazard analysis (PHA) and hazard analysis (HA), subsystem HA (SSHA) and operation and support hazard analyses (O&SHA) can be used to determine for instance potential disruptions as well as their effects on resilience performance functions. In each case, the modifications, tailoring and amendments determine the effectiveness of the method. For instance, a resilience analysis inspired by a hazard analysis could replace hazard source types by disruption types and assess the associated risks on system level considering the counter measures in all resilience management phases. Such hazard analysis methods can be used to generate tables that are useful in implementing process-based semi-quantitative analytical assessment approaches.	Häring et al. (2009), Committee on Increasing National Resilience to Hazards and Disasters, Committee on Science, Engineering, and Public Policy (2012), Ouyang and Dueñas-Osorio (2012), Linkov et al. (2013a), Schäfer et al. (2014), and Cutter (2016)
19. Modified deductive system analysis methods: deductive resilience assessment methods Examples include fault tree analysis (FTA) and time dependent or dynamic FTA (DFTA). A possible starting point are double-failure matrix (DFM) and higher order failure combinations. Such analytical assessments can be used to determine analytically and quantitatively the effect of combinations of events on overall system resilience. At typical tailoring is for instance the attack tree analysis, where the top event of an FTA is an event relevant for resilience assessment.	Ouyang and Dueñas-Osorio (2012), Shirali et al. (2013), Laprise et al. (2015), Ouyang and Wang (2015), and Renger et al. (2015)
20. Flow simulations Flow simulation cover a wide range of systems, e.g. traffic simulation, electric grid alternating current (AC) power flow simulations, water and sewage grid simulations, gas and oil pipeline grid simulations. Such simulations are also provided within GIS environments, e.g. within the ESRI tool suite see e.g. (Benda et al. 2007; Procter et al. 2010; Allegrini et al. 2015).	Hollnagel et al. (2010), The city of New Castle (2010), Vugrin et al. (2011), Kerner (2014), Antenucci and Sansavini (2016), Li and Sansavini (2016), and Nan and Sansavini (2017)
21. Modified event analysis and all hazard approach: Disruptions analysis, all disruptions approach This includes approaches to identify all possible disruption events relevant for assessing the resilience of systems. This can become input for a modified all hazard analysis, which starts out from threat and disruption events and takes account of technical resilience capabilities and their possible failure. If extended to all possible disruptions and their combinations, similar to all hazards approaches, an all resilience approach is obeyed.	Rose and Liao (2005), Jackson (2010), Jenelius and Mattsson (2012), Burgholzer et al. (2013), Sterbenz et al. (2013), and Hamilton et al. (2016)
22. Operational research models and simulations Operational research models and simulations can be used for top level system modelling as well as for extracting for instance economic information from cyber-physical models relevant for decision making of agents.	Linkov et al. (2013b), DiMase et al. (2015), Sahebjamnia et al. (2015), and Aven (2016)

(continued)

Table 2.2 (continued)

Title of method; short description	References
23. Data-based models, data-mining methods Data-based methods apply simple data analytics up to data mining, learning and deep learning algorithms methods to extract information of interest for assessing the resilience response from various data sources, e.g. real time sensor data, social media data, or operational data. Rather established sample methods include time series analysis methods, trigger event detection, anomaly detection, and knowledge mining.	Sturrock and Shoub (1982), Enders et al. (1992), Bloomfield (2000), Box et al. (2008), Larisch et al. (2011), Henry and Ramirez-Marquez (2012), Bellini et al. (2014a), Faturechi and Miller-Hooks (2014), National Consortium for the study of Terrorism and Responses to Terrorism (2014), Stergiopoulos et al. (2015), Bellini et al. (2016b), and Grasso et al. (2016)
24. Experimental methods Experimental methods can be defined to comprise a wide range of scaled, real size, partial, laboratory and free field experiments to assess aspects of resilience, for example in the structural-engineering domain. Field tests involve real environments, e.g. operator or situation awareness rooms. Experimental methods in the presented definition do not include empirical field studies.	Law et al. (2006), Schrenk (2008), Fischer and Häring (2009), Larisch et al. (2011), and Sterbenz et al. (2013)
25. Table top exercises, red teaming/penetration tests, serious gaming This method category comprises a wide range of approaches from table top exercises, red teaming/penetration tests to serious gaming. Each of them by now has been described and applied in very different contexts, in particular in the civil security domain for assessing and identifying improvement needs for the resilience of systems. Examples include the application of all of these methods to airport checkpoint security questions which can easily be framed as resilience engineering challenges.	Mansfield (2010), Renger et al. (2015), Siebold et al. (2015), and van der Beek and Schraagen (2015)
26. Decision support methods Decision support methods are used for decision making taking into account the context and multiple diverse criteria in a rational way. Often rather fast ad hoc methods like multi criteria decision making are preferred. Also prospect theory approaches can be applied to rationalize decisions.	Arboleda et al. (2006), Falasca et al. (2008), Greene (2011), and Larkin et al. (2015), Bellini et al. (2016a)
27. Expert estimates, expert elicitation This comprises structured approaches to extract information from expert opinions including associated uncertainties, e.g. technical capabilities believed to be relevant for prevention, protection, response and recovery, as well as number estimates e.g. for frequency of disruptions, damage effects on system, etc. Sample methods include Delphi method variants.	Bologna et al. (2016)

(continued)

Table 2.2 (continued)

Title of method; short description	References
28. Functional Resonance Method (FRAM) FRAM is a heuristic method to model and understand functional properties of systems and their subsystems, in particular to model the effect of disruptions. A close link to more established system models, in particular static and dynamic graphical semi-formal models, e.g. of SysML, is not yet established.	(Bellini et al., Rose and Liao (2005), Hollnagel (2009), Jackson (2010), Hollnagel (2011), Jenelius and Mattsson (2012), and Burgholzer et al. (2013)
29. Resilience score cards Typically ad hoc criteria or generic resilience criteria like robustness, rapidity, redundancy, resourcefulness, etc. are used. However, they rather should be the outcome of a resilience assessment process than being input right at the beginning, because not in all cases for instance redundancy is the best option. Such resilience criteria can be used for identifying possible resilience objectives in early phases of procedural assessment methods.	Bruneau et al. (2003), Chang and Shinozuka (2004), MCEER (2006), O'Rourke (2007), Tierney and Bruneau (2007), Rose (2009), Baird (2010), Cimellaro et al. (2010), Størseth et al. (2010), Dorbritz (2011), Renschler et al. (2011), and Tamvakis and Xenidis (2013)
30. System modelling languages Examples for graphical and semi-formal system modelling languages that can be used for a wide range of technical, socio-technical and social systems (e.g. organizations) are the Unified Modelling Language (originally developed for the software domain, UML) and the systems modelling language (SysML) for systems engineering across disciplinary domains. Such models can be extended (using extensions) and restricted to allow for formal models.	Weilkiens (2007), Friedenthal et al. (2011), Object Management Group (2012), Delligatti (2014), and Renger et al. (2015)

This recommendation of methods for each resilience level is made independent of potential application systems. In case of (strongly) not recommended (strong) arguments have to be given if the method (class) is selected. In case of (strongly) recommended (strong) arguments have to be given if the method (class) is not selected.

Figure 2.5 shows the assessment of three different sample methods for each resilience level and for each resilience management step. In Fig. 2.5a, the level of recommendation for qualitative/(semi-) quantitative analytical resilience assessment approaches is shown. It can be seen, that for nearly all resilience levels the resilience management steps 4–8 show high values meaning that these methods support those resilience management steps strongly. On the other hand, small values are indicated for the context analysis and the system definition.

Figure 2.5b shows, that physical-engineering (multi-domain, combinational, coupled) simulations based on 2D, 3D, CAD, GIS data models show advantages in the disruptions identification and the measure selection. Small values are shown for the context analysis. Of course their main strength is in the detailed resilience

Table 2.3 Suitability of each method or method class for each resilience quantification and development process step (resilience management step)

Methods for resilience quantification	Resilience Level (RL)	Resilience management phases								
		(1) Context analysis	(2) System definition	(3) Performance function identification	(4) Disruptions identification	(5) Pre-identification of critical combinations of functions and disruptions	(6) Overall resilience quantification	(7) Resilience/cost evaluation	(8) Measure selection	(9) Measure development and implementation
1. Qualitative/ (semi-) quantitative analytical resilience assessment approaches	Level 1	2	2	3	3	5	5	5	4	3
	Level 2	3	3	4	4	5	5	5	5	3
	Level 3	2	3	5	5	5	5	5	5	4
2. Resilience order expansion approaches and their quantification using statistical and probabilistic approaches	Level 1	3	2	3	3	4	4	3	3	2
	Level 2	4	3	4	4	4	5	4	4	2
	Level 3	4	3	4	5	5	5	5	4	2
3. Resilience trajectory/ propagation/ transition matrix/ dynamic approaches	Level 1	2	2	2	3	3	3	3	3	2
	Level 2	2	2	2	3	3	4	4	4	2
	Level 3	2	3	3	3	3	5	5	5	2
4. Socio technical cyber physical-engineering system modelling, simulation and analysis	Level 1	3	3	2	3	3	3	2	3	3
	Level 2	4	4	3	4	4	4	3	4	4
	Level 3	5	5	5	5	5	5	4	5	5
5. Network/Grid models and simulation	Level 1	3	3	3	4	4	4	3	3	3
	Level 2	4	4	4	4	4	4	4	4	3
	Level 3	4	5	5	5	5	5	4	4	4
6. Physical-engineering (multi-domain, combinational, coupled) simulations based on 2D, 3D, CAD, GIS data models	Level 1	2	2	2	2	2	3	2	3	2
	Level 2	2	3	3	3	3	4	3	4	3
	Level 3	2	4	4	4	4	5	4	5	4
7. Cyber logic/ layer modelling and simulation	Level 1	2	2	3	3	3	4	2	3	2
	Level 2	3	4	4	4	4	4	3	4	3
	Level 3	4	4	5	5	5	5	4	5	4

(continued)

Table 2.3 (continued)

8. Procedure and process modelling and analysis	Level 1	4	4	3	3	4	4	2	3	3
	Level 2	4	4	4	4	4	5	3	4	4
	Level 3	5	5	5	5	5	5	4	5	5
9. Human factor approaches, human-machine-modelling, mental modelling technologies	Level 1	2	2	2	3	3	4	2	3	3
	Level 2	3	3	3	4	4	5	3	4	4
	Level 3	3	4	4	5	4	5	4	5	4
10. State machine modolling and simulation, including Boolean failure state evaluation, forward and backward simulation	Level 1	1	2	3	3	3	3	3	3	2
	Level 2	2	4	4	4	4	4	4	4	3
	Level 3	3	5	5	5	5	5	5	5	4
11. Domain specific models and simulations for infrastructure types	Level 1	2	2	2	3	3	3	3	3	3
	Level 2	3	3	3	4	4	5	4	4	4
	Level 3	4	4	4	5	5	5	5	5	4
12. Resilience and risk visualization	Level 1	2	2	1	3	3	4	4	3	3
	Level 2	3	3	2	4	4	5	5	4	4
	Level 3	3	3	2	5	5	5	5	4	5
13. Interoperability models, Input-Output models	Level 1	2	3	3	3	3	3	2	4	3
	Level 2	3	4	4	4	4	4	3	4	4
	Level 3	4	4	4	4	4	5	4	5	5
14. Probabilistic and stochastic approaches, Markov processes, probabilistic network approaches	Level 1	2	2	2	2	2	3	2	3	2
	Level 2	3	3	3	3	3	4	3	4	3
	Level 3	3	4	4	4	4	5	4	5	4
15. Empirical and field studies	Level 1	4	4	4	4	4	4	2	3	2
	Level 2	5	4	5	5	5	4	3	4	3
	Level 3	5	5	5	5	5	5	4	5	4
16. Engineering approaches	Level 1	2	3	3	3	3	4	3	4	3
	Level 2	3	4	4	4	4	5	4	4	4
	Level 3	4	5	4	4	4	5	4	5	5
17. Modified inductive system analysis methods: inductive analytical resilience assessment	Level 1	3	3	4	4	4	5	4	5	4
	Level 2	3	4	4	4	4	5	5	5	4
	Level 3	3	5	5	5	5	5	5	5	5
18. Modified Classical hazard analysis methods: analytical disruptions analysis	Level 1	4	4	4	4	4	4	4	4	4
	Level 2	4	4	4	5	4	5	4	5	4
	Level 3	5	5	5	5	5	5	5	5	5

(continued)

Table 2.3 (continued)

19. Modified deductive system analysis methods: deductive resilience assessment methods	Level 1	3	3	4	4	4	4	4	4	3
	Level 2	3	3	4	5	4	5	4	5	4
	Level 3	4	4	5	5	5	5	4	5	4
20. Flow simulations	Level 1	2	2	2	2	3	3	1	2	2
	Level 2	3	3	3	3	4	4	2	3	3
	Level 3	4	4	4	4	5	5	3	4	4
21. Modified event analysis and all hazard approach: Disruptions analysis, all disruptions approach	Level 1	3	3	3	4	4	3	4	4	3
	Level 2	4	4	4	5	4	4	4	4	3
	Level 3	5	5	5	5	5	4	5	5	4
22. Operational research models and simulations	Level 1	1	2	2	2	2	3	2	2	2
	Level 2	2	3	3	3	3	4	3	3	3
	Level 3	3	4	4	4	4	5	4	4	4
23. Data-based models, data-mining methods	Level 1	2	2	2	3	3	3	2	3	3
	Level 2	3	3	3	4	4	4	3	4	4
	Level 3	5	4	5	5	5	5	5	4	5
24. Experimental methods	Level 1	1	2	3	3	2	3	2	3	2
	Level 2	2	3	4	4	3	4	3	4	3
	Level 3	2	4	5	5	4	5	4	5	4
25. Table top exercises, red teaming, penetration tests, serious gaming	Level 1	2	2	2	3	3	4	2	3	2
	Level 2	3	3	3	4	4	5	3	4	3
	Level 3	4	4	4	5	5	5	4	5	4
26. Decision support methods	Level 1	3	3	4	3	4	4	2	4	4
	Level 2	4	4	3	4	5	5	3	5	5
	Level 3	5	5	4	5	5	5	3	5	5
27. Expert estimates, expert elucidation	Level 1	3	3	4	4	4	4	4	4	3
	Level 2	4	4	5	5	5	5	4	5	4
	Level 3	5	5	5	5	5	5	5	5	5
28. Functional Resonance Method (FRAM)	Level 1	3	3	4	3	3	3	3	3	3
	Level 2	4	4	4	4	3	3	3	3	3
	Level 3	5	5	5	3	3	3	3	2	2
29. Resilience score cards	Level 1	3	2	3	3	2	2	2	5	4
	Level 2	3	2	3	3	2	2	2	5	4
	Level 3	3	2	3	3	2	2	3	5	4
30. System modelling languages	Level 1	3	4	3	3	3	4	2	4	4
	Level 2	4	5	5	4	5	5	3	5	5
	Level 3	5	5	5	5	5	5	3	5	5

The following scale set is used: strongly not recommended, not recommended, no recommendation, recommended, strongly recommended. This covers the recommended use of the method for the first time as well as if the method has already been used in earlier phases, e.g. as in the case of system modeling approaches which are typically used in many phases

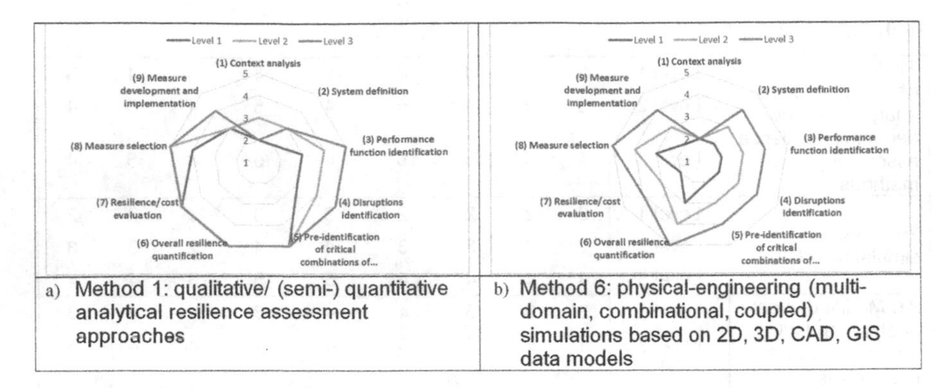

Fig. 2.5 Visualization of recommendation level in spider diagrams for two different method classes for each resilience quantification and generation phase and each resilience level 1 to 3. (**a**) Method 1: qualitative/(semi-) quantitative analytical resilience assessment approaches. (**b**) Method 6: physical-engineering (multi-domain, combinational, coupled) simulations based on 2D, 3D, CAD, GIS data models

quantification and overall resilience quantification, which needs to take account of a lot of detailed information.

Typically, it is found that qualitative methods are useful for more resilience management steps than methods that are more specific, e.g. engineering methods or infrastructure-specific models. Further is observed, that for higher resilience levels more methods are used. Most classical engineering methods are only prepared for in other phases.

Than step 6 in terms of building up the model and actually employed only in resilience management phase 6: detailed resilience quantification and overall resilience quantification.

2.7 Sample Cases for Resilience Level Determination and Selection of Methods for Case Studies

In a sample application, three case studies are considered:

(A) **Urban transport resilience improvement using real time data for the town of Florence**
(B) **Electrical high voltage grid vulnerability and resilience assessment of Swiss high voltage grid**
(C) **Province of Ontario resilience model**

Below, for each case study the following is given:

(i) A short description of the sample system considered, delivered by a short plain text.

(ii) The resilience level obtained for few (one or two) selected resilience system performance functions taking account of a defined set of potential disruptions. This is delivered by a qualitative description of the system performance functions and by giving their resilience levels.
(iii) The methods (classes) used throughout the resilience quantification and development process without distinguishing between resilience management steps. This includes a recommendation of their usage for the sample cases taking into account the highest recommendation only. This is delivered by attributing a value to each method (class) for each case study according to Eq. (2.5) as delivered in Table 2.4 for each case study.
(iv) Selected few resilience improvement efforts conducted and their estimated resilience level.

Regarding (ii), (iii) and (iv), it is noted again, that for simplicity and consistency the same resilience level scale of Table 2.1 is used for the assessment of critical combinations of system functions and disruptions (phase 5), for the selection of methods for refined for resilience quantification (step 6), and for resilience development and implementation, respectively.

(A) Urban transport resilience improvement using real time data for Florence

(i) Enhancing resilience in urban transport systems (UTS) is considered imperative for two main reasons: such systems provide critical support to every socio-economic activity and are currently themselves one of the most important economic sectors and secondly, the paths that convey people, goods and information, are the same through which also many risks are propagated (Taylor et al. 2006; Faturechi and Miller-Hooks 2014; Demirel et al. 2015; Hughes et al. 2015). The UTS in Florence is characterized by several drawbacks (Bellini et al. 2016b).

Here below we provide some of the relevant data able to characterize the UTS in Florence:

- the density of cars is 2.159 car/km^2–205,650 vehicles and 71,167 motorbikes (914.925 in total for the province);
- the UTS has 700 Km or streets in the urban area most of them are situated in the historical/mediaeval area (Restricted Traffic zone: Zona a Traffico Limitato, ZTL, Controlled parking zone: Zone a Controllo di Sosta, ZCS) where the dimensions are scarce (usually one way street);
- streets/bridges/rails etc. are managed by several authorities. Including metropolitan area there are: 1439 km under regional/provincial control, 114 KM under state control (National Autonomous Roads Corporation: Azienda Nazionale Autonoma delle Strade, ANAS), and 95 km highway under Autrostrade s.p.a. several urban underpasses are managed by Trenitalia spa, Florence Metro is managed by the public company Gestione Servizio tramviario (GEST), etc.;
- 70% of the street victims occur in urban area the rest extra-urban;
- the tourism pressure in Florence is about 10.000.000 of non-residential persons each year and usually concentrated in specific periods of the year;

Table 2.4 Methods and methods classes used within the three case studies, including their maximum recommendation level considering all resilience management phases

Resilience assessment, quantification and development methods and method classes (see Table 2 for detailed descriptions)	**Maximum recommendation for method (class) in at least one resilience management phase**		
	Case study		
Resilience quantification: short description including application examples	**Case study A: Urban transport resilience improvement using real time data for Florence**	**Case Study B: Electrical high voltage grid vulnerability and resilience assessment of Swiss grid**	**Case Study C: Province of Ontario resilience model**
1. Qualitative/ (semi-) quantitative analytical resilience assessment approaches	4	2	5
2. Resilience order expansion approaches and their quantification using statistical and probabilistic approaches	3	4	5
3. Resilience trajectory/ propagation/ transition matrix/ dynamic approaches	4	5	5
4. Socio technical cyber physical-engineering system modelling, simulation and analysis	5	5	5
5. Network/Grid models and simulation	4	5	5
6. Physical-engineering (multi-domain, combinational, coupled) simulations based on 2D, 3D, CAD, GIS data models	4	3	5
7. Cyber logic/ layer modelling and simulation	3	4	5
8. Procedure and process modelling and analysis	5	5	5
9. Human factor approaches, human-machine-modelling, mental modeling technologies	5	4	5
10. State machine modeling and simulation, including Boolean failure state evaluation, forward and backward simulation	3	4	5
11. Domain specific models and simulations for infrastructure types	4	4	4
12. Resilience and risk visualization	5	5	5
13. Interoperability models, Input-Output models	3	3	5
14. Probabilistic and stochastic approaches, Markov processes, probabilistic network approaches	4	4	5
15. Empirical and field studies	4	2	5

(continued)

Table 2.4 (continued)

16. Engineering approaches	3	5	5
17. Modified inductive system analysis methods: inductive analytical resilience assessment	4	3	4
18. Modified Classical hazard analysis methods: analytical disruptions analysis	4	3	4
19. Modified deductive system analysis methods: deductive resilience assessment methods	4	4	4
20. Flow simulations	4	3	5
21. Modified event analysis and all hazard approach: Disruptions analysis, all disruptions approach	5	5	5
22. Operational research models and simulations	0	4	0
23. Data-based models, data-mining methods	5	2	4
24. Experimental methods	5	3	3
25. Table top exercises, red teaming/penetration tests, serious gaming	3	3	3
26. Decision support methods	5	4	4
27. Expert estimates, expert elucidation	5	3	5
28. Functional Resonance Method (FRAM)	5	3	3
29. Resilience score cards	3	2	3
30. System modelling languages	5	3	4

- 150 K of commuters every day that arrive with cars, trains, buses;
- the average number of passengers in a car is 1.7, thus an inefficient usage;
- the just-in-time inventory management strategy of the Florence downtown shops requires a continuous provision;
- the level of particles on the order of 10 micrometers or less (PM10) in Florence tends to go over the national average and sometimes goes over the legal limit of 40 μg/m^3 causing traffic stops for days;
- More than 80% of the streets are at flooding risk.

(ii) The threats addressed in Florence are river flooding and flash flooding. According to the historical records and hydrogeological risk maps published on the City council open data platform, the 80% of the city area can be considered at risk. Typical system resilience performance functions include the "adaptive provision of mobility of citizens as organized by multi-modal public transport" in the advent of minor up to major disruptions like flooding but also accidents, persons blocking roads or railways, technical failures, strong rainfall, strikes, terror alerts. The range of resilience level is typically from 1 to 2, possibly 3.

Another resilience performance system function is related to the "intelligent early warning" where population is timely advised with context aware messages elaborated through a data driven situational awareness system and delivered through different communication channels as mobile phones, variable message panels, radios, TVs, etc. The range of resilience level is around 2.

Every resilience performance function is designed according to the evidence driven adaptive cycle as presented in Fig. 2.6 that requires to collect a huge amount of heterogeneous data from the technical systems as well as the human beings (UTS users). The scope is to continuously monitor the adaptive and buffer capacity of the UTS thus monitoring its resource availability in order to support real time decision making. In fact, the UTS relies on human actors to deal with dynamics, complexity and uncertainty (Norros 2004) that cannot be controlled on the basis of fixed rules and procedures. For this purpose, the user requires tools and an organizational con-

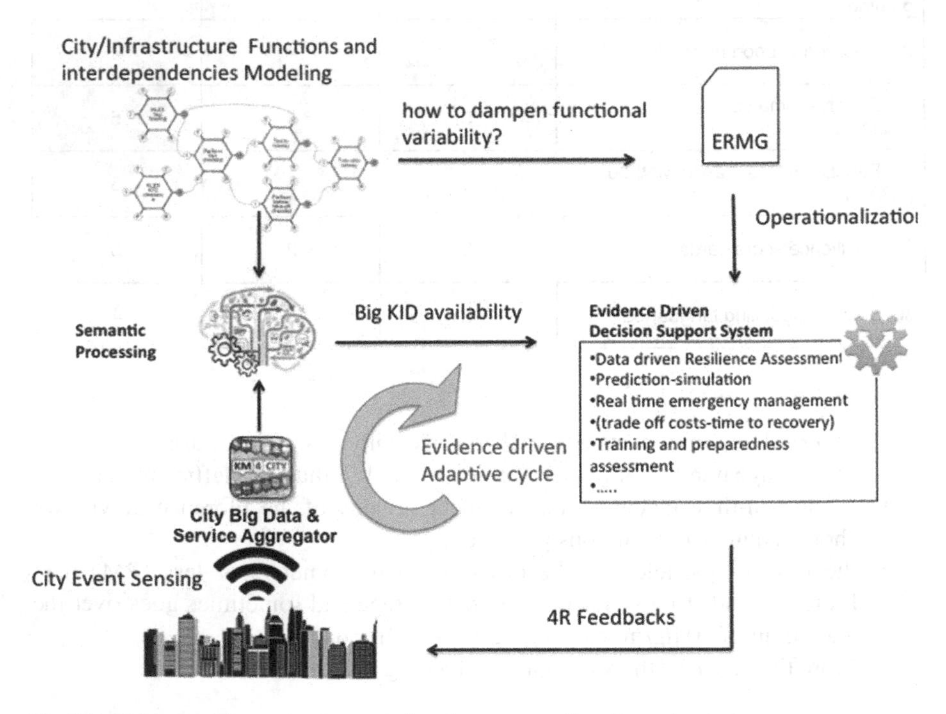

Fig. 2.6 Evidence driven adaptive cycle in urban context. See discussion in text

text that allows doing so efficiently. In order to operationalize such a model, in Florence we worked with Big Data and Internet of Everything (IoE) moving from simulated data to real data generated by the smart city.

Such data are input of the Collaborative Resilience Assessment and Management Support System (CRAMSS), a data driven tool able to support decisions for multiple decision makers in a complex environment. There are three types of data being collected and managed by the Big Data Management Platform (2017a) developed within the EU project RESOLUTE (2015) and used by the CRAMSS: urban data, human behavior data and social network data. In particular:

- **Urban data** include municipality open data, such as: structure of the city, seismic risk maps, hydrological risk maps, services, statistics, time series of major disasters, descriptors of structures such as schools, hospitals, streets, river level, weather conditions, position of Wi-Fi access points, locations of people aggregation facilities (such as: gym, schools, mall, social house, theatres, stadium, hospital). Some of these issues generate real time data such as the emergency triage status of hospitals, environmental sensors, parking areas availability, metro status and position, traffic flow information, origin destination matrices for cars, traffic flow movements (Bellini et al. 2016c).
- **Human behavior data** may be either individual or group-based and include activity related and behavioral personal or collective profiles addressing psychological, habitual and cognitive aspects. These profiles may be extracted based on different kinds of sensors: Wi-Fi network, Bluetooth servers, traffic flow sensors as spires (wearable activity trackers), TV-cameras, mobile cells from telecom operators, mobile Apps, etc., by using data mining and data analytics techniques. All these multidisciplinary and multimodal raw data need to be integrated in a common comprehensive format towards discovering meaning-bearing annotations (Bellini et al. 2014a).
- **Social networks data** are tweets, comments, posts, likes, etc. A social network crawler can be exploited to manage and analyze all real-time data streaming from the citizens and the city infrastructure (Grasso et al. 2016). The crawler should be language independent utilizing multilingual thesaurus. Text processing and knowledge mining techniques should be used to discover hidden information, to identify sentiments, trends, influencers, to detect events or to predict flows. In addition to the dynamic data, an interoperable knowledge base contains cross sectors data that can be used to provide services to help the environment to become more efficient in disaster situations. Furthermore, the activities of data analytics and semantic reasoning are used to generate new knowledge that can be integrated into the interoperable knowledge base where cross sectors data are used to help improve resilience in situations of danger (e.g. data ingestion, mining and algorithms, computing models and recommendations).

These heterogeneous datasets have to be accommodated in a scalable and interoperable Knowledge Base (Bellini et al. 2014a), which contains cross-sectors data that can be used to provide services to help the environment become more efficient in the event of a disaster. Furthermore, the Data Analytics Semantic

Computing layer computes several elaborations to generate new knowledge (such as: extraction of typical human trajectories in the city, computation of origin destination matrices at different time slots and week days, computation of predictions about eventual city dysfunctions, computation of sentiment analysis with respect to major city services) that can be integrated into the Knowledge Base, where cross-sectors data are used to help improving resilience in situations of danger (e.g., data ingestion, mining and algorithms, computing models and recommendations).

(iii) In the following the resilience quantification methods as described in Sect. 2.6 Table 2.2 are used for resilience quantification in example A as well as for the identification of the best resilience improvement measures, see also Table 2.4 second column:

- Method 2: to identify the UTS threats;
- Method 4: to consider human, technology and organization as assets in UTS;
- Method 9: Human factor approaches, human-machine-modeling, and mental modeling technologies, to model human behavior and movement at city level;
- Method 12: Data visualization for risk and resilience understanding and decision making;
- Method 13: to use a semantic approach to fuse heterogeneous data;
- Method 15: to collect process data from the ground to extract meaningful information about the capacity of the system of coping with changing and unexpected conditions;
- Method 21: to calculate in real time the damages according to the area affected, the magnitude of the phenomena and its dynamics measured through sensors (user as a sensors; environmental sensors, etc.);
- Methods 20 and 22: Simulative approaches, to model metro, road traffic dynamics, to define and share strategies in the decision support system;
- Method 26: Data driven decision support systems has been implemented to support decision makers in applying;
- Method 28: To describe and understand the UTS complexity and interdependency and to drive an ERMG definition and big data platform implementation.

The project RESOLUTE is answering the need for improving the resilience of critical infrastructure, in particular for UTSs, by conducting a systematic review and assessment of the state of the art of the resilience assessment and management concepts, as a basis for the deployment of an European Resilience Management Guide (ERMG). The guide also takes into account that resilience is not only about the performance of individual system elements but rather the emerging behavior associated with intra and inter system interactions. Hence the project focuses on a process for the identification of system performance functions that are paramount for resilient system behavior in case of disruptions. Thus it considers a broad set of methods and method classes with a strong focus on data-driven methods which help to support and implement such adaptive resilience assessment processes and resilience generation processes. Hence all methods of Table 2.2 could be relevant for the short case study A.

(iv) The EU project RESOLUTE (2015) is based on the vision of achieving higher sustainability of operations in European UTSs. This requires overall heightened operational efficiency, mainly by optimizing the allocation and utilization of available resources (organizational, technical and human), whilst striving to continuously minimize any source of environmental pollution as well as any disruptive events like incidents, accidents and other operational failures. Within this context, RESOLUTE considers resilience as a useful management paradigm, within which the consideration of adaptability capacities is paramount, in particular as driven by real-time (observational) data.

Organizations must generate the ability to continuously adjust to ever-changing operational environments. This requires rapid resilient response in case of disruptive events rather than inefficient built-up of redundancies.

The resilience improvement measures that will be results as project outcomes include:

- Data driven Risk and resilience assessment of UTS;
- Reduction of the consequences of events, enabling multi operator coordination and stakeholder awareness;
- Reduction of likelihood of high-consequence events through technical and/or organizational means using data generated by many sources (e.g. environmental sensors, user as a sensor, traffic data);
- Complex system definition, UTS function and interdependencies identification;
- Prevention of upstream and downstream propagation of functional variability and resonance quantification, i.e. prevention of cascading effects managing resources availability and allocation;
- Mitigation of critical event effects on population, e.g. by optimizing emergency response and evacuation through preparation, early warnings and real time re-routing through mobile apps;
- Release of an European Resilience Management Guide (ERMG). The guide also takes into account that resilience is not only about the performance of individual system elements but rather the emerging behavior associated to intra and inter system interactions.
- Collaborative Resilience Assessment and Management Support System (CRAMSS), that adopts a highly synergic approach towards the definition of a resilience model for the next-generation of collaborative emergency services and decision making process.

In summary, the example case study A on urban transport resilience assessment and improvement using real time data focuses on a broad understanding of urban transport systems and subsystems including their interfacing and management, the identification of key transport functionalities and related performance measures, the accessibility of data-driven indicators of performance and for disruptions identification as well as the data-guided selection of efficient response strategies in the advent of disruptions.

For all steps already existing solutions are taken into account and in parts significantly extended. So far rather generic transport system performance functions have been identified. However, the project assumes that a strong focus should be on the generic capability of transport systems to recover in case of disruptions and hence related system performance measures and the employment of appropriate methods. Furthermore, even if data-driven approaches rely on substantial technical systems, resilience quantification and generation methods that take account of the human, organizational and societal factors have been identified to be critical for successful resilience assessment and improvement, in particular in ongoing crises.

(B) Electrical high voltage grid vulnerability and resilience assessment of Swiss grid

(i) The high-voltage electric power supply system (EPSS) consists of three interdependent subsystems arranged in three different layers, i.e. System Under Control (SUC), Operational Control System (OCS), and Human Operator Level system (HOL). The SUC represents a technical part of the EPSS, its components include transmission lines, generators, busbars and relays. It is a time-stepped system, i.e. the time scale has a strong influence on its functionalities.

The OCS also represents the technical part of the EPSS. Its major responsibility is to control and monitor the coupled SUCs. Compared to the SUC, the OCS is an event-driven system, i.e. its functionalities are mainly influenced by events rather than by the time scale. The Supervisory Control and Data Acquisition (SCADA) system represents the OCS, i.e. is a major part of the OCS. Components of the SCADA include field instrumentation and control devices (FIDs and FCDs), remote terminal units (RTUs), communication units (CUs), and master terminal unit (MTU).

Finally, the HOL represents a non-technical part of the EPSS, which is related to human and organizational factors influencing the overall system performance. The HOL is responsible for monitoring and processing generated alarms, switching off components at remote substations and sending commands to remote substations. In order to achieve a high-fidelity modeling of SUC and OCS, both functionality (physical laws) and structure (topology) should be considered. Furthermore, the model for OCS needs to be able to process messages among components.

An agent based model (ABM) is selected to combine all these systems in a single modelling approach. This approach intends to represent the whole system by dividing it into interacting agents. Each agent is capable of modifying its internal status, behaviors and adapts itself to environmental changes. ABM is a bottom-up approach and each component is represented as an agent (Tolk and Uhrmacher 2009; Chappin and Dijkema 2010).

The model for the HOL should be able to quantify the effects of human performance. Human Reliability Analysis (HRA) is suitable to this aim, and provides a way to assess human performance in either qualitative or quantitative ways. Qualitative methods focus on the identification of events or errors, and quantitative methods focus on translating identified events/errors into Human Error Probability (Sharit 2012).

(ii) In the following, it is motivated that winter storms are a natural threat and potential disruption of strong interest for power supply grids in Switzerland. It is also exemplarily listed which system performance measures are of interest in this case. The system performance measure "actual power demand served" is selected. Depending on the households, infrastructure (e.g. hospitals) and industry that is supplied, resilience levels of 2–3 or more can be attributed to this power grid system functionality. In each case, the system resilience performance measure could further be refined for applications e.g. "actual power demand served, i.e. power losless than 4 h and less than 3 losses per week", etc.

Historical records reveal that hazards such as earthquakes and winter storms were the cause of significant damage in at least nine events over the past 1000 years in Switzerland (Bilis et al. 2010). According to (Raschke et al. 2011), the estimated frequency of natural hazards, i.e. winter storms, which have the potential of resulting in the simultaneous disconnection of 20 transmission lines is in the range of $6 \cdot 10^{-4}$ to $7 \cdot 10^{-4}$ per year. In this resilience assessment experiment, it is assumed that a natural hazard, i.e. winter storm or ice rain, impacts the central region of Switzerland, where power transmission lines are located; as a result, about 17 power transmission lines are disconnected.

Several system measures of performance (MOP) quantify the response of the EPSS to the disruptive event, which focus on different characteristics, examples include:

(1) MOP_{SUC1}, the number of available transmission lines (topology related),
(2) MOP_{SUC2}, actual power demand served (functionality related).

One MOP is selected for the SCADA:

(3) MOP_{OCS}, the number of available RTUs (topology related).

The multiplicative metric proposed to quantify general resilience, GR_{SUC}, integrates the various measures of resilience capabilities, i.e. robustness, recovery speed, recovery ability, performance loss and loss speed, and allows comparisons among different systems and system configurations (Nan and Sansavini 2017). Strategies focusing on the enhancement of a specific system the resilience capability can be tested.

(iii) In the following the resilience quantification methods as described in Sect. 2.3 are used for resilience quantification as well as for the identification of the best resilience improvement measure.

Relating to Table 2.2, the following methods can be identified (see also Table 2.4):

- Method 2: Resilience order expansion approaches and their quantification using statistical and probabilistic approaches, to identify the threats and their frequency;
- Method 4: Socio technical cyber physical- engineering system modelling, simulation and analysis, including agent-based, for the overall modelling approach of the EPSS;
- Method 7: Cyber logic/layer modelling and simulation, to model the OCS;
- Method 9: Human factor approaches, human-machine-modelling, and mental modeling technologies, to model the HOL;

- Method 11: Domain-specific models, to model the SUC;
- Method 16: Engineering approaches, to determine the physical SUC behavior in case of line interruptions;
- Methods 20 and 22: Simulative approaches, to model OCS and SUC;

(iv) Next, some possible resilience improvement measures or options are discussed:
- Strategy 1: The improvement of the efficiency of line reparation enhances the restorability capability during the recovery phase, i.e. the mean time to repair *MTTR*.
- Strategy 2: The improvement of the human operator performance enhances the adaptive capability during the response and recovery phase, i.e. the human error probability threshold HEP_A.
- Strategy 3: The improvement of RTU battery capacity enhances the absorptive capability during the disruptive phase.

The target system for each strategy also varies: SUC is the target system for Strategy 1 and 2, and SCADA is the target system for Strategy 3.

Figure 2.7 illustrates the value of GR_{SUC} as defined in Eq. 2.6 below, i.e. the multiplicative resilience metric for SUC to the disruptive event, with respect to Strategy 1 and 2 using MOP_{SUC2} as introduced in section B.ii. When both strategies are

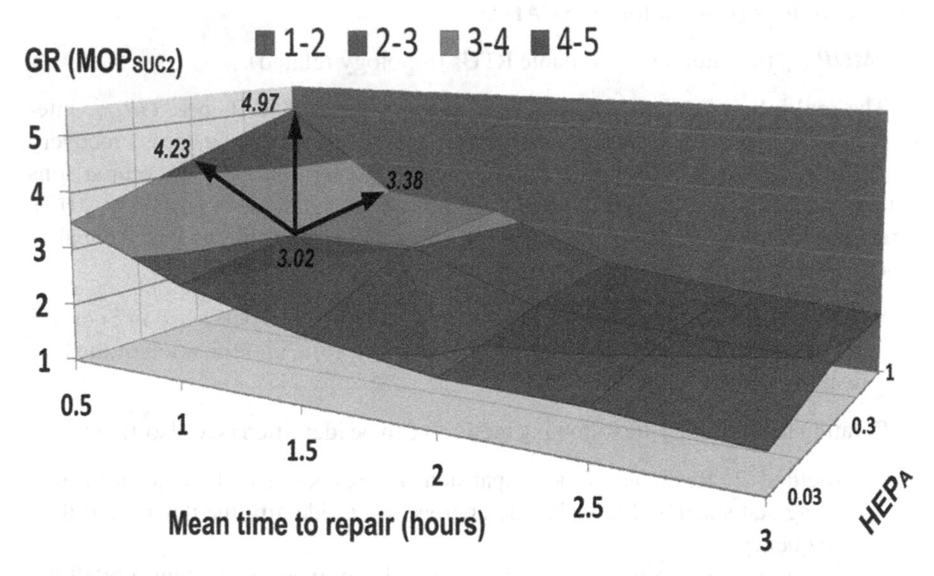

Fig. 2.7 Example for overall resilience quantification and improvement of an energy high voltage grid system in a context where two system parameters can be changed and a quantitative multiplicative resilience system performance measure of interest for optimization has been selected. The overall system resilience quantity is $GR_{SUC}(MOP_{SUC2})$, the actual power demand served depending on the mean time to repair (MTTR) and the human error probability threshold (HEP_A). Values for different simulation scenarios implementing Strategy 1 and 2 are given

implemented simultaneously, the resilience of SUC is enhanced significantly, see Fig. 2.7.

Furthermore, the values of GR_{SUC} allow comparing the relative benefits of different improvement strategies. $GR_{SUC}(MOP_{SUC2}) = 3.02$ when $MTTR = 1$ h and $HEP_A = 0.3$; at this point, if the efficiency of reparation is further improved, i.e. $MTTR = 0.5$ h, $GR_{SUC}(MOP_{SUC2})$ indicates 40% resilience increase. On the other hand, if the human operator performance is further improved, i.e. $HEP_A = 1$, the $GR_{SUC}(MOP_{SUC2})$ metric indicates 12% resilience increase. If both strategies are implemented, $GR_{SUC}(MOP_{SUC2})$ metric indicates 64% resilience increase. Hence the best selection of improvement strategies can be determined based on GR and on the implementation costs.

In summary, the short example case study B on the electrical high voltage grid vulnerability and resilience assessment of the Swiss grid reveals that for this domain specific socio-technical system and the disruption set "storm and ice-rain" a multiplicative system performance measures of interest can be identified to quantify resilience. In this case, the pre-quantification of resilience was conducted using statistical-historical data which revealed that the loss of the identified system performance functions in the case of the selected disruptions are of interest. In addition, the resilience quantification allowed to compare different resilience improvement measures.

(C) The Province of Ontario critical infrastructure resilience model

(i) The Province of Ontario is Canada's most densely populated and highly industrialized Province, with a concomitant high concentration of and dependence on industrial and information age critical infrastructure. As Canada's manufacturing, governance and financial center, Ontario is responsible for approximately 36% of Canada's GDP (Status of 2014; e.g. data from Statistics Canada (2014). Consequently, disruptions in any of the critical infrastructure sectors that interdict commercial or governance operations can have a disproportionate effect on the national economy.

The highly dependent nature of all commercial and governance operations on the densely concentrated infrastructures and the inter-dependent nature of the critical infrastructure sectors themselves demands a network approach to the analysis, assessment and improvement of Provincial resilience. To this end a network based approach, employing path analysis in a graph theory based tool set was used to identify pathways of exposure to risk, risk effects, pathways of consequence and the feasibility of various proposed mitigations (see also (2017b).

The discrete directed graph model is made up of nodes and edges. Nodes depict an operation, an asset or an enabler and each one is assigned two values on a 0–10 scale. One for impact on the system and another for likelihood of failure for a list of identified possible hazard and disruptive events. Impact is derived from expert elicitation and mental modeling and/or from physical system evidence. Likelihood of failure is likewise derived from expert elicitation, system design parameters and operational history evidence.

Edges depict a level of dependence between nodes using the same 11-integer scale. A node can have an upstream or downstream relationship with another node or both; not necessarily of the same value. This is modeled using (possibly multiple) directed edges between nodes. Upstream and downstream is always dependent on the disruptive event considered and how its effects can propagate through the graph model and its system (of systems) layers. Dependency relationship scoring is derived from expert data elicitation and documented systems evidence. When a scenario is introduced the three scores are manipulated to reflect the state change in each entity. The path analysis algorithms then calculate the impact and return the results in an adjusted risk index that reflects the new exposures or vulnerabilities.

The resulting model includes more than 1300 infrastructure entities in a network control framework where dependencies are not only understood within their critical infrastructure sectors but are also understood cross-sector. The model is dynamic in that multiple timely ordered hazard scenarios can be tested against the control framework individually or in combination to reveal the pathways of exposure to risk for each scenario. The upstream and downstream propagation of the disruption and damage also takes into account how long the propagation takes. This allows for an absolute dynamic sequencing of cascading effects in case the propagation times are known. In case that propagation time information is not available only relative dynamic sequencing is available.

Where costs are known they can be calculated and aggregated along a consequence chain to reflect the whole cost of a given scenario over time in terms of fixed and variable costs to auditable standards. Where node locations are known, risk effects can be represented in GIS or 3D constructive simulation and fused with other relevant geospatial data allowing advanced geospatial analysis of operational risks. In a similar fashion, SCADA systems are mapped to the nodes they control and the consequence of cyber attacks can be understood.

This approach provides for the emergency management agency a living model of Provincial Infrastructure to plan and exercise with. It is credited with reducing the effects of the 2013 floods in Toronto and several other Provincial level natural disaster events. The model is gradually being used by other Canadian Provincial government departments, not specifically mandated for resilience for planning and response management. Such an approach, while comprehensive and informative, depends on human understanding and skill to drive it and maintain it, and this is proving its greatest challenge.

(ii) Typical system performance functions of interest are the provision of the services of the respective infrastructure grids or industrial sites in case of advert events. In particular, the capability of industrial facilities to cope with risk and disruption events that are assessed to be critical and that cannot be hedged by societally acceptable insurance coverage.

More precisely a single such system performance function is the "sufficient drinking water supply of the population" in case of flooding and cyber-attacks. This are examples for natural and man-made malicious events, respectively, the latter possibly even supported by external governmental resources. In this case, the resilience level of the performance function and risk event combination would be assessed to be 2–3 or higher.

(iii) For an overview of the methods used within the modelling approach, see Table 2.4.

Main methods used within the approach include:

- Graph models: Method 5: Network/Grid models and simulation; Method 12: Resilience and risk visualization
- GIS Based Models: Method 6: Physical-engineering (multi-domain, combinational, coupled) simulations based on 2D, 3D, CAD, GIS data models
- Simulation: Method 20: Flow simulations; Method 22: Operational research models and simulations; Method 5: Network/Grid models and simulation; Method 11: Domain specific models and simulations for infrastructure types
- Mental mapping: Method 9: Human factor approaches, human-machine-modelling, mental modeling technologies
- Engineering approaches: Method 16: Engineering approaches
- Expert elucidation: Method 27: Expert estimates, expert elucidation
- Human factors: Method 9: Human factor approaches, human-machine-modelling, mental modeling technologies
- Cyber systems modelling: Method 7: Cyber logic/layer modelling and simulation

(iv) Typical resilience improvement measures that can be inferred and tested by the Province of Ontario resilience model include:

- Risk and resilience assessment of several design options of industrial sites or infrastructure designs
- Local increase of robustness, mainly regarding the modeled nodes, i.e. reducing the consequences of events, e.g. mechanical retrofit, change of requirements regarding the fulfillment of building codes depending on threat levels
- Reduction of likelihood of high-consequence events through technical and/or organizational means, e.g. using video surveillance and early detection of internal and external disruptions to improve prevention
- Prevention of upstream and downstream propagation of events, i.e. prevention of cascading effects within grids or systems or beyond them, e.g. using smart grids that are locally self-sufficient on demand
- Reduction of damage event effects on on-site personnel and the population, e.g. by optimizing emergency response and evacuation through preparation and technical alerting systems

Table 2.4 gives an overview of the methods used for each case study as well as their maximum level of recommendation.

2.8 Discussion of Relation of Framework Approach to Existing Resilience Concepts, Definitions and Quantifications and Improvement Processes

In the following, some representative existing resilience quantification and/or development approaches are discussed regarding their relation to the presented approach. In each case at least the resilience concept used is named, the (implicitly or explicitly used) definition operationalized for resilience quantification, the actually used resilience quantification method or methods and possibly further methods typically used within this strand of research.

2.8.1 Quantification of Resilience Using a System Performance Function Based Measure

Using the time-dependent system performance diagrams of Figs. 2.3 or 2.4, a single additive resilience metric can be constructed which is different from the often used resilience triangle loss quantity, see e.g. (Häring et al. 2016a). Possibly several such resilience metrics can be weighted and combined. Such a resilience quantification is one of the options for the "overall resilience quantification" in phase 6 of the resilience management cycle.

This also holds true for the multiplicative overall general resilience measures as used within Case study B in the present text, see also (Nan et al. 2016). The motivation for this measure is given below.

Resilience can be quantified by considering resilience capabilities of systems (i.e. absorptive, adaptive and restorative capability) in different phases (i.e. original steady, disruptive, recovery and new steady phase) and integrating them into a unique resilience metric.

The selection of the appropriate time-dependent system performance function or system measure of performance (MOP), as conducted in the resilience management phase 3 "identification of system performance functions", depends on the specific service provided by the infrastructure under analysis.

Referring to Fig. 2.3, in the original steady phase, the system performance is within its target value. In the disruptive phase, the performance drops until reaching the lowest level. During this phase, the system absorptive capability can be assessed by *Robustness* (*R*), which quantifies the minimum performance level. This measure is able to identify the maximum impact of disruptive events; however, it is not sufficient to reflect the ability of the system to absorb the impact.

To this aim, two complementary measures are further employed, i.e. *Rapidity* ($RAPI_{DP}$) and *Performance Loss* (PL_{DP}) in the disruptive phase (DP). *Rapidity* can be approximated by the average slope of the system performance function, and captures the speed of change in the system performance during the disruption phase and the recovery phase. In this way, $RAPI_{DP}$ and $RAPI_{RP}$ can be defined. *Performance loss* in the disruptive phase (PL_{DP}) can be quantified as the area of the region bounded by the MOP curve with and without occurrence of the disruptions effects, i.e. the area between the continued original steady curve and the performance curve in case of a disruption till the lowest performance level.

Alternatively, the *time averaged performance loss* (*TAPL*) can also be used, which considers the time of appearance of negative effects due to disruptive events up to full system recovery, and provides a time-independent indication of adaptive and restorative capabilities in response to the disruptive events. It is obtained by dividing the performance loss by the time length of the just described time span of system performance drop and increase.

The last phase is the new steady state of the system performance level, which may equal the previous steady level, reach a lower level or may even be at a higher level than the original one. In order to take this situation into consideration, a simple relative quantitative measure *Recovery Ability* (*RA*) is considered, which measures the new steady state performance level in units of the maximum loss, i.e. is greater than unity if the system bounces back better, equal to unity in case of full recovery and less than unity if it is worse than unity.

An integrated metric with the ability of combining these capabilities can quantify system resilience with an overall perspective and allow comparisons among different systems and system configurations. A general resilience metric (*GR*) is proposed (Nan and Sansavini 2017), which integrates the measures of the resilience capabilities:

$$GR = f\left(R, RAPI_{DP}, RAPI_{RP}, TAPL, RA\right) = R \times \left(\frac{RAPI_{RP}}{RAPI_{DP}}\right) \times \left(TAPL\right)^{-1} \times RA. \tag{2.6}$$

The measure *GR* assumes that robustness *R* (here defined as the lowest performance level reached), recovery speed $RAPI_{RP}$ and recovery ability *RA* have a positive effect on resilience, i.e. are direct proportional to resilience. Conversely, the measure GR assumes that time-averaged performance loss *TAPL* and loss speed $RAPI_{DP}$ have a negative effect, i.e. are indirect proportional to resilience.

No weighting factor is assigned to the various terms so that no bias is introduced, i.e. they contribute equally to resilience. *GR* is dimensionless and is most useful in a comparative manner, i.e. to compare the resilience of various systems to the same disruptive event, or to compare resilience of the same system under different disruptive events. This approach of measuring system resilience is neither model nor domain specific. For instance, historical data can also be used for the resilience analysis. It only requires the time series that represents system output during a time period that covers at least one disruptive event. In this respect, the selection of the MOP is very important.

This performance based overall resilience quantification is an example that resilience quantities for application domains have already been defined and applied (Sansavini 2015).

2.8.2 FRAM Analysis

According to the last development in the functional resonance analysis (FRAM) method (Bellini et al. 2016a), resilience quantification of socio-technical systems like urban transport systems (UTS) can be approached connecting real data to the models. In FRAM there are several dimensions which qualitatively characterize the variability of the output of a FRAM function. However, even when considering invariant the variability of such a FRAM function, the impact of this variability may vary based on the function dampening capacity (FDC) of the downstream FRAM functions. The function damping capacity FDC is defined as the capability in a certain instant of the downstream function of absorbing the variability of the incoming input I (changing conditions) maintaining its output O within acceptable/expected variability.

The formalization proposed is to quantify the amount of performance variability in input exceeding the function dampening capacity (FDC) of the receiving functions. In particular we call this matching the Variability Rate (VR). The VR expresses the amount of input variability still dampened or absorbed by all the downstream functions avoiding effects on their subsequent outputs. In (Bellini et al. 2016a), the FDC expresses the limits of this distribution imposed by the destination functions.

As well known, the Z-score represents a normalization of the distance of an observation from the means in a distributions. The basic Z -score formula is expressed as the ratio between the distance of a value from the mean and the standard deviation: $z = (x - \mu)/\sigma$. Thus the FDC can be represented by the Z-score (ZFDC) value reflecting the limits of the values that can be dampened by the downstream functions. The Variability Rate (VR) in percent reads

$$VR_{R,t}\% = P\left(Z_{FDC_{(R,t)}}\right)\cdot 100 \tag{2.7}$$

where R is a FRAM relationship (see e.g. (Bellini et al. 2016a) and references therein), t is the instant considered and P is the probability associated with the Z-score.

Moreover, since the variability in how a function is carried out may show itself by the variability of the output, the evaluation of the FDC of a function can be done only after the evaluation of its output variability. However, thanks to the Big Data approach, the possibility to determinate FDC in advance and predict the subsequent output variability in order to apply decisions to enhance the FDC through the increase of resource availability, is complex but not complicated.

The focus is on monitoring the resource availability of each function defining firing and variability acceptance thresholds for each of them. This is what is done within the RESOLUTE project. It is connecting all needed information coming from the smart city to the FRAM model, in order to continuously compute and estimate the VRs and connect the resulting scores to a data driven next generation of Decision Support system (Bellini et al. 2016d).

In summary, the FRAM approach helps to identify which quantitative thresholds for functional performance function resilience of the system are of interest, which possibly also can be attributed to single or few components or subsystems. This allows to identify related data sources for the empirical-statistical determination of the thresholds.

2.8.3 Network Models

Network models in resilience management are used as a first order abstract representation for interconnected systems. Here, network nodes (or vertices) are used to represent infrastructure or system units with certain properties and internal states, while network edges (or links) represent dependency relationships between the units.

Depending on specifics of the system of interest, different types of networks can be used. To this end, the most basic model is the undirected network, where links have no preferential direction and equal weight. Undirected networks are suitable representation for e.g., social networks, and certain types of communication networks, where communication between the nodes is reciprocal.

The straightforward generalizations are weighted and directed network models that can be used to represent systems with unequal link weights and/or directional links. Examples of such systems span power grids, transportation networks, and communication networks, e.g., the Internet. For comprehensive review of network models see, e.g., (Newman 2010).

Many interconnected systems are coupled and should not be analyzed separately. Paradigmatic examples are power-grids that are controlled through the Internet, social networks where interactions can be naturally categorized as professional, friendship, romantic, etc., and airline transportation networks, consisting of multiple layers, each corresponding to a distinct airline carrier. To properly model interdependent systems, the wealth of multilayered networks has been developed in recent years (see e.g. Boccaletti et al. 2014).

Despite the success of network models in the analysis of complex systems, resilience approaches to interconnected systems are still at their infancy and a unified approach to resilience in networked systems is yet to be formulated. In recent years there has been several works aiming to quantify resilience in a variety of networked systems. One example is the work of (Sterbenz et al. 2013), describing a methodology to evaluate network resilience using topology generation, analytical, simulation, and experimental emulation techniques.

Another notable contribution is the work of (Zhang et al. 2015) offering a systematic numerical analysis of resilience in a number of different network structures including the grid, ring, hub-and-spoke, complete, scale-free and small-world networks.

The work of (Ganin et al. 2016) is the first to propose quantitative methodologies for engineering resilience in directed graphs and interdependent coupled networks.

The work of (Gao et al. 2016) proposes a method to project coupled linear differential equations describing network dynamics onto a single differential equation quantifying network resilience. Finally, the work by (Vlacheas et al. 2013) attempts to unify the concept of network resilience by identifying principal network resilience concepts and describing the interactions between them.

In summary, network approaches have been used successfully to qualitatively, topologically and quantitatively assess resilience of abstract models of systems, layered systems and systems of systems. However, there is not yet a generally accepted overall approach to the quantification of resilience based on graph models, including multilayered models.

As the wide fields of applications of graph-based methods reveal, it is questionable whether a unique and equally relevant resilience quantification is feasible. This latter statement fits nicely in the approach of the present work which claims that resilience quantification and improvement should occur in a context aware way, which in particular includes the objectives of the resilience quantification, as well as to be well aware of the system definition and boundaries, see Fig. 2.1.

2.9 Summary and Conclusions

The present work addresses the strong need for a generic and tailorable resilience framework and process that covers resilience management and improvement, in particular resilience quantification, development and implementation. This has been attacked by identifying general requirements of such an approach and process requirements, by defining nine process steps (see Fig. 2.1) and most importantly by proposing a wide range of methods and method classes (see Table 2.2) that allow to implement the process and its process steps.

A further main focus of the present work was the definition and quantification of resilience levels (see Table 2.1). Inter alia, they are used

(i) to assess the criticality of combinations of system performance functions and potential (several) disruptions
(ii) to determine the level of rigor of the resilience quantification effort, e.g., for determining the necessary reliability of system resilience functions in the response and recovery phases,
(iii) to determine the effort necessary for development and implementation of such system resilience functions.

By introducing a rich ontology of methods and method classes consisting of 30 categories, mainly covering resilience quantification, it is shown which methods are deemed most relevant for which phase of the resilience management and quantification process phases as well as resilience level (see Table 2.3). This supports to select the most relevant methods and their combinations for each step when assessing and implementing system resilience functions.

The focus of methods, techniques and measures for resilience generation is on resilience assessment and quantification based on appropriate system understanding, modelling and simulation, since resilience development and improvement can resort to existing standards of reliable and dependable system development, as soon as the resilience functions are well defined.

The resilience assessment and improvement process is expected to be reproducible, certifiable and auditable. In particular, it is shown how it links and extends classical risk management, functional safety as well as emerging business continuity approaches.

The resilience management framework is demonstrated by three different case studies, where exemplary system functions deemed relevant for sufficient system resilience have been identified and the used resilience quantification methods and in some cases also the resilience improvement and generation methods have been identified (see Table 2.4). It was shown that the process and methods cover current sample resilience quantification and improvement efforts.

Major advantages of the proposed resilience framework and management process have been discussed, including but not limited to (i) the strong take up of the system context, (ii) the explicit requirement to understand the system and its main (critical) functions and services, (iii) to cover known threats and hazards as well as potential disruptions, (iv) to ask for a complete pre-screening and semi-quantitative identification of critical combinations of system functions and disruptions, (v) the verbose and quantitative definition of resilience system functions, (vi) the resilience level driven selection of resilience quantifications approaches, (vii) the explicit resilience evaluation (decision making) step, (viii) the explicit resilience improvement measure selection step, and (ix) the compact resilience improvement and development step that strongly resorts to standard system (domain-specific) approaches.

With the chosen sample cases as well within the overall presentation of the method it becomes obvious that the presented resilience framework, technical resilience quantification and generation process (in summary generic resilience management process) in particular covers cyber-physical socio-technical systems, non-linear system behavior, snowball and cascading effects. Furthermore, the approach covers physical security, societal security, technical safety, cyber and IT safety, as in particular relevant for internet of things (IoT) developments as assessed relevant for system resilience, respectively.

It is expected that the presented approach is suitable as a starting point for a technical science driven resilience management and improvement. The approach embraces resilience management standardization that takes up inter and intra-disciplinary needs of a wide range of technical and social science domains, in particular the science, technology, engineering and mathematics (STEM) disciplines as well as social sciences, psychology, ecology and economy sciences (e.g. banking and insurance), ethics and political sciences.

2.10 Outlook and Research Needs

The following main future research needs have been identified:

- Refinement of definitions and terminology introduced and their relation to existing frameworks and terms
- Refinement of identified generic and process specific requirements
- Refinement of process step requirements, in particular input and output of each step and more detailed proposal of methods
- Improvement of the completeness and orthogonality of the methods ontology, possibly introduction of hierarchies, e.g. explicit distinction between method classes and methods
- Refinement of the introduced quantitative resilience levels, in particular their role within further resilience management phases, i.e. beyond the resilience management phases 5, 6 and 9
- The suitability of methods within the resilience management process could also take into account the rigor and depth with which the methods are used
- Adaption, tailoring, amendment and extension of existing methods for the proposed resilience management and engineering approach
- Development of novel methods for supporting the proposed process, in particular in the modelling, simulation and resilience assessment of coupled network systems
- Application of the proposed resilience management and engineering process ex post and ex ante to existing and future systems, respectively
- Sharpening the added and orthogonal value when compared with classical (lived, implemented) risk management
- Identification and generation of further engineering/technical science driven resilience quantities that support the proposed resilience management and engineering process
- Complete case studies that apply the proposed resilience management, quantification and implementation process
- Studies on the relation of the proposed approach to existing and emerging standards
- Development of specific methods that are capable
  - to deal with unknown (unknown) disruptions,
  - anomalies of systems,
  - emergent system behavior,
  - uncertainties in all varieties,
  - sensitivity of resilience quantities and
  - capabilities that suffice for fast (near) real-time resilience prediction.

Acknowledgements The research leading to these results has in parts received funding from the European Union's Horizon 2020 Research and Innovation Programme, under Grant Agreement no 653260. The contributions of all RESILENS consortium members are gratefully acknowledged. This research has also partly received funding from the project "Resilience Indicators for

Optimization of Technical Systems" (Number: 181226) of the Freiburg Sustainability Center. The Freiburg Sustainability Center is a cooperation of the Fraunhofer Society and the Albert-Ludwigs-University Freiburg. It is supported by grants from the Baden-Württemberg Ministry of Economics and the Baden-Württemberg Ministry of Science, Research and the Arts.

Further Suggested Readings

Abdul-Rahman A (ed) (2016) Advances in 3D geoinformation, 1st edn. Lecture Notes in Geoinformation and Cartography. Springer International Publishing

AIRMIC, Alarm, IRM (2002) A risk management standard

Aitsi-Selmi A, Egawa S, Sasaki H, Wannous C, Murray V (2015) The sendai framework for disaster risk reduction: renewing the global commitment to people's resilience, health, and well-being. Int J Disaster Risk Sci 6(2):164–176. doi:10.1007/s13753-015-0050-9

Alberts DS, Hayes RE (2003) Power to the edge: command, control in the information age, Information age transformation series. CCRP Publication Series, Washington, DC

Allegrini J, Orehounig K, Mavromatidis G, Ruesch F, Dorer V, Evins R (2015) A review of modelling approaches and tools for the simulation of district-scale energy systems. Renew Sust Energ Rev 52:1391–1404. doi:10.1016/j.rser.2015.07.123

Antenucci A, Sansavini G (2016) Security analysis of the operations of coupled electric and gas network. J Risk Reliab. (to appear soon)

Arboleda CA, Abraham DM, Richard JP., Lubitz R (2006) Impact of interdependencies between infrastructure systems in the operation of health care facilities during disaster events. Joint International Conference on Computing and Decision Making in Civil and Building Engineering, Montreal, Canada

Augustinus C (2003) Handbook on best practices, security of tenure, and access to land: implementation of the Habitat Agenda. United Nations Human Settlements Programme, Nairobi

Australian Government (2010) Critical infrastructure resilience strategy http://www.emergency.qld.gov.au/publications/pdf/Critical_Infrastructure_Resilience_Strategy.pdf. Accessed 5 Aug 2016

Aven T (2016) Risk assessment and risk management: review of recent advances on their foundation. Eur J Oper Res 253(1):1–13. doi:10.1016/j.ejor.2015.12.023

Baird (2010) The phases of emergecy management

Barker K, Ramirez-Marquez JE, Rocco CM (2013) Resilience-based network component importance measures. Reliab Eng Syst Saf 117:89–97. doi:10.1016/j.ress.2013.03.012

Baroud H, Barker K, Ramirez-Marquez JE, Rocco SCM (2014a) Importance measures for inland waterway network resilience. Transport Res Part E Logist Transport Rev 62:55–67. doi:10.1016/j.tre.2013.11.010

Baroud H, Ramirez-Marquez JE, Barker K, Rocco CM (2014b) Stochastic measures of network resilience: applications to waterway commodity flows. Risk Anal Off Public Soc Risk Anal 34(7):1317–1335. doi:10.1111/risa.12175

Barthélemy M (2011) Spatial networks. Phys Rep 499(1–3):1–101. doi:10.1016/j.physrep.2010.11.002

Baumann D, Häring I, Siebold U, Finger J (2014) A web application for urban security enhancement. In: Thoma K, Häring I, Leismann T (eds) 9th future security: Berlin, September 16–18, 2014; proceedings, pp 17–25. Fraunhofer-Verlag, Stuttgart

Bellini P, Benigni M, Billero R, Nesi P, Rauch N (2014a) Km4City ontology building vs data harvesting and cleaning for smart-city services. J Vis Lang Comput 25(6):827–839. doi:10.1016/j.jvlc.2014.10.023

Bellini P, Nesi P, Simoncini M, Tibo A (2014b) Maintenance and emergency management with an integrated indoor/outdoor navigation support. J Vis Lang Comput 25(6):637–649

Bellini E, Gaitanidou E, Ferreira P (2015) D3.5 European resilience management guidelines – H2020 RESOLUTE. www.resolute-eu.org
Bellini E, Nesi P, Pantaleo G, Venturi A (2016a) Functional resonance analysis method based-decision support tool for urban transport system resilience management. In: Proceedings of the second IEEE International Smart Cities Conference (ISC2), Trento (Italy)
Bellini E, Ceravolo P, Nesi P (2016b) Quantify resilience enhancement of UTS through exploiting connect community and internet of everything emerging technologies. ArXiv (Pre-print)
Bellini E, Nesi P, Ferreira P, Simoes A, Candelieri A, Gaitanidou E (2016c) Towards resilience operationalization in urban transport system: the RESOLUTE project aApproach. In: Walls L, Revie M, Bedford T (eds) European safety and reliability conference (ESREL), Glasgow, 25–29.09. Taylor & Francis Group, London
Bellini P, Cenni D, Nesi P (2016d) AP positioning for estimating people flow as origin destination matrix for smart cities. In: The 22nd international conference on distributed multimedia systems, DMS 2016, Italy
Benda L, Miller D, Andras K, Bigelow P, Reeves G, Michael D (2007) NetMap: a new tool in support of watershed science and resource management. For Sci 53(2):206–219
Bilis E, Raschke M, Kröger W (2010) Seismic response of the swiss transmission grid. In: Ale, B.J.M. (ed) Reliability, risk and safety: back to the future; ESREL (European Safety and Reliability), Island of Rhodes, 5–9 September. CRC Press, London, pp 5–9
Bloomfield P (2000) Fourier analysis of time series: an introduction, 2nd edn. Wiley series in probability and statistics. Applied probability and statistics section. Wiley, New York
Boccaletti S, Latora V, Moreno Y, Chavez M, Hwang D-U (2006) Complex networks: structure and dynamics. Phys Rep 424(4–5):175–308. doi:10.1016/j.physrep.2005.10.009
Boccaletti S, Bianconi G, Criado R, del Genio CI, Gómez-Gardeñes J, Romance M, Sendiña-Nadal I, Wang Z, Zanin M (2014) The structure and dynamics of multilayer networks. Phys Rep 544(1):1–122. doi:10.1016/j.physrep.2014.07.001
Bologna S, Carducci G, Bertocchi G, Carrozzi L, Cavallini S, Lazari A, Oliva G, Traballesi A (2016) Guidelines for critical infrastructures resilience evaluation
Box GEP, Jenkins GM, Reinsel GC (2008) Time series analysis: forecasting and control, 4th edn. Wiley series in probability and statistics. John Wiley, Hoboken
Boyd J (1995) The essence of winning and losing: a five slide set by Boyd
Bruneau M, Chang SE, Eguchi RT, Lee GC, O'Rourke TD, Reinhorn AM, Shinozuka M, Tierney K, Wallace WA, von Winterfeldt D (2003) A framework to quantitatively assess and enhance the seismic resilience of communities. Earthq Spectra 19(4):733–752. doi:10.1193/1.1623497
Burgholzer W, Bauer G, Posset M, Jammernegg W (2013) Analysing the impact of disruptions in intermodal transport networks: a micro simulation-based model. Decis Support Syst 54(4):1580–1586. doi:10.1016/j.dss.2012.05.060
Cavallo A, Ireland V (2014) Preparing for complex interdependent risks: a system of systems approach to building disaster resilience. Int J Disaster Risk Reduc 9:181–193. doi:10.1016/j.ijdrr.2014.05.001
Chang SE, Shinozuka M (2004) Measuring improvements in the disaster resilience of communities. Earthq Spectra 20(3):739–755. doi:10.1193/1.1775796
Chappin EJL, Dijkema GPJ (2010) Agent-based modelling of energy infrastructure transitions. Int J Crit Infrastruct 6(2):106–130
Christmann GB (2014) Investigating spatial transformation processes. An ethnographic discourse analysis in disadvantaged neighbourhoods. Hist Soc Res 2:235–256
Cimellaro GP, Reinhorn AM, Bruneau M (2010) Framework for analytical quantification of disaster resilience. Eng Struct 32(11):3639–3649. doi:10.1016/j.engstruct.2010.08.008
Committee on Increasing National Resilience to Hazards and Disasters, Committee on Science, Engineering, and Public Policy (2012) The national academies: disaster resilience: a national imperative. National Academies Press, Washington, DC
Cooper CHV, Chiaradia AJ (2015) sDNA: how and why we reinvented spatial network analysis for health, economics and active modes of transport. GIS Research UK (GISRUK)

Cutter SL (2016) The landscape of disaster resilience indicators in the USA. Nat Hazards 80(2):741–758. doi:10.1007/s11069-015-1993-2
Dekker S, Hollnagel E, Woods D, Cook R (2008) Resilience engineering: new directions for measuring and maintaining safety in complex systems: final report
Delligatti L (2014) SysML distilled: a brief guide to the systems modeling language. Addison-Wesley, Upper Saddle River
Demirel H, Kompil M, Nemry F (2015) A framework to analyze the vulnerability of European road networks due to Sea-Level Rise (SLR) and sea storm surges. Transp Res A Policy Pract 81:62–76. doi:10.1016/j.tra.2015.05.002
DiMase D, Collier ZA, Heffner K, Linkov I (2015) Systems engineering framework for cyber physical security and resilience. Environ Syst Decis 35(2):291–300. doi:10.1007/s10669-015-9540-y
Djordjević S, Butler D, Gourbesville P, Mark O, Pasche E (2011) New policies to deal with climate change and other drivers impacting on resilience to flooding in urban areas: the CORFU approach. Environ Sci Pol 14(7):864–873. doi:10.1016/j.envsci.2011.05.008
Dorbritz R (2011) Assessing the resilience of transportation systems in case of large-scale disastrous events. In: Cygas D, Froehner KD, Breznikar A (eds) Environmental engineering: selected papers. VGTU Press "Technika" scientific book, No 1867-M, pp 1070–1076. Vilnius Gediminas Technical University press "Technika", Vilnius
Edwards C (2009) Resilient nation. Demos, London
Enders W, Parise GF, Sandler T (1992) A time-series analysis of transnational terrorism: trends and cycles. Def Econ 3(4):305–320. doi:10.1080/10430719208404739
Esmiller B, Curatella F, Kalousi G, Kelly D, Amato F, Häring I, Schäfer J, Ryzenko J, Banaszek M, Katzmarekt KU (2013) FP7 integration project D-Box: comprehensive toolbox for humanitarian clearing of large civil areas from anti-personal landmines and cluster munitions. In: International symposium humanitarian demining, Šibenik, 23–25.04, pp 21–22
Falasca M, Zobel CW, Cook D (2008) A decision support framework to assess supply chain resilience. In: Friedrich F, van de Valle B (eds) Proceedings of the 5th international ISCRAM conference, Washington, DC, pp 596–605
Faturechi R, Miller-Hooks E (2014) Travel time resilience of roadway networks under disaster. Transp Res B Methodol 70:47–64. doi:10.1016/j.trb.2014.08.007
Fenwick M, Hisatake M, Vermeulen EPM (2016) Intelligent cars INC. – Governance principles to build a disruptive company. SSRN J. doi:10.2139/ssrn.2823006
Finger J, Hasenstein S, Siebold U, Häring I (2016) Analytical resilience quantification for critical infrastructure and technical systems. In: Walls L, Revie M, Bedford T (eds) European safety and reliability conference (ESREL), Glasgow, 25–29.09. Taylor & Francis Group, London, pp 2122–2128
Fischer K, Häring I (2009) SDOF response model parameters from dynamic blast loading experiments. Eng Struct 31(8):1677–1686. doi:10.1016/j.engstruct.2009.02.040
Fischer K, Siebold U, Vogelbacher G, Häring I, Riedel W (2014) Empirische Analyse sicherheitskritischer Ereignisse in urbanisierten Gebieten. Bautechnik 91(4):262–273. doi:10.1002/bate.201300041
Fischer K, Häring I, Riedel W (2015) Risk-based resilience quantification and improvement for urban areas. In: Beyerer J, Meissner A, Geisler J (eds) Security research conference: 10th future security. Fraunhofer, Stuttgart, pp 417–424
Fischer K, Häring I, Riedel W, Vogelbacher G, Hiermaier S (2016) Susceptibility, vulnerability, and averaged risk analysis for resilience enhancement of urban areas. Int J Protect Struct 7(1):45–76. doi:10.1177/2041419615622727
Fortunato S (2010) Community detection in graphs. Phys Rep 486(3–5):75–174. doi:10.1016/j.physrep.2009.11.002
Fox-Lent C, Bates ME, Linkov I (2015) A matrix approach to community resilience assessment: an illustrative case at Rockaway Peninsula. Environ Syst Decis 35(2):209–218. doi:10.1007/s10669-015-9555-4

Francis R, Bekera B (2014) A metric and frameworks for resilience analysis of engineered and infrastructure systems. Reliab Eng Syst Saf 121:90–103. doi:10.1016/j.ress.2013.07.004
Friedenthal S, Moore A, Steiner R (2011) A practical guide to SysML: the systems modeling language, 2nd edn. Morgan Kaufmann, Amsterdam/Boston
Ganin AA, Massaro E, Gutfraind A, Steen N, Keisler JM, Kott A, Mangoubi R, Linkov I (2016) Operational resilience: concepts, design and analysis. Sci Rep 6:19540. doi:10.1038/srep19540
Gao J, Barzel B, Barabasi A-L (2016) Universal resilience patterns in complex networks. Nature 530(7590):307–312. doi:10.1038/nature16948
Gertsbakh IB, Shpungin Y (2011) Network reliability and resilience. SpringerBriefs in electrical and computer engineering. Springer, Heidelberg/New York. doi: 10.1007/978-3-642-22374-7
Gopalakrishnan K, Peeta S (eds) (2010) Sustainable and resilient critical infrastructure systems: simulation, modeling, and intelligent engineering. Springer, Berlin/Heidelberg
Grasso V, Crisci A, Nesi P, Pantaleo G, Zaza I, Gozzini B (2016) Public-crowd-sensing of heatwaves by social-media data. 16th EMS annual meeting & 11th European conference on applied climatology (ECAC), 12–16 September 2016 Trieste, Italy CE2/AM3 delivery and communication of impact and based forecasts and risk based warnings
Greene R (2011) GIS-Based multiple-criteria decision analysis. Geography Compass 5:412–432
Gröger G, Plümer L (2012) CityGML – interoperable semantic 3D city models. ISPRS J Photogramm Remote Sens 71:12–33. doi:10.1016/j.isprsjprs.2012.04.004
Hamilton MC, Lambert JH, Connelly EB, Barker K (2016) Resilience analytics with disruption of preferences and lifecycle cost analysis for energy microgrids. Reliab Eng Syst Saf 150:11–21. doi:10.1016/j.ress.2016.01.005
Häring I, Schönherr M, Richter C (2009) Quantitative hazard and risk analysis for fragments of high-explosive shells in air. Reliab Eng Syst Saf 94(9):1461–1470. doi:10.1016/j.ress.2009.02.003
Häring I, Ebenhöch S, Stolz A (2016a) Quantifying resilience for resilience engineering of socio technical systems. Eur J Secur Res 1(1):21–58. doi:10.1007/s41125-015-0001-x
Häring I, Scharte B, Stolz A, Leismann T, Hiermaier S (2016b) Resilience engineering and quantification for sustainable systems development and assessment: socio-technical systems and critical infrastructures. In: Resource guide on resilience. EPFL international risk governance center, Lausanne
Havran V, Prikryl J, Purgathofer W (2000) Statistical comparison of ray shooting effiency schemes. Institute of Computer Graphics, Vienna University of Technology
Henry D, Ramirez-Marquez JE (2012) Generic metrics and quantitative approaches for system resilience as a function of time. Reliab Eng Syst Saf 99:114–122. doi:10.1016/j.ress.2011.09.002
Holling CS (1973) Resilience and stability of ecological systems. Annu Rev Ecol Syst 4:1–23
Hollnagel E (2009) The four cornerstones of resilience engineering. In: Nemeth CP, Hollnagel E, Dekker S (ed) Resilience engineering perspectives, vol 2: preparation and restoration. Ashgate studies in resilience engineering, Chapter 6, pp 117–134. Ashgate
Hollnagel E (2011) Prologue: the scope of resilience engineering. Resil Eng Pract Guidebook
Hollnagel E, Tveiten CK, Albrechtsen E (2010) Resilience engineering and integrated operations in the petroleum industry. IO-center (SINTEF) report, SINTEF A16331. Center for Integrated Operations in the Petroleum Industry, Trondheim
Holme P, Saramäki J (2012) Temporal networks. Phys Rep 519(3):97–125. doi:10.1016/j.physrep.2012.03.001
Hosseini S, Barker K (2016) Modeling infrastructure resilience using Bayesian networks: a case study of inland waterway ports. Comput Ind Eng 93:252–266. doi:10.1016/j.cie.2016.01.007
Hughes BP, Newstead S, Anund A, Shu CC, Falkmer T (2015) A review of models relevant to road safety. Accid Anal Prev 74:250–270. doi:10.1016/j.aap.2014.06.003
Index to Qualitative Health Research (2000) Qual Health Res 10(6):855–860. doi:10.1177/104973200129118778
Ip WH, Wang D (2011) Resilience and friability of transportation networks: evaluation, analysis and optimization. IEEE Syst J 5(2):189–198. doi:10.1109/JSYST.2010.2096670

ISO 31010 (2009) Risk management—risk assessment techniques. International Organization for Standardization, Genf

Jackson S (2010) Architecting resilient systems: accident avoidance and survival and recovery from disruptions. Wiley series in systems engineering and management. Wiley, Hoboken

Jendrian K (2014) Der Standard ISO/IEC 27001: 2013. Datenschutz und Datensicherheit-DuD 38(8):552–557

Jenelius E, Mattsson L-G (2012) Road network vulnerability analysis of area-covering disruptions: a grid-based approach with case study. Transp Res A Policy Pract 46(5):746–760. doi:10.1016/j.tra.2012.02.003

Jenelius E, Mattsson L-G (2015) Road network vulnerability analysis: conceptualization, implementation and application. Comput Environ Urban Syst 49:136–147. doi:10.1016/j.compenvurbsys.2014.02.003

Jing Y, Ahn G-J, Zhao Z, Hu H (2014) RiskMon: continuous and automated risk assessment of mobile applications. In: Bertino E, Sandhu R, Park J (eds) CODASPY'14: proceedings of the 4th ACM conference on data and application security and privacy, March 3–5, 2014, San Antonio, Texas, pp 99–110. Association for Computing Machinery, New York

Johansson, ACH, Svedung I, Andersson R (2006) Management of risks in societal planning – an analysis of scope and variety of health, safety and security issues in municipality plan documents. Safe Sci 44(8):675–688. doi:10.1016/j.ssci.2006.03.001

Joint International Conference on Computing and Decision Making in Civil and Building Engineering (2006) Impact of interdependencies between infrastructure systems in the operation of health care facilities during disaster events

Kaufman J, Häring I (2011) Functional safety requirements for active protection systems from individual and collective risk criteria. In: Soares C (ed) Advances in safety, reliability and risk management. CRC Press, pp 133–140

Kaufmann R, Häring I (2013) Comparison of 3D visualization options for quantitative risk analyses. In: Steenbergen RDJM (ed) Safety, reliability and risk analysis: beyond the horizon: proceedings of the European Safety and Reliability Conference, ESREL 2013, Amsterdam, 29 September–2 October. A Balkema book. CRC Press, Boca Raton, pp 2019–2026

Kerner BS (2014) Three-phase theory of city traffic: moving synchronized flow patterns in under-saturated city traffic at signals. Phys A Stat Mech Appl **397**:76–110. doi:10.1016/j.physa.2013.11.009

Keybl M, Fandozzi J, Graves R, Taylor M, Yost B (2012) Harmonizing risk and quantifying preparedness. In: IEEE conference on technologies for Homeland Security (HST), 2012: 13–15, November 2012, Waltham, Massachusetts, USA. IEEE, Piscataway. doi:10.1109/THS.2012.6459845

Khademi N, Balaei B, Shahri M, Mirzaei M, Sarrafi B, Zahabiun M, Mohaymany AS (2015) Transportation network vulnerability analysis for the case of a catastrophic earthquake. Int J Disaster Risk Reduc 12:234–254. doi:10.1016/j.ijdrr.2015.01.009

Khakzad N, Reniers G (2015) Using graph theory to analyze the vulnerability of process plants in the context of cascading effects. Reliab Eng Syst Saf 143:63–73. doi:10.1016/j.ress.2015.04.015

Khaled AA, Jin M, Clarke DB, Hoque MA (2015) Train design and routing optimization for evaluating criticality of freight railroad infrastructures. Transp Res B Methodol 71:71–84. doi:10.1016/j.trb.2014.10.002

Km4City major tools. (2017) http://www.km4city.org

Koulakezian A, Abdelgawad H, Tizghadam A, Abdulhai B, Leon-Garcia A (2015) Robust network design for roadway networks: unifying framework and application. IEEE Intell Transport Syst Mag 7(2):34–46. doi:10.1109/MITS.2014.2386654

Kovalenko T, Sornette D (2013) Dynamical diagnosis and solutions for resilient natural and social systems. Planet@Risk 1(1):7–33

La Porte TR (1996) High reliability organizations: unlikely, demanding and at risk. J Conting Crisis Manag 4(2):60–71

Laprise M, Lufkin S, Rey E (2015) An indicator system for the assessment of sustainability integrated into the project dynamics of regeneration of disused urban areas. Build Environ 86:29–38. doi:10.1016/j.buildenv.2014.12.002

Larisch M, Siebold U, Häring I (2011) Safety aspects of generic real-time embedded software model checking in the fuzing domain. In: Soares C (ed) Advances in safety, reliability and risk management. CRC Press, pp 2678–2684

Larkin S, Fox-Lent C, Eisenberg DA, Trump BD, Wallace S, Chadderton C, Linkov I (2015) Benchmarking agency and organizational practices in resilience decision making. Environ Syst Decis 35(2):185–195. doi:10.1007/s10669-015-9554-5

Law C-W, Lee C-K, Tai M-K (2006) Visualization of complex noise environment by virtual reality technologies. Environment Protection Department (EPD), Hong Kong. http://www.science.gov.hk/paper/EPD_CWLaw.pdf. Access Date: March 2008

Li B, Sansavini G (2016) Effective multi-objective selection of inter-subnetwork power shifts to mitigate cascading failures. Electr Power Syst Res 134:114–125

Linkov I, Eisenberg DA, Bates ME, Chang D, Convertino M, Allen JH, Flynn SE, Seager TP (2013a) Measurable resilience for actionable policy. Environ Sci Technol 10:108–10110. doi:10.1021/es403443n

Linkov I, Eisenberg DA, Plourde K, Seager TP, Allen J, Kott A (2013b) Resilience metrics for cyber systems. Environ Syst Decis 33(4):471–476. doi:10.1007/s10669-013-9485-y

Linkov I, Bridges T, Creutzig F, Decker J, Fox-Lent C, Kröger W, Lambert JH, Levermann A, Montreuil B, Nathwani J, Nyer R, Renn O, Scharte B, Scheffler A, Schreurs M, Thiel-Clemen T (2014) Changing the resilience paradigm. Nat Clim Chang 4(6):407–409. doi:10.1038/nclimate2227

Lu L, Becker T, Löwner MO (2016) 3D complete traffic noise analysis based on CityGML. In: Abdul-Rahman A (ed) Advances in 3D geoinformation, 1st edn. Lecture notes in geoinformation and cartography. Springer International Publishing

Mansfield J (2010) The nature of change or the law of unintended consequences: an introductory text to designing complex systems and managing change. Imperial College Press; Distributed by World Scientific Pub, London/Singapore/Hackensack

Mattsson L-G, Jenelius E (2015) Vulnerability and resilience of transport systems – a discussion of recent research. Transp Res A Policy Pract 81:16–34. doi:10.1016/j.tra.2015.06.002

McDaniels T, Chang S, Cole D, Mikawoz J, Longstaff H (2008) Fostering resilience to extreme events within infrastructure systems: characterizing decision contexts for mitigation and adaptation. Glob Environ Chang 18(2):310–318. doi:10.1016/j.gloenvcha.2008.03.001

MCEER (2006) MCEER's resilience framework

MIT (2006) Global airline industry programm – airline industry overview. http://web.mit.edu/airlines/analysis/analysis_airline_industry.html. Accessed 16 Sept 2014

Mock RG, Lopez de Obeso L, Zipper C (2016) Resilience assessment of internet of things: a case study on smart buildings. In: Walls L, Revie M, Bedford T (eds) European safety and reliability conference (ESREL), Glasgow, 25–29.09. Taylor & Francis Group, London, pp 2260–2267

Naim Kapucu N, Hawkins CV, Rivera FI (eds) (2013) Disaster resiliency. Interdisciplinary perspectives. Routledge, London

Nan C, Sansavini G (2017) A quantitative method for assessing resilience of interdependent infrastructures. Reliab Eng Syst Saf 157:35–53

Nan C, Sansavini G, Kröger W (2016) Building an integrated metric for quantifying the resilience of interdependent infrastructure systems. In: Panayiotou CG, Ellinas G, Kyriakides E, Polycarpou MM (eds) Critical information infrastructures security. Lecture notes in computer science. Springer International Publishing, Cham, pp 159–171

National Consortium for the study of Terrorism and Responses to Terrorism (2014) Global terrorism database. http://www.start.umd.edu/gtd/. Accessed 28 Aug 2013

Newcastle City-wide Floodplain Risk Management Study and Plan (2010) The city of New Castle

Newman M (2010) Networks: an introduction. Oxford University Press, Oxford

Nnorom IC, Osibanjo O (2009) Toxicity characterization of waste mobile phone plastics. J Hazard Mater 161(1):183–188. doi:10.1016/j.jhazmat.2008.03.067

Norros L (2004) Acting under uncertainty The core-task analysis in ecological study of work. Espoo, VTT, Finland

O'Rourke TD (2007) Critical infrastructure, interdependencies, and resilience. The Bridge

Object Management Group (2012) OMG systems modeling language (OMG SysML™): Version 1.3. http://www.omg.org/spec/SysML/1.3/. Accessed 15 Oct 2014

Oliveira EDL, Portugal LDS, Porto Junior W (2016) Indicators of reliability and vulnerability: similarities and differences in ranking links of a complex road system. Transport Res Part A Policy Pract 88:195–208. doi:10.1016/j.tra.2016.04.004

Ontario critical infrastructure resilience model (2017b) http://www.risklogik.com/ontario-critical-infrastructure-resilience-model

Osinga FPB (2007) Science, strategy and war: the strategic theory of John Boyd. Strategy and history, vol 18. Routledge, London/New York

Ouyang M (2014) Review on modeling and simulation of interdependent critical infrastructure systems. Reliab Eng Syst Safe 121:43–60. doi:10.1016/j.ress.2013.06.040

Ouyang M, Dueñas-Osorio L (2012) Time-dependent resilience assessment and improvement of urban infrastructure systems. Chaos (Woodbury, NY) 22(3):033122. doi:10.1063/1.4737204

Ouyang M, Wang Z (2015) Resilience assessment of interdependent infrastructure systems: with a focus on joint restoration modeling and analysis. Reliab Eng Syst Saf 141:74–82. doi:10.1016/j.ress.2015.03.011

Pant R, Barker K, Ramirez-Marquez JE, Rocco CM (2014) Stochastic measures of resilience and their application to container terminals. Comput Ind Eng 70:183–194. doi:10.1016/j.cie.2014.01.017

Paul Seidenstat, FX Splane (ed) (2009) Protecting airline passengers in the age of terrorism. Praeger Security International, Santa Barbara

Pearl TH (2016) Fast & Furious: the misregulation of driverless cars. SSRN J. doi:10.2139/ssrn.2819473

Perrow C (2011) Normal accidents: living with high risk technologies. Princeton University Press, Princeton

Podofillini L, Sudret B, Stojadinović B, Zio E, Kröger W (eds) (2015) Safety and reliability of complex engineered systems. In: Proceedings of the 25th European safety and reliability conference, ESREL 2015, Zürich, 7–10 September. CRC Press, Boca Raton

Procter JN, Cronin SJ, Platz T, Patra A, Dalbey K, Sheridan M, Neall V (2010) Mapping block-and-ash flow hazards based on Titan 2D simulations: a case study from Mt. Taranaki, NZ. Nat Hazards 53(3):483–501

Purdy G (2010) ISO 31000: 2009—setting a new standard for risk management. Risk Anal Off Pub Soc Risk Anal 30(6):881–886. doi:10.1111/j.1539-6924.2010.01442.x

Ramirez-Marquez JE, Rocco CM, Moronta J, Gama Dessavre D (2016) Robustness in network community detection under links weights uncertainties. Reliab Eng Syst Saf 153:88–95. doi:10.1016/j.ress.2016.04.009

Raschke M, Bilis E, Kröger W (2011) Vulnerability of the Swiss electric power transmission grid against natural hazards. Appl Stat Probab Civil Eng:1407–1414

Reggiani A (2013) Network resilience for transport security: some methodological considerations. Transp Policy 28:63–68. doi:10.1016/j.tranpol.2012.09.007

Renger P, Siebold U, Kaufmann R, Häring I (2015) Semi-formal static and dynamic modeling and categorization of airport checkpoints. In: Nowakowski T, Mlynczak M, Jodejko-Pietruczuk A, Werbinska-Wojciechowska S (eds) Safety and reliability: methodology and applications; [ESREL 2014 conference, held in Wrocław, Poland]. CRC Press, London, pp 1721–1731. doi:10.1201/b17399-234

Renschler CS, Fraizer AE, Arendt LA, Cimellaro G-P, Reinhorn AM, Bruneau M (2011) A framework for defining and measuring resilience at the community scale. The PEOPLES Resilience Framework

Resolute – RESOLUTE – Home (2015) www.resolute-eu.org
Riedel W, Fischer K, Stolz A, Häring I, Bachmann M (2015) Modeling the vulnerability of urban areas to explosion scenarios. In: Stewart M, Netherton M (eds) 3rd International Conference on Protective Structures (ICPS3), Newcastle, Australia, 03–06.02
Riedel W, Niwenhuijs A, Fischer K, Crabbe S, Heynes W, Müllers I, Trojaberg S, Häring I (2014) Quantifying urban risk and vulnerability – a toolsuite of new methods for planners. In: Thoma K, Häring I, Leismann T (eds) 9th future security: Berlin, September 16–18, 2014; proceedings. Fraunhofer-Verlag, Stuttgart, pp 8–16
Righi AW, Saurin TA, Wachs P (2015) A systematic literature review of resilience engineering: research areas and a research agenda proposal. Reliab Eng Syst Saf 141:142–152. doi:10.1016/j.ress.2015.03.007
Rose A (2004) Defining and measuring economic resilience to disasters. Disaster Prev Manag 13(4):307–314. doi:10.1108/09653560410556528
Rose A (2009) Economic resilience to disasters
Rose A, Liao S-Y (2005) Modeling regional economic resilience to disasters: a computable general equilibrium analysis of water service disruptions*. J Reg Sci 45(1):75–112. doi:10.1111/j.0022-4146.2005.00365.x
Sahebjamnia N, Torabi SA, Mansouri SA (2015) Integrated business continuity and disaster recovery planning: towards organizational resilience. Eur J Oper Res 242(1):261–273. doi:10.1016/j.ejor.2014.09.055
Sansavini G (2015) Vulnerability and resilience of future interdependent energy networks
Satumtira G, Dueñas-Osorio L (2010) Synthesis of modeling and simulation methods on critical infrastructure interdependencies research. In: Gopalakrishnan K, Peeta S (eds) Sustainable and resilient critical infrastructure systems: simulation, modeling, and intelligent engineering. Springer, Berlin/Heidelberg, pp 1–51
Schäfer J, Kopf N, Häring I (2014) Empirical risk analysis of humanitarian demining for characterization of hazard sources. In: Thoma K, Häring I, Leismann T (eds) 9th future security: Berlin, September 16–18, 2014; proceedings. Fraunhofer-Verlag, Stuttgart, pp 598–602
Schoppe C, Häring I, Siebold U (2013) Semi-formal modeling of risk management process and application to chance management and monitoring. In: Steenbergen RDJM (ed) Safety, reliability and risk analysis: beyond the horizon: proceedings of the European Safety and Reliability Conference, ESREL 2013, Amsterdam, 29 September–2 October. A Balkema book. CRC Press, Boca Raton, pp 1411–1418
Schoppe C, Zehetner J, Finger J, Baumann D, Siebold U, Häring I (2015) Risk assessment methods for improving urban security. In: Nowakowski T, Mlynczak M, Jodejko-Pietruczuk A, Werbinska-Wojciechowska S (eds) Safety and reliability: methodology and applications; [ESREL 2014 Conference, held in Wrocław, Poland]. CRC Press, London, pp 701–708
Schrenk M (ed) (2008) Mobility nodes as innovation hubs: REAL CORP 008; 13th international conference on urban planning, regional development and information society, Vienna International Airport, 19–21.05. Selbstverl. des Vereins CORP – Competence Center of Urban and Regional Planning, Schwechat-Rannersdorf
Sharit J (2012) Human error and human reliability analysis. Handbook of human factors and ergonomics, 4th edn. pp 734–800
Shirali G, Mohammadfam I, Ebrahimipour V (2013) A new method for quantitative assessment of resilience engineering by PCA and NT approach: a case study in a process industry. Reliab Eng Syst Saf 119:88–94. doi:10.1016/j.ress.2013.05.003
Siebold U (n.d.) Untersuchung statistischer Auswerteverfahren zur Analyse sicherheitsrelevanter Ereignisse: Diplomarbeit
Siebold U (2013) Identifikation und Analyse von sicherheitsbezogenen Komponenten in semiformalen Modellen. Schriftenreihe Epsilon – Forschungsergebnisse aus der Kurzzeitdynamik, vol 25. Fraunhofer Verlag, Stuttgart
Siebold U, Hasenstein S, Finger J, Häring I (2015) Table-top urban risk and resilience management for football events. In: Podofillini L, Sudret B, Stojadinović B, Zio E, Kröger W (eds)

Safety and reliability of complex engineered systems: proceedings of the 25th European safety and reliability conference, ESREL 2015, Zürich, 7–10 September. CRC Press, Boca Raton, pp 3375–3382

Sivaharan T, Blair G, Friday A, Wu M, Duran-Limon H, Okanda P, Sorensen CF (eds) (2004) Cooperating sentient vehicles for next generation automobiles. ACM/USENIX MobiSys 2004 International Workshop on Applications of Mobile Embedded Systems (WAMES 2004 online proceedings)

Statistics Canada (2014) Canada's National Statistical Agency. http://www.statcan.gc.ca/eng/start

Steenbergen RDJM (ed) (2013) Safety, reliability and risk analysis: beyond the horizon. In: Proceedings of the European safety and reliability conference, ESREL 2013, Amsterdam, 29 September–2 October. A Balkema book. CRC Press, Boca Raton

Sterbenz JPG, Çetinkaya EK, Hameed MA, Jabbar A, Qian S, Rohrer JP (2013) Evaluation of network resilience, survivability, and disruption tolerance: analysis, topology generation, simulation, and experimentation. Telecommun Syst. doi:10.1007/s11235-0119573-6

Stergiopoulos G, Kotzanikolaou P, Theocharidou M, Lykou G, Gritzalis D (2015) Time-based critical infrastructure dependency analysis for large-scale and cross-sectoral failures. Int J Crit Infrastruct Prot. doi:10.1016/j.ijcip.2015.12.002

Størseth F, Tinmannsvik RK, Øien K (2010) Building safety by resilient organization – a case specific approach. In: Briš R (ed) Reliability, risk and safety: theory and applications; proceedings of the European safety and reliability conference, ESREL 2009, Prague, Czech Republic, 7–10 September 2009. CRC Press/Balkema, Leiden

Stouffer K, Pillitteri V, Lightman S, Abrams M, Hahn A (2015) Guide to Industrial Control Systems (ICS) security. National Institute of Standards and Technology

Sturrock PA, Shoub EC (1982) Examination of time series through randomly broken windows. ApJ 256:788. doi:10.1086/159951

Suter M (2011) Resilience and risk management in critical infrastructure protection policy: exploring the relationship and comparing its use. Focal Report, 7. Center for Security Studies (CSS), ETH Zurich, Zurich

Tamvakis P, Xenidis Y (2013) Comparative evaluation of resilience quantification methods for infrastructure systems. Procedia Soc Behav Sci 74:339–348. doi:10.1016/j.sbspro.2013.03.030

Taylor MAP, Sekhar SVC, D'Este GM (2006) Application of accessibility based methods for vulnerability analysis of strategic road networks. Netw Spatial Econ 6(3):267–291. doi:10.1007/s11067-006-9284-9

Thoma K (ed) (2011) European perspectives on security research. Acatech diskutiert. Springer, Berlin/Heidelberg

Thoma K (2014) Resilien-Tech:» Resilience by Design «: a strategy for the technology issues of the future. Herbert Utz Verlag

Thoma K, Scharte B, Hiller D, Leismann T (2016) Resilience engineering as part of security research: definitions. Concept Sci Approach Eur J Sec Res 1(1):3–19. doi:10.1007/s41125-016-0002-4

Tierney K, Bruneau M (2007) Conceptualizing and measuring resilience: a key to disaster loss reduction. TR News 250:14–18

Tochtermann K, Scharl A (eds) (2006) EnviroInfo 2006: managing environmental knowledge. In: Proceedings of the 20th international conference "Informatics for Environmental Protection", Graz (Austria). [Berichte aus der Umweltinformatik]. Shaker, Aachen

Tolk A, Uhrmacher AM (2009) Agents: agenthood, agent architectures, and agent taxonomies. Agent-directed simulation and systems engineering. pp 75–109

UN General Assembly (2015) Integrated and coordinated implementation of and follow-up Integrated and coordinated implementation of and follow-up to the outcomes of the major United Nations conferences and summits in the economic, social and related fields. Follow-up to the outcome of the Millennium Summit, New York van der Beek D, Schraagen JM (2015) ADAPTER: Analysing and developing adaptability and performance in teams to enhance resilience. Reliab Eng Syst Saf 141:33–44. doi:10.1016/j.ress.2015.03.019

van Someren MW, Barnard YF, Sandberg JAC (1994) The think aloud method: a practical guide to modelling cognitive processes. Academic Press, London

Vlacheas P, Stavroulaki V, Demestichas P, Cadzow S, Ikonomou D, Gorniak S (2013) Towards end-to-end network resilience. Int J Crit Infrastruct Prot 6(3–4):159–178. doi:10.1016/j.ijcip.2013.08.004

Vogelbacher G, Häring I, Fischer K, Riedel W (2016) Empirical susceptibility, vulnerability and risk analysis for resilience enhancement of urban areas to terrorist events. Eur J Secur Res 1(2):151–186. doi:10.1007/s41125-016-0009-x

Voss M, Häring I, Fischer K, Riedel W, Siebold U (2012) Susceptibility and vulnerability of urban buildings and infrastructure against terroristic threats from qualitative and quantitative risk analyses. In: 11th international probabilistic safety assessment and management conference and the annual European safety and reliability conference 2012: (PSAM11 ESREL 2012). Curran Associates, Inc., New York, pp 5757–5767

Vugrin ED, Turnquist MA, Brown NJ (2014) Optimal recovery sequencing for enhanced resilience and service restoration in transportation networks. IJCIS 10(3/4):218. doi:10.1504/IJCIS.2014.066356

Vugrin ED, Warren DE, Ehlen MA (2011) A resilience assessment framework for infrastructure and economic systems: quantitative and qualitative resilience analysis of petrochemical supply chains to a hurricane. Proc Safety Prog 30(3):280–290. doi:10.1002/prs.10437

Walker B, Carpenter S, Anderies J, Abel N, Cumming G, Janssen M, Lebel L, Norberg J, Peterson GD, Pritchard R (2002) Resilience management in social-ecological systems: a working hypothesis for a participatory approach. Conserv Ecol 6(1)

Weilkiens T (2007) Systems engineering with SysML/UML: modeling, analysis, design. The OMG press. Morgan Kaufmann, Burlington

Zhang X, Miller-Hooks E, Denny K (2015) Assessing the role of network topology in transportation network resilience. J Transp Geogr 46:35–45. doi:10.1016/j.jtrangeo.2015.05.006

Zobel CW (2011) Representing perceived tradeoffs in defining disaster resilience. Decis Support Syst 50(2):394–403. doi:10.1016/j.dss.2010.10.001

Chapter 3
Redesigning Resilient Infrastructure Research

Thomas P. Seager, Susan Spierre Clark, Daniel A. Eisenberg, John E. Thomas, Margaret M. Hinrichs, Ryan Kofron, Camilla Nørgaard Jensen, Lauren R. McBurnett, Marcus Snell, and David L. Alderson

Abstract Despite federal policy directives to strengthen the *resilience* of critical infrastructure systems to extreme weather and other adverse events, several knowledge and governance barriers currently frustrate progress towards policy goals, namely: (1) a lack of awareness of what constitutes resilience in diverse infrastructure applications, (2) a lack of judgement about how to create resilience, (3) a lack of incentives that motivate resilience creation, and (4) obstacles that prevent action or reform, even where incentives exist, within existing governance systems. In this chapter, we describe each of these barriers in greater detail and provide a catalog of theories for overcoming them. Regarding awareness, we contrast four different characterizations of resilience as rebound, robustness, graceful extensibility, and sustained adaptability. We apply Integral Theory to demonstrate the necessity of integrating multiple investigative perspectives. Further, we illustrate the importance of recognizing resilience as a set of processes, in addition to resources and outcomes, and the difficulty of measuring quality and quality of resilience actions. Regarding judgement, we position infrastructure as the principal mechanism by which human rights are realized as human capabilities, and propose applying theories of human development such as Maslow's hierarchy of needs to identify the most critical infrastructure in terms of the services they provide to end users. Regarding a lack of incentives, we examine the modes and tools of financial analysis by which investments in resilience infrastructure may be prioritized and find two failings: the difficulty of estimating the monetary value of optionality, and the problem of exponential discounting of future cash flows. Regarding obstacles to action, we describe a hierarchy of adaptive actions applicable to physical infrastructure and

T.P. Seager (✉) • S.S. Clark • D.A. Eisenberg • J.E. Thomas • M.M. Hinrichs • R. Kofron
C.N. Jensen • L.R. McBurnett • M. Snell
School of Sustainable Engineering & The Built Environment, Arizona State University, Tempe 85287, AZ, USA
e-mail: thomas.seager@asu.edu

D.L. Alderson
Center for Infrastructure Defense, Operations Research Department, Naval Post Graduate School, Monterey 93943, CA, USA

© Springer Science+Business Media B.V. 2017

I. Linkov, J.M. Palma-Oliveira (eds.), *Resilience and Risk*, NATO Science for Peace and Security Series C: Environmental Security,
DOI 10.1007/978-94-024-1123-2_3

the essential dimensions of organizational maturity that determine how these adaptive actions might be initiated. Additionally, we discuss the difficulty of education and training for resilient infrastructure systems and propose simulation gaming as an integrative research and education approach for capturing lessons learned from historical catastrophes, play-testing scenarios, sharing knowledge, and training a workforce prepared for the challenges of the post-industrial infrastructure age. Finally, we suggest establishing a National Network for Resilient Infrastructure Simulation to coordinate research and practice focused on interactive case studies in resilient infrastructure systems.

Keywords Critical infrastructure • Adaptive governance • Resilience engineering • Resilience processes • Socio-technical systems integration • Resilience economics • Organizational resilience • Human resilience development • Integral theory • Resilient infrastructure education

3.1 Introduction

Policy objectives sometimes outpace the science and governance mechanisms necessary to achieve them (Seager et al. 2017). In examining why, Flynn (2016) identifies four primary knowledge impediments to infrastructure resilience, which we organize here according to a model of moral capacities established by Hannah et al. (2011):

- **Awareness:** We remain unaware of how poorly we are prepared.
- **Judgement:** We lack the capacity to formulate preferential alternatives.
- **Motivation:** We lack the incentives necessary to motivate resilience.
- **Action:** Existing governance frameworks face barriers to action that could create resilience.

In this chapter, we propose an integrated research agenda for creating the knowledge necessary to overcome these four barriers, and propose simulation games as an effective pedagogical strategy for Flynn's fifth impediment: education and training of the workforce that must apply the resilience knowledge in action.

3.2 *Awareness:* Recognizing Resilience

The rapid growth of the term "resilience" in a diverse set of academic and popular writings has been "astonishing" (Sage and Zebrowski 2016, See Fig. 3.1). Perhaps to the dismay of scholars who seek greater specificity in definition of the term, resilience has become a "hyper-popular" buzzword (Woods 2015). Nevertheless, Alexander (2013) traces the etymological origins and finds that usage in fields such as law, mechanics, social science, business, and the natural sciences dates back at

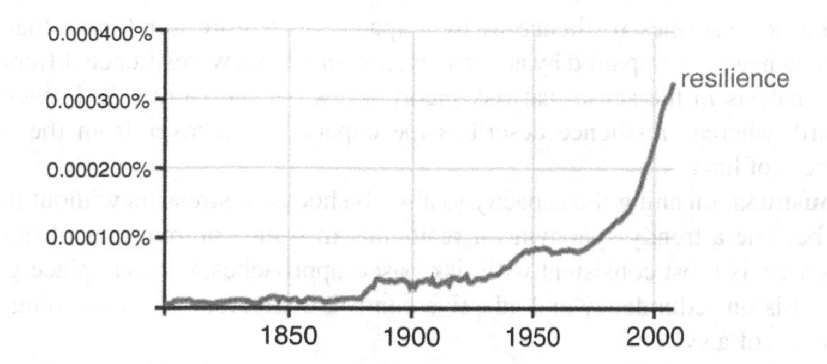

Fig. 3.1 Use of the word "resilience" in English-language books has grown exponentially since Holling (1973) popularized the term in the natural sciences (Google Ngram 2017a)

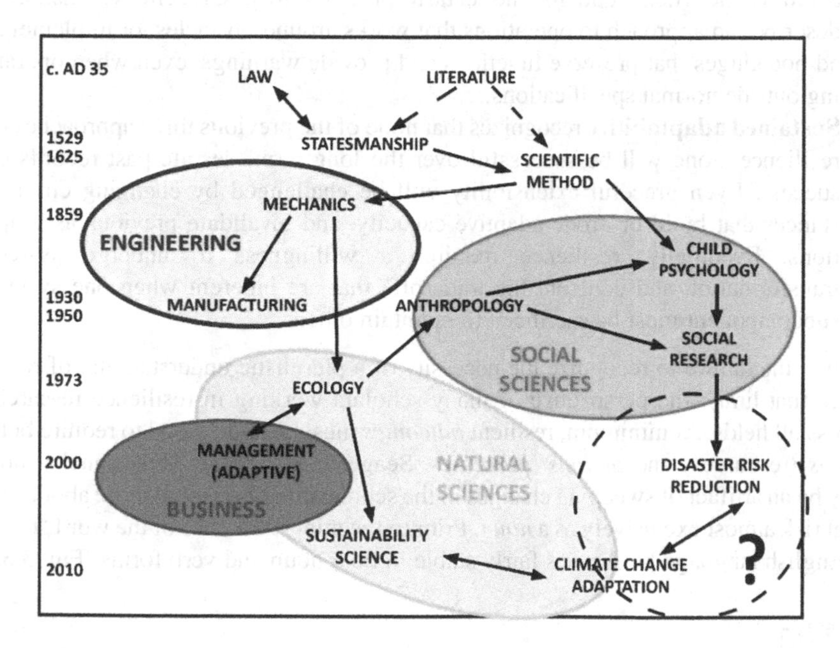

Fig. 3.2 The etymological evolution of resilience can be traced back at least several centuries. Approximate disciplinary boundaries have been added to a diagram reproduced from Alexander (2013, licensed via CC-BY). The dashed line indicates the potential emergence of new discipline

least two centuries (Fig. 3.2). The legacy of this twisting evolution is a confusing and disjointed landscape of scholarship that has failed to result in a generalizable theory.

In infrastructure applications, resilience is often conflated with risk, although some scholars have taken pains to attempt to draw distinctions (e.g., Park et al. 2013; Linkov et al. 2014). Given the explosion of recent interest, it must be recognized that multiple conceptions or perspectives have been brought to bear. For example, Seager (2008) identifies four different understandings of adaptation for sustainability, all of them extant in recent literature and policy documents, that roughly correspond to the four understandings of resilience more recently described by Woods (2015):

1. **Rebound** describes resilience as the capacity to restore conditions that have been damaged or impaired by adverse events. In this view, resilience differs from risk analysis in the sense that risk incorporates the probability and severity of hazard, whereas resilience describes the capacity to recover from the consequences of hazard.
2. **Robustness**, meaning the capacity to absorb shocks or stressors without failing, has become a trendy synonym for resilience in some communities of practice. This view is most consistent with risk-based approaches, but may place greater emphasis on redundancy and adaptive control, compared with hardening static elements of a system.
3. **Graceful extensibility** recognizes surprise as an inevitable feature of complex systems, and thus seeks to manage the modes and consequence of surprise to avoid brittle (i.e., sudden) and catastrophic failures. Graceful extensibility describes an approach to operations that works around obstacles, or implements ad hoc kluges that preserve functions and provide warnings, even when operating outside normal specifications.
4. **Sustained adaptability** recognizes that none of the previous three approaches to resilience alone will be successful over the long term, despite past records of success. Even graceful extensibility will be challenged by changing circumstances that build or erode adaptive capacity, and invalidate previous assumptions. Eventually, resilience requires a willingness to undergo system transformation and confront the trade-offs that are inherent when one system subcomponent must be sacrificed to maintain others.

It is the failure to recognize the necessity of a pluralistic understanding of resilience that limits the perspective of many scholars working in resilience research, across all fields. At minimum, resilient *outcomes* must be understood to require both *things* (resources) and *actions* (processes, Seager et al. 2007). This tunnel vision may be an artifact of sweeping changes in the scholarship of risk that write about and treat risk almost exclusively as a *noun*. Prior to the mid-1960s, use of the word "risk" in English language books was fairly stable in both noun and verb forms (Fig. 3.3).

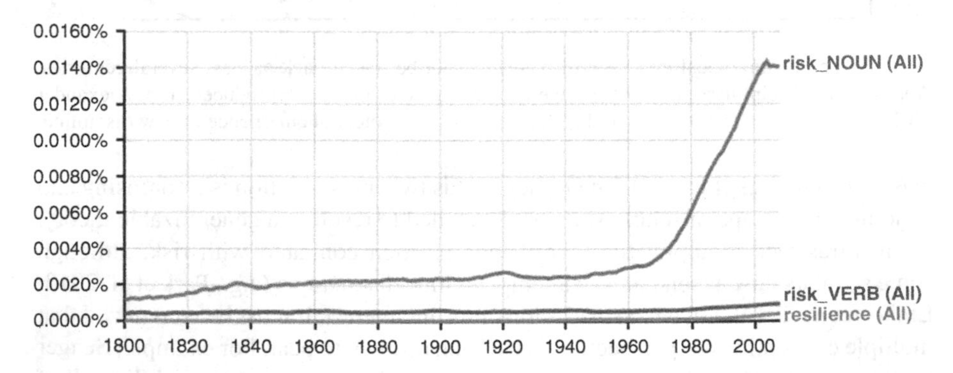

Fig 3.3 The concept of 'risk' has changed since the early 1970s to emphasize the noun form, rather than the verb. Without analogous verb forms for resilience, researchers may neglect to understand resilience actions, and focus instead exclusively on resources (Google Ngram 2017b)

However, the 1970s marked an exponential increase in the treatment of risk in exclusively as a noun. During this period, scholars emphasized framing risk as something that could be determined objectively and mathematically, and thought about "rationally" (Hamilton et al. 2007). However, unlike "risk" there is no dual usage of "resilience" in both noun and verb forms. Thus, coming to an understanding of resilience as a set of things a system *does* rather than as something a system *has* would run counter to the trend toward objectification already represented in risk research.

3.2.1 Conceptualizations of Resilience as Action Verbs

Rooting resilience in *action* helps distinguish resilience from robustness. Because quantifying the quality or capacity for adaptive action is harder than measuring resources (e.g., backup generators), and outcomes (e.g., restoration of electric power), resilience actions are easy to overlook. Nevertheless, processes are the mechanisms by which resources are deployed to result in outcomes. From this perspective, infrastructure resilience is neither found in portable, decentralized electric power generation, nor in robust, centralized and efficient grid-based electric power, but in the capacity to switch between the two.

In thinking of resilience action, there are at least four verbs necessary for every complex system to be resilient – sensing, anticipating, adapting, and learning or SAAL (Park et al. 2013). *Sensing* is the ability to recognize and incorporate new and changing stressors on a system into our understanding. *Anticipating* is used to foresee possible threats given our sense and understanding of new stressors. *Adapting* reflects the actions taken and changes made in response to what has been sensed and anticipated. *Learning* is the capacity to create, share, and apply new knowledge. These actions can happen simultaneously, but they must all be alert to enhance adaptive capacity. This continued alertness prevents stagnation and complacency, and promotes thinking that goes beyond risk analysis (Hollnagel 2014; Hollnagel et al. 2007; Madni and Jackson 2009).

The compartmentalization of science into disciplines and sub-disciplines frustrates infrastructure resilience research. While PPD-21 explicitly calls for a "holistic" approach, and it is widely recognized that resilience is a systems concept requiring integrative approaches to creation and organization of knowledge, academic traditions and bureaucratic incentive structures work against this goal. The failure of scholars of one subdiscipline to recognize the necessity, potential contributions, and opportunity to learn from others presents an obstacle to realization of a holistic research agenda that must accommodate multiple perspectives, systems, definitions, dimensions, and methods (Esbjörn-hargens and Zimmerman 2009).

3.2.2 *Application of Integral Theory for Organizing Holistic Awareness*

One way to generate greater awareness of resilience knowledge is through application of Integral Theory, which presents a holistic framework for organizing and comparing different epistemologies (i.e., ways of knowing) representing how knowledge perspectives are conceptualized and presented. Integral Theory provides a logical structure for integrating resilience knowledge from multiple disciplines. Wilber (2001) notes that the vast majority of human languages have some form of first-, second-, and third-person view or perspective of the world. These perspectives are present in linguistic structures, which are used to formulate, communicate, and interpret meaning, knowledge, and experience. The perspectives identified by Integral Theory – represented in the English language by the pronouns I, we, and it – each enable a unique domain of investigation (Esbjörn-hargens 2010). Considering how a third person perspective may be either singular ("it") or plural ("its"), Integral Theory suggests there are at least four irreducible perspectives: (1) the subjective "I" encompassing experience, (2) the intersubjective "we" encompassing culture, (3) the objective "it" encompassing behavior, and (4) the interobjective "its" that captures the interaction of singular subcomponents as a system (Fig. 3.4). Where resilience awareness favors one domain over another, scholars are at risk of offering partial solutions. An Integral Theory approach aims to include as many knowledge domains and perspectives as possible to enable more holistic and comprehensive solutions.

The perspectives illustrated in Fig. 3.4 may be viewed as four distinct but interrelated epistemological orientations that offer structure for organizing awareness. The left-hand quadrants are the interior subjective domains, and the right hand quadrants are the exterior, objective domains. The upper two quadrants are the singular perspectives and the lower two are the group or collective perspectives. Thus each quadrant contributes a unique orientation, described as follows:

- *Experience*. The upper left quadrant of the Integral map corresponds to the individual interior, which is a first-person "I" perspective of experience. The individual interior is subjective in nature and includes factors like personal values and beliefs that underpin a person's experience of their environment. Example epistemologies in this quadrant include cognitive, affective, moral, and psychological development, capacities, and dispositions of an individual person. Resilience research corresponding to the experience quadrant is found in psychology (Bonanno 2004; Noltemeyer and Bush 2013) and psychiatry (Connor 2006), and is concerned with the adaptive and maladaptive response of individuals to stress.
- *Culture*. The lower left quadrant corresponds to the collective (i.e., social) interior, which is a second-person "we" perspective of culture. The collective interior is intersubjective in nature and includes factors like ethics, shared values, collective meaning-making, and worldviews of a collective or group of people representing a social experience. Example epistemologies include factors such as

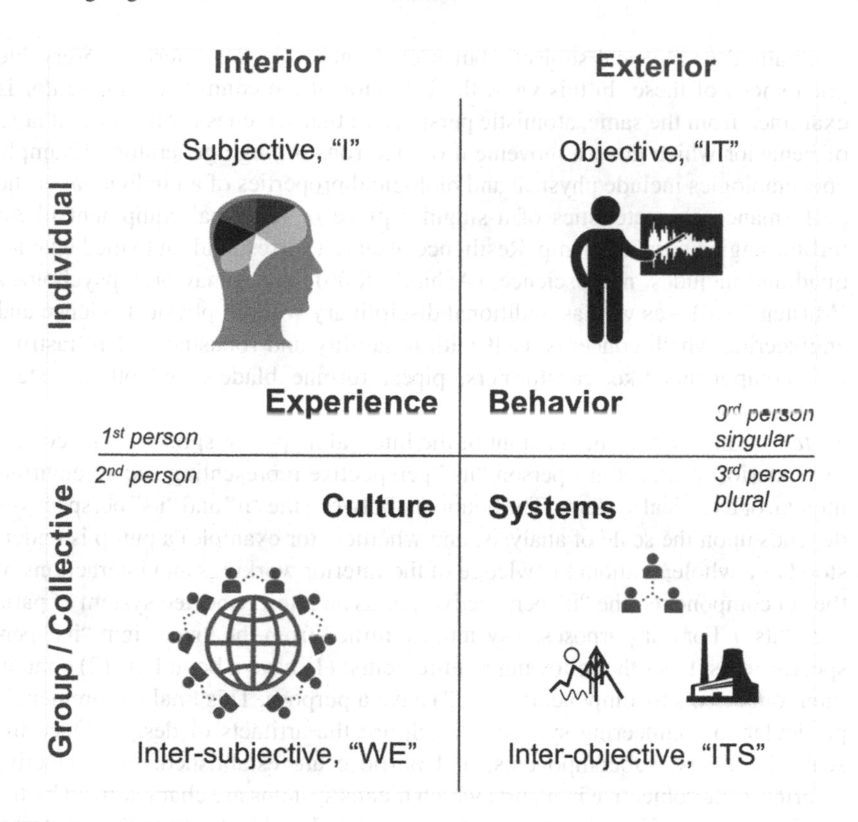

Fig. 3.4 **Integral Map**. Each quadrant corresponds to human perspectives representing different ways of knowing: (1) **experience** refers to the individual interior, which is a 1st person, subjective perspective characterized by the pronoun "I" and includes the cognitive, affective, and instinctive dimensions; (2) **behavior** refers to the individual exterior, which is a 3rd person (singular) objective perspective represented by the pronoun "IT" and includes the individual characteristics or actions of a person, object, or an artifact; (3) **culture** refers to the collective interior, which is a 2nd person intersubjective perspective represented by the pronoun "WE" and includes factors like shared values, ethics, and worldviews corresponding to groups, organizations, and other affiliations; (4) **systems** refers to the collective exterior, which is a 3rd person (plural) inter-objective perspective characterized by the pronoun "ITS" and includes complex interdependent social, environmental, and technological systems and the relationships among them (Adapted from Esbjörn-hargens 2010; Wilber 2001)

social cohesion, community efficacy, and the ability of a group to cope with adversity. Resilience research corresponding to the culture quadrant is found in the sociology and community resilience literature (Zautra et al. 2008; Norris et al. 2008; Berkes and Ross 2012).

- *Behavior*. The upper right quadrant corresponds to an individual exterior third-person "it" perspective representing behavioral phenomena and interactions in the physical environment. The individual exterior is objective in nature and considers factors like the characteristics and measures of a person or physical objects in addition to their actions and behaviors. This view incorporates both human and natural or physical subjects, but it examines only the exterior, observable phenomena of these. In this view, the behavior of the control room operator is examined from the same, atomistic perspective that searches for the natural laws of behavior which govern movement of electrons through generators. Example epistemologies include physical and biological properties of an individual or the performance characteristics of a singular piece of technical equipment like a turbine engine or water pump. Resilience research corresponding to the behavior quadrant includes neuroscience (Achard 2006) and behavioral psychology (Masten 2014) – as well as traditional disciplinary work in physical science and engineering which concerns itself with reliability and robustness of infrastructure components like transformers, pipes, turbine blades, and other system subcomponents.

- *Systems*. The lower right quadrant of the Integral map corresponds to the collective exterior (plural) third-person "its" perspective representing social, environmental, or technical systems. The boundary between the "it" and "its" perspectives depends upon the scale of analysis, and whether (for example) a pump is understood as a whole, without knowledge of the interior workings and interactions of the subcomponents (the "it" perspective), or as an interconnected system of parts (i.e., "its"). For our purposes, a system examined from the lower right "its" perspective must meet the following requirements: (1) have a boundary, (2) contain interconnected subcomponents, and (3) have a purpose. This final requirement is particular to engineering systems, which are the artifacts of design. Once the scale, boundary, subcomponents, and purpose are established, the collective exterior is interobjective in nature, which means systems are characterized by the empirical relationships between and among people, objects, and other systems. Recent examples of the systems perspective in resilient infrastructure research include examinations of fuel and transportation (Spierre Clark and Chester 2016) and network-based analyses of interdependent infrastructure systems (Chen et al. 2014; Zhang et al. 2015) and myriad other examples including research examining the system of generation, transmission, distribution and consumption of electric power, and interaction of electric power systems with water or agricultural systems.

Taken together, these quadrants clarify the epistemological perspectives that might be missing from certain knowledge claims and thus, application of Integral Theory allows more complete awareness of the knowledge necessary to understand and pursue holistic approaches to infrastructure resilience research that broaden awareness.

3.3 *Judgement*: Resilience Preferences

The second knowledge barrier to resilience relates to judgement, or the capacity to formulate decision alternatives and rank-order them from most to least preferable. That is, even if we knew what resilience was in all its different noun and verb forms and we obtained awareness of all four of its irreducible perspectives, we still would not know which processes are most worthy of scarce resources, and the outcomes that are most important to achieve. The inevitability of failure in complex systems requires a recognition that not all system functions or components can be protected, and yet there is no agreement across sectors and networks on what to prioritize. Without prioritization, subsystems naturally prioritize their own interests – even in cases where pursuit of subcomponent interests causes collapse of the whole (Sadowski et al. 2015).

3.3.1 Continuity of Operations, Mission Assurance, Critical Infrastructure

Within the U.S. Government, agencies within the Executive Branch have adopted different guidance documents intended to inform resilience judgement. These include:

- **Continuity of Operations (COOP).** The White House National Security Council (NSC) defines *continuity of operations* as "an effort within individual executive departments and agencies to ensure that Primary Mission Essential Functions (PMEFs) continue to be performed during a wide range of emergencies, including localized acts of nature, accidents and technological or attack-related emergencies" (HSC 2007). PMEFs are "Those department and agency Mission Essential Functions… which must be performed in order to support the performance of [eight National Essential Functions, including the continued function of our form of government under the Constitution] before, during, and in the aftermath of an emergency." A National Continuity Policy Implementation Plan (NCIP) defines the eight NEFs, the corresponding PMEFs, along with specific requirements for timely recovery of PMEFs after an event (NSC 2007).
- **Mission Assurance (MA).** Within the Department of Defense (DoD), the notion of a *mission* is important for focusing attention and prioritization of infrastructure. Directive 3020.40 defines *mission assurance* as, "A process to protect or ensure the continued function and resilience of capabilities and assets, including personnel, equipment, facilities, networks, information and information systems, infrastructure, and supply chains, critical to the execution of DoD mission-essential functions in any operating environment or condition." Thus, the concept of mission assurance is used "as a process to protect or ensure the continued function and resilience of capabilities and assets by refining, integrating, and synchronizing the aspects of the DoD security, protection, and risk-management programs that directly relate to mission execution" (DoD 2016).
- **Critical Infrastructure.** To provide guidance on those operations that should take precedence over others, the Department of Homeland Security (DHS) has identified 16 critical infrastructure (CI) sectors that are considered "so vital to the United States that the incapacity or destruction of such systems and assets would have a debilitating impact on security, national economic security, national public health or safety, or any combination of those matters" (DHS 2013). To maintain the function of these sectors, the 2007 National Strategy for Homeland

Security notes the importance of *operational resilience* as "an approach that centers on investments that make the system better able to absorb the impact of an event without losing the capacity to function." (HSC 2007), while Alderson et al. (2015) present quantitative models for assessing and improving the operational resilience of infrastructure systems using game-theoretic models that identify worst-case disruptions to system operation.

The emphasis in these guidance documents reveals a view of resilience that emphasizes robustness, and is most consistent with risk-based approaches to prioritize infrastructure assets within each government agency or sector based on the likelihood of threats and infrastructure vulnerabilities, as well as the potential consequences the nation would face if it were to fail (GAO 2013). This compartmental approach has resulted in inconsistencies among risk assessment tools, areas assessed for vulnerability, and the detail of information collected that has inhibited integration and coordination of prioritization efforts (Larkin et al. 2015; GAO 2014). Moreover, the DHS approach to critical infrastructure suffers from a misplaced emphasis on the physical condition of the infrastructure, rather than the services provided. A more complete resilience approach would alternatively recognize multiple adaptive pathways to provide end-users with the function of the infrastructure, as well as the capacity of any sectors to substitute for, reinforce, or pose a threat to other sectors (e.g., Ganin et al. 2016).

3.3.2 A Capabilities Approach Emphasizes the Function of Infrastructure

The view of infrastructure as a service begs the question, "What are the infrastructure services valued most in the United States?" To develop resilience judgement requires examination of the foundational values that form the basis for civil society. In the US, these are codified in the Declaration of Independence, the Constitution, and specifically the Bill of Rights, among other documents, which describe the inalienable human rights the US government is bound to protect. Such rights include the freedom of speech, freedom to bear arms, the right to privacy, and the right to peacefully assemble, among others. However, in a modern, interdependent world protection of these rights is an empty, philosophical gesture without the affordances that acknowledge these rights as capabilities. That is, formal rights require infrastructure and other factors to facilitate the transformation of rights into effective freedoms (Sen 1999a, b). For example, access to education does not happen without affordable and adequate institutions of education. Health care cannot take place without access to medicine and medical equipment. A capability for freedom of speech cannot exist without the technological platforms that make exercise of these fundamental human rights possible.[1]

---

[1] In Thomas Jefferson's time, the printing press was a technology platform considered so essential to freedom of speech that it was written directly into the 1st amendment. In today's age, this right might be expressed as freedom to tweet.

Table 3.1 An abbreviated summary of Nussbaum's (2003) list of central human capabilities

Human Capability	Being able to…
Life	Live to the end of a human life of normal length
Bodily health	Have good health, nourishment and shelter
Bodily integrity	Move freely from place to place, be secure against violent assault, and choice in matters of reproduction
Senses, imagination, and thought	Use the *senses*; being able to *imagine*, to *think*, and have adequate education
Emotions	Have attachments to things and persons outside ourselves
Practical reason	Engage in critical reflection about the planning of one's own life
Affiliation	Live for and in relation to other human beings
Other species	Live with concern for and in relation to animals, plants, and nature
Play	Laugh, to play, to enjoy recreational activities
Control over one's political environment	Participate effectively in political choices that govern one's life; having the rights of political participation, free speech and freedom of association
Control over one's material environment	Hold property (both land and movable goods) and seek employment on an equal basis with others

The capabilities approach (CA) is the framework used by the United Nations Development Program to understand the underlying basis of how rights become realized. The CA is founded on the claim that the achievement of human well-being is of primary moral importance and emphasizes that the freedom to achieve well-being is understood in terms of capabilities, or real opportunities to be and do what people value (Sen 1999a, b; Nussbaum and Sen 1992; Nussbaum 2000, 2006). The framework emphasizes the role of conversion factors, including personal, social or environmental characteristics, public policies and social norms, as well as available institutions and infrastructure which enable rights and resources to become capabilities. Thus, from a capabilities perspective the most critical infrastructure can be understood as those which are vital for protecting or providing essential human capabilities.

Nussbaum (2003) provides a list of ten capabilities that she claims are important because the activities and freedoms they enable are central to a life that is truly human (Table 3.1).[2] She defends these capabilities as being the moral entitlements of every human being on earth. The list specifies the minimum entitlements a citizen should be guaranteed by their governments and relevant international institutions. Nussbaum formulates the list at a general, legislative level and advocates that the translation to implementation and policies should be done at a local level, taking into account local differences. However, within this set of capabilities, critics of the CA argue that not all capabilities can be protected at all times. Thus, operational guidance is required to create a hierarchy of capabilities for prioritizing infrastructure services under times of scarcity or stress.

[2] Other multidimensional lists and conceptions of human well-being have been generated and vary according to the questions that each author seek to address and the context of operation; see Alkire (2002) and Hall et al. (2010) for a discussion and comparison of different approaches.

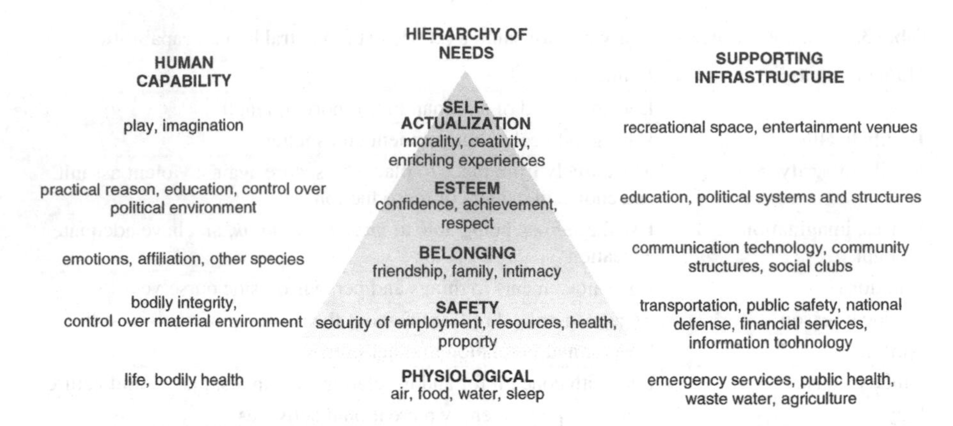

Fig. 3.5 Nussbaum's central capabilities (*left*) and supporting critical infrastructures (*right*) mapped onto Maslow's hierarchy of needs (*center*)

3.3.3 Development Hierarchies for Prioritizing Infrastructure

One approach to accomplish prioritization is to employ a hierarchical theory of human development such as Maslow's hierarchy of needs (Maslow 1943). The model is illustrated as a pyramid with the most urgent survival and safety needs at the base, followed by less urgent needs including belonging, esteem, and ultimately self-actualization (Fig. 3.5).[3] Maslow explains that there are preconditions for the basic needs in his model that include things like freedom to express one's self, freedom to seek information, and freedom to act without harm, as well as justice, order in the community, and fairness, which are not considered ends in themselves but important for achieving basic satisfactions. According to Maslow, the higher needs on the pyramid require more preconditions, or better external conditions, for achievement whereas the lower needs are more tangible, localized, and limited. Moreover, Maslow claims that relative to superficial and conscious desires that are impacted by one's culture, the basic needs represented in his theory of motivation are more universal and common among all humans.[4]

[3] Maslow acknowledges that his hierarchy suggests a degree of fixity, even though some people will be motivated by needs in a different order. He discusses how the hierarchy does not usually occur in a step-wise fashion as the pyramid implies. A more realistic description of the hierarchy is decreasing percentages of satisfaction as one moves up the pyramid.

[4] Maslow's hierarchy has resonated across many disciplines, from psychology, to education, business, engineering, and technology because it organizes a very complex topic into a cognitively appealing and intuitive model. Its popularity stems from the model's relative simplicity and hierarchical nature which allows for more practical application, yet these characteristics are also heavily criticized. Alkire (2002) argue that dimensions of human development should be *nonhierarchical* because what seems most important to an individual will change over time, depending on the situation and context. Others contend that people are capable of higher order needs such as love and belonging, even if their basic psychological needs are unmet.

Figure 3.5 demonstrates how Maslow's hierarchy provides a framework from which we can begin to prioritize infrastructure, according to the role that infrastructure systems play in enabling or supporting basic human needs. Categorizing the human capabilities identified by Nussbaum into a tier on Maslow's hierarchy is useful for teasing out particular infrastructure sectors that are important for each. For example, our most urgent physiological needs are closely related to the capability of being able to live a life of normal length and having good health, nourishment and shelter. The corresponding critical infrastructure systems would be things like emergency services, public health, water and wastewater, as well as food and agriculture. The next level of safety needs relates to the capabilities of bodily integrity and control over one's material environment. These capabilities are described as being able to move freely from place to place, be secure against violent assault, and choice in matters of reproduction as well as the ability to hold property (both land and movable goods) and seek employment on an equal basis with others. Infrastructure that supports these values are transportation systems, public safety (e.g., police and fire protection), national defense, financial services, information technology, and other government facilities. Further up Maslow's pyramid, we find the sense of belonging that can be enabled by communication technologies, schools and community structures, as well as other social clubs and institutions that relate directly to Nussbaum's idea of affiliation. Next to last is 'esteem' or confidence, which is enabled through education, participating in political choices, as well as freedom of speech. Finally, Maslow lists at the apex of his pyramid certain qualities of "self-actualization" like creativity and a capacity for moral judgement – activities that are related to Nussbaum's identification of "play" as a fundamental human capability (Selinger et al. 2015, 2016; Sadowski et al. 2013, 2015; Clark et al. 2015).

Maslow's implication is that human development needs at the base of the pyramid must be met before those at higher levels can be realized. Thus, Maslow provides a system by which infrastructure systems that realize services related to physiological needs and safety must be met before those that realize a sense of belonging or self-actualization. Nonetheless, not every tractor or farm must be prioritized during a disaster. In fact, a resilient food system would not be dependent on any one particular food source or facility to maintain access to nutritious food. For water, even if the pipelines or water treatments fail to deliver clean water, having the ability to boil water or truck in water from other locations would still satisfy basic needs. The latter example illustrates the ability to sacrifice part of the system but still maintain the supply of basic services to people in need. It also shows how the interdependent nature of critical infrastructure could be an asset for alleviating failures or disruptions in other areas, which supports the need to move away from the current sectoral approach and toward a more holistic and systems approach to critical infrastructure resilience.

3.4 *Motivation:* Incentivizing Resilience

The third barrier is a lack of incentives which *motivate* those resilience actions identified by resilience judgement. Obstacles to motivation include the perception of high costs, entrenched self-protective interests, a lack of urgency, a lack of investment capital, and problems of moral hazard and moral luck (Nagel 1993).[5] Whatever the reason, all too often disincentives to resilience outnumber or outweigh motivations.

Infrastructure typically requires large, upfront capital investments which create long-lasting benefits and maintenance requirements and infrastructure finance takes many forms, including property, income, special sales taxes, or user fees such as tolls and metered rates. Given the complicated finance and ownership structures, it is no wonder that incentive structures for building resilience may be difficult to decode.

Current practices in the United States are to discount the future value of benefits and liabilities using an exponential discount factor that corresponds to the inverse of compound interest. The result is that future events appear inconsequential from a present value perspective. Hyperbolic discounting provides a better description of real human and animal behavior, and places greater value on future events.

3.4.1 *The Myopia of Finance*

In many cases, such as regulated utilities, investment incentives are legislatively reduced to financial measures or constraints that mandate actions with certain pecuniary consequences. Thus, the financial models that estimate infrastructure costs and return on investment have regulatory compliance status in ways that constrain or mandate action. In these cases, where finance is the principal driving incentive structure, resilience presents a special type of problem. Figure 3.6 represents a hypothetical frequency plot showing the likelihood of disaster compared to its severity. The graph assumes the most serious conditions exist at the far right, and occur most infrequently. By contrast, normal operating environments are found at the left, and occur most frequently. This is the typical type of frequency diagram that is associated with storm hydrographs, where more severe storms are expected with longer return periods (such as the 100-year flood).

[5] Moral hazard refers to the externalization of risks to third parties, even in the absence of the intent to cause harm (Pauly 1968). Where the benefits of risk taking accrue to those making the decision to take risks, but the downsides accrue to others, the distortion causes decision-makers with "skin in the game" (Taleb 2012) to place irresponsible bets with poor expected social outcomes. The complementary concept is moral luck (Nagel 1993), which refers to the tendency to judge the moral worthiness of actions by their outcome, rather than intent. Because in complex systems, outcomes will never perfectly align with intentions, judging exclusively on the basis of outcome leaves open the possibility that some poor, or irresponsible decisions may nevertheless be judged morally worthy simply out of good fortune. A more complete description of the relationship between these two fascinating philosophical concepts and resilience must be left for some future publication.

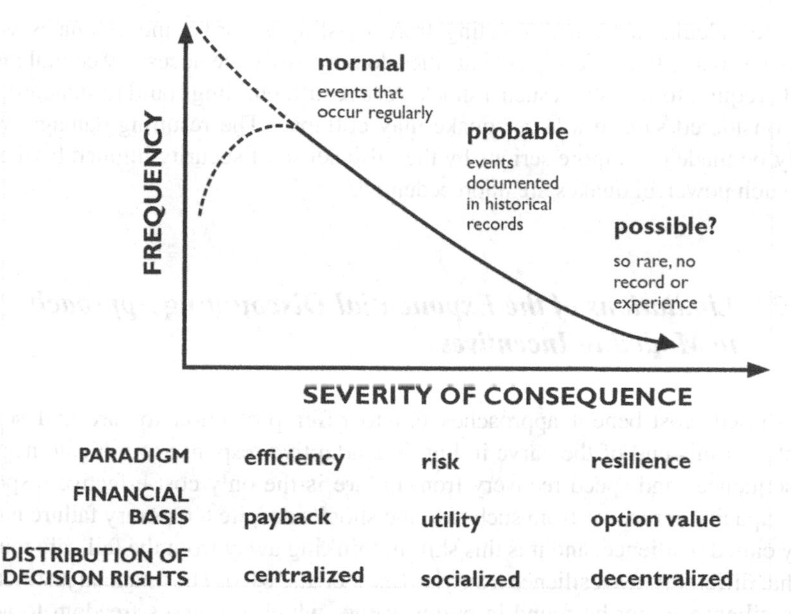

Fig. 3.6 Less frequent events may have more severe consequences. In these case, response demands optionality. However, financial models of option value are insufficient to incorporate the value of future optionality in present value terms

Financial justification for investments at the left edge of the frequency plot are typically predicated on the expectation of improved economic efficiency, as measured by payback period (for example). Because these operating conditions are normal, there are readily available datasets that describe this shape of the frequency diagram, and investment expectations can be computed reliably. Less frequent events are found in the middle of the diagram. They occur with enough frequency (in the historical record) to estimate a probability of occurrence. In this region, financial justification is based upon risk-adjusted, or probability-weighted utilities. Although exact returns to capital investment are contingent on uncertain future events, a probabilistic cost-benefit optimization is calculable, and over a large enough portfolio of projects, good and bad luck can be expected to even out.

The difficulties in financial modeling occur at the right-most edge of the frequency diagram, where events are so rare that there is no historical record. Consequently, probability estimates require an unreliable extrapolation from observed data. In fact, events are so infrequent that improbable in theory may be treated as impossible in practice – partly because such events have never been experienced before. For example, earthquakes that register above 8.0 on the moment magnitude or Richter scales are extremely rare. There are fewer than half a dozen earthquakes in the global historical record that register above a 9.0, and none that registered above 9.5. Therefore, it may seem impossible to designers or planners to anticipate an earthquake that registers 10.0 (which would be at least 5 times more powerful than the strongest quake ever recorded). The probability of this event can only be calculated by extrapolating from existing records, and estimates would likely be so low that risk-adjusted utilities do not justify the necessary capital investments require to withstand such a shock. As a result, buildings built to standards that are considered safe in a lesser quake may collapse. The resulting damage would likely be made even more serious by the false sense of security implied by the fact that such powerful quakes are unprecedented.

3.4.2 Limitations of the Exponential Discounting Approach to Modeling Incentives

Risk-based, cost-benefit approaches fail to offer protection to rare and serious events. At this end of the curve in Fig. 3.6, adaptive response to mitigate negative consequences and speed recovery from failure is the only cost-effective response. The capacity to recover from such extreme shocks despite temporary failure is typically called resilience, and it is this shift in thinking away from the fail-safe mentality that differentiates resilience from risk (Park et al. 2013). The financial justification for resilience might be found in option value, which preserves freedom to adapt. While there is no doubt that optionality has economic value that can be represented in financial models, there is no consensus on the proper way to do it. Even in financial markets, where contracts, dates, and prices are well known, options are notoriously criticized for being mispriced. The difficulties of pricing option value in infrastructure must be orders of magnitude more serious than for financial contracts, and thus are likely to be neglected altogether.

Even where future liabilities can be well described, there remain difficulties of assessing intertemporal trade-offs. Because so much infrastructure is long-lived, decisions made during design and early stages of operation can have consequences lasting for decades, if not a century. Given the long lifetimes of typical civil engineering systems, determining costs is an uncertain proposition that involves calculated forecasts of their replacement and maintenance needs. It requires amortization of initial capital outlays, setting aside money for future expenditures, and managing emergencies and unforeseen events. Financial models must compare alternatives with different future cash flow consequences, such as sizing of constructed components and maintenance or replacements schedules. For example, where the expense of replacement exceeds the expected future expense of repair or replacement (when discounted to present value), the financially rational argument is to defer replacement.

In California confusion over infrastructure financing has been especially problematic. In 1996, Proposition 218 modified the state constitution to require that voters approve utility rate increases. A series of court challenges over the next 20 years have failed to clarify the implications of the law for infrastructure resilience (Stranger 2013). For example, a 2015 ruling effectively makes it illegal for utilities to charge more for water than "… the actual costs of providing water" (Munoz 2015). Because the cost of water provision must include expected depreciation and replacement, financial models must take into account issues such as infrastructure ageing and the discounting of repairs and maintenance to present value to estimate the actual fiscal burden of maintaining infrastructure. However, traditional investment forecasting techniques used by analysts and decision makers are flawed by misconceptions in infrastructure economics. For years, economists have used an exponential discounting methodology as a basis for modeling time preferences functions (Ayres and Axtell 1996).

The exponential discounting approach is justified by the opportunity costs of foregone investment interest. That is, the future value of any cash transaction may be expressed as the present worth equivalent that would grow to the future cash amount via compound interest. Growth of an investment account under continuously compounded interest is an exponential of the form e^{rt} where r represents the rate of interest and t represents time. Typically, discount rates are based on investment returns (i.e., rates of interest) considered risk free, such as US Treasury bonds. Nonetheless, large infrastructure investments, such as dams, bridges, highways or even power plants, often outlast the currency systems and debt structures that finance them. Consequently, "risk free" is more a point of comparison more than it is an absolute guarantee of financial security. The exponential approach is problematic in the long term, in that it rapidly compounds present values to massive future expectations, or (conversely) diminishes future events to the point of insignificance in the present.

The exponential discount model is "time insensitive" in that it is typically applied with a constant discount rate assumed to be applicable over the entire discounting period. Under truly "risk free" conditions, a time insensitive model might be justifiable. However, resilience cautions against the illusion that *any* prospect can be made free of risk. Without such an assumption, the rational approach in finance might be to apply higher discount rates longer-term time horizons – exactly as evidenced by market rates in bond yields, which typically apply higher rate expectations to bonds with longer-term maturities. (The opposite, where short-term rates are higher than long-term is called an *inverted* yield curve).

3.4.3 Strengths of the Hyperbolic Discounting Curve for Infrastructure Investment

However, real human decision making does not conform to the expectations of the normal, positive yield curve. In dozens of human and animal studies, real behavior conforms to a view of the long-term future that is more patient than exponential expectations predict (e.g., Hayden 2016; Winkler 2006). Empirical evidence shows that a *hyperbolic discount* function (Fig. 3.7, below) describes and predicts choices better than the exponential curve.

Compared to exponential discounting, the hyperbolic curve discounts near-term events more, and extremely long-term events less. In this way, the hyperbolic view is presbyopic (i.e, far-sighted) compared to the myopic, exponential alternative. For example, the hyperbolic view of time preferences explains procrastination, whereas the time insensitive exponential model predicts that everyone would get up in the morning at the same time for which they had set their alarms the night before.

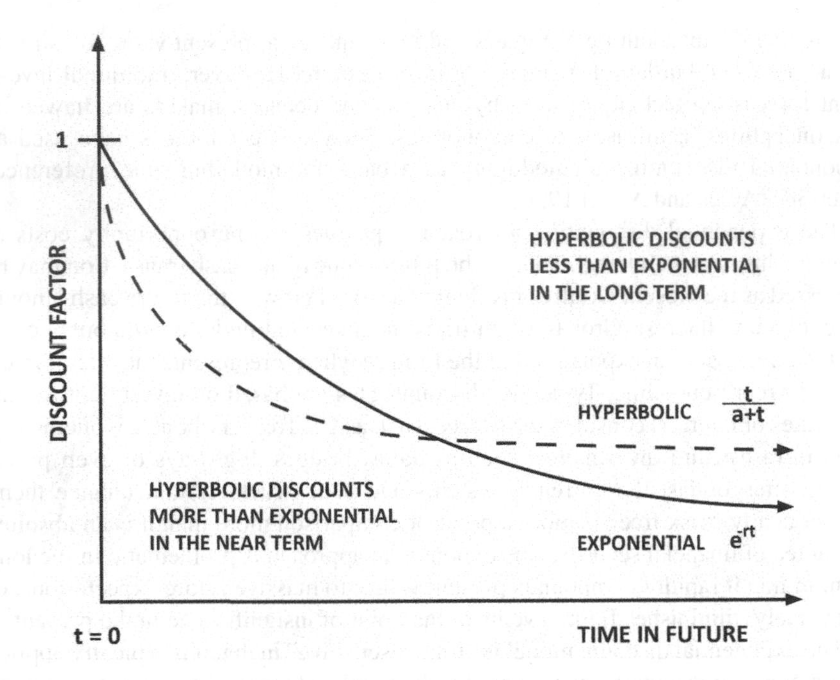

Fig. 3.7 Hyperbolic discounting values long-term consequences more than exponential, and thus provides a rational for long-term infrastructure investments that is absent from existing financial models

The implications of hyperbolic discounting for infrastructure investment may be extraordinary, insofar as the calculation of present worth expectations of long-term liabilities such as deferred maintenance may cause reallocation of resources away from short term financial benefits in favor of long-term structural integrity. The time-sensitive nature of hyperbolic discounting incentivizes a continuous reassessment of priorities, and may avoid the trap of putting off expenditures until a future date. This is especially important over the long term, as infrastructure ageing, replacement needs, or even the risk of collapse of entire segments of the system, accelerates with time. In fact, hyperbolic discounting is already the standard for assessment of infrastructure projects under certain conditions in the European Union.

3.5 *Action*: Reorganizing for Resilience

The fourth barrier is the lack of organizational and governance structures that enable adaptive action across infrastructures and services. Infrastructure in the United States is owned, financed, operated, and reconstructed by a myriad of different private and public organizations with overlapping jurisdictions. In some cases, these jurisdictions ascribe to political boundaries such as city, county, or state lines – but in other cases infrastructure managed by pseudo-governmental authorities that transcend these boundaries. As different infrastructure systems provide a diverse array of services such as highway, air, and water transportation, electric power, communications, potable water and sewage, flood management, and others, the design, operation and adaptation of these systems are often incompatible with one another. Likewise, policies and protocols for handling system shocks and tradeoffs among infrastructure providers are inherently different. Consequently, the organizational and governance structures that support resilience action are currently incompatible. Even within a single critical infrastructure system, different operational goals, management structures, political or geographic boundaries, and governance systems exist. These differences can prevent even the most aware, motivated, and incentivized organizations from enacting resilience.

In general, infrastructure governance refers to the combination of laws, protocols, and norms that dictate decision-making activities taken for service provision. Each policy on its own prescribes the roles, authority, expectations, and liability of individuals and organizations within infrastructure systems. The combination of these policies generates a "functional layering" (Gim et al. 2017) of individuals and organizations into specific administrative structures for managing normal operations and crisis events. Although infrastructure governance is instituted through individual policies, we focus attention on this functional layering among organizations that dictates coordinated actions to manage infrastructure crises.

3.5.1 Infrastructure Governance Through Horizontal and Vertical Structures

Infrastructure governance in practice is characterized by the establishment of one of two administrative structures: horizontal and vertical (Kapucu and Garayev 2014; Kapucu et al. 2013). Horizontal structures focus on grouping agencies of similar function together to into streamlined units, emphasizing related operations and operational goals. Current practices in the National Infrastructure Protection Plan (DHS 2013) emphasize this form of crisis response, grouping infrastructure managers within a single sector, defined by function (e.g., energy, transportation, water, communications). The sectoral groupings form multi-agency groups that serve specific emergency support functions (ESF) within physical systems that bear technical similarity, but may serve different geographic regions or market segments (DHS 2013). This administrative structure is often characterized by flexible response activities that can change according to crisis needs.

In contrast, vertical administrative structures focus on having a standard operational command system for managing all incidents, epitomized by incident command systems (ICS) employed across the United States by first responder agencies (Kapucu and Garayev 2014). ICSs function by designating a single incident commander for overall management and decision-making of all infrastructure sectors, in any crisis event. This approach is an attempt to solve problems of coordination across jurisdictional, agency, and technical boundaries, by including joint decision-making processes by representatives from multiple infrastructure systems (Kapucu and Garayev 2014).

Infrastructure systems are governed by a hybrid of these two administrative structures depending on the scale and level of decision-making required. Individual infrastructure installations (e.g., power plants, refineries, dams, airports, or data server farms) follow a vertical, ICS structure, with on-site commanders and control rooms organizing all failure response and recovery activities (Fox-Lent et al. 2015). Larger, interconnected infrastructure systems (e.g., power grids, highways, pipelines) follow a horizontal, ESF structure, where the provision of specialized services have separate control rooms that coordinate in case of emergency. These separations differ for different infrastructure systems depending on the manner in which services are provided. For example, electric power transmission and distribution may be provided by the same entity but managed by different control rooms due to geographic scale (long-range transmission vs. local distribution circuits) and the physics of electricity itself (single vs. three-phase models). Traffic and water distribution within a single metropolitan area is often managed by multiple, distributed control centers due to shifting jurisdictional boundaries across interconnected infrastructures. Within a single transportation system, multiple modes of transit like trains and buses require different control centers due to operational differences, yet seek the same overarching goal of mobility. Divides in infrastructure management also occur depending on economic structures across jurisdictions. For example, power grids link both vertically integrated utilities that own and operate their own generation, transmission, and distribution infrastructure as well as horizontally integrated systems that own and operate these infrastructures with separate entities.

3.5.2 Governance Failures in Cascading Crisis Scenarios

While this complex web of governance frameworks may succeed in normal operations, large-scale crisis scenarios often reveal mismatches in decision-making authority and expertise that exacerbate negative consequences. In crisis scenarios, localized failures can cascade across multiple infrastructures, cities, states, and countries as services become unavailable (Clark et al. 2017). Unforeseen cascading failure events are further complicated when losses cross ownership, operational, and regulatory boundaries, and crisis response requires the coordination of dissimilar organizations that may have never previously worked together. Existing policies and protocols for these interactions require local expertise within horizontal governance systems to yield to bureaucratic, ICS-based processes for information-sharing and decision-making across infrastructures and sectors. Where even electric power, transportation, and water providers may have difficulty understanding the full range of policies governing their own systems, misunderstandings are amplified across distinct services. Thus, joint decision-making provided by vertical systems may still remain ineffective for crisis response by slowing the capacity of expert organizations to act. Similar to how failures cascade across built infrastructures, the inability of individuals, organizations, and industries to cope with uncommon, inter-organizational communication and coordination demands can both amplify damage and slow recovery.

Electric power grids provide an important example in which maladaptive coordination activities can lead to cascading losses. Power grids are pervasive infrastructure systems that connect electricity generation to point of use across multiple cities, regions, countries, and continents. Despite many power systems having standards for their design, operation, and use, interconnected jurisdictions often have different laws, protocols, and norms. These differences in policies lead to both a varied landscape of technologies and governance frameworks for operations and management. Technological and social differences alike have historically exacerbated losses. For example, the 2003 US Northeast blackout included a combination of infrastructure, control system, and decision-making failures that led to cascading damages (Pourbeik et al. 2006). And, the 2000–2001 rolling blackouts in California that brought a premature end to Governor Gray Davis' term were the result of failure to understand the consequences of policy reforms that left the California power supply and distribution system vulnerable to manipulation when hydropower availability was curtailed by lower instream flows (Navvaro 2004).

Because other critical infrastructure systems are dependent on electricity and vice versa, communication and decision-making failures in both cases are not isolated to electric power utilities, or affected water and transportation systems (Zimmerman and Restrepo 2006). Thus, miscommunication and lack of integrated planning typically leads to slowed recovery of all infrastructure services. For example, since 2003, post-mortem analysis of several major blackout events continue to identify improved communication within and across governing organizations to enhance blackout response (Adibi and Fink 2006; Andersson et al. 2005; Kirschen and Bouffard 2009). Moreover, an increase in studies on interdependent infrastructure also indicate the potential for losses in water and transportation systems to affect power grids (Clark et al. 2017).

3.5.3 Connecting Crisis Coordination and Infrastructure Governance

In these unforeseen and cascading situations where resilience is most pertinent, interacting and incompatible governance frameworks manifest as failures in coordination among disparate entities. Mismanaging crisis coordination exacerbates losses and slows recovery by causing duplication of work, hindrance of first responders, delays due to misunderstanding, and misallocation of resources (Petrenj et al. 2012). A recent review by Petrenj et al. (2013) outlines 17 issues that affect infrastructure crisis coordination, eight of which are directly associated with the organizational and governance structures of infrastructure providers, including:

- a lack of incentives to share information, resulting in failures of information flows
- incompatibility of crisis management processes and procedures, resulting in failures of joint activities or planning
- differences in organizational structure
- unbalanced distribution of workloads
- role ambiguity
- mismatch between goals and interests

Both horizontal and vertical governance frameworks can be rendered impractical by these issues. Successful crisis response by horizontal emergency support structures are more likely impaired by lack of information sharing, incompatibility of processes, and mismatched goals. Vertical incident command systems are impaired by differences in organizational structures, incomplete information flows, unbalanced workloads, and role ambiguity. Thus, even harmonized management and knowledge systems across multiple infrastructures would not necessarily be successful at solving infrastructure governance problems. As Flynn (2016) argues, there simply are no governance frameworks that build resilience in critical infrastructure.

In lieu of a perfect approach, a first step toward more effective infrastructure governance is modelling knowledge from multiple industries together to identify mismatches in policies prior to inevitable infrastructure and coordination failures. Figure 3.8 illustrates the current division of expertise among interacting critical

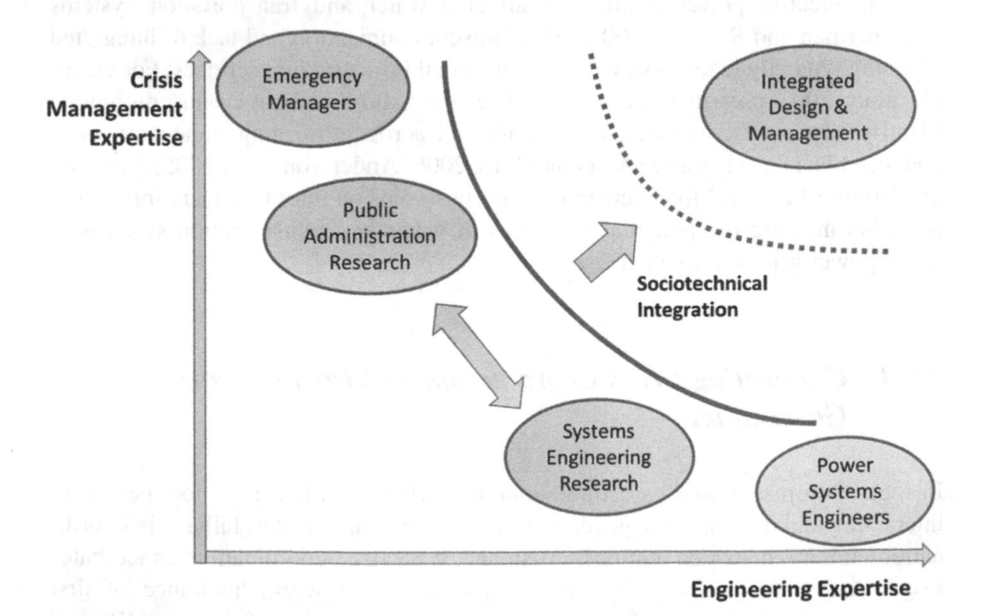

Fig. 3.8 Segregation of knowledge among engineering experts and governance experts means that technical experts may harbor misconceptions about governance, while crisis management experts harbor misconceptions about the workings of technical systems

services, including knowledge of relevant infrastructure governance frameworks. Because knowledge sharing between these two specialized groups is expensive and difficult, especially under the pressure of crisis. Knowledge held by power system infrastructure providers (operators, managers, owners) is largely unknown by those who provide crisis management support during system failures. Likewise, knowledge of crisis support activities, needs, and operations are unknown by infrastructure providers. Successful failure response requires coordination across these disparate groups that synthesizes different types and sources of knowledge. Where existing governance frameworks rely on rigid, bureaucratic processes, they may fail in response to unforeseen events because mechanisms of knowledge sharing are inadequate for adaptive response. Instead, governance that supports collaborative, adaptive, creative, and innovative action across these groups must be developed to enable resilient action. These likely begin by establishing improved methods for communication – suggesting that a common basis for modeling both systems might be an effective approach for integration of knowledge and correction of misconceptions.

Similar divisions in expertise exist across all 16 critical infrastructure sectors. Current practice establishes policies for public administration and power grid systems absent from each other – the physical limitations of electric power transmission are not considered in emergency response protocols, and crisis management roles are not reflected in infrastructure design. For example, technical models of electric power provision focus exclusively on the structure and function of the infrastructure itself to understand the physical limitations of built systems and potential failure modes. However, the complex physics of power generation and transmission systems requires simplifications that highlight some features and mask others. Choices made in technical modeling may have repercussions in the form of governance decisions, such as investment in redundancy, or allocation of authority among the emergency managers and power grid engineers required to coordinate actions when natural disasters, terrorist attacks, or other threats overwhelm existing infrastructures, automatic controls, and security. Because only large-scale failures bring these experts together, there is scant empirical data demonstrating whether coordinated blackout response policies will be adaptive or maladaptive. By contrast, integrated models of both governance and power grid physics may enable study of the interactions between them during crises. Linking knowledge across both forms of expertise may reveal how physical infrastructure function relates to socially constructed institutions which establish the regulations, protocols, and norms to prepare for and manage unforeseen events.

3.5.4 Four Fundamental Elements of Governance Models for Adaptive Capacity

Several modelling approaches exist which can support improved understanding across layered governance frameworks in critical infrastructure systems. In interdependent infrastructure models, governance frameworks and their respective policies are reduced to "logical dependencies" across individuals and organizations (Petit et al. 2015; Rinaldi et al. 2001). Ouyang (2014) recently reviewed different modeling approaches for logical dependencies in interdependent critical infrastructure systems and the potential ways in which they can inform resilient system design. The most relevant modelling methods for developing critical infrastructure resilience strategies are from operations research (agent-based, systems dynamics, and network-based models) rather than those normally used by social scientists (empirical and economic models). Moreover, hybrid approaches that link agent-, systems dynamics-, and network-based models together may be capable to address technological and governance strategies together (Ouyang 2014; Eisenberg et al. 2014).

While model choice influences the governance and coordination solutions offered, agent-, systems dynamics-, and network-based models all offer four broad solution types for overcoming governance barriers and enabling resilience actions. These operations research approaches each have idiosyncratic structures and functions, suggesting they lend themselves to modelling some dependencies better than others (Eisenberg et al. 2014). However, the three modelling approaches all require consideration of at least four fundamental elements in their construction that enable adaptive capacity:

- **Design variables** can be adjusted in response to changing stressors or boundary conditions – often, automatically. For example, some highways use adaptive speed limits, ramp meters, or toll lanes to meter traffic loads. Air travel systems typically make adjustments in gate assignments, landing or takeoff queues, and even destinations to adapt to changing weather, equipment or other conditions. Water distribution systems may adjust water sources, or distribution pressures to adapt to changing precipitation or demand conditions.
- **System constraints** limit the feasible region in which design variables may fluctuate. These may include regulatory or procedural constraints, such as the airline crew service limitations, water quality treatment standards, or electric power reserve margins. In times of crises, a temporary relaxation of, or work around, constraints may allow operation that avoids catastrophic collapse (i.e., graceful extensibility).
- When adjustment of design variables and relaxation of constraints fails, adjustments may need to be made that alter the **relationships** between design and system variables. System performance is typically judged by state variables that describe the quantity and quality of system end functions, such as water pressure, voltage, or passenger-miles traveled. For example, deployment of microgrid power transmission architecture changes the structural relationship between power generation, transmission, and consumption by introducing new design variables. Thus voltage may be maintained within the microgrid by different mechanisms than grid-dependent systems. Adaptation of the fundamental relationship between design and state variables may require longer lead times and larger capital investment than adjustment of design variables or relaxation of constraints. For example, demand-adjusted toll lanes may mitigate traffic congestion by making travel more expensive during peak travel periods – an example

of adaptive capacity introduced by adjustment of design variables (e.g., tolls). However, introduction of public transport alternatives changes the fundamental relationship between demand for travel and highway congestion by introducing new choices – at the expense of investment in purchase of buses, rail systems or technologies of increasing passenger densities.

- Finally, the ultimate source of adaptive capacity from an operations research perspective is **transformation** of the entire system itself. Here, wholesale replacement of design and state variables make the old system obsolete. The boundary between adaptation of system relationships and system transformation may depend upon the perspective of the observer. For example, the canal system of the nineteenth century resulted in a transformation of bulk goods transport in the US that made possible the accumulation of capital in cities like New York, and the economic development of the northern Midwest for farming and immigration. However, the canal era was short-lived as railroad technology advanced and offered new choices to accomplish the same task. In modern times, email could represent a fundamental transformation in information flows, or merely a more efficient fax machine. Typically, transformative leaps forward in technology may be deployed in the service of incremental improvements in design and state variable relationships before the full potential of the transformation is realized.

Taken together, these elements form operational models useful for understanding how both infrastructure failure and recovery and governance processes influence service provision (Alderson et al. 2014, 2015). Still, the above operations research actions are strategies for exercising adaptive capacity in physical and economic systems and do not speak to the decision-making processes or authority necessary for making adjustments. Those processes that lead to choices for changing decision variables, modifying constraints, changing relationships, or transformation are the purview of governance systems, rather than operations systems. Strategies for changing relationships and system transformation have no direct analogs in governance systems. In particular, deployment of any of the operations research strategies for adaptive capacity requires some expression of initiative that depends on resilience awareness (of an unsatisfactory condition), resilience judgment (formulation and preferential ordering of alternatives to the status quo), motivation (stimulus to employ an alternative), and action.

3.5.5 The Need for Human Ingenuity in Enacting Resilience Governance

The first two operations research sources of adaptive capacity (adjustment of design variables and constraints) are amenable to automated control systems that combine sensors and automated algorithms to execute the sensing and adaptation resilience processes. For example, in automobiles advancements in adaptive cruise control, crash detection, and autonomous or assisted driving rely on sensing and adjustment of design variables (e.g., acceleration, braking, deploy of airbags, parking controls, or fully autonomous driving) to avoid collisions. To some extent, artificial intelligence may even enhance anticipation and learning – the resilience processes which are most difficult to automate. Nevertheless, the last two operations research sources of adaptive capacity (changing relationships and system transformation), require greater human ingenuity.

To assess capacity to enact resilience governance we must conduct a critical examination of an organization's maturity (Alberts et al. 2010), where maturity is comprised of three dimensions:

- Patterns of interaction,
- Allocation of decision rights, and
- Distribution of information.

The three dimensions of organizational maturity outlined above are critical to resilience governance because the patterns of interaction, allocation of decision rights, and distribution of information among people who comprise an organization dictate that organization's capacity for sensing, anticipating, adapting, and learning in the face of change. Even where decision rights are allocated to automatic control systems, these controls must have access to information (e.g., sensing), allocation of decision rights (e.g., to adapt design variables), and function according to algorithms which dictate the patterns and policies which regulate interaction of system components, both technical and social.

Higher order adaptive strategies of changing fundamental relationships and system transformation place a greater burden on human imagination in anticipation and learning processes. Because these strategies operate outside existing systems constraints, they require an irreducible human component. Further, adaptive capacity depends not only on execution of the four operations research strategies, but also on collective human capacity to adjust the three essential aspects of organizational maturity: patterns and policies of interaction (who reports to whom, how, and when?), allocation of decision rights (who or what decides?), distribution of information (who or what knows what?).

Thus, resilience is not intrinsic to any design, adaptive strategy, or state of organizational maturity. Rather, resilience is the capacity to execute the processes of sensing, anticipation, adaptation, and learning (SAAL) that deploy the adaptive strategies and levels of organizational maturity appropriate to the specific stressor. Resilient systems will carry out the SAAL resilience processes and adjust organizational maturity of governance systems which then select adaptive strategies from the four operations research options detailed above. Thus, the SAAL resilience processes are essential to both adaptive capacity in governance, and adaptive capacity in the physical systems that control the processes which convert resources to desirable outcomes under conditions of non-stationary stress.

3.6 Cultivating Resilience Action at Individual and Collective Levels

The first four knowledge barriers provided in Flynn (2016) correspond exactly to the model of moral development described by Hannah et al. (2011) as moral conation.[6] Working at the scale of individual human development (the upper left quadrant of Fig. 3.4, above), the concept of moral conation describes the development of moral awareness, moral judgement, moral motivation, and the courage required to act morally in the world. Here, we thread together the analogous concepts in resilience awareness, resilience judgement, resilience motivation, and the courage to take the initiative to create resilience action applied at both the individual and collective scales.

Nonetheless, a fifth barrier to resilience action – characteristically different than the first four summarized above – is the problem of cultivating resilience capacities through education and training "that draw on the kind of interdisciplinary collaboration across technical, non-technical, professional and research programs that is required to advance a comprehensive approach to building resilience" Flynn (2016). Resilient infrastructure systems will require a workforce with the capacity to operate as resilience experts. Given the paucity of education programs specific to resilient infrastructure systems, and the traditional paradigmatic obstacles to integration of knowledge across disciplinary domains within existing educational intuitions, creating a workforce of resilience experts will require novel pedagogical strategies addressing both explicit and tacit knowledge. In this section, we consider the challenge of developing and teaching capacity for resilience action by integrating capacities for resilience awareness, resilience judgement, resilience motivation, and resilience action at both the individual human and collective organization scales.

3.6.1 The Limits of Risk-Based Approaches

The failures of risk analysis in the face of complexity are already well-chronicled. For example, in *Normal Accidents* (Perrow 1984), Yale sociologist Charles Perrow describes the counter-intuitive phenomenon in which the addition of emergency backup systems, additional controls, and well-intentioned interventions in complicated technological systems paradoxically increased vulnerabilities to the very risks they were intended to mitigate. Perrow's study was motivated partly by the question, "How could safety systems make us more vulnerable?" The answer, he argued, was *complexity*.

Since Perrow a series of additional works have reinforced his thesis. For example, the Logic of Failure (Dörner 1996) summarizes a series of studies that revealed

[6] Conation is an obscure word that describes volition, or willful action. Conation describes behavior that is purposeful striving, rather than recreational or hedonic.

only about 10% of human subjects possessed the complex systems reasoning skills to manage non-linear feedback loops in simulation game environments. Meanwhile, books like *Fooled By Randomness* (Taleb 2005) and *The Failure of Risk Management* (Hubbard 2009) provide scathing critiques of data fluency and statistical reasoning skills among managers responsible for complex, interdependent technological systems. Despite these insights, the US is still experiencing increasing insurance losses and worse catastrophic outcomes from changing stress conditions with which decision-makers have little prior experience, including California droughts, Louisiana floods, and San Diego power outages.

Given the accelerating pace of environmental change, surprise is inevitable. In contrast to robustness approaches that rely on risk analysis to identify hazards and reduce the probability of failure, the more dynamic resilience approaches (recovery, graceful extensibility, and sustained adaptability) emphasize rapid recovery from failures and adaptation to surprise. For example, faced with record instream flows during the 2011 Mississippi river floods, the US Army Corp of Engineers took the unprecedented action of dynamiting levees in eastern Missouri, inundating a region called the New Madrid Floodway (Olson and Morton 2012). Although the floodway had been created by a Congressional Act almost 80 years earlier for the purpose of relieving swollen rivers (Barry 1997), the usual response to river flooding had always been attempts to build protective levees higher – with disastrous consequences that were predictable in retrospect.

Nonetheless, instances like the 2011 Mississippi case in which federal agencies play a direct role in proactive, adaptive management of interstate infrastructure systems are exceedingly rare. The traditional role of the federal government has been to provide funding and research while design, construction, operation, and adaptation of critical infrastructure systems is delegated to an array of private companies and local, county, and state government, or trans-boundary special authorities. The Salt River Project in Arizona is a quintessential example (Gim et al. 2017) Therefore, we must find a way to engage hundreds of thousands of managers, engineers, technicians, leaders, and other stakeholders in diverse organizations to establish and strengthen capacities for dynamic resilience action.

The difficulties posed to traditional education and training by the problem of infrastructure *complexity* are formidable. While there is no doubt that infrastructure is complicated, existing education programs already do a good job training for complicated tasks. Here, we use the term complexity to describe the interconnected feedback loops which make infrastructure systems behave in surprising ways that obscure discovery of root causes (Alderson and Doyle 2010). At least four concepts are essential to understand:

- **Interdependency**. The sub-systems which comprise a complex system are interdependent in that they are mutually reliant. Thus, in addition to understanding how water, power, and transportation systems operate in isolation, people must also understand the relationships between each of these systems, and the ways in which these interdependencies comprise the larger, complex system which we refer to as infrastructure.

- **Feedback Loops**. A feedback loop is a specific type of interdependency wherein the output of one subsystem becomes input for others that ultimately feed information or resources back into the first subsystem. For example, increased traffic congestion may result in demand for construction of new highways. However, construction of new highways that relieve traffic congestion consequently induce greater demand for the privately owned vehicles that result in increased traffic congestion. The result is a positive feedback loop of seemingly unending increases in highway construction.
- **Nonlinearity**. Where linear relationships exist, human intuition is effective at extrapolating from observation and experience to anticipate events for which there are few or no observations. However, where relationships are non-linear, human intuition fails. Nonlinear systems are consequently unpredictable because small changes in interpretation of existing datasets can create enormous difference in anticipation.
- **Stochasticity**. Complex systems are subject to irregular and random phenomena. Such stochastic events are unpredictable, and subject to fallacies of oversimplification. For example, it is characteristic of human bias to confuse rare or unprecedented events with the impossible. Infrastructure systems subject to extreme weather events such as floods, earthquakes, and heat waves can cause catastrophic infrastructure failures like power outages, water main bursts, and structural collapses that managers erroneously thought were so unlikely that they could or should be ignored (Clark et al. 2017).

3.6.2 Education Theory for Resilient Infrastructure

The Kolb Learning cycle provides a theory applicable a holistic program of education and training for resilient infrastructure expertise. Regardless of the particular methods of teaching, it is now understood that learning requires at least four activities: abstraction, experimentation, experience, and reflection (Kolb 2014):

- **Abstraction** is the process of building representations of reality (e.g., equations, models) that highlight some features for while masking others. The advantage of abstraction is that it yields generalizable knowledge, typically stripped of context, with broad applicability. Thus, abstract knowledge (e.g., theory) can be long-lasting and cost effective in the sense that investigation of models and equation (e.g., virtual) space is far less costly than investigations which require physical space.
- **Experimentation** is the process of manipulating independent variables for the purpose of observing changes in dependent variables. When experiments operate on highly abstracted representations (e.g., idealized laboratory equipment), they ideally reveal reliable, reproducible, and empirically verifiable relationships. Logical positivist expressions of the scientific method are based upon iterative processes of abstraction (e.g., mathematical representations) which result in

predictions that are either verified or falsified in idealized laboratory environments. Typically, these methods rely upon reductionism – or a narrow definition of system boundaries – which removes confounding variables and improves reproducibility of results. Such approaches are problematic for complex systems, as it is difficult to identify which are the essential variables critical to reproduce in the laboratory, and which can be held constant or otherwise ignored. All infrastructure systems, when examined at scales large enough, are complex. Therefore, the disadvantage of relying solely on the iterative abstraction/experimentation loop in infrastructure education is that the resulting generalizable knowledge can never be more than an approximation of real systems. In some cases, these approximations may be so poor as to result in serious misconceptions. Thus, empirical verification of abstract knowledge in complex systems is never fully realizable, and ultimately experimentation must take place in the real world where unpredictable, costly, and even unethical consequences may result.

- **Experience** refers to the knowledge acquired through direct, sensory engagement with an activity. Unlike experimentation, experience may or may not involve the deliberate manipulation of independent variables and conscious monitoring of dependent variables. The advantage of experience is that knowledge is rich in context, but the disadvantage is that it may not be generalizable. Also, experiential knowledge is notoriously difficult to assess, and expensive to share without first being made explicit. The importance of experience in acquisition of expert knowledge is illustrated in foreign language learning. Despite mastering a foreign alphabet, word definitions, sentence structure, idioms, and culture, effective foreign language training typically resorts to some sort of immersion to gain tacit knowledge – such as jokes – which can only come with experience.
- Lastly, **reflection** is the process of giving serious consideration to or examining experience. As abstraction works with experimentation, so does reflection work in concert with experience. Through the processes of observing ourselves and recalling memory of our experiences, reflection can make connections between different experiences which strengthen the applicability of knowledge to new contexts. Reflection helps retain the salient aspects of an experience, make sense of them, and sometimes play out in the imagination alternative experiences which might result from different choices.

Different academic disciplines emphasize iteration between different aspects of the learning cycle. For example, the predominant focus in traditional engineering pedagogy is on cognitive learning objectives and outcomes using abstract conceptualization and active experimentation, both of which are activities for mastering generalizable concepts about infrastructure. However, the holism required to create resilient infrastructure cannot be achieved without strengthening aspects of the Kolbe Cycle that are absent from the disciplines essential to education and training of a resilient infrastructure workforce. For example, the lack of concrete experience in engineering education represents a deficiency that is especially problematic for building adaptive capacity in the context of extraordinarily rare events (Fig. 3.9).

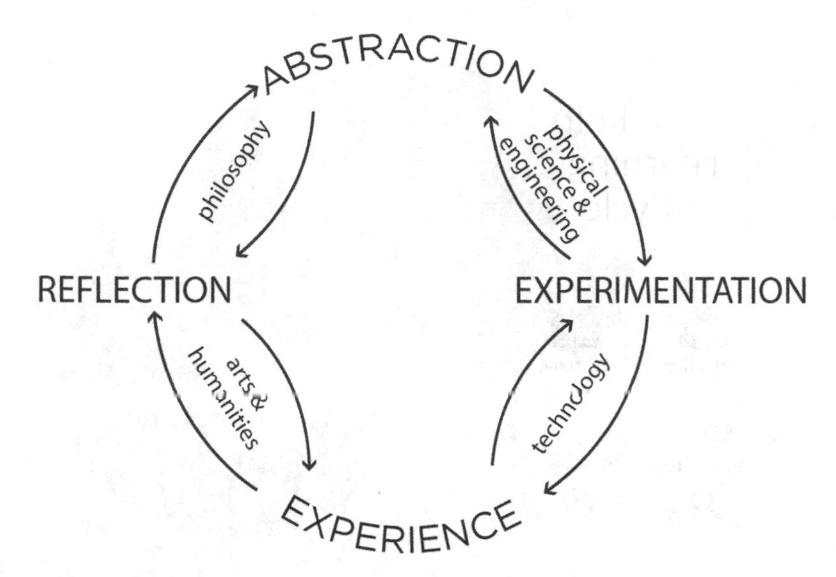

Fig. 3.9 The Kolb Learning Cycle describes the minimum activities necessary for effective education and training. Some disciplines adopt pedagogical approaches that emphasize some stages of the cycle, and neglect others

3.6.3 Simulation Games: A Vehicle for Demonstrating Infrastructure Complexity

Because no single discipline contains both the technical knowledge and the pedagogical strategies necessary for the entire task, a synthesis of methods is required. In particular, the necessity of *experience* presents a significant barrier. Partly this may be because experience is expensive, but it is likely also due to the fact that knowledge acquired through experience is difficult to assess. Knowledge of facts, figures, equations, and book-learning is typically standardized and amenable to comparative assessment. However, this type of knowledge alone is insufficient for the SAAL resilience processes, which require both the acquisition and interpretation of information (i.e., sensing) and a continuous creation of the new knowledge necessary for anticipation, and learning. While knowledge that is standardized and codified is easy to share, experiential knowledge is not. The former is called explicit knowledge, and it is reducible to symbolic language such as software code, design standards, data tables, operation manuals, mathematical equations, and it can be embodied in materials such as machinery, robots, circuit boards, and other physical objects. Explicit knowledge, once created, is typically easy to duplicate and share. The latter (knowledge from experience) is called tacit and it because it is gained exclusively through experience, it is difficult and expensive to share (Grant 1996; Grant et al. 2010). Expertise requires both explicit and tacit knowledge, and thus typically requires both rapid recall of facts and data, and the synthesis of these into

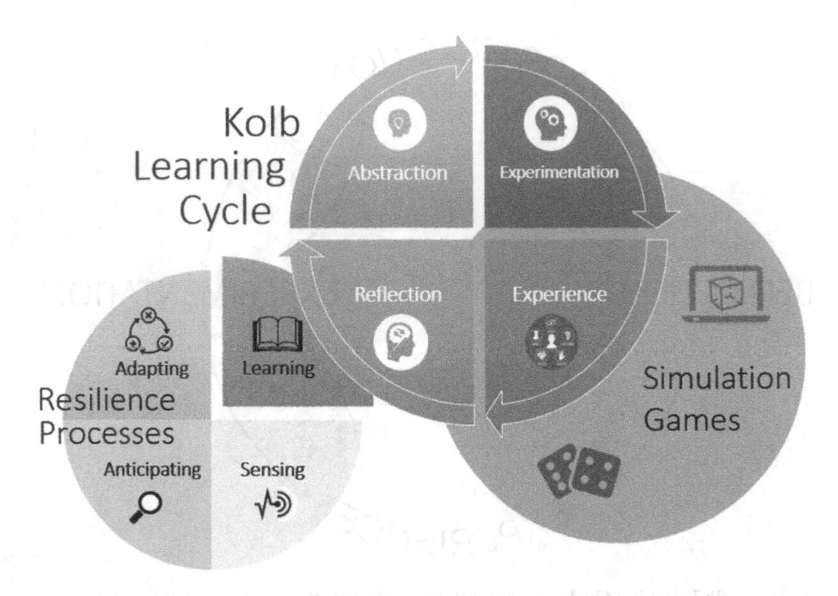

Fig. 3.10 While the Kolb Learning Cycle provides a theory of resilience learning, existing educational programs lack experiential instruction in resilient infrastructure. Simulation games may fill this gap by providing hands-on tools for building tacit knowledge

action informed by experience – i.e., without the necessity of calculation or rational reasoning (Fig. 3.10).

Simulation games provide several affordances that overcome obstacles to teaching the complex systems reasoning skills required to create resilience action (Abt 1970). First, they provide an environment in which trainees can explore historical failures, or build future scenarios without incurring the cost of drills, pilot projects, or catastrophe. Such experiences speak directly to building resilience awareness, as trainees can compress decadal timeframes into hours or minutes to observe the power of non-linearity and explore a more complete range of stochasticity. Participants can practice working in teams and experiment with different governance structures that correspond to different assignments of information, decision rights, and patterns of interaction.

Secondly, trainees can experiment with different adaptive strategies and build the judgement necessary to form more reliable intuitions about preferable alternatives, as well as practice the deliberative skills necessary to identify and choose between or construct the values necessary to make tradeoffs about which components or functions of an infrastructure system must be compromised to maintain graceful extensibility when robustness and recovery fail. Third, good games provide an emotional experience that may motivate trainees to take action prior to catastrophic events, rather than ex ante, as has been the case with the major policy reforms and investments that followed television images of the devastation of New Orleans by Hurricane Katrina. Simulation of the emotional experience may prompt pre-emptive action to mitigate adverse negative consequences in a way that failed to materialize in New Orleans.

Whether simulation games are effective at building courage for resilience action remains an open question. Nevertheless, simulation games have a long history of success in organizations that build and typically rely upon the courage of their members, most notably:

- the military's extensive use of war games (Brewer and Shubik 1979),
- the Red Cross use of simulation games for disaster response training,
- flight simulators for pilot training, and
- simulation in medical training with cadavers, dummies, and virtual reality.

3.7 Conclusion

3.7.1 A National Network for Resilient Infrastructure Case Studies and Simulation

Despite the increasing frequency and insurance losses of extreme weather events experienced nationwide, serious catastrophes such as tornadoes, ice storms, floods, and hurricanes remain rare events. Consequently, communities with little experience with extreme weather often find themselves ill-prepared for even mild stressors. For example, a 1 inch dusting of snow paralyzed traffic in Washington DC in January 2016. After the storm, Mayor Muriel Bowser offered an apology via twitter, admitting that "the District failed to deploy the necessary resources in response to the snow" (Bowser 2016). But the dire consequences of just a few inches of snow should not have been mysterious to District officials, given the experience of Atlanta almost exactly two years earlier, when less than three inches of snow accumulation caused 13 deaths, complete gridlock, and declaration of a state of emergency.

A key contributor to the problem is that we are not learning effectively from past disasters or adverse events on a national scale. Toft and Reynolds (2016) explain that this problem stems from 'hidden influences' that impact the perception of risk and the decision-making ability of both individuals and groups of people. These influences include individual and organizational heuristics that are biased by past experiences (or the lack of), a sense of personal invulnerability or overconfidence about risk, and an aversion to learning from negative events. For critical infrastructure systems, organizations such as water departments and power regulators are additionally challenged by the increasing and complex interdependencies of the systems they govern, that can amplify the consequences of poor decision-making and/or disruptions during an extreme event (Clark et al. 2017).

To integrate and build capacities for resilience awareness, resilience judgement, resilience motivation, and resilience action, we propose a National Network for Resilient Infrastructure Case Studies and Simulation. Networks are fast becoming the future of scientific societies which leverage information communication technology to accelerate collaboration and co-production of knowledge (Seager and Hinrichs 2017). We envision the Network as a platform for capture, creation, sharing, and dissemination of lessons learned from previous disasters, and for offering experiential training and education through simulation of future infrastructure scenarios. The Network would enable knowledge from past experiences to be transferred to other locations and situations, and simulations allow stakeholders to safely experience infrastructure failures as well as reflect on those experiences by playing out multiple scenarios and alternative endings, therefore motivating change. These cases and simulation experiences will influence decision-making heuristics to help address common but inhibitive tendencies for understanding risk.

The case study method is widely used for learning business, law and medical disciplines as well as teaching science (Yadav et al. 2007; Irwin 2000), and organizational learning (Thomas et al. 2001; March et al. 1991). Nonetheless, historical accounts alone are insufficient for providing a meaningful, personal experience necessary for changes in behavior (Seager et al. 2010). Simulations compliment case studies by engaging both affective and cognitive processes through experiential learning (Tennyson and Jorczak 2008) and creating spaces for participants to embody new forms of leadership necessary for creative problem solving in diverse, multi-sectoral teams (Hinrichs et al. 2017). Simulations allow opportunities to practice decision-making related to crises of complex infrastructure systems, which would be immoral in the lived, physical world (Sterman 1994). Therefore, an essential aspect of the proposed approach would be to augment historical case studies with computer simulations that allow interaction with the case in ways that emulate the choices and stressors faced by real infrastructure managers. Well documented disasters like Hurricanes Katrina and Sandy, the 2011 San Diego Blackout, the 2011 New Madrid Flooding Event, and the current Los Angeles water infrastructure crises offer content rich cases from which we can learn about successes and failures of our infrastructure systems and their governance structures during and after an extreme event.

Simulation-based learning is common for military skills training, such as flight simulator training (Cioppa et al. 2004; Hays et al. 1992). Healthcare professionals also use simulations to learn and practice skills without risking the health of lives of patients (Hammond 2004; Kneebone 2003; Ziv et al. 2000), including the American Red Cross, which employs simulation learning for emergency healthcare training. In a similar fashion, we can use simulations to learn about how our systems fail and how they might respond to anticipated changes in the future. Together, case studies and simulations offer a promising strategy for fostering a more adaptive approach to infrastructure management that benefits from past infrastructure experiences as well as considers longer-term implications of design strategies. Just as the Harvard Business Review offers a collection of case studies for business management education, the Federal Aviation Administration provides lessons from airplane accidents (see lessonslearned.faa.gov), and the Center for Army Lessons Learned (CALL) collects and disseminates lessons from past training and operation experiences (Thomas et al. 2001), we imagine a continuously growing database of infrastructure case studies that allow for more effective creation knowledge from past and simulated infrastructure failures, overcoming Flynn's (2016) four knowledge barriers discussed at the outset of the chapter. Through repeated simulation, participants may experiment with different strategies and reflect upon the outcomes, develop reliable theories and test new practices that investigate resilient infrastructure systems as whole. Moreover, we propose that the simulations be used as exercises for students and practitioners working in municipalities, businesses, utilities, engineering firms, research institutes, and urban planners to overcome Flynn's fifth barrier, education and training that prepares the workforce necessary to resolve the impending infrastructure crisis.

Acknowledgements The authors acknowledge the guidance provided in extended conversations with Drs. Stephen Flynn of Northeastern University, David Woods of The Ohio State University, and Igor Linkov of the US Army Engineer and Development Center. Also, Dr. David Alexander of University College London provided feedback and encouragement in the development of Fig. 3.2. This material is based upon work supported by the National Science Foundation awards 1,441,352 and 1,360,509, the Department of Defense (DOD) Grant11967796-ONR-Navy Enterprise Partnership Teaming with Universities for National Excellence (NEPTUNE), and the DOD Defense Threat Reduction Agency award HDTRA1-11-16-BRCWMD-PerE). The content does not necessarily reflect the position or the policy of the Federal Government, and no official endorsement should be inferred.

Further Suggested Readings

Abt C (1970) Serious games. Viking Press, New York

Achard S (2006) A resilient, low-frequency, small-world human brain functional network with highly connected association cortical hubs. J Neurosci 26(1):63–72

Adibi MM, Fink LH (2006) Restoration from cascading failures. IEEE Power and Energ Mag 4(5):68–77

Alberts D, Huber RK, Moffat J (2010) NATO NECe C2 Maturity Model, Available at: http://oai.dtic.mil/oai/oai?verb=getRecord&metadataPrefix=html&identifier=ADA555717. Accessed 15 Feb 2017

Alderson DL, Doyle JC (2010) Contrasting views of complexity and their implications for network-centric infrastructures. IEEE Trans Syst ManCybernPart A Syst Hum 40(4):839–852

Alderson DL, Brown GG, Carlyle WM (2014) Assessing and improving operational resilience of critical infrastructures and other systems. In: Bridging Data and Decisions, INFORMS pp 180–215

Alderson DL, Brown GG, Carlyle WM (2015) Operational models of infrastructure resilience. Risk Anal 35(4):562–586

Alexander DE (2013) Resilience and disaster risk reduction: an etymological journey. NatHazards Earth Syst Sci 13(11):2707–2716. Available at: http://www.nat-hazards-earth-syst-sci.net/13/2707/2013/. Accessed 30 Jan 2014

Alkire S (2002) Dimensions of human development. World Dev 30(2):181–205

Andersson G et al (2005) Causes of the 2003 major grid blackouts in North America and Europe, and recommended means to improve system dynamic performance. IEEE Trans Power Syst 20(4):1922–1928

Ayres RU, Axtell R (1996) Foresight as a survival characteristic: when (if ever) does the long view pay? Technol Forecast Soc Chang 51(3):209–235

Barry J (1997) Rising tide: the great Mississippi flood of 1927 and how it changed America. Simon & Schuster, New York

Berkes F, Ross H (2012) Community resilience: toward an integrated approach. Soc Nat Resour 26(1):5–20
Bonanno GA (2004) Loss, truama, and human resilience: have we underestimated the human capacity to thrive after extremely adverse events? Am Psychol 59(1):20–28
Bowser M (2016) Last night the district failed to deploy the necessary resources in response to the snow – for that I am sorry. Twitter 21 January 2016. Available at: https://twitter.com/mayorbowser/status/690196235593080832. Accessed 13 Feb 2017
Brewer G, Shubik M (1979) The war game: a critique of military problem solving. Harvard University Press, Cambridge, MA
Chen Y-Z, Huang Z-G, Zhang H-F, Eisenberg DA, Seager TP, Lai Y-C (2014) Extreme events in multilayer, interdependent complex networks and control. Sci Rep 5:17277
Cioppa TM, Lucas TW, Sanchez SM (2004) Military applications of agent-based simulations. In: Proceedings of the 2004 Winter Simulation conference. IEEE, pp 165–174
Clark SS, Chester MV (2016) A hybrid approach for assessing the multi-scale impacts of urban resource use: transportation in phoenix, Arizona. J Ind Ecol 21(1):136–150
Clark SS, Berardy A, Hannah M, Seager TP, Selinger E, Mikanda JV (2015) The role of tacit knowledge in facilitating globally distributed virtual teams: lessons learned from using games and social media in the classroom. Connexions 3(1):113–151
Clark SS, Chester MV, Seager TP (2017) The vulnerability of interdependent urban infrastructure systems to climate change: could phoenix experience a Katrina of extreme heat? Climactic Change. Under review. Prepublication draft. Available at https://awsum.box.com/v/2017Clark-KatrinaExtremeHeat. Accessed 15 Feb 2017
Connor K (2006) Assessment of resilience in the aftermath of trauma. J Clin Psychiatry 67(2):46–49
Department of Defense (DoD) (2016) DoD directive 3020.40: mission assurance, Office of the Under Secretary of Defense for Policy, Washington, DC, Effective: November 29, 2016
Department of Homeland Security (DHS) (2013) National Infrastructure Protection Plan (NIPP): partnering for critical infrastructure security and resilience. Available at: https://www.dhs.gov/national-infrastructure-protection-plan. Accessed 15 Feb 2017
Dörner D (1996) The logic of failure: recognizing and avoiding error in complex situations. Basic Books, New York
Eisenberg, D.A., Park, J., Kim, D., Seager, T.P. (2014). Resilience analysis of critical infrastructure systems requires integration of multiple analytical techniques. Figshare. Available at https://doi.org/10.6084/m9.figshare.3085810.v1. Accessed 15 Feb 2017
Esbjörn-hargens S (2010) An overview of integral theory. In: Esbjörn-hargens S (ed) Integral theory in action: applied, theoretical, and constructive perspectives on the AQAL model. State University of New York Press, Albany, pp 33–61
Esbjörn-hargens S, Zimmerman MA (2009) Integral ecology, uniting multiple perspectives on the natural world. Integral Books, Boston
Flynn SE (2016) Five impediments to building societal resilience. In IRGC Resource Guide on Resilience pp 1–4. Available at https://www.irgc.org/wp-content/uploads/2016/04/Flynn-Five-Impediments-to-Building-Societal-Resilience.pdf. Accessed 14 Feb 2017
Fox-Lent C, Bates M, Linkov I (2015) A matrix approach to community resilience assessment: an illustrative case at Rockaway Peninsula. EnvironSyst Decis 35:209–218
Ganin AA, Massaro E, Gutfraind A, Steen N, Keisler JM, Kott A, Mangoubi R, Linkov I (2016) Operational resilience: concepts, design and analysis. Sci Rep 6:19540
GAO (2013) Critical infrastructure protection: DHS list of priority assests needs to be validated and reported to congress. United States Government Accountability Office Report to Congressional Requesters. Available at: http://www.gao.gov/assets/660/653300.pdf
GAO (2014) Critical infrastructure protection: DHS action needed to enhance integration and coordination of vulnerability assessment efforts. United States Government Accountability Office Report to Congressional Requesters. Available at: http://www.gao.gov/assets/670/665788.pdf
Gim C, Kennedy EB, Spierre Clark S, Millar CA, Ruddell B (2017) Institutions in critical infrastructure resilience. J Infrastruct Syst Under review. Prepublication draft available at https://awsum.box.com/v/2017Gim-InstitutionalLayering. Accessed 15 Feb 2017

Google Ngram (2017a) https://books.google.com/ngrams/graph?content=resilience&year_start=1800&year_end=2015&corpus=15&smoothing=3&share=&direct&direct_url=t1%3B%2Cresilience%3B%2Cc0. Accessed 14 Feb 2017
Google Ngram (2017b) https://books.google.com/ngrams/graph?content=risk_NOUN%2Crisk_VERB%2Cresilience&year_start=1800&year_end=2015&corpus=15&smoothing=3&share=&direct_url=t1%3B%2Crisk_NOUN%3B%2Cc0%3B.t1%3B%2Crisk_VERB%3B%2Cc0%3B.t1%3B%2Cresilience%3B%2Cc0. Accessed 15 Feb 2017
Grant R (1996) Toward a knowledge-based theory of the firm. Strateg Manag J 17:109–122
Grant GB, Seager TP, Massard G, Nies L (2010) Information and communication technology for industrial symbiosis. J Ind Ecol 14(5):40–753
Hall J et al (2010) A framework to measure the progress of societies, OECD statistics working papers, 2010/05. OECD Publishing. http://dx.doi.org/10.1787/5km4k7mnrkzw-en
Hamilton C, Adolphs S, Nerlich B (2007) The meanings of "risk": a view from corpus linguistics. Discourse Soc 18(2):163–181
Hammond J (2004) Simulation in critical care and trauma education and training. Curr Opin Crit Care 10(5):325–329
Hannah ST, Avolio BJ, May DR (2011) Moral maturation and moral conation: a capacity approach to explaining moral thought and action. Acad Manag Rev 36:663–685
Hayden BY (2016) Time discounting and time preference in animals: a critical review. Psychon Bull Rev 23:39–53
Hays RT et al (1992) Flight simulator training effectiveness: a meta-analysis. Mil Psychol 4(2):63–74
Hinrichs MM, Seager TP, Tracy SJ, Hannah MA (2017) Innovation in the knowledge age: implications for collaborative science. Environ Syst Decis In press. doi:10.1007/s10669-016-9610-9
Holling CS (1973) Resilience and stability of ecological systems. Annu Rev Ecol Syst 4:1–23
Hollnagel E (2014) Resilience engineering and the built environment. Build Res Inf 42(2):221–228
Hollnagel E, Woods DD, Leveson N (2007) Resilience engineering: concepts and precepts. Ashgate Publishing Ltd., Aldershot
Homeland Security Council (HSC) (2007) National strategy for homeland security. The White House, Washington, DC
Hubbard DW (2009) The failure of risk management: why it's broken and how to fix it. John Wiley & Sons, Hoboken
Irwin AR (2000) Historical case studies: teaching the nature of science in context. Sci Educ 84(1):5–26
Kapucu N, Garayev V (2014) Structure and network performance: horizontal and vertical networks in emergency management. Adm Soc 48(8):931–961
Kapucu N, Garayev V, Wang X (2013) Sustaining networks in emergency management. Public Perform Manag Rev 37(1):104–133
Kirschen D, Bouffard F (2009) Keeping the lights on and the information flowing. IEEE Power and Energ Mag 7(1):50–60
Kneebone R (2003) Simulation in surgical training: educational issues and practical implications. Med Educ 37(3):267–277
Kolb DA (2014) Experiential learning: experience as the source of learning and development. FT press
Larkin S, Fox-Lent C, Eisenberg DA, Trump BD, Wallace S, Chadderton C, Linkov I (2015) Benchmarking agency and organizational practices in resilience decision making. Environment Systems and Decisions 35(2):185–195
Linkov I, Bridges T, Creutzig F, Decker J, Fox-Lent C, Kröger W, Lambert JH, Levermann A, Montreuil B, Nathwani J, Nyer R (2014) Changing the resilience paradigm. Nat Clim Chang 4(6):407–409
Madni AM, Jackson S (2009) Towards a conceptual framework for resilience engineering. IEEE Syst J 3(2):181–191
March JG, Sproull LS, Tamuz M (1991) Learning from samples of one or fewer. Organ Sci 2(1):1–13

Maslow AH (1943) A theory of human motivation. Psychol Rev 50(4):370
Masten AS (2014) Ordinary magic: resilience in development. The Guilford Press, Upper Saddle River
Munoz G (2015) Capistrano Taxpayers Association, Inc. v. City of San Juan Capistrano, Court of Appeal of the State of California Fourth Appellate District Division Three, Super. Ct. No. 30-2012-00594579. Available at http://wilcoxbenumof.blazonco.com/files/articles/G048969.pdf. Accessed 15 Feb 2017
Nagel T (1993) Moral luck. In: Russell P, Deery O (eds) Essential readings from the contemporary debates. Oxford University Press, Oxford
National Security Council (NSC) (2007) National Continuity Policy Implementation Plan. The White House, Washington, DC
Navvaro P (2004) On the political economy of electricity deregulation—California style. Electr J 17(2):47–54
Noltemeyer AL, Bush KR (2013) Adversity and resilience: a synthesis of international research. Sch Psychol Int 34(5):474–487
Norris FH, Stevens SP, Pfefferbaum B, Wyche KF, Pfefferbaum RL (2008) Community resilience as a metaphor, theory, set of capacities, and strategy for disaster readiness. Am J Community Psychol 41(1–2):127–150
Nussbaum M (2000) Women and human development: the capabilities approach. Cambridge University Press, Cambridge
Nussbaum M (2003) Capabilities as fundamental entitlements: Sen and social justice. Fem Econ 9(2–3):33–59
Nussbaum M (2006) Frontiers of justice: disability, nationality, species membership. Harvard University Press, London
Nussbaum M, Sen A (1992) The quality of life. Oxford University Press, Oxford
Olson K, Morton L (2012) The impacts of 2011 induced levee breaches on agricultural lands of Mississippi River valley. J Soil Water Conserv 67(1):5A–10A
Ouyang M (2014) Review on modeling and simulation of interdependent critical infrastructure systems. Reliab Eng Syst Saf 121:43–60
Park J, Seager TP, Rao PSC, Convertino M, Linkov I (2013) Integrating risk and resilience approaches to catastrophe management in engineering systems. Risk Anal 33(3):356–367
Pauly MV (1968) The economics of moral hazard: comment. Am Econ Rev 58(3, Part 1):531–537
Perrow C (1984) Normal accidents: living with high-risk technologies. Basic Books, New York
Petit F et al (2015) Analysis of critical infrastructure dependencies and interdependencies. Argonne National Lab Report ANL/GSS-15/4. Available at: http://www.ipd.anl.gov/anlpubs/2015/06/111906.pdf. Accessed 15 Feb 2017
Petrenj B, Lettieri E, Trucco P (2012) Towards enhanced collaboration and information sharing for critical infrastructure resilience: current barriers and emerging capabilities. Int J Crit Infrastruct 8:107
Petrenj B, Lettieri E, Trucco P (2013) Information sharing and collaboration for critical infrastructure resilience – a comprehensive review on barriers and emerging capabilities. Int J Crit Infrastruct 9(4):304–329
Pourbeik P, Kundur PS, Taylor CW (2006) The anatomy of a power grid blackout. IEEE Power Energ Mag 4(5):22–29
Rinaldi SM, Peerenboom JP, Kelly TK (2001) Identifying, understanding, and analyzing critical infrastructure interdependencies. IEEE Control Syst Mag 21(6):11–25
Sadowski J, Seager TP, Selinger E, Spierre SG, Whyte KP (2013) An experiential, game-theoretic pedagogy for sustainability ethics. Sci Eng Ethics 19(3):1323–1339
Sadowski J, Spierre SG, Selinger E, Seager TP, Adams EA, Berardy A (2015) Intergroup cooperation in common pool resource dilemmas. Sci Eng Ethics 21(5):1197–1215
Sage D, Zebrowski C (2016) Resilience and critical infrastructure: origins, theories and critiques. In: Dover R, Huw D, Goodman MS (eds) The Palgrave handbook of security, risk and intelligence. Palgrave MacMillan, London

Seager TP (2008) The sustainability spectrum and the sciences of sustainability. Bus Strateg Environ 17(7):444–453

Seager TP, Hinrichs MM (2017) Technology and science: innovation at the international symposium on sustainable systems and technology. Environ Syst Decis In press. doi:10.1007/s10669-017-9630-0

Seager TP et al (2007) Typological review of environmental performance metrics (with illustrative examples for oil spill response). Integr Environ Assess Manag 3(3):310–321

Seager TP, Selinger E, Whiddon D, Schwartz D, Spierre S, Berardy A (2010) Debunking the fallacy of the individual decision-maker: an experiential pedagogy for sustainability ethics. In: Sustainable Systems and Technology (ISSST), 2010 IEEE international symposium on IEEE, pp 1–5

Seager TP, Gisladottirb V, Mancillas J, Roege P, Linkov I (2017) Inspiration to operation: securing net benefits vs. zero outcome. J Clean Prod 148:422–426

Selinger E, Sadowski J, Seager TP (2015) Gamification and morality. In: Waltz SP, Deterding S (eds) The gameful world: approaches, issues, applications. MIT Press, Cambridge, MA, pp 371–392

Selinger E, Seager TP, Spierre, S, Schwartz D (2016) Using sustainability games to elicit moral hypotheses from scientists and engineers. Rethinking climate change research: clean technology, culture and communication, Routledge, pp 117–128

Sen A (1999a) Commodities and capabilities. Oxford University Press, Oxford

Sen A (1999b) Development as freedom. Anchor Books, New York

Sterman J (1994) Learning in and about complex systems. Syst Dyn Rev 10(2–3):291–330

Stranger W (2013) California wastewater rates: life after Bighorn v Verjil. California Water Law J. Available at: http://waterlawjournal.com/california-wastewater-rates-life-after-bighorn-v-verjil/

Taleb NN (2005) Fooled by randomness: the hidden role of chance in life and in the markets. Random House, New York

Taleb NN (2012) Antifragile: things that gain from disorder. Random House, New York

Tennyson RD, Jorczak RL (2008) A conceptual framework for the empirical study of instructional games. In: O'Neil HF, Perez RS (eds) Computer games and team and individual learning. Elsevier, Amsterdam, pp 39–54

Thomas JB, Watts Sussman S, Henderson JC (2001) Understanding "Strategic Learning": linking organizational learning, knowledge management, and sensemaking. Organ Sci 12(3):331–335

Toft B, Reynolds S (2016) Learning from disasters. Springer, New York

Wilber K (2001) A theory of everything: an integral vision for business, politics, science, and spirituality. Shambhala Publications, Boston

Winkler R (2006) Does 'better' discounting lead to 'worse' outcomes in long-run decisions? The dilemma of hyperbolic discounting. Ecol Econ 57(4):573–582

Woods DD (2015) Four concepts for resilience and the implications for the future of resilience engineering. Reliab Eng Syst Saf 141:5–9

Yadav A et al (2007) Teaching science with case studies: a national survey of faculty perceptions of the benefits and challenges of using cases. J Coll Sci Teach 37(1):34

Zautra A, Hall J, Murray K (2008) Community development and community resilience: an integrative approach. Community Dev 39(3):130–147

Zhang S-P, Huang Z-G, Dong J-Q, Eisenberg DA, Seager TP, Lai Y-C (2015) Optimization and resilience of complex supply-demand networks. New J Phys 17(6):063029

Zimmerman R, Restrepo CE (2006) The next step: quantifying infrastructure interdependencies to improve security. Int J Crit Infrastruct 2:215–230

Ziv A, Small SD, Wolpe PR (2000) Patient safety and simulation-based medical education. Med Teach 22(5):489–495

Chapter 4
Designing Resilient Systems

Scott Jackson and Timothy L.J. Ferris

Abstract This chapter describes a method to approach the design of systems to ensure resilience. The state machine model describes a set of states in which a system may be situated and a set of transitions between those states which represent the response of the system to either events of threat applied to the system or restorative, maintenance or management actions taken in interaction with the system. The method is based on the analysis of systems proposals using a state machine description of resilience which is presented in the first major section of this chapter. Systems are developed to provide specific capabilities, usually a set of cognate capabilities that are either, or both, capabilities which belong together as a set or which are usefully grouped together to provide improved value from the system compared with only building the system to provide for its primary purpose. The design approach described in this chapter extends the normal design activities required to design a system to provide the specified capability with design and analysis activities required to ensure that the system provides the required resilience characteristics.

Keywords Resilience • States • Transitions • Events • Principles • Design • Processes • Systems • Decisions

4.1 Introduction

The state machine model described in this chapter identifies seven possible system states and 31 transitions between those states which result from particular events which may occur to the system or actions deliberately performed on the system to change its state. Each state is the result of the system as designed and the history of

---

S. Jackson (✉)
Burnham Systems Consulting, Irvine, CA, USA
e-mail: jackson@burnhamsystems.net

T.L.J. Ferris
Centre for Systems Engineering, Cranfield University, Defence Academy of the United Kingdom, Shrivenham SN6 8LA, UK
e-mail: timothy.ferris@cranfield.ac.uk

© Springer Science+Business Media B.V. 2017

I. Linkov, J.M. Palma-Oliveira (eds.), *Resilience and Risk*, NATO Science for Peace and Security Series C: Environmental Security,
DOI 10.1007/978-94-024-1123-2_4

threats which have impacted and actions performed upon the system. Each transition between states is the result of either a threat impacting the system or a deliberate action performed on the system. The design of systems with the intention of ensuring resilience is, we argue, appropriately guided by the application of a set of 34 principles described in Jackson (2016b). The choice of the system architecture proposed in the design process is a result of the combination of the architectural demands necessary for satisfying the requirements upon the whole system to enable delivery of the intended capability, and the demands to enable the specified resilience characteristics to be delivered. The architecture determined as the appropriate means to achieve the necessary resilience characteristics is most likely different from that which may have been chosen if resilience had not been considered. The specific architectural differences that support resilience, compared to the architecture which would have been chosen if resilience had not been considered, are the result of implementation of particular resilience principles which will enable the achievement of the particular resilience goals needed for the system under development in a manner which is appropriate for the specific system. The resilience principles which guide the design reasoning associated with enabling a desirable solution for each of the transitions may be different across the set of transitions, and the particular choice of principle that provides the most desirable predicted outcome for each transition may be a different member of the set of principles.

The specification of resilience objectives includes determination of the level of performance of each function of the system that is considered appropriate given particular classes of threat event that may occur. Through the life of the system it will be affected by a number of threat events, some of which will cause transition to states for which a restoration of management remedy is required with the result that through the life cycle a system may pass through the possible states and transitions described in the state machine model in any combination permitted by the structure of the model. The design of a system with resilience as a factor of interest is governed by the clear determination of the resilience objectives and the use of design principles selected from a set of principles that provide means to improve resilience through methods that have been demonstrated as helpful to improve achievement of specific resilience related objectives.

4.2 State Machine Model

In Fig. 4.1 we show a generic state machine model of system states and transitions that describes the relationships related to resilience. The state machine model includes all the states and transitions which could be meaningful in any system. In practice, in some specific systems, because of the nature of those systems some of the states or transitions may not be meaningful. Reasons for this include which states are meaningful for particular systems. Similarly, only some of the possible transitions described in Fig. 4.1 may be practical or desirable for particular systems. During a system development, the people responsible for design have the

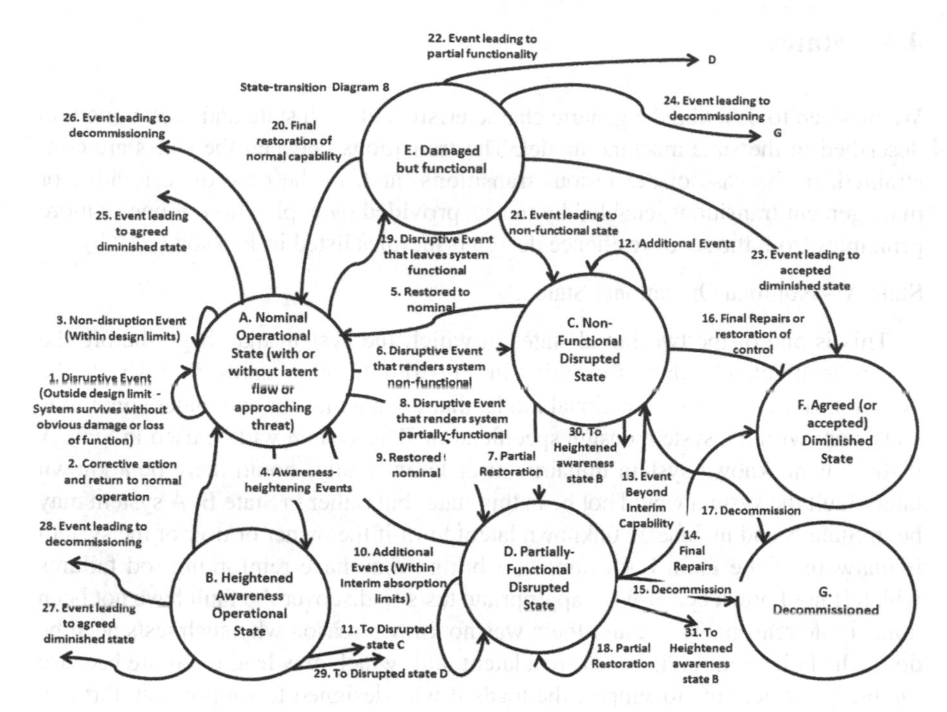

Fig. 4.1 – State-transition diagram (Adapted from Jackson et al. 2015)

responsibility to determine specification description of each of the states, that is, detail that instantiates the desirable condition and performance of the system when it is in each of the states described. In the specification process it may be decided that there is no meaningful description of a particular state. For example, some of the interim states representing partial impairment of the system may be determined to be not meaningful because a system that suffers such an impairment may be transitioned directly to decommissioned.

Similarly, the designers of a particular system must consider which transitions are appropriate for the specific system. Some transitions occur because the system is affected by a threat applied to it, and in this case the transition must be described in the system specific model. The transitions which occur as a result of the restorative or management actions of system owners and managers are discretionary and, depending on the nature of the system, it may be appropriate to support, or not support, particular discretionary transitions. The benefit of the state machine model is that it provides system designers with a tool to provide a reasoned specification of the characteristics to be enabled by the system with respect to both its primary capability and its resilience characteristics.

4.3 States

We proceed to describe the generic characteristics of each state and each transition described in the state machine model. The transitions between the states are constrained, in the case of deleterious transitions, and, in the case of restorative or management transitions, enabled by means provided by application of one or more principles from the set of resilience design principles listed in Jackson (2016b).

State A – Nominal Operational State

This is one of the two initial states in which the system may begin before the system is affected by threats resulting in transition to other states. In this state the system is used in a usual functional mode, that is, in a manner and under conditions consistent with the system design specification. The system will be used in State A if there is no known system impairment or latent fault. Should there be a known latent fault the system would not be in this state, but rather in State B. A system may be in State A and include an unknown latent fault if the owner or user of the system is unaware of the fault. For example, a bridge may have reinforcing rod failures which if not known because the appropriate tests to discover the fault have not been done, quite reasonably because there was no known reason why such tests need be done, the failed reinforcing rods are a latent fault which may lead to failure because the bridge is not able to support the loads it was designed to support but, through deterioration is no longer able to support. The system operators are not aware of the impending threat.

State A, as the nominal operational state, includes the system operating in an operating condition within the operational envelope of the system. If the normal operational envelop includes a condition in which the system is switched off, either before first operation or at any time thereafter, then State A includes the system in a design compliant switched off state. In such a switched off condition the system is in a state consistent with one of the normal conditions for which it was designed, an inactive state in which it is either switched off awaiting an occasion of use or is switched off for any other reason that belongs within the normal, not awaiting repair, reasons a system may be switched off. These might include storage or shipping, or some other system relevant purpose for being switched off.

State B – Heightened Awareness Operational State

This is the second of the two states in which the system may begin before it is subjected to additional threat events. The difference between states A and B is that in State B the system operators are aware of one or more impending threats. An example of such a system may be the bridge with compromised reinforcing rods. Awareness of the weakened state of the bridge may lead to analysis to determine the extent of the weakness and the heightened awareness of the operating condition may lead to operational constraints to only allow a reduced loading of the bridge in use. Change in the operational environment may lead to a similar situation. There are a number of stone bridges across the Thames River in Oxfordshire which were built to enable horse drawn vehicle traffic. The bridges remain in a state in which they could support their design traffic loading but because of changes in the traffic possibilities resulting from modern vehicles and traffic densities have been reassessed and now have a limited vehicle loading rating. This is an example of a State B, heightened awareness state in which the original design state is the cause of what is a current latent weakness.

State C – Non-functional Disrupted State

In this state the threat, or threats, have rendered the system completely non-functional. The accumulation of threats that have impacted on the system have done damage to the system that results in the system not operating. In this state the system has been disabled, and so is not available for immediate use, but the management decision concerning whether to restore or decommission the system has not yet been made. An example of a system in State C is a vehicle which has been involved in a significant collision which has disabled the vehicle before the assessment has been made that leads to the decision to either decommission the vehicle or to repair it. The outcomes of the assessment decision may be restoration to full functionality (State A), to partial functionality (State D), to an acceptable diminished state (State F) or to complete decommissioning (State G), as shown by transitions shown in Fig. 4.1. The decision depends on an overall assessment of the available choices in the context of the specifics of the system. The USS Cole, described by the US Department of Defence (2001), is an example of a high value system in State C in which it made financial and strategic sense to restore to State A.

State D – Partially Functional Disrupted State

In this state the threat has rendered the system partially functional. The diminishment of function may be in the form of complete removal of certain functions or the reduction in performance levels achievable in one or more functions. The effect of the diminishment is that the system cannot perform all design functions at the design specification but the system retains some functionality, and might continue to have capacity to provide useful services with the remaining capacity. An example of loss of a function may be a vehicle suffering loss of communications services, in which case the work around is to revert to following pre-departure instructions for what to do in the event of loss of communications and the capacity that provides for intra-journey update of instructions or situation awareness. An example of impairment of a function is the capacity to continue driving, with some restrictions, by the 'run flat' tyre fitted to some models of car. On the occasion of a puncture the vehicle can continue driving for a useful distance prior to effecting a full repair avoiding the risks associated with an enforced tyre change at the site of the puncture event. US Airways Flight 1549 encountered a flock of birds causing loss of both engines as described by Pariès (2011) forcing the aircraft to depend only on auxiliary power, which was sufficient to enable control by the pilot which resulted in ditching into the Hudson River. The long-term result was rescue of all occupants, no collision with anything else, but total loss of the aircraft. The capacity for limited function in that event provided the opportunity for a skilled operator to attempt remedial action reducing the effect of the original loss of function.

State E – Damaged but Functional State

In this state the threat has damaged the system, but the system continues to maintain full functionality at least in the near term. One mechanism typically used to achieve this state is system design with sufficient redundancy to continue function even though part of the system is damaged. An example is the design requirement for commercial aircraft to have sufficient engine capacity to enable continuation of a flight and landing with the loss of all but one engine. This system requires both the technical capacity of the equipment to perform the functions necessary and also that the system has the necessary support, such as pilot training, to appropriately respond to such events. More generally systems may be designed to support functions by methods, either similar to the primary method, or markedly differently but supporting the same outcome action. An example of a different method is seen if a customer service computer system also has a manual backup method available to address the event of failure of the computing system.

State F – Agreed Diminished State

This is one of two final states in the state-transition analysis. In this state the system is damaged in a way that leads to a management decision that the effective solution is to decide to modify the concept of the system to be less capable than it was originally intended to be. The management decision may be based on any, or a combination of, factors including the economics of cost of repair and benefit of the investment in repairs, availability of repair parts and materials, and the impact of disruptions associated with performing remedial work, and potentially other factors. Transition to an agreed diminished state may involve some work to modify the system from its impaired state to the agreed end state. This choice is likely in several situations: the system is near end of life and restoration is not practical or is excessively expensive for the anticipated benefit; or the system can be made to provide a useful service in the current context without restoration to its original condition.

State G – Totally Decommissioned State

This is the second of two final states in the state-transition analysis. In this state the system is removed from service, and in most situations is scrapped. In this state there is no way to bring the system back into service. If the service that was provided by the system continues to be necessary the only method to provide that service is to use a different system. The system may be transitioned into State G as a result of a catastrophic event which destroys the original system, or an event which causes sufficient damage that a management decision is made that it is not worthwhile attempting to restore the system to either full or partial functionality for either cost or practical reasons.

4.4 Transitions

Transitions are the paths the system passes through between the states. It is necessary to understand the nature of the transitions in order to understand the impact of the resilience state model on system architecture for design purposes and to assist system management.

Transitions are initiated by events. There are three classes of events, each of which can be the trigger for a state transition. The class of events are:

1. Threat events. These are occasions on which either an external or latent threat materialises to present an actual threat to the system. These may be of kinds and magnitudes that are anticipated, and may be specified as events with which the system must deal, or of unanticipated kinds or magnitudes exceeding explicit design margins. Threats may be generated as a result of a physical effect, whether arising from the environment or from the internal properties of a system, or human caused, whether by outsider or insider parties and potentially accidental or malicious, or the interaction of and management to address predictable events, the possibility of unanticipated and unanticipatable events demands that resilience be framed in a specific cause agnostic manner. This has profound impacts on how resilience is conceptualised and addressed.
2. Restorative activities. These are actions performed by system managers to perform action to restore the system from a damaged state to either a like original condition or a condition that is agreed to be a suitable repaired state.
3. Management activities. These are actions performed by system operators or managers in which a decision about whether, or what kind of, restoration, or other management action is appropriate to best achieve the objectives of the system owner.

The combination of event described above can interact. In particular, it is possible that a system operator or manager action in response to a first threat event may interact with the characteristics of the system situated in its environment to introduce a new threat. To illustrate, a common event on country roads in Australia is a car roll-over involving the event sequence of the car drifting off the paved roadway on the passenger side, the driver, on realising the error, pulls hard on the steering to the right to return to the car to the paved roadway, the tires bit hard on the pavement and the car over-corrects and progresses across the road and off the other side, leading to a high-speed rollover on the opposite side of the road than the original travel. The car rolls with consequent vehicle damage, injury and possibly death for the occupants. The driver response to the original error was inappropriate and led to the catastrophe.

4.4.1 Transition Descriptions

We discuss the 31 possible transitions between states to explain the type of event which initiates the transition. The transitions initiated by events can result either in preservation of system functionality, restoration of functionality, or the managed transition to lower functionality, or in the absence of desirable results the threat induced diminution of functionality.

Transition 1 – Disruptive Event outside Design Limit (State A to State A)

This transition describes the situation in which a system returns to its normal operating state after the impact of a threat which is greater than the specified capacity of the system to absorb a threat of the type. For example, in the aviation industry a design margin of 50% for most structural components so threats up to these limits are within the design capacity of the aircraft. These margins allow the aircraft to absorb disruptions such as gust loads. Events which exceed design limits may have no effect on immediate functionality, but they may cause damage that may accumulate over time through effects such as fatigue which may reduce the margin level and potentially require repair at a later time. A specific instance of this transition may be initiated by an event such as the gust load discussed above. The design action to pre-empt this threat may be effected by implementation of the *absorption* principle and the *margin* support principle as described in Table 4.1.

Transition 1 occurs in response to a threat event.

Transition 2 – Return to Normal Operational State (State B to State A)

This transition occurs if the threat detection which resulted in the system being in State B has been neutralised or avoided. Upon confirmation that the threat is no longer present the system can transition from State B to State A, which happens when any heightened awareness measures are discontinued. Transition 2 is also enabled if it is confirmed that the threat which was discovered, and resulted in the system being in State B, is within design limits, thus enabling reversion to State A.

Transition 2 is initiated by a management decision that the threat has gone and therefore the special measures of State B can be relieved.

Transition 3 – Non-disruptive Event Inside Design Limit (State A to State A)

This scenario can be called the nominal case because it is a threat event within the limits for which the system is built. The threat is recognised here as a threat because it is between the nominal design load and the standard margin of safety loading. However, for example, a building designed to withstand a 9.0 Richter earthquake should easily withstand a 5.0 or 6.0 quake. Yet, even the lower value threats may cause minor damage which needs repair. The earthquake is the initiating event. The system design approach to address the example scenario is to implement the *absorption* principle described in Table 4.1.

Transition 3 occurs as a result of a threat event.

Table 4.1 Resilience principles, support principles and their sources

Top-level principle	Support principle
Absorption – The system shall be capable of withstanding the design level disruption. Derived from Hollnagel et al. (2006)	*Margin* – The design level shall be increased to allow for an increase in the disruption. Hollnagel et al. (2006). Hollnagel lists this as a top-level principle
	Hardening – The system shall be resistant to deformation. Source: Richards (2009)
	Context Spanning – The system shall be designed for both the maximum disruption level and the most likely disruption. Source: Madni (2008)
	Limit Degradation – The absorption capability shall not be allowed to degrade due to aging or poor maintenance. Source: Derived; Jackson and Ferris (2013)
Restructuring – The system shall be capable of restructuring itself. Hollnagel et al. (2006)	*Authority escalation* – Authority to manage crises shall escalate in accordance with the severity of the crisis. Maxwell and Emerson (2009)
	Regroup – the system shall restructure itself after an encounter with a threat. Raveh (2008)
Reparability – The system shall be capable of repairing itself. Richards (2009)	
Drift Correction – When approaching the boundary of resilience, the system can avoid or perform corrective action; action can be taken against either real-time or latent threats. Hollnagel et al. (2006)	*Detection* – The system shall be capable of detecting an approaching threat. Derived: Jackson and Ferris (2013)
	Corrective Action – The system shall be capable of performing a corrective action following a detection. Source: Derived: Jackson and Ferris (2013)
	Independent Review – The system shall be capable of detecting faults that may result in a disruption at a later time. Derived: Haddon-Cave (2009)
Cross-scale Interaction– Every node of a system should be capable of communicating, cooperating, and collaborating with every other node. Source: Hollnagel et al. (2006)	*Knowledge Between Nodes* – All nodes of the system should be capable of knowing what all the other nodes are doing. Billings (1997)
	Human Monitoring – Automated systems should understand the intent of the human operator. Billings (1997)
	Automated System Monitoring – the human should understand the intent of the automated system. Billings (1997)
	Intent Awareness – All the nodes of a system should understand the intent of the other nodes. Billings (1997)
	Informed Operator – the human should be informed as to all aspects of an automated system. Billings (1997)
	Internode Impediment – There should be no administrative or technical obstacle to the interactions among elements of a system. Derived from case studies; Jackson (2010)

(continued)

Table 4.1 (continued)

Top-level principle	Support principle
Functional Redundancy – There should be two or more independent and physically different ways to perform a critical task. Leveson (1995), Leveson uses the term *design diversity*	
Physical Redundancy – The system should possess two or more independent and identical legs to perform critical tasks. Leveson (1995) Leveson uses the term *design redundancy*	
Layered Defence – The system should be capable of having two or more ways to address a single vulnerability. Derived from reason (1997)	
Neutral State – Human agents should delay in taking action to make a more reasoned judgement as to what the best action might be. Madni and Jackson (2009)	
Human in the loop – there should always be human in the system when there is a need for human cognition. Madni and Jackson (2009)	*Automated function* – It is preferable for humans to perform a function rather than automated systems when conditions are acceptable. Billings (1997)
	Reduce Human Error – Standard strategies should be used to reduce human error. Derived from Billings (1997) and Reason (1990)
	Human in Control – Humans should have final decision making authority unless conditions preclude it. Billings (1997)
Complexity Avoidance – The system should not be more complex than necessary. Madni and Jackson (2009), derived from Perrow (1999)	*Reduce Variability* – The relationship between the elements of the system should be as stable as possible. Marczyk (2012)
Reduce Hidden Interactions – Potentially harmful interactions between elements of the system should be reduced. Derived from Leveson (1995) and Perrow (1999)	
Modularity. Madni and Jackson (2009), Perrow (2011)	
Loose Coupling – The system should have the capability of limiting cascading failures by intentional delays at the nodes. Perrow (1999)	*Containment* – The system will assure that failures cannot propagate from node to node. Derived; Jackson and Ferris (2013)
Neutral State- the system should be in a neutral state following a disruption, Madni (2008)	

From Jackson (2016b)

Transition 4 –Awareness Heightening Event (State A to State B)

This transition is initiated by the discovery of an impending threat. The threat could take the form of discovery of a threat in the physical environment, discovery of a human originated threat, discovery of a flaw in the system such as early warning signs of an impending failure, or discovery of drift from the proper operating condition. The origin of the threat is not important but the fact that a threat is discovered while it has still not become manifest is the important issue. The effect of discovery of the threat is that the manner of use of the system is modified to a more defensive, than usual, pattern to reduce the risk associated with encounter with the threat. Design of a method to discover an impending threat would normally rely on application of the drift correction principle, Table 4.1, an example of such a threat would be an approaching train for which positive train control (PTC) is designed as described in the Metrolink report of the National Transportation Safety Board (NTSB 2010).

Transition 4 occurs as a result of a threat event, as either, or a combination of, an automatic or management action.

Transition 5 – Restored to Nominal (State C to State A)

Transition 5 is the transition from State C, a system which has been rendered non-functional is restored to normal operation, State A. The action to effect Transition 5would normally be a repair activity. An example of such a transition is the USS Cole restoration by the DoD (2001). The ship was rendered non-functional by a terrorist attack, and had to be retrieved using heavy ship salvage equipment. The transition 5 trigger event was the management decision to restore the ship to full operational capability, which was then implemented in an appropriate shipyard. The design principle applied in this scenario was *reparability*, described in Table 4.1. The system design implementation of reparability demanded certain design characteristics in the ship platform and also the design of the support systems around the platform to provide means to perform repair.

Transition 5 is an action of repair initiated by a management decision to proceed with the repair.

Transition 6 – Disruptive Event that Renders the System Non-Functional (State A to State C)

Transition 6 is an event which does something that disables a system in a way that results in the system being unable to perform any of its functions forthwith. In the case of USS Cole the attack damaged the ship sufficiently that it was unable to perform any of its functions, even for a short time, leaving the only opportunity of remedy to begin with a salvage retrieval operation. In general transition 6 events are the result of the effect of a threat of a kind and magnitude that the system is immediately disabled. Such threats may be of any threat type, originating either external or internal to the system. The design approach to manage risks associated with transition 6 type events may include the *margin* support principle described in Table 4.1, which would increase the magnitude of threat the system would withstand, or the *modularity* principle which may enable the continuation of some functions of the system. Clearly the principles which it is practical to apply, and the effectiveness of design solutions, depend on the nature of the specific system and its context.

Transition 6 is the result of a threat event.

Transition 7 – Partial Restoration (State C to State D)

If a system is completely dysfunctional, as it is in State C, it may be advantageous to restore the system to a partially functional state, State D. The initiating event would be the decision to perform a partial restoration. State D is an inherently interim state. Following the shrot-term restoration to a partially functional state, which enables partial use of the system until a further decision, or action which can only be performed after some lead-time, where it is advantageous to have partial capability during the lead-time interval, where the later action may result in transition to any of States A, F or G. The interval of partial functionally enabled by State D may be useful to reduce disruptions that would otherwise be cause by complete loss of all the system functionality. The design principles of *modularity*, *physical redundancy* and *reparability*, Table 4.1, are examples of principles which may provide opportunity for transition 7.

Transition 7 is an action of partial repair initiated by a management decision to proceed.

Transition 8 –Disruptive Event that Renders the System Partially Functional (State A to State D)

A disruptive event, whether originated externally or internally to the system, may cause damage which results in partial loss of functionality, whether that loss is loss of a complete function or an impairment of the performance of a particular function. In the event that part of the function is lost the remaining functions continue to be available, and if a suitable set of functions remain it may be possible for the system to continue to perform some of its function or to provide opportunity for an orderly shut-down to be effected. Design action to enable State D may include application of the *defence in depth* and the *absorption* and the *functional redundancy* principle. At least in some systems, as exemplified by US Airways Flight 1549 (Pariès 2011) the *human in the loop* principle may enable the system to be guided into a satisfactory outcome through human action which may enable sensible response to situations for which it is impractical to determine automated responses.

The use of US Airways 1549 as an example indicates that the state machine model time intervals are concerned with the instantaneous state of the system and that a major event may involve a sequence of events wherein the system progresses through more than one transition and state from the pre-event state before arriving in the settled post-incident state. In a later part of the US Airways 1549 the aircraft ditched, with complete loss of function and the platform being declared unrepairable, but the partial functionality was important for achieving a satisfactory outcome.

Transition 8 is the result of a threat event.

Transition 9 – Restored to Nominal (State D to State A)

In this transition the system is restored from a state of partial functionality (State D) to a fully functional state (State A). As Table 4.1 shows, all the Adaptability group of principles can be used to achieve this goal. The adaptability group of principles includes *drift correction*, *reparability*, and *human in the loop* as described in Table 4.1. The design approach to enabling transition 9 involves identifying a design proposal that implements one or more of these principles and analysing the impact of the design proposals incorporating instantiations of these principles with a view to determining the proposal which best achieves the resilience characteristics required. In most situations the achievement of transition 9 requires a management decision to proceed with remedial action. The remedial action is made possible by the affordances provided through the design which has incorporated particular resilience principles in particular ways.

Transition 9 is an action of repair initiated by a management decision to proceed.

Transition 10 – Additional Events (Within Interim Absorption Capability) (State D to State D)

This transition describes a situation similar to that described by transition 3. In both cases a threat is encountered, acts upon the system and the system remains in a state catagorised by the same state in the state machine model. However, whereas in the case of State A and transition 3 the overall effect is that the resultant State A has no impairment of the system, in the case of transition 10 the State D condition of the system after the transition may be a further diminishment of the system condition compared with its state prior to the event. Design approaches to support transition 10 use the Robustness principles listed in Table 4.1, including *absorption*, *physical redundancy*, and *functional redundancy*.

Transition 10 is the result of a threat event.

Transition 11 – Transition from the Heightened Awareness state to the Non-functional State (State C)

This transition involves a threat event which occurs when the system is in a state of heightened awareness which, in turn, arises when there is awareness of possible threat events. The threat which causes the event leading to this transitions may originate external or internal to the system. A typical military scenario for State B is one where incoming missiles are detected and the system operates in some modified way, the heightened awareness state. But a missile penetrates the protective methods employed by the system, creating a disruptive event. A resilience perspective in this transition implies a controlled or constrained degradation to a lower level of functionality in contrast to a non-resilience perspective in which the system suffers degradation without any design method applied in the attempt to manage the degradation. Design approaches include selection from the Tolerance principles, including modularity, *neutral state*, *loose coupling*, *complexity avoidance*, *hidden interaction avoidance*, *defence in depth*, and *functional redundancy*, depending on the specific system details.

Transition 11 is the result of a threat event.

Transition 12 – Additional Events (State C to State C)

This transition describes an event that impacts on a non-functional system and results in the system being non-functional. The system commences in a non-functional state and ends in a non-functional state which will have detail differences than the original state. The cause of the threat may arise external or internal to the system and takes the form of a threat additional to any. Where a threat acts on a system which is non-functional because of some earlier event, the outcome is a system with additional damage which in turn, impacts the practical and management issues associated with future repair or management decisions associated with the system.

Transition 12 is the result of a threat event.

Transition 13 – Event beyond Interim Capability (State D to State C)

This transition occurs when an event, originating internal or external to the system, occurs that renders a partially functional system non-functional. A system in State D has already suffered some impairment compared with its design capacity and this transition is triggered by some additional threat event which results in total loss of functionality. Desirable behaviour of the system during the degradation of transition 13 is achieved through the design choice to implement resilience principles already identified in relation to transitions 6, 8 and 11.

Transition 13 is the result of a threat event.

Transition 14 – Final Resolution (State D to State F)

This transition represents the path from a partially functional state (State D) to a final agreed to diminished state (State F). This transition is the result of a system management decision that the system in State D, which is viewed as an impaired version of what it should be, is either declared to be a system with reduced functionality matching the physical state that it was in at the time, or to modify the original system in State D so that it becomes a system with different functionality. In most cases effecting transition 14 involves work that converts the system from one form and purpose to something different. Where such work is done, the manager would often perform modifications to suit the environment current at the time of the transition, and the environmental changes may contribute to the decision of perform a transition type 14 rather than to repair to return to original configuration, transition 9.

Transition 14 is an action performed through a management decision.

Transition 15 – Decommission (State D to State G)

If, after review of the situation of a system in State D, it is decided that it is not worthwhile to perform work to either restore original or modified capacity of the system a management decision may be made to decommission the system. Such a system is withdrawn from service and the appropriate disposal actions are performed.

Transition 15 is an action performed through a management decision.

Transition 16 – Agreed repairs or restoration of control (State C to State F)

This transition represents the state of affairs when a system which has already been rendered non-functional is either partially repaired, or modified, so that it becomes a functioning system with different characteristics than those displayed by the original system. This transition is similar to transition 14, described above, with the difference being in the originating state. This transition involves a management decision to modify the system and work which performs the system modification.

Transition 16 is an action of partial repair initiated by a management decision to proceed.

Transition 17 – Decommission (State C to State G)

This transition represents the case in which a system, already non-functional, is decommissioned to State G. The initiating event is the recognition that the system is non-functional and that there is no further use for it, or that any possible work to repair or modify it would not provide an adequate return on investment. A common example of transition 17 is the decision to scrap a motor vehicle after an accident in which damage that would cost more to repair than the market value of the vehicle has been sustained.

Transition 17 is an action performed through a management decision.

Transition 18 – Partial Restoration (State D to State D)

This transition describes a situation in which a partially functional system, State D, is worked upon with a short-term remedy that partially fixes a problem. The remedy is valuable because it improves the system condition, possibly by restoring some, but not all of the lost functionality, and by making the system condition more stable than in the original State D condition. Partial restoration is useful in situations where the system performs a critical task and the improvement achieved through the partial restoration provide a valuable improvement and enables delay of the final restoration. Transition 18 requires a management decision that it is beneficial to perform a partial restoration when a full restoration is not achievable quickly, and is often used to afford time during the lead time to obtain parts or perform work.

Transition 18 is an action of partial repair initiated by a management decision to proceed.

Transition 19 – Disruptive Event that leaves system damaged but functional (State A to State E)

This transition is initiated when a system is operating in its normal operational state, State A, encounters a threat and is damaged but continues to operate in a fully functional capacity. This situation can occur either when a component not essential to system functionality is damaged or one branch of a redundant pathway is disabled. The usual design approach to support transition 19 is to design with physical or functional redundancy. Such a system is impaired by the threat it has encountered making the redundancy principles useful for enabling graceful degradation on loss of part of the system, thus enabling safer outcomes upon failure of parts of the system.

Transition 19 is the result of a threat event.

Transition 20 – Final Restoration to a Normal Operational State (State E to State A)

Transition 20 is initiated when the system is restored to a fully operational state following damage to one or more components. This transition is usually effected through the repair or restoration of the system by a repair action. Transition 20 is enabled by a design decision to make the system repairable.

Transition 20 is an action of repair initiated by a management decision to proceed.

Transition 21 – Event leading to non-functional state (State E to State C)

Transition 21 is caused by an additional threat event impacting a system which is already in State E and resulting in a loss of function. The design method to enable the transition 21 is to use any of the Tolerance principles as listed in Table 4.2.

Transition 21 is the result of a threat event.

Transition 22 – Event leading to partial functionality (State E to State D)

This transition is initiated when another disruptive event occurs to the system that is already damaged but functional in State E. The additional disruptive event leaves the system partially functional, in State D. The design approach to address this scenario is to use any of the Tolerance principles listed in Table 4.2.

Transition 22 is the result of a threat event.

Transition 23 – Event leading to agreed diminished state (State E to State F)

Transition 23 is triggered by a decision to perform work which modifies the system from its original design to a different configuration which provides satisfactory functionality in the modified environment of the system at the time of this transition. In State E the system had been damaged by prior threat events. Transition 23 is similar to transitions 14 and 16 except that the origin state is different.

Transition 23 is an action initiated by a management decision to proceed and may involve some work to repair or modify the system.

Transition 24 – Event leading to decommissioning (State E to State G)

Transition 24 is initiated when a decision is made to decommission a partially functional system. The motivation for decommissioning a partially functional system is that it is judged to be not worthwhile to do work to restore the system to full functionality nor to modify it for agreed different functionality.

Transition 24 is an action performed through a management decision.

Transition 25 – Event leading to agreed diminished state (State A to State F)

This transition occurs when a decision is made to transition a fully functional system in State A to a diminished state, F. This transition is rare, most likely resulting from some of the functionality of the system becoming not necessary or too

Table 4.2 Groupings of principles

Grouping	Top-level principles	Usefulness of groupings
Robustness	*Absorption*	Particularly useful in withstanding the initial disruption caused by a threat
	Physical redundancy	
	Functional redundancy	
Adaptability	*Drift correction*	Useful in assisting the system in restoring a level of functionality which has previously been degraded
	Restructuring	
	Reparability	
	Human in the loop	
Tolerance	*Modularity*	Particularly useful in guiding the system to a lower level of functionality from any given level
	Neutral state	
	Loose coupling	
	Complexity avoidance	
	Hidden interaction avoidance	
	Defence in depth	
	Functional redundancy	
Integrity	*Internode interaction*	Applies to all systems and transitions since this grouping and the associated principle assures that the system remains a system throughout the encounter with the threat

expensive to support, whilst other functionality of the system remains worthwhile maintaining. This transition results from a management decision, like the other transitions into state F, transitions 14, 16 and 23.

Transition 25 is an action initiated by a management decision to proceed and may involvesome work to repair or modify the system.

Transition 26 – Event leading to decommissioning (State A to State G)

Like Transition 25, Transition 26 is very rare. Transition 26 happens when a decision is made to decommission a system that is in normal operating condition. This decision would be made when the anticipated cost of maintaining the system is greater than the value the system provides or where the context has changed making the system no longer valuable. Note that transition 26 does not describe a situation of significant breakage of the system because a threat induced decommissioning sequence has at least two transitions, the first caused by the significant threat event which results in the system transitioning to a degraded state and the second, or later, transition being the decision to decommission the system.

Transition 26 is an action performed through a management decision.

Transition 27 – Event leading to agreed diminished state (State B to State F)

Like Transition 25 the system, in this case in State B, encounters a disruptive event leaving it non-functional in State F. Although the system in State B, the origin state of Transition 27, was in a state of heightened awareness of a threat, either the magnitude of the threat was too great for the system to respond in a protective manner or the threat which caused the failure was of a kind different than that for which the heightened awareness actions were watching. The design approach to assist control in Transition 27 is likely one of the Tolerance principles in Table 4.2.

Transition 27 is an action initiated by a management decision to proceed and may involve some work to repair or modify the system.

Transition 28 – Event leading to decommissioning (State B to State G)

This transition differs from Transition 27 only in that the damage caused by the disruptive event is not reparable. The system can be controlled to State G by any of the principles in the Tolerance grouping.

Transition 28 is an action performed through a management decision with an initial event caused by a threat event.

Transition 29 – Transition from the Heightened Awareness state to the Partially Functional Disrupted state (State B to State C)

This transition involves a threat event which occurs when the system is in a state of heightened awareness which, in turn, arises when there is awareness of possible threat events. The threat which causes the event leading to this transition may originate external or internal to the system. Design approaches include selection from the Tolerance principles, including modularity, *neutral state*, *loose coupling*, *complexity avoidance*, *hidden interaction avoidance*, *defence in depth*, and *functional redundancy*, depending on the specific system details.

Transition 29 is the result of a threat event.

Transition 30 – Transition from the Non-Functional State to the Heightened Awareness state (State C to State B)

This transition involves an action of repair to a system in a non-functional state so that the system is returned to a normal operational capacity while there is a reason for the system to operate with heightened awareness of potential threats.

Transition 30 is the result of a repair activity.

Transition 31 – Transition from the Partially Functioning Disrupted State to the Heightened Awareness state (State D to State B)

This transition involves an action of repair to a system in a partially functional state so that the system is returned to a normal operational capacity while there is a reason for the system to operate with heightened awareness of potential threats.

Transition 31 is the result of a repair activity.

4.5 An Approach to Design for Resilience

The engineer's task is to design a system which performs its intended functions in the manner intended and is inherently as resilient as possible and practical in the circumstances, and demonstrating the resilience characteristics that provide the desired system characteristics under the stresses imposed by threats of kinds or magnitudes beyond the specified design loadings of the system. The design process, after the system requirements have been identified, follows the steps:

1. Propose one or more candidate system architectures capable of satisfying the set of requirements and applying established good design practice as determined in the specific domain.
2. For each candidate architecture review the resilience implications. The particular concern is to review the impact on system capacity to perform its function if subsystems identified in the architecture fail to perform their function. At this stage the cause of failure is irrelevant because the analysis concerns the impact on the system if the particular subsystem does not operate as intended and therefore does not contribute the services into the system that it has been configured to perform. This analysis will enable determination of the failures which lead to each of the kinds of degraded state. The knowledge of the system State resulting from failure of particular subsystems will assist in determining the impact of loss of each subsystem.
3. It is now necessary to identify what possible threats could lead to loss of each subsystem. This investigation results in knowledge of the probability of each threat type and magnitude beyond the subsystem failure threshold, assuming normal design precautions are incorporated, enabling construction of two types of conjoint representations of the system. One of the conjoint measures links magnitude of threat event of each type and the loss of system capability, indicating the impact of various levels of each type of threat. The other representation concerns the threat magnitude of each type and the probability of loss of function of each subsystem.
4. The designer then identifies any deficiencies in the resilience of the current system design proposal compared with the desired characteristics. The resilience design principles listed in Table 4.1 are reviewed to suggest strategies to improve the resilience of the system.
5. The designer then proposes methods of incorporating one or more resilience principles to address the particular concerns identified in the candidate architecture. It is likely that multiple methods could be effective ways to improve system resilience. Each method proposed must be analysed to enable selection of a method that provides the best combination of resilience characteristics, function performance, cost and delivery schedule.
6. Each candidate architecture must be developed and analysed in order to determine the most appropriate system architecture and resilience strategy in order to best satisfy the overall system objectives.

By following this approach, particularly focusing on the impact of loss of particular subsystems on the performance of the system, the designer is guided towards ensuring the resilience and adequate behaviour under the influence of a diversity of threats, rather than focusing on particular threats. The question of particular threats is a second order question associated with determining the value to be gained through hardening particular subsystems against the effect of a particular threat type beyond the inherent resistance of the normal production standard.

4.6 Resilience Principles to be Used in Design

We list a set of resilience design principles in Table 4.1. These principles have been collected from resilience literature. Various contributors to the engineering concept of resilience have identified a number of resilience related principles, often with different names and slightly variant descriptions, and each has listed only some of the whole set listed in Table 4.1. The subsets of principles found in each of the earlier authors have significant overlaps with other contributors but none listed the whole set.

Each of these principles has been demonstrated as valuable in (Jackson 2016a). The method used to demonstrate the value of the principles was to find support for each principle in the recommendations in reports of relevant authorities into a set of system failures chosen from various domains of engineered systems. The reports were official investigations into particular disasters. In each case the circumstances of the disaster interacted with specific aspects of the system involved with the result that the disaster took the form of a particular threat manifestation which interacted with a particular weakness, or combination of weaknesses, with the outcome that the system failed. Past experience of design in many domains has led to design guidance, or required practices, which implement many of the resilience principles. In such cases, the design rules have been enacted and enforced because of historic experience. The result is system designs which are unlikely to fail for a reason that would be remedied by that principle. Therefore, in such cases, principles already embedded in the normal design practice of a domain are unlikely to be implied by the recommendations of a report into a disaster because the causal sequence is more likely to involve a different issue. However, the set of resilience design principles were all each supported by the recommendations of at least one disaster report from a set of ten disasters representing four domains.

Table 4.1 divides the principles into Top-level Principles and Support Principles. Top-level Principles express a primary concept of an approach to some aspect of resilience. The Support Principles express methods of implementation of the Top-level Principles. As such the primary choice in the design process is to choose to apply a Top-level Principle, which will address one of the primary perspectives on the nature of resilience. The secondary decision concerns methods of implementing the Top-level Principle, and these are indicated through the Support Principles.

4.7 Groupings

Jackson (2016a) concluded that the principles listed in Table 4.1 can be organised into four groups each identifying an attribute of resilience. These attributes derive from the multiple aspects of the concept of resilience, concerning the resistance of the system to degradation caused by threats, the enablement of recovery after a threat event has caused diminishment of the system, managing the diminishment of system in the immediate aftermath of a threat event and ensuring the integrity or coherence of the system at all times. The four groups and the associated top-level principles are shown in Table 4.2. It will be noticed that the *functional redundancy* principle applies to two groupings.

4.8 Mapping of Principles

In a previous work Jackson (2016a) mapped the principles to the transitions using the rules shown in Table 4.3. These rules are based on whether the transition will result in the system having partial functionality, functionality lower than the pre-transition state, return to the pre-transition state, restoration of part or all the functionality, involves humans in the loop, or retains the integrity of the system. The rules, which identify principles which are likely to be useful in designing solutions that assist implementation of particular transitions in a manner that satisfied the resilience objectives relevant to the system. The principle or principles that are appropriate, and therefore required the qualification "likely" above, depend on the specific system context which may permit or prohibit certain principles. In addition,

Table 4.3 Rules for mapping transitions and candidate resilience principles

	Rule	Applicable transitions
1.	Any transition resulting in a partially functional state ⇨ *modularity, physical redundancy, functional redundancy*	7, 8
2.	Any transition to a lower (less functional or non-functional) state ⇨ all Tolerance principles.	6, 8, 11, 13, 14, 15, 16, 17, 19, 21, 22, 23, 24, 25, 26, 27, 28, 29
3.	Any transition to the same state ⇨ all Robustness principles	1, 3, 10, 12, 18
4.	Any transition resulting in an increase or restoration in functionality ⇨ all Adaptability principles	2, 5, 9, 20
5.	Any transition resulting in heightened awareness ⇨ *drift correction, human in the loop*	4
6.	The *human in the loop* principle can execute the following other principles: *drift correction, neutral state, loose coupling, functional redundancy*	All relevant transitions for the named principles
7.	The *internode interaction* and *defense in depth* principles apply to all transitions ⇨ All Integrity principles	All transitions

design architecture choices may point to the appropriateness or inappropriateness of particular principles, particularly because of the cost of implementation of solutions based on one principle or the alignment of a principle already used for another purpose which may have incidental benefit in achieving another aspect of resilience.

4.9 Conclusion

The state machine formulation of the problem of resilience is a useful tool to assist reasoning about the acceptable resilience characteristics of a particular system. Systems will be subjected to events arising from external sources which are outside the conditions which they were specified and designed to withstand. In addition, any system is subject to threats arising internally, either as a result of latent faults that escaped detection in the testing process, or as a result of some action or lack of action of a person inside the system which results in things going awry, some of which may be accidental and some malicious. The result of this is that systems cannot be built to withstand all possible bad events without any risk of degradation.

The state machine model assists the systems engineer to separate the cause of challenges to the system, the specific form of threats, from the effect on the system by emphasising the effects on the system, through the various states in which the system could be situated. The states are important because they represent the various states of completeness of the system function and therefore analysis of the system performance when in each state enables discovery of the impact on the whole system functionality when particular subsystems have failed.

There are two philosophies for achieving resilience. One focuses on the causes of system degradation. Where the cause philosophy dominates the attempt to create resilience will be focused on toughening the system and its components against the types of threats identified and at levels which represent a reasonable balance of cost and benefit, since it may be impossible to identify the highest possible level of a threat or it may be impractical to build the system to withstand that extreme level.

The second philosophy focuses on the effect of failures of components or subsystems in the system. In this case the first analysis of a system architecture is to posit the "what if" questions associated with parts of the system failing or functioning in a manner different than intended, and analysing the system to determine the impact of those failures on the whole system performance. This enables clear understanding of the effect of the loss of things that could be lost through the effect of any threat. Further analysis of the possible causes of failure of the particular components can be performed to determine the impact on whole system performance of any of the possible threats. Understanding of the impact on whole system performance opens the designers thinking to a wider range of strategies to achieve improved resilience of the system and frames the matter on achievement of the desired system effectiveness outcomes rather than the survivability of particular components of the system.

A systems engineer working with the second, effects, philosophy will be free to look at approaches to resilience that include providing the service by alternative methods and by hardening particular components against specific threat causes, possibly to different magnitudes of threat for different components because the impact of loss of some is greater than for others. The effects philosophy also focuses on considering what are appropriate levels of performance of system functions when the system is affected by threats and the decision may depend on the scenario leading to the impairment of the system.

The authors have presented a comprehensive multi-faceted approach to designing a resilient system. This approach can be applied to many system types, many threat events, and many scenarios because it is focused on understanding the effects of threat events rather than focusing on the causes and mechanisms of threats to the system. This approach is presented to help the reader develop more refined approaches to the challenge of designing resilient systems.

Further Suggested Readings

Billings C (1997) Aviation automation: the search for human-centered approach. Lawrence Erlbaum Associates, Mahwah

DoD (2001) USS Cole Commission report. Department of Defense, Washington, DC

Haddon-Cave C (2009) An independent review into the broader issues surrounding the loss of the RAF nimrod MR2 aircraft XV230 in Afganistan in 2006. The House of Commons, London

Hollnagel E, Woods DD, Leveson N (eds) (2006) Resilience engineering: concepts and precepts. Ashgate Publishing Limited, Aldershot

Jackson S (2010) Architecting resilient systems: accident avoidance and survival and recovery from disruptions. Wiley series in systems engineering and management. John Wiley & Sons, Hoboken

Jackson S (2016a) Evaluation of resilience principles for engineered systems. Thesis, University of South Australia, Adelaide, Australia

Jackson S (2016b) Principles for resilient design – a guide for understanding and implementation. In: Linkov I (ed) IRGC Rresource Guide on Resilience. International Risk Governance Council (IRGC), University of Lausanne, Lausanne

Jackson S, Ferris T (2013) Resilience principles for engineered systems. Syst Eng 16(2):152–164

Jackson S, Cook SC, Ferris T (2015) Towards a method to describe resilience to assist in system specificaion. Paper presented at the IS 2015, Seattle, 15 July

Leveson N (1995) Safeware: system safety and computers. Addison Wesley, Reading

Madni A (2008) Suggested heuristics and validation. Various suggested resilience heuristics and their validation. This item covers several conversations. edn., Los Angeles, CA

Madni A, Jackson S (2009) Towards a conceptual framework for resilience engineering. IEEE Syst J 3(2):181–191

Marczyk J (2012) Complexity reduction. Email communication edn, Como, Italy

Maxwell J, Emerson R (2009) Observations of the resilience architecture of the firefighting and emergency response infrastructure. Insight International Council on Systems Engineering

NTSB (2010) Collision of Metrolink train 111 with union Pacific train LOF65–12 Chatsworth, California, 12 Sep 2008. National Transportation Safety Board, Washington, DC

Pariès J (2011) Lessons from the Hudson. In: Hollnagel E, Pariès J, Woods DD, Wreathhall J (eds) Resilience engineering in practice: a guidebook. Ashgate Studies in Resilience Engineering. Ashgate Publishing Limited, Farnham

Perrow C (1999) Normal accidents: living with high risk technologies. Princeton University Press, Princeton
Perrow C (2011) Modular power grids. New Haven
Raveh A (2008) Regroup heuristic. Comment during tutorial on resilience edn., Utrecht, The Netherlands
Reason J (1990) Human error. Cambridge University Press, Cambridge
Reason J (1997) Managing the risks of Organisational accidents. Ashgate Publishing Limited, Aldershot
Richards MG (2009) Mult-attribute tradespace exploration for survivability. Dissertation, Massachusetts Institute of Technology, Cambridge, MA

Part III
Infrastructure

Chapter 5
Infrastructure Resilience Assessment, Management and Governance – State and Perspectives

Hans R. Heinimann and Kirk Hatfield

Abstract Rapid urbanisation worldwide has created a host of fundamental challenges, which in conjunction with natural or man-made disasters, now threaten the resilience of communities, cities, and mega-cities. Consider the first challenge of climate change which has undermined the fundamental assumptions used to design the engineered systems that currently define the critical functionalities of cities today, and for tomorrow will demand innovative paradigms and new assumptions for designing resilient cities of the future. Another challenge concerns the damages directly or indirectly associated with natural and man-made disasters. These damages are expected to escalate as long as the value at risk continues to increase. And then there is the challenge of surging connectivity within and between critical infrastructure systems which has left such systems interdependent and vulnerable to cascading failures and regime shifts which foment ill-defined changes in system functionality. Changes that include but are not limited to emergent disruptions of critical services, system damages and even system-wide failures. The purpose of this chapter is to propose a framework for understanding and assessing critical infrastructure system resilience, to introduce a vision of resilient governance, and to propose a framework for harnessing knowledge transfer and continuous learning as required of policymakers seeking to elucidate and promote best practices that shape desired behaviour from individuals, social systems, stakeholders and communities.

Significant findings are the following. First, a set of 10 questions (deca-tuple set) is formulated to frame resilience assessment and management concepts. The approach taken is analogous to that of the triplet-question set of risk management. The deca-tuple question set serves to guide the work of resilience evaluation and analysis and even resilience building. The process of building resilience is, in fact,

---

H.R. Heinimann (✉)
ETH Zurich, Future Resilient Systems at the Singapore-ETH Centre (SEC), Singapore, Singapore
e-mail: Hans.heinimann@frs.ethz.ch

K. Hatfield
Engineering School of Sustainable Infrastructure and Environment, Department of Civil and Coastal Engineering, University of Florida, Gainesville, FL, USA

© Springer Science+Business Media B.V. 2017

I. Linkov, J.M. Palma-Oliveira (eds.), *Resilience and Risk*, NATO Science for Peace and Security Series C: Environmental Security,
DOI 10.1007/978-94-024-1123-2_5

a collective action of public and private stakeholders responding to infrastructure disruptions. We propose a resilience assessment framework consisting of 5 phases: pre-assessment, appraisal, characterization and evaluation, management and communication. This framework follows that of the risk governance framework of the International Risk Governance Council (IRGC). Building resilience requires continuous learning and adaptation from individuals, teams, organisations, governance and government systems. We propose three levels: knowledge transformation for policymaking, building best practices, and adapting individual behaviour. Authors identified several knowledge gaps for future scientific investigation in the domains of: context and framing, disruption identification, biophysical resilience analysis, cognitive resilience analysis, resilience evaluation, and building system resilience.

Keywords Resilience framework • Biophysical resilience • Cognitive resilience • Resilience functions • Resilience governance

5.1 Introduction

Urbanisation is an aggregation process that fosters the emergence of urban clusters where much of the world population resides and the magnitude of human activities prevail. This process continuously changes the very spaces that humans occupy and their contiguous environments. By 2030 the world will have 41 megacities with more than 10 million inhabitants (UN 2014). In 2014, 1 out of 8 inhabitants worldwide were living in one of 28 megacities. Several trends have come to characterise the urbanisation process. First, the density of infrastructure per unit of area continues to increase and with it the value at risk which results in higher expected damages, even if the hazard profile remains constant. Second, the urban metabolism – the flow of goods, services, information, and people – continues to expand dramatically, and is borne by a stratified systems of coupled infrastructure systems that enable different kinds of flows. Third, cities of today represent the nascent skeleton of the cyborg ("cybernetic organism") cities of the future (Swyngedouw 2006), cities that constitute ever sophisticated and interdependent mosaic of advancing infrastructure systems, enabling technologies, green spaces and social systems. The flow of materials, energy, information, people, and resources is converging wherever critical services are provided and essential functions are executed. It is within this context that natural ecosystems, social organisations, and constructed infrastructure networks and assets merge into complex interdependent socio-ecological-technological systems, which epitomise zones of urban activity. This rapid urban development has brought to forefront several fundamental challenges. For example and as previously indicated, expected damages due to disruptions will increase because the value at risk has grown. Second, climate change is augmenting changes in the spatial and temporal variability and uncertainty of environmental conditions, which will in turn undermine the very assumptions used to design the systems that currently define our urban environment. Third, the coupling strength between and within connected infrastructure systems continues to grow and under the right conditions will foment cascading system failures or even regime shifts that precipitate emergent system disruptions and damages never before experienced.

Risk management has been a successful approach in dealing with the future and its associated uncertainties as long as hazards, system responses, actions, and mechanical properties of engineered structures can be characterised using stationary probability distributions over the parameters of interest. The trends mentioned above will yield a growing fraction of ambiguous or weak signals that emerge into disruptions with unexpected outcomes. It is difficult at best to design and implement preventive measures if critical event drivers and system consequences are vague or impossible to predict. Current risk-based engineering approaches treat hazard based loss mitigation as a problem that can be solved with classic infrastructure hardening or building an increasingly hazard resistant infrastructure. This approach, has proven to be costly and ineffective against extreme or unexpected events (Francis and Bekera 2014, Olsen 2015, Park et al. 2013). This dilemma calls for a new paradigm for system design and management that consider the inability to fully define and estimate the uncertainties affecting the system (Olsen 2015). Traditional engineering design pursues essentially a preventive strategy of balancing the actions on a structure with structural resistance or one of enabling a structure to resist a whole set of actions based on assumptions. Biological and Ecological Systems possess this type of "resistance" but have additional, postevent strategies. The wound healing process of organisms is an example. Immediately after wounding, the inflammation mechanism aims to maintain critical body functions (stop bleeding, the breakdown of necrotic tissue), followed by rebuilding the wound and rearranging affected body structures and functions to normalcy. Resilience is a concept that stems from system's resistance, and mimics postevent recovery functions of natural systems. Although the concept has been around for a while, a comprehensive generic concept is missing that is applicable across multiple disciplinary domains.

In this chapter, we first propose a framework to understand and assess the resilience behaviour of critical infrastructure systems, which is based on a brief review of the state-of-the-art in the relevant fields. Recognising that building resilience is a collective action problem, we then sketch a vision of a resilience governance system – predicated on the IRGC's risk governance framework. Assuming resilience building requires changes in both technical and social spheres, we propose a framework for using knowledge transfer and continuous learning in policymaking, defining and promoting best practices, and shaping the behaviour of individuals, social systems and communities.

5.2 Understanding the Behaviour of Infrastructure Systems Under Multi-hazard Scenarios

Our infrastructure systems and the environment in which they operate, are getting more and more complex and interdependent. Our ability to understand them with common sense or with the aid of multifaceted models and tools has not kept pace with this growing complexity and interdependence. Engineered infrastructure systems are based on specific assumptions about the properties and behaviours of these systems and their surrounding environments. Unexpected changes in the environment or inadequate understanding of interactions within and between systems may result in unexpected system behaviours that range from normal to disrupted operation to failure. The vulnerability of modern infrastructure systems to emergent and cascading failures calls for novel approaches that foster the design, analysis, monitoring, and management of engineered systems, which are more resilient.

5.2.1 *Need for a New Infrastructure Assessment Approach*

The crucial issue –as presented in Fig. 5.1 on the x-axis – is essentially the quality of available knowledge on infrastructure system behaviour under different endogenous and exogenous conditions. Many engineering approaches, for example, those identifying optimal solutions with mathematical methods, rely on *certainty* assumptions; that is, they assume complete knowledge of the system and the environment and the classic model of human behaviour (Homo Economicus). The introduction of *uncertainty* enables the relaxation of these assumptions through the use of expectation values, which are the products of probabilities and various metrics of consequences. Probabilistic risk analysis, which is the backbone of risk management since the 1950s is a stream of thinking, which still represents the state-of-the-art (ISO 2009). Further relaxation of what is known about a target system is engendered in the concept of *ambiguity* (Renn and Klinke 2004) which assumes

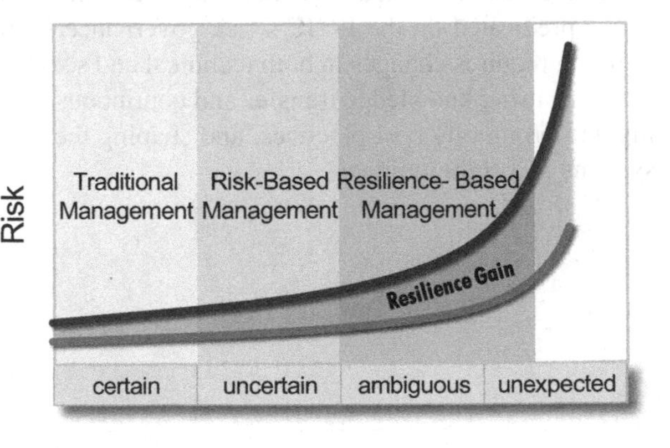

Fig. 5.1 The influence of knowledge domains and a new design and management paradigm on risk (Adapted from Murray et al. 2013)

there are observations on some phenomena which proffer several legitimate interpretations of meaning. The Zika virus is an excellent example of ambiguity; its existence has been known for some time, but the sheer diversity of interpretations regarding threats, actions and consequences confounded efforts to forecast the ensuing epidemic. In air traffic control, ambiguity is termed as "weak signals", and most often, there is no finite interpretation of meaning. There is yet another level of system knowledge, which is characterised as "unexpected" or "unknown". The black swan concept (Taleb 2007) is a metaphor for this category, which assumes there are events that just happen for which there are no means to anticipate or predict them. The "dragon king" concept (Sornette 2009) assumes – and successfully demonstrates – that there are options to anticipate or predict "unknown events" if we have real-time information regarding system behaviour with an adequate time granularity (Sornette and Ouillon 2012).

The y-axis of Fig. 5.1 presents the expected adverse consequences for the different categories of knowledge and different management strategies for a system. Up to the 1950s, approaches to design and manage infrastructure systems were based on certainty assumptions. With the emergence of probabilistic risk analysis in the 1950s (Keller and Modarres 2005), risk management approaches – aiming to eliminate or prevent unacceptable risks – improved the overall safety and security of our systems tremendously. Today's constructed infrastructure systems are more complex and highly connected; these changes have pushed a large fraction of urban infrastructure systems into the "ambiguous" and "unexpected" knowledge domains, and in turn increased the risk of adverse and ill-defined consequences (upper line, Fig. 5.1). Ambiguity or lacking sufficient knowledge about a system means that communities, engineers, and governance have to cope with invalid assumptions (Day et al. 2015), either due to unexpected changes in the environment or due to emerging and unexpected behaviours in critical infrastructure systems (Haimes et al. 2008). Communities must acquire the capacity to ensure infrastructure systems continue to provide critical services and support essential function whenever and wherever invalid assumptions reside (Day et al. 2015). The capacity of a system to cope with invalid assumptions is the hallmark of a resilient system, which unfortunately does not appear in most resilience definitions. Alternatively, if the assumptions about the system and its environment are constant, knowledge is '*certain*', and system resilience is not in question. Enabling a system to behave resiliently lowers the adverse consequences (Fig. 5.1, lower curve) and the difference between the "risk management" and the "resilience management" curves results in a resilience gain, creating economic benefits that must be evaluated against the additional costs of resilience measures.

5.2.2 What Are the Main Gaps in Infrastructure System Assessment?

The knowledge domains of Fig. 5.1 correspond to three types of systems: deterministic, probabilistic, and non-deterministic (Dove et al. 2012). The systems engineering community characterise non-deterministic systems as those requiring novel (resilience) approaches including complex-adaptive, autonomous, chaotic, agile, etc. This type of behaviour is often the result of two drivers, an increase in coupling strength and a decrease in heterogeneity both within and between systems (Osorio et al. 2010). If the coupling strength of a system increases and its heterogeneity concurrently decreases, the system may enter a critical state where a transition to a new regime ("regime shift") occurs and system behaviour changes.

Since 1978, it was known that probabilistic risk assessment, a well-established approach for probabilistic systems [systems of the 2nd kind], suffered major shortcomings as a tool for assessing of non-deterministic systems [systems of the 3rd kind] (Lewis et al. 1978). It was argued then that it was conceptually impossible to construct event-trees or fault-trees to completeness and that probabilistic risk assessment models do not account for unexpected failure modes. More recently, Leveson emphasised that this type of "chain-of-event" analysis is not be capable of accounting for indirect, nonlinear, and feedback relationships that govern the behaviour of complex systems (Leveson 2004). She claimed that traditional approaches performed poorly in modelling human action behaviour by often reducing human factors to deterministic operator models that were based on human reliability assumptions. Aware of these challenges, the systems engineering community-initiated research on "systems of systems engineering" (Deiotte 2016) and on "non-deterministic systems" (Dove et al. 2012). They found serious systems engineering gaps arising from the application of from previous concepts, approaches, and tools that were not appropriate to tackle these new challenges (Dove et al. 2012). A gap closing endeavour should start with analysing differences between previous systems and "systems of the 3rd kind" and the requirements to be fulfilled by novel approaches and tools.

5.2.3 Infrastructure Resilience Assessment

A useful approach to coping with infrastructure assessment challenges is to look outwards from where we are; where the phase, "where we are" is taken to mean the standard approach to assessing systems for adverse consequences, which is risk management. The ISO standard 31,000 (ISO 2009) defined a risk management and assessment framework that is widely accepted. Accordingly, the risk management process consists of five activities: (1) establishing the context, (2) risk identification, (3) risk analysis, (4) risk evaluation, and (5) risk treatment. Below, we characterise

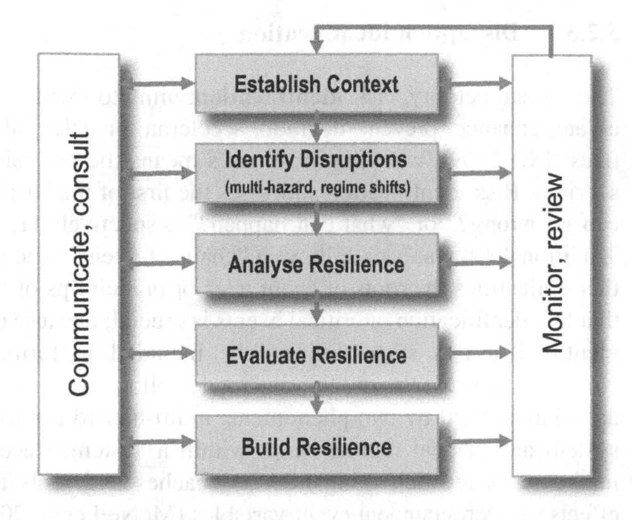

Fig. 5.2 Resilience management framework, adapted from the ISO risk management framework {ISO 2009 #26}

each of those activities and explore directions for advancing each but from the perspective of resilience management and assessment (Fig. 5.2).

The scope of the analysis is comprehensive; it looks at an infrastructure system as an integration of engineered components, the operating organisation, and user subsystems.

5.2.3.1 Establishing the Context

The first activity, **establishing the context**, aims to evaluate and understand the environment and internal governance arrangements, the latter of which defines human interactions and operational procedures within any organisation. Non-deterministic systems in need of resilience management must be viewed beyond local system boundaries. Ecologists have long recognised that phenomena associated with complex systems often extend across different spatiotemporal scales (Holling 2001). They were among the first to demonstrate the value of interrogating complex systems at temporal and spatial scales both above and below the target system of interest. In an ecological context, the analysis of a single plant would require an investigation of the pertinent plant's ecological community at the upper level and plant's organs at a lower. For a social-technical system, such as a manufacturing plant, the analysis would include the supply chain at the upper scale and production cells at a lower. This is multi-scale perspective supports Leveson's view that human actions cannot be reduced to a simple operator that follows a deterministic human reliability model. For example, explaining and understanding the multitude of human factor issues that existed before and then evolved during and after the Fukushima nuclear power plant accident is only possible, if the control room, the plant management, and the plant owner are viewed concurrently as distinct interdependent levels of a larger complex system.

5.2.3.2 Disruption Identification

The second activity, **risk identification**, aims to identify those events that "might create, enhance, prevent, degrade, accelerate or delay" the achievement of objectives (ISO 2009), which in most cases means the provision of a single or a set of services. Risk identification addresses the first of the "triplet idea" questions "what can go wrong?" or "what can happen?" respectively (Kaplan and Garrick 1981). Traditional risk analysis follows a "chain-of-event" type of analysis, which means that it identifies the roots of event trees or branch tips of failure trees. It is obvious that the identification of critical events is crucial, because critical events that are not identified at this stage will not be included in further analysis (ISO 2009). Nondeterministic systems failures rarely follow "chain-of-event" type patterns but are characterised by two phenomena: multi-hazard events in the environment of a system and critical regime shifts within a system. Recent advances in financial mathematics are delivering new approaches and tools to model the cumulative effects of interdependent event variables (McNeil et al. 2015). Compound distributions can model the cumulative effect of two or more variables, for example, a first variable describing the frequency distribution of an event in time and a second variable describing the frequency distribution of the event magnitude. A so-called compound Poisson distribution (McNeil et al. 2015) is one example of such an approach that could be used for infrastructure system events such as the modelling rock fall. Copula models provide a more comprehensive approach to describing the dependence within a set of random variables that is based on the assumption that any multivariate joint distribution can be represented regarding univariate marginal distribution functions and a copula, which describes the dependence structure between the variables (McNeil et al. 2015). Copulas possess considerable potential in modelling multi-hazard events, but calibration requires good data. Regime shifts are another challenge, for which traditional risk management approaches do not provide methods and tools. A regime shift is a large, abrupt, persistent change in the structure and functions of a system, which produces an overall change in system behaviour. Hence, an analysis of system regime shifts focuses on gathering and interpreting evidence of internal system disruptions and change. The existence of system regime shifts presents two challenges in achieving resilience: first, there is a need to understand system-specific critical states, their existence, and the conditions that induce their emergence. Second, there is a need for tools to detect or anticipate critical regime shifts. Dragon King theory (Sornette and Ouillon 2012) provides the basis and the tools to detect a considerable fraction of Dragon King events when sufficient time series data exists to characterise the state of the system of interest.

5.2.3.3 Resilience Analysis

The third activity, **risk analysis**, intends to develop a thorough understanding of context-specific risks. Following the "triplet idea" set of questions (Kaplan and Garrick 1981), it tries to answer the second and the third question, "how likely will an event happen?", Moreover, "if it does happen, what are the consequences?". Ideally, we have a high number of scenarios, representing the whole range between "high probability – low consequence" and "low probability – high consequence," which would allow us to derive an empirical distribution function over the range of possible consequences. Risk management professionals have been presenting this type of empirical distribution functions in double logarithmic plots with a consequence metric on the x- and the exceedance frequency on the y-axis. These plots are known as frequency-consequence diagrams, in short FC-diagrams, that emerged at the end of the 1960s (Farmer 1967) and are now standard to present the probabilistic results of risk analysis. FC-diagrams are easy to understand but limited by the number of scenarios considered for analysis and by the "chain-of-events" approach that cannot account for indirect, nonlinear, and feedback relationships that govern the behaviour of complex systems (Leveson 2004).

5.2.3.4 Deca-Tuple Resilience Question Set

The question for this audience is in what direction risk analysis should evolve to address current needs for the analysis of resilience in urban socio-ecological-technological systems and in particular constructed infrastructure. The "triplet idea" concept (Kaplan and Garrick 1981), defining three key questions, has shaped our conceptual understanding of risk analysis. In analogy, a first step is to further develop the "triplet idea" in such a way that it captures the essential functionality or behaviour of resilient systems. Although many resilience definitions have been around, a consistent description based on generic system functions is still lacking. In this paper, resilient system behaviour is framed within the context of three classes of generic functions: biophysical core functions, enabling functions, and cognitive functions (Fig. 5.3).

The *biophysical core functions* characterise the "bathtub" behaviour of systems, triggered by some shock, absorbing the shock, recovering and adapting, which can be restated in a "quadruplet idea" question set (Fig. 5.3). (1) What is a systems' ability to resist within acceptable degradation (Haimes 2009)?; (2) How can we best re-establish a systems' key functionalities or re-stabilize its behaviour?; (3) How can we best re-build a systems' performance up to normalcy?; and (4) How can we best change the biophysical architecture/topology of the system to make it more fault-tolerant? *Enabling functions* support and amplify biophysical core functions and are expressed in a double question set: (1) Does preparation increase the capability to cope with unexpected disruptions?; and (2) Does an emergency response of a system significantly reduce degradation?

Cognitive functions comprise the capabilities of individuals and organisations to perceive the state of a system and the environment in which it is operating, understanding its significance and meaning, to retrieve or develop courses of actions, and to select and release the most meaningful action. Cognitive sciences and psychology have made significant progress in understanding how both cognitive and affective appraisal processes operate (Moors et al. 2013). Cognitive functions are

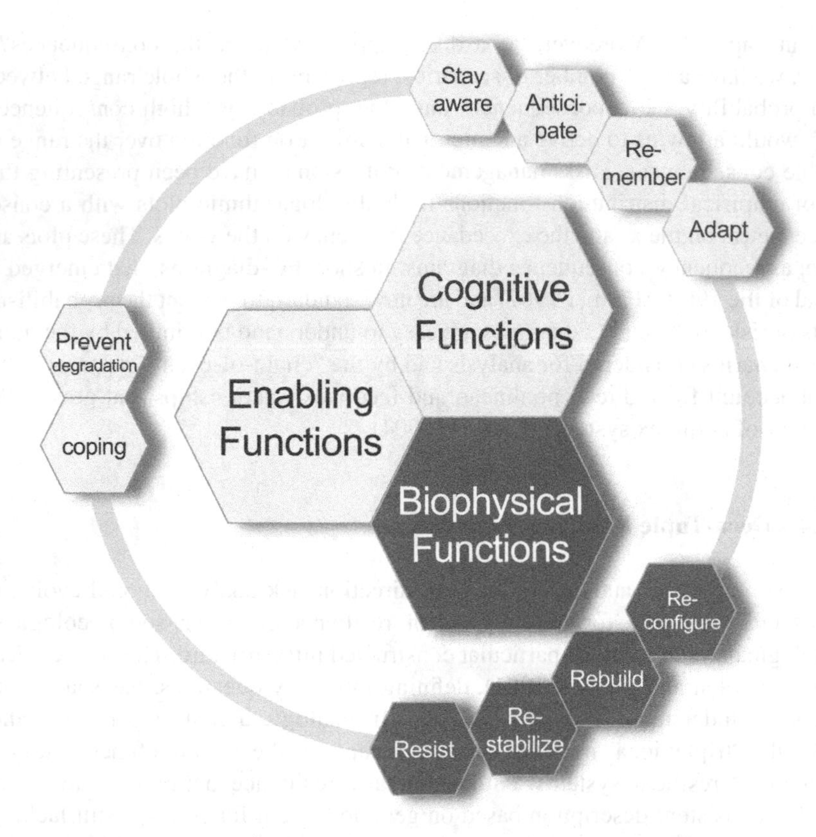

Fig. 5.3 Resilient System Behaviour, framed with three classes of functions: (1) biophysical, (2) cognitive, and (3) enabling. The three classes split into ten resilience functions

embedded within higher order organisms, storing sets of meaningful actions in the repository and having mechanisms to perceive threats from past observations. If the cognitive part of resilient system behaviour were articulated as a quadruplet set of questions, it would look as follows (Fig. 5.3). (1) "How can we keep the perceptive awareness of a system alive over a long term?"; (2) "How can we best detect or anticipate unexpected, critical events, or regime shifts?"; (3) How can we best remember – store and retrieve – actions that successfully dampened past critical events or mitigated regime shifts?"; and (4) "How can we continuously adapt the behaviour of individuals and organisations to increase their capability to cope with unexpected events?". Following this logic, answers to 10 questions are needed to acquire a comprehensive resilience assessment, which is – compared to the triple risk management question set – a challenge. A pragmatic approach is to start with the quadruple biophysical core function set, that characterises the "bathtub" behaviour at the aggregated level of systems or the system of a systems perspective, as illustrated in Fig. 5.4 for power grid behaviour in areas affected by super storm Sandy (Henry and Ramirez-Marquez 2016).

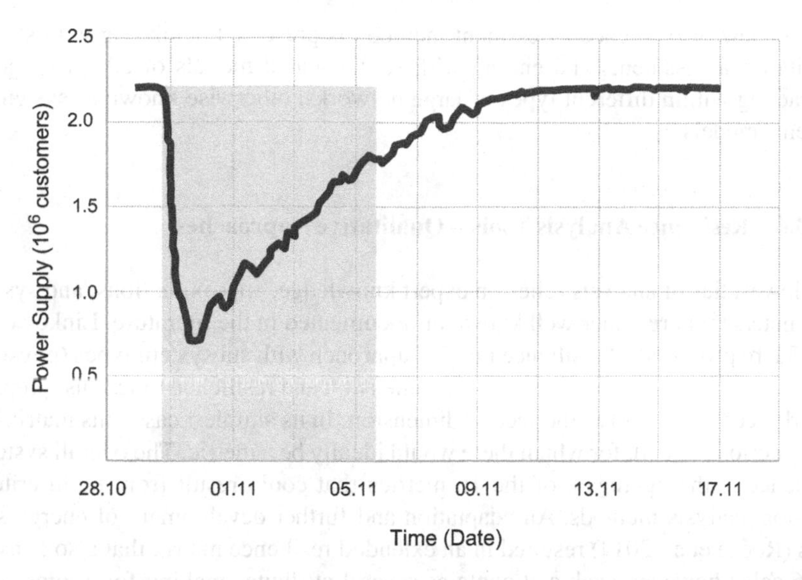

Fig. 5.4 Power supply performance after the disruption of Hurricane Sandy (Adapted from Henry and Ramirez-Marquez 2016)

The immediate power system response to the shock of Hurricane Sandy was the collapse of performance to about 30% of the pre-event level. This phase equals the "resist" function, which has to make sure that critical systems stay within an acceptable range of functionality. After reaching the bottom, the "re-stabilize" function ensures critical system functionality survives, which in the Sandy case took a couple of hours. The third function, "re-build" allows the system to recover to the pre-event performance level, which took about eight days. Figure 5.4 does not enable one to make any statements regarding the "reconfigure" function or feasible changes that may have occurred with the architecture of system components and system topology. This nicely documented real-world case illustrates that what we learn from beyond the typical approach to "resilience analysis" which answers only a subset of the Deca-tuple resilience questions, and it demonstrates the value ofr a more comprehensive approach to resilience analysis.

5.2.3.5 Resilience Analysis Tools

Previous sections defined ten dimensions of analysis that provide quantitative data on infrastructure systems resilience. Standardised analysis and assessment approaches often follow a tiered approach consisting of several levels of analysis. Tiered approaches have been used to complete typical environmental risk assessments (Backhaus and Faust 2012) and in the assessment of risks associated with nuclear waste repositories (Harvey 1980). Analyses can vary with analytical granularity, data availability, model sophistication, etc. In the following paragraphs, a brief review of resilience assessment methods is presented beginning with simple qualitative assessments and ending with sophisticated models of adaptive agents interacting within different types of large networks, otherwise known as system of systems models.

5.2.3.6 Resilience Analysis Tools – Qualitative Approaches

The lowest tier of analysis relies on expert knowledge, approximations, and system parameters that are either well known or documented in the literature. Linkov et al. (2013a, b) proposed a "resilience matrix" approach with subsystem types (physical, informational, cognitive, social) as one dimension and resilience functions (prepare, absorb, recover, adapt) as the second dimension. In its simplest case, this matrix has one criterion per cell, for which there would ideally be a metric. The overall systems resilience is the aggregate of the 16 metrics that could result from multi-criteria decision analysis methods. An adaptation and further development of energy systems (Roege et al. 2014) resulted in an extended resilience matrix that also consists of 16 cells; however, each cell contains several attributes making for a total of 93 attributes. The authors claim that this approach covers several phases of an event and integrates different system properties. Because of its simplicity and focus on qualitative assessments, this matrix approach seems appealing; however, its scientific defensibility and reproducibility are low, and it does not capture interconnectedness within and between systems that are often the main cause for critical regime shifts, cascading failure, and emergent system behaviour.

Higher tiered approaches often follow systems analysis, a process to develop a sound technical understanding of the system predicated on a rigorous basis of data and information (Walden et al. 2015). Experts have used quantitative modelling, analytical modelling and simulations to study various levels of granularity and complexity, and the results serve as inputs into various technical design and decision processes. There are two classes of analytical approaches introduced and explored below, (1) low-dimensional models with a few interacting components (Gao et al. 2016), and (2) multi-dimensional models consisting of a large number of components that interact through a complex network.

5.2.3.7 Resilience Analysis Tools – Low-Dimensional Models

Low-dimensional models approximate the behaviour of a complex system with a small set of non-coupled equations. Figure 5.4 illustrates the aggregated behaviour of a power grid system consisting of tens-of-thousands of interacting components. Accordingly, it critical that we can describe the performance curve for Fig. 5.4 mathematically. The earthquake engineering community proposed a framework to quantify this aggregate behaviour (Cimellaro et al. 2010) by modelling the first three biophysical core functions, resist, re-establish, and rebuild. They simulated the "resist" function with loss and fragility models, and the "re-establish and rebuild" functions with recovery models. Since recovery models are essential for resilience analysis, the authors introduced three types of functions: linear, exponential, and trigonometric. Although this type of aggregated analysis is useful in gaining a first-order understanding of problem, there is a need for approaches that treat multi-dimensional systems with large numbers of interacting components.

5.2.3.8 Resilience Analysis Tools – Flows on Networks Models

Infrastructure systems are networks that are expected to provide specific levels of services. The flow of goods and services is the essential property of infrastructure systems. Consequently, there is an essential need to simulate the flow of those goods and services over different system states and during disruptions and provide adequate representation of corresponding physical and dynamical responses. The field of operations research has brought advances in graph theory, network analysis and a host of mathematical tools and algorithms to represent flows on networks and identify the best possible flow patterns for different requirements. Typical types of problems are maximum flows, minimum cost flows, assignments, matchings, minimum spanning trees, etc. Pertinent methods are useful in any study of critical infrastructure systems where the flow of energy, communications, or materials can be represented as graphical problems and where there is a need to identify the shortest paths and the minimum spanning tree or to reroute traffic flows after sections of a network have lost functionality (Ahuja et al. 1993). Power flow studies are essential in power engineering to numerically describe the flow of electric power in interconnected systems, given total system losses and individual line losses. In a system with N buses and R generators, there are 2 (N- 1)(R-1) unknowns. Due to the non-linear nature of this problem, electrical engineers use numerical methods to obtain solutions that are within an acceptable tolerance. There is a standard textbook describing the whole body of knowledge and providing numerical tools for power system analysis (Saadat 2010).

5.2.3.9 Resilience Analysis Tools – Complex Network Approaches

The scientific study of complex networks emerged after the appearance of two seminal papers before the end of the 1990s (Barabási 2009, Barabási and Albert 1999, Watts and Strogatz 1998). Prior to these papers, the distribution of connections from nodes was either assumed to be completely regular or completely random (Watts and Strogatz 1998), and in then 1999 Barabasi discovered that many real-world networks were something in between, which may be characterized by a power law distribution (Barabási and Albert 1999, Watts and Strogatz 1998). They found this type of so-called scale-free networks to be remarkably robust. Research revealed that there are three principal approaches to increase robustness: (1) to increase the fraction of autonomous nodes, (2) to design dependence links such that they connect to nodes with similar degrees, and (3) to protect high-degree nodes against attack (Gao et al. 2011)(Barabási and Albert 1999, Gao et al. 2011, Watts and Strogatz 1998). The theory of complex networks provides a comprehensive approach to studying generic properties of networks, such as robustness. Nodes on a graph – representing physical facilities such as power stations, network routers, etc. – are considered either operable or inoperable. Percolation theory has been using as an approach to model disruptions of complex networks (Callaway et al. 2000, Hu et al. 2011, Parshani et al. 2010). This approach provides the advantage that it can yield explicit solutions for some networks, such as random networks. It is surprising that only a few real-world networks were studied with percolation approaches, among which was the 2003 Italian power grid blackout is prominent (Parshani et al. 2011, Schneider et al. 2011). However, complex network theory is predicated on strong assumptions that do not hold for many real-world networks: (1) behavior of a single node is rather continuous than discrete between operable and inoperable, and the response is time-delayed, resulting in oscillating system behavior; (2) single nodes contribute differently to the overall system performance because of varying physical performance of facilities; (3) system disruptions may not follow a percolation approach because of spatial constraints and targeted attack; (4) engineered systems often demonstrate connectivity degree distributions that do not follow power law assumptions; and, (5) Up to now, there are no studies yielding information that is common in the risk management community, particularly frequency distributions over expected consequences.

5.2.3.10 Resilience Analysis Tools – Multiphase Models

Moving up in a tiered resilience assessment system, there is a body of models that integrate different types of behaviour, such as the flow of a specific service, the vulnerability of system components, characterization of disruption magnitudes, etc. Complex network theory depends on the following assumptions: (1) disruptions of nodes or edges follow a stochastic process (e.g. Poisson or percolation), (2) fragility of components is discrete (a disrupted component will fail), and (3) components do not exhibit a dynamic recovery behaviour. Research is producing new approaches to overcome these shortcomings, and a recent resilience investigation on the British power grid is an illustrative example (Espinoza et al. 2016). As a first step, the authors simplified the grid to 29 nodes and 98 transmission lines and then modelled power flow changes triggered by different kinds of disruptions, whereby both, DC and AC assumptions applied. As a second step, they modelled the frequency of windstorm magnitudes using an extreme value distribution. A third step consisted of developing storm fragility functions for transmission lines and transmission line towers, whereas in a fourth and final step they modelled the recovery of transmission lines and towers using exponentially distributed meantime to repair (MTTR) functions. The study demonstrated that the risk of blackouts significantly increased when average wind speeds doubled. An assessment of recovery behaviour is an essential component of resilience analysis. In this power grid study, the authors explicitly mentioned that much of the required data was not available and as a result, assumptions had to be made. Nevertheless, the study provide a framework for advancing the way forward.

5.2.3.11 Resilience Analysis Tools – Agent-Based Modelling Systems

Agent-based modelling systems (ABMS) is a relatively new approach to modelling complex, adaptive systems that consists of interacting, autonomous "agents" (Macal and North 2010) with the capability to adapt at the individual or population level. Therefore, they are an excellent tool to study phenomena such as self-organization, adaptation, or response emergence. Macal and North described essential characteristics as follows: (1) self-contained, (2) autonomous and self-directed, (3) state variables vary in time, and (4) social, having dynamic interactions with other agents that influence behaviour. ABMS is mostly used to study phenomena in which adaptation and emergence are important, which is the case for social and human components of Complex-Interactive (CI) systems. Amin advocated CI systems composed of many agents, mainly in the form of decision-making and control units distributed across physical, financial and operational subsystems (Amin 2000, Amin 2001, 2002, Amin and Giacomoni 2012). By the end of the 1990s, two trends emerged that made power grid's more complex, the availability of very high voltage active control devices that made distributed control possible, and the substitution of centralised control by individual agents that cooperate in a market setting (Wildberger 1997). Those trends heralded a new modelling approach for representing complex, distributed systems. Traditional control systems could handle faults; whereas the power system "dispatcher" had to manually handle the recovery from faults and the relocation of power generation (Wildberger 1997). Agent-based modelling for electric power used the following base-classes of entities: (1) generation unit agents, (2) transmission system agents, (3) load agents, and (4) corporate agents. In 1999, the complex interactive networks/systems initiative – a joint initiative of the US Electric Power Research Institute and the US Department of Defence – was established to promote research on the self-healing of infrastructure systems after threats, failures, or other disruptions (Amin 2000). The initiative proposed multi-agent models for power grid systems that were organised in layers (Amin 2000). The base layer consisted of the physical structure of the power grid; whereas, agents performed self-healing actions organised in a reactive layer. Agents that prioritise disruptive events and continuously update real-world systems model were organised in a coordination layer and cognitive agents strived for viability and robustness in a deliberate environment. The appeal of agent-based modelling rests with its ability to represent the behaviour of agents, whereas the representation of physical agents seems to be somewhat limited. ABMS was also used successfully to model infrastructure interdependencies (Becker et al. 2011, Cardellini et al. 2007, Casalicchio et al. 2007, Laprie et al. 2008) over a range of different system scopes and boundaries.

5.2.3.12 Resilience Analysis Tools – System of System Models

The increasing complexity of infrastructure systems has exceeded system engineers' ability to predict their behaviour and ensure control (Pennock and Wade 2015). As mentioned earlier, this calls for a novel approach that explicitly looks for interconnections within and between systems, the so-called "system of systems approach" A system of systems model must provide platforms and tools to represent and dynamically update interconnections from a bottom-up approach to imitate patterns of behavioural at the macro scale. Additionally, they have to adequately describe a system of systems properties, such as (1) decentralised authority and control; (2) network-centric structure that is dynamically supplied by the constituent systems; and (3) indeterminable and often unpredictable behaviour (Gorod et al. 2008). A system of systems model usually consist of systems from different scientific disciplines, including an operating organisation, a user subsystem, an SCADA system, and an engineered system, and what the resilience matrix approach aims to represent in the Y-dimension. Macal and North presented one of the first infrastructure systems of system models (Macal and North 2002) that consisted of five subsystems: (1) a natural gas physical infrastructure layer; (2) a natural gas business layer; (3) an electric power physical infrastructure layer; (4) an electric power business layer; and (5) a consumer layer. Whereas many traditional models focused on physical systems, this model included complete operating organisations and consumers linked together by market mechanisms. The model mapped interdependencies between disparate flows of: (1) electricity, (2) gas, (3) money, and (4) directives (i.e., information). The authors emphasised that at that time detailed physical models of the power grid behaviour could not be integrated within the agent-based model system and that therefore power flow had to be approximated. A comprehensive review of CI modelling systems (Pederson et al. 2006) reported several ABMS CI modelling systems, among which the majority were developed at US national laboratories – Argonne, Los Alamos, SANDIA – to study the operation of complex power systems, financial system infrastructure, and aggregate behaviour of road users. An Italian-based approach, CISIA, was used to model a heterogeneous "system-of-systems" context (Panzieri et al. 2004); it aimed to study fault propagation across heterogeneous infrastructure systems. Although the system of systems approach has been around for a while, the scope of pertinent investigations was limited to two or three interacting systems alone. Examples are the interaction of a power system and an SCADA (supervisory control and data acquisition)-system (Eusgeld et al. 2011), power system and an operator (Schläpfer et al. 2012), or power system with operator and SCADA system (Nan and Sansavini 2017). A recent study, proposed an integrated modelling framework for infrastructure system of systems simulation which only confirms that the field is evolving.

5.2.3.13 Resilience Evaluation

Risk evaluation takes the outcomes of risk analysis, compares them with tolerability criteria, identifies eventual safety gaps, and prioritises needs for risk treatment. In analogy, resilience evaluation compares resilience assessment results with the lowest acceptable resilience metrics to identify gaps that call for action. By adopting the deca-tuple question set, a comprehensive assessment considers at least ten resilience metrics – assuming each question is characterised by one metric – that are then aggregated into one comprehensive assessment. There is no single optimum for this type of multidimensional problem, but only a Pareto frontier that defines the efficiency limit. The decision of which solution should be selected from the Pareto frontier always depends on the value preferences of decision-makers.

5.2.3.14 System Treatment for Resilience

Risk treatment aims to modify risks by selecting and implementing a single or a set of actions, such as removing the risk source, modifying the likelihood, modifying the consequences, sharing the risk with other parties, etc. In analogy, resilience treatment aims to identify the set of actions that increases system resilience in a cost-effective way. Haimes et al. (Haimes et al. 2008) suggested that there are two basic options for action: (1) protecting system assets, and (2) adding resilience to systems. Whereas risk management mainly focuses on protection measures, resilience management balances protection against adding resilience. Although the Haimes' framework of balancing protective and resilience actions is an excellent starting point, there is still no comprehensive approach of identifying and prioritising resilience actions against cost-effectiveness. A possible approach could consist of walking through the deca-tuple question set, and asking what options are available, and how do they compare from the perspective of cost-effectiveness (Haimes 2016). Below, is the deca-tuple question set, adapted for the identification of resilience actions:

Biophysical core functions: what are possible options to:

- Improve the system's ability to resist within acceptable degradation?
- Improve the system's ability to re-establish key functionality and stable behaviour?
- Improve the cost-effectiveness of rebuilding system performance to the level of normalcy?
- Change the biophysical architecture/topology of the system to make it more fault-tolerant?

Enabling functions: what are possible options to:

- Increase the capability to cope with unexpected disruptions?
- Reduce degradation with appropriate emergency responses?

Cognitive functions: what are possible options to:

- Retain long term perceptive system awareness?
- Improve the detection or anticipation of unexpected, critical events, or regime shifts?
- Remember – store and retrieve – actions that successfully dampened past critical events or mitigated regime shifts?
- Speed up the ability of individuals or an organisation to adapt their behaviours, and in turn, better cope with unexpected events?

5.2.4 *Infrastructure Resilience Research Challenges*

After sketching a vision of resilience assessment in Sect. 5.2.3, we now shift the discussion to the topic of knowledge gaps which must be addressed to develop a comprehensive resilience assessment framework. We begin with an alignment of these gaps along the lines of 5 main functions: (1) establishing the context, (2) disruption identification, (3) resilience analysis, (4) resilience evaluation, and (5) system treatment for resilience. It is emphasised that is not possible to anticipate every challenge or threat, but that additional knowledge gaps will be discovered in the process.

- **Context and framing**. The key issue in framing the resilience assessment problem is to identify essential flows of services within a system and the exchange of flows between systems of lower- and higher-levels. In this emerging world of cyber-physical systems that are driving the "smart system" movement, the management of dependencies, heterogeneity and uncertainty are the new challenges that traditional engineering methodologies fail to address (Friedenthal and Burkhart 2015). To tackle this issue, there is an ongoing effort to develop a system of systems engineering approach that is, for the most part, model-driven (Fitzgerald 2016). System of systems engineering as a paradigm emerged at the end of the 1990s (Fuchs 2015), out of which modelling languages, such as SysML and COMPASS emerged (Fitzgerald 2016) to ensure consistency and traceability across all lifecycle phases of the system modelling. Although these platforms were developed from a design perspective, they offer the potential for a system of system framing, definition, and innovative analyses. Whereas the modelling of system architectures is quite mature, the modelling of interdependencies remains a major challenge (Fitzgerald 2016); because, subdomains rely on distinct terminologies. Hence, the use of ontological engineering approaches possess the potential for consistent communication and reasoning not unlike a natural language. Moreover, these approaches present opportunities for modelling interdependencies in a system of system setting.
- **Disruption identification**. Traditional, chain-of-event type analysis has looked at a single disruption for each scenario. Hence, multi-peril disruptions are the first challenge that must be considered in a system of systems context. Financial

engineering approaches, such as copulas or compound Poisson processes (MCNEIL et al., 201 5) are possible approaches to tackling this challenge but have not been extensively explored. Critical regime shifts are a second challenge that is characteristic of complex systems. This endogenous shift in system functions and services to a "new normal" is an emergent phenomenon that occurs with a change in significant external drivers or critical system components; it is difficult to anticipate without complete knowledge of the system and its environment. Hence, there is a need to explore and understand system-specific conditions which foment incipient regime shifts. The field of transportation offers a well-studied example of a regime shift phenomenon to illustrate the difficult nature of the problem at hand, and that would be the transition of traffic from steady flows to congestion.

- **Biophysical Resilience Analysis**. This analysis aims to understand, quantify and assess the performance of biophysical, cognitive and enabling functions that characterise resilience behaviours in systems. The biophysical function set (see 2.3) – resist, restabilize, rebuild, reconfigure – has been the core of physical resilience engineering, that focused on draw-down (resisting) and draw-up (restabilizing and rebuilding) behaviours; but, for the most part, neglected the aspect of system reconfiguration. Drawdown behaviour covers the range between full functionality to collapse, which can be modelled as system fragility with S-shaped functions such as Log Normal or other distribution functions. Draw-up (restoring) behaviour can be approximated with various maintenance functions which characterise the probability of completing system repairs over a given period. Log Normal and other distribution functions have been widely used for modelling maintenance functions. When data are limited, the maintenance function is approximated with an exponential distribution, which alone requires an estimate of the meantime to repair (MTTR). Draw-down, as well as draw-up phenomena, reflect system interdependencies which will complicate the application of simple distribution functions. However, we are at the very beginning of understanding draw-down/draw-up behaviours regarding mathematical functions for a single system component much less an entire complex system. This is why scorecard approaches, such as the "resilience-matrix" approach (Linkov et al. 2013a, Linkov et al. 2013b) are attractive and effective for initiating the salient first step of organising information pertinent to advancing resilience; however, at this stage their capacity to consider interdependencies is limited. System reconfigurability, the fourth biophysical function, is a systemic property that depends on system topology. To the best of our knowledge, reconfigurability analysis is at the nascent stages of exploration and development. Graph theory offers connectivity measures that are candidates for quantifying reconfigurability. Metrics for characterising the global connectivity of graphs are available (Newman 2010); but, many lack immediate utility for network analyses pertinent to the critical infrastructure systems discussed here.
- **Cognitive Resilience Analysis**. Cognitive resilience functions (2.3) had been the focus of cognitive resilience engineering and included functions that retained system awareness, anticipate disruptions, and recall successful actions that

dampened or mitigated past disruptions, and adaptive behaviours. Cognitive resilience (see also 4.2) stems from the concept of distributed cognition, which instigated research on "resilience engineering" (Hollnagel 2006) and "high-reliability organisations" (Weick et al. 1999). Weick proposed an assessment framework of mindfulness to assess and measure cognitive resilience functions. This initial endeavour deployed a scorecard ("resilience analysis grid") to characterise cognitive resilience performance (Hollnagel 2012). The analysis grid consisted of four functions, the ability to respond, to monitor, to learn, and to anticipate. Beyond this effort, there exists another but different question-based assessment method (Sutcliffe et al. 2016). In viewing both methods, there exists an opportunity to merge both and forge a "distributed cognition anvil".

- **Resilience Evaluation**. Resilience assessment compares actual system resilience with a required or desired resilience standard to determine the need for action. We are at the very beginning of quantifying system resilience in a comparable, meaningful, repeatable, and comprehensive way. We are far from expressing metrics or standards that define the minimum resilience such systems must provide. Since resilience performance is always based on multi-dimensional systems, the problem is one of developing a global metric. Methods of multi-criteria decision-making are quite popular for this kind of problem, but they rely on strong assumptions, such as the independency of individual metrics, reliable estimates of preferences for single metrics, and linear aggregation. In other fields, global metrics are calculated with geometric means and data envelope analysis (DEA) as another approach. There is a need for a resilience evaluation framework that covers both the technical aspects and the assessment and decision-making process.
- **Building System Resilience**: In the paragraphs, on system treatment for resilience in Sect. 5.2.3 we outlined what building system resilience means: a balance between protective and resilience actions (Haimes et al. 2008). Considering our 10 resilience functions – 4 biophysical, 4 cognitive, 2 enabling – the problem is how to allocate resources across the 10 functions, constrained by a budget of financial, physical, human, and temporal resources, such that the overall resilience is maximised. This is combinatorial optimisation problem of which the simplest case (2 alternatives for each function – no action, action) produces 1024 combinations. If we increase the number of options to 3 per function, the problem grows to 59,049 combinations. These examples illustrate the complexity of these problems, which easily extend beyond the cognitive capability of humans and calls for the development of support tools. Operations research is providing a wealth of approaches to solving such problems, among which robust optimisation methods are particularly appealing because they can handle uncertainties. Up to now, resilience building is a field largely unexplored but offering many opportunities for future research.

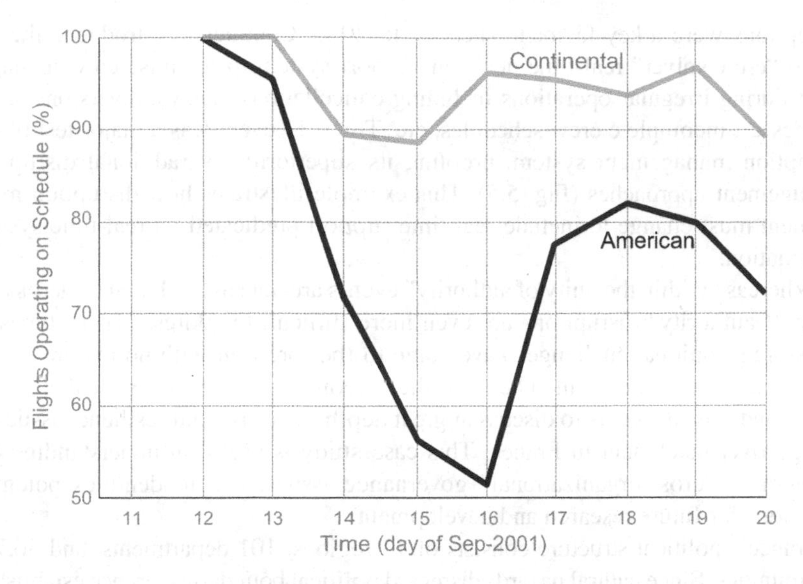

Fig. 5.5 Re-establishment of operational performance of two airlines after the 9/11 grounding in the US (Adapted from Yu and Qi 2004)

5.2.5 What Organisational Changes Are Required for Successful Implementation?

Events that demand resilience-based management have some common characteristics. First, they precipitate unanticipated consequences of significant magnitude. Second, these events cross public and private borders of authority. Third, they induce systems to transition into alternative states or regimes where the assumptions that underpin risk-management plans are no longer valid. While methods for designing optimal operations and risk mitigation plans have achieved a certain level of maturity, methods for real-time disruption management are rare (Yu and Qi 2004). Figure 5.5 illustrates how two US airlines re-established their operational functionality after 9/11.

Whereas Continental was able to re-establish and maintain the ratio of flights operating on schedule at 90% and above. For comparison, American's operational performance dropped after three days to about 50% and recovered to around 75% (Yu and Qi 2004). Although the starting conditions were the same for all the airlines, one observed different patterns of disruption and the recovery of system functionalities. After the storm of the "century" in 1993 – also called the "93 super storm" and the "great blizzard of 1993" –US aircraft along the eastern seaboard were grounded four days, and the Newark airport had to be closed for almost two days (Yu and Qi 2004). Snow covered airplane identification numbers; crews were dispersed among different hotels around airports, and Continental completely lost operational control (Yu and Qi 2004). After 1994, Continental went through a major transformation process; the management realised that crewing flights after major disruptions were a key factor to success. In 2000, Continental introduced the so-called "crew solver" real-time decision support system to optimise crew management during irregular operations including cancellations, delays, diversions, crew sicknesses, incomplete crew schedules, etc. The 9/11 event was a major test of the disruption management system, proofing its superiority to traditional disruption management approaches (Fig. 5.5). This example illustrates how disruption management must change to include real-time support predicated on real-time system information.

Whereas "within the unity of authority" events are not easy to handle, "across the unity of authority" disruptions are even more difficult to address. These types of cross-organizational challenges have come to the forefront with no mature strategies to handle them. Presented below is the outcome of a recent exercise that forced public and private agents to discuss at great depth flood risk and resilience issues in a large river catchment in France. This case study is useful in understanding the problem of "cross-organizational" governance issues, and it identifies potential directions for future research and development.

France's political structure consists of 13 regions, 101 departments, and 36,700 communities. Since natural hazards disregard political boundaries, France established seven defence and safety zones to manage disruptions that extend beyond the jurisdictions and capabilities of departments. The "Paris Defence and Safety Zone" is one such unit responsible for (1) coordinating crisis responses between public and private entities, (2) managing information gathering and distribution, (3) managing operational resources, and (4) providing communication beyond the departmental level.

Following the subsidiarity principle, the European Union established a mechanism to support and complement member states in cases that push them beyond their capabilities. Additionally, the EU Directive on assessment and management of flood risks (2007/60/EC) was introduced to establish a framework for the assessment and management of flood risks aimed at the reduction of the adverse consequences to human health, the environment, cultural heritage and the economic activity within the EU. The directive requires member states to perform preliminary flood risk assessments for each river basin within their territory. The assessment should consist of river basin maps, descriptions of previous flooding events, descriptions of extreme flooding events with significant adverse consequences, and assessments of potential adverse consequences for future floods. In March 2016, the "Sequana" exercise took place to test the operational effectiveness of this European civil protection mechanism as outlined above. The exercise focused on crisis response activities in four areas:

- The coordination of actions between public and private bodies;
- The processing, interpretation and feeding back of information;
- The activation of resources within and beyond the "Paris Defence and Safety Zone" to support departmental authorities; and
- The communication beyond the capabilities of departmental authorities.

More than 90 public and private entities and business sectors participated in the Sequana exercise including: insurance, banking, telecommunications, transport, sanitation, electrical power, gas, museums, and hospitals. The exercise revealed the crucial role of the utility operators. An extreme flood in the Seine catchment could potentially result in a disruption of the electricity grid, affecting about 1.2 million consumers, and a disruption of both the Metro and RER (regional transport network), leaving only Metro line 2 fully functional and closing RER lines.

Sequana looked at resilience as the capability to maintain critical system functions in the face of disruptive events and to recover to normalcy following such events. Business Continuity Plans and Municipal Safeguarding Plans were enabling processes of firms and authorities respectively, which aimed to prepare, maintain or manage information, and align expected resources to meet expected disruption intervention requirements.

The Sequana exercise yielded the following lessons learned:

- The preparation for main flooding events in the Seine basin requires public and private actors to coordinate, and in particular to identify upstream interdependencies.
- A shared understanding of the situation is crucial for effective flood management coordination, which was enabled in this case by geographical information systems providing data on networks of vital importance.
- System recovery, performed by work undertaken following the event and by the coordination of restoration efforts, needs further attention to improve system resilience.
- Communication among public and private actors, as well as affected citizens is crucial in building adaptive behaviours to cope with flooding events.

5.3 Infrastructure Resilience Governance

Building resilience is a collective action problem because infrastructure disruptions affect both public and private sectors, as well as society as a whole. Related problems and activities are embedded in a socio-economic context that varies in both time and space and must be considered to gain broader acceptance of efforts to enhance community resilience. For a long time, command-and-control arrangements were in use to frame problems, to solve them, and to identify and implement effective, efficient actions. More recently, governance approaches became popular, extending traditional processes by using negotiation, dialogue, deliberation and consensus. A recent article (Amsler 2016) defined collaborative governance as the set of structures and processes that enable collaborations among (1) public agencies, (2) private sector agents, (3) civil society, and (4) the general public, aiming to solve problems, to design and implement programs, plans, and projects, and to enforce policies.

Here, we focus on government agencies, for which programs, plans and projects are the entities of management and action. Designing this type of actions is mainly solving problems and making decisions, which may be characterized by a set of activities, in particular: (1) choosing issues that require attention; (2) setting goals;

(3) finding or designing suitable courses of action; (4) evaluating actions; and (5) selecting actions (Simon et al. 1986). While decision-making addresses activities (4) and (5), problems solving in the narrow sense usually focuses on activities (1) to (3). The research briefing panel on decision-making and problem-solving concluded that the very first step, choosing and characterising issues that require attention is the least understood (Simon et al. 1986). It emphasised that once a problem is identified, it should be represented in a way that facilitates its solution, and that representation highly affects the quality of the solutions that will be found. The panel called the problem representation process "framing," emphasising that it was even less understood than agenda setting. Logic calls first for some sequential problem-solving and decision-making process, and second for tools, that support and improve the "problem framing" activity. Introduced below is the risk governance process which is modified and proffered here as a resilience governance process, and mental model concept that helps to frame the problem and to develop a shared understanding of the problem.

Resilience assessment and building as a collaborative action requires a purposeful process that enables and fosters collaboration among participating agents and related problem-solving and decision-making issues. Below, we are proposing (1) a collaborative resilience assessment process, and introducing (2) tools that support joint sense-making and understanding.

Systems Engineering has developed frameworks for coordinating interactions between engineering specialists, other engineering disciplines, system stakeholders, operators and manufacturers (Walden et al. 2015). Systems Engineering activities start from a problem that has been identified and then expend considerable time and effort to frame and understand the problem within the context of risks and resilience before solutions are developed, selected and implemented. The International Risk Governance Council (IRGC) developed a process that can be adapted to solve resilience assessment and management issues. Convinced, that emerging systemic risks are often uncertain or even ambiguous, IRGC experts came to the conclusion that a more comprehensive assessment and management approach is needed because systemic, emerging issues are typically embedded in a much larger political, social, economic and environmental context that is characterized by plural value systems and distributed authority (IRGC 2012). Figure 5.6 illustrates the IRGC framework that consists of five processes: (1) pre-assessment, (2) appraisal, (3) characterization and evaluation, (4) management, and (5) communication. The approach follows Simon's "intelligence-design-choice" logic (Simon 1960), whereby the intelligence phase splits into three, and the design of choice phases merge into "management." The fact that most resilience and emerging risk issues are vague, ill-structured, differently perceived across stakeholders calls for a more comprehensive "intelligence" phase that must go further to understanding how "resilience issues are seen". In the following paragraphs, we introduce and explain the 5 phases of resilience assessment and management, which directly evolve from the IRGC framework, but are adapted to fit a resilience perspective.

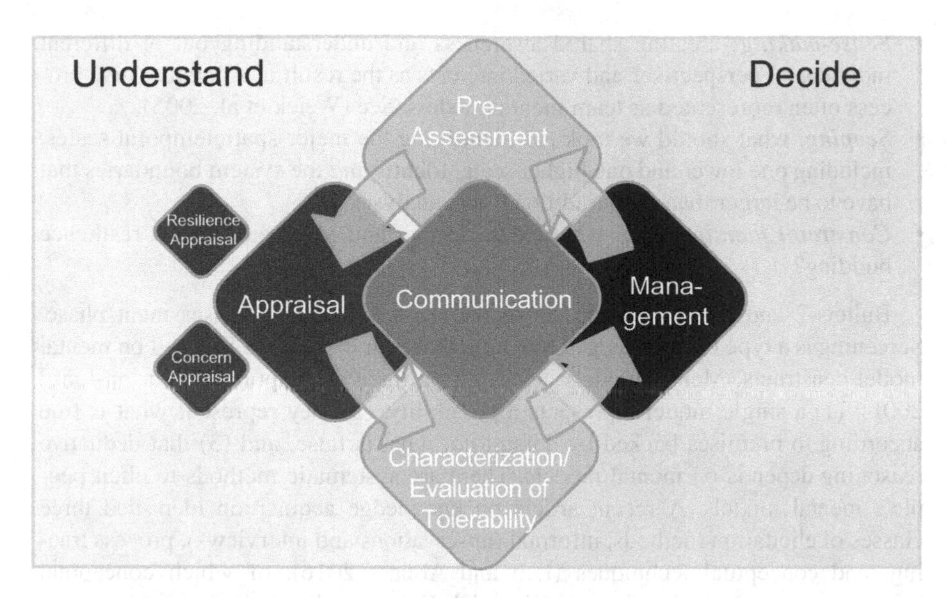

Fig. 5.6 Resilience assessment and management framework (Adapted from IRGC's Risk Governance Framework IRGC 2012)

5.3.1 Pre-assessment

Resilience pre-assessment (Fig. 5.6, top) aims to develop a joint view and a joint meaning of resilience issues across experts, stakeholders and citizens. The backbone of infrastructure systems that supported modern societies were first built in the nineteen century, these included railway networks, water and wastewater pipelines, roads, and electric power grids, and the social embeddedness of the environment was quite homogeneous regarding values, beliefs and experiences. In the last decades, human value, belief and experience systems have become more heterogeneous, which is often called "value plurality.". A second trend is one that amplifies this plurality and is the consequences of complex systems spreading across boundaries of public and private authority. Non-experts perceive issues in their environment by creating "small-scale models of reality" that are used to anticipate events and to reason (Johnson-Laird 2001). It has been a well-supported hypothesis among psychologist that values, beliefs, goals and experiences shape those small-scale "mental models" (Craik 1943). Consequently, there is a need for collaboration that reveals the variety of extant mental models, brings them together in a joint group mental model that defines the issues that are likely to be important, and defines the scope of the appraisal phase. Five key issues critical in the pre-assessment phase include:

- ***Selecting***: what expert, government agency, private and citizens should be invited to participate in the pre-assessment phase?
- ***Screening***: Revealing the variety of views on the issue, encapsulated in "small-scale models of reality" (Johnson-Laird 2001), often called mental models.

- ***Sense-making***: creating shared awareness and understanding out of different individuals' perspectives and varied interests as the result of a collaborative process often represented as team mental models. See (Weick et al. 2005).
- ***Scoping***: what should we look at? Identifying the major spatiotemporal scales, including one lower and one higher scale. Identifying the system boundaries that have to be larger than with traditional risk analysis.
- ***Constraint identification***: what extant formal and legal rules affect resilience building?

Bullets 2 and three from the above list are key in the pre-assessment phase. Screening is a type of knowledge elicitation that is in most cases is based on mental model constructs. Mental models are based on three assumptions (Johnson-Laird 2001): (1) a single model represents a possibility, (2) they represent what is true according to premises backed by default not what is false, and (3) that deductive reasoning depends on mental models. There are systematic methods to elicit people's mental models. A recent article on knowledge acquisition identified three classes of elicitation methods, informal (observations and interviews), process tracing, and conceptual techniques (Leu and Abbass 2016), of which conceptual approaches are appropriate for mental model elicitation. An elicitation process typically starts by establishing a conceptual model of the resilience, based on literature review. This resulting conceptual model provides the framework to design structured interviews, which will take place in a one-to-one mode with each of the selected explored, government agencies, industrial, and citizen.

The whole set of interview-extracted mental models has to be integrated, which requires interpretation and structured interrelated representations of the relevant concepts (Leu and Abbass 2016) within a resilience domain under consideration. Facilitated workshops are a meaningful approach for concept interpretation and structuring; that aim to integrate concepts with some formal representation methods, such as conceptual graph analysis (concept maps), diagramming methods (influence diagrams), networks (Petri Nets), etc. The investigation of the Challenger and Columbia accidents at NASA was one of the first attempts to systematically look at risk from the systemic, emerging point of view, which required studying the behaviour of the NASA system as a compound of technical and organisational subsystems. Leveson and her collaborators represented the body of elicited mental models and data with casual loop diagrams (Dulac et al. 2007, Leveson 2008) that can represent positive (reinforcing) and negative (damping) feedback loops, which were the main drivers for system failures. Structure-function diagrams, as they have been in use for representing the functional architecture of systems (Walden et al. 2015), or flow diagrams representing the physical flows of people, goods, services, information, etc., might eventually be converted, See (Harris 1999). The choice of the specific representation approach should consider its transition into an expert model. Leveson's casual loop diagrams are easily transformed into a system dynamics model representation and used for quantitative analysis.

The bullet, sense making, is probably the most crucial. Whereas a mental model is a construct that is encapsulating sense making for a single individual, team mental models, which were earlier called shared or group mental models refer to overlapping representations of reality that teams share to anticipate events and to reason (Mohammed et al. 2010). They allow team members (1) to view incomplete information in a similar manner, (2) share expectations concerning future events, and (3) to develop similar causal accounts for specific situations, which can also be summarised as "description, prediction, and explanation" (Mohammed et al. 2010). If the degree of shared vision is high, it is assumed shared sense-making is indeed significant, and has resulted in shared cognition. However, metrics of shared cognition appear to be underdeveloped, which makes assessments of the extent to which team mental models are shared across team members difficult (Cannon-Bowers and Salas 2001).

5.3.2 Appraisal

Resilience appraisal (Fig. 5.6, left) aims to reveal, analyse and integrate the best possible scientific knowledge to characterise and understand relevant resilience issues for a system under consideration. Following IRGC's framework, resilience appraisal has to engender an appraisal of concerns. As mentioned earlier, it is assumed individual and team mental models represent what is true, but by default not what is false (Johnson-Laird 2001). This is why a concern assessment has to systematically analyse the perceptions of different stakeholder with the objective of revealing issues that have a negative valence. The IRGC framework seeks to solve this societal dilemma: the systematic deviations of analytic risk assessment on the one hand and the intuitive risk perception on the other. We assume that resilience assessment faces the same challenge.

Resilience assessment is a systematic process that was described in Sect. 5.2.3 "infrastructure resilience assessment," and for which there is a whole set of scientific methods from "export assessment" to "system of system models." Risk analysis' triple question set, as first outlined in 1981 (Kaplan and Garrick 1981) was an important step that made numerous analysis studies more coherent and scientifically reproducible. In analogy, we propose that resilience analysis should follow the proposed deca-tuple question set, which was outlined in Sect. 5.2.3. Those questions are:

How do biophysical functions perform and what are opportunities for improvement, in particular."

(1) What is a systems' ability to resist within acceptable degradation (Haimes 2009)?
(2) What is a systems' capability to re-establish its key functionality to re-stabilize its behaviour?
(3) What is a systems' capability to re-build its performance up to normalcy?
(4) How can we best change the biophysical architecture/topology of the system to make it more fault-tolerant?

How do enabling functions perform, in terms of aiming to support und amplify the biophysical core functions, and what are the opportunities for improvement, in particular:

(5) Does the level of preparation increase the capability to cope with unexpected disruptions?
(6) Does the level of emergency response measures significantly reduce system degradation immediately after an event?

How do cognitive functions perform and what are the opportunities for improvement, in particular:

(7) To what degree is a system able to keep perceptive awareness alive over a long term?
(8) What are the capabilities of a system to detect or anticipate unexpected, critical events, or regime shifts?
(9) To what extend can a system remember – store and retrieve – actions that successfully dampened past critical events or mitigated regime shifts?
(10) What is the capability of a system to continuously adapt the behaviour of its social subsystems (individuals, organisations) to increase the capability to cope with unexpected events?

Informed problem solving and assessment is predicated on scientific information and the balancing of risks with benefits or cost with benefits respectively. Laypersons intuitively build perceptual "small-scale models" of an issue (mental models, (Johnson-Laird 2001)) that are guiding their preference to accept or to reject an issue. Informed and perceptual assessments of risk result in most cases within an acceptance gap, which has been a societal dilemma since the 1980s (Slovic et al. 1981). IRGC's risk governance framework proposed a "concern assessment" process that uses the best scientific knowledge from social sciences to elicit, structure and assess public concerns. A concern assessment raises questions such as (IRGC 2012):

(1) What are the public's concerns and perceptions of a resilience issue?
(2) What are the main drivers in shaping public concerns (organisations, opinion leaders, media, NGOs, etc.)?
(3) Are informed and perceptual assessments connected or disconnected?
(4) Is there a possibility that the resilience issue will result in political mobilisation or potential conflict?

5.3.3 Characterization and Evaluation

Resilience characterization and evaluation (Fig. 5.4, bottom) aims to identify and describe gaps across ten resilience functions that call for action. The first step, characterization of resilience performance, is an outcome of a resilience appraisal. It has to integrate scientific evidence, public perceptions, and perceived knowledge gaps to provide a comprehensive description of performance under each of the ten resilience functions. The characterization of gaps, a result of "situational awareness" (McKinney et al. 2015) as in understanding a system under investigation, and gaps in the verification and validation of the underlying scientific knowledge are essential ingredients of resilience-oriented approaches. A tolerability judgment is the second step, aims to make a balanced assessment of whether performances under each of the ten resilience functions are acceptable or not. This results in a resilience profile that illustrates the level of targeted resilience performance under all ten functions. The appraisal of the leverage effect of those functions is the third step, resulting in a prioritisation of the resilience functions to maximise the affectivity-effort ratio.

5.3.4 Management

Management (Fig. 5.4, right) transforms the requirements (resilience gaps in the evaluation phase) into a system of objectives, defines constraints, designs solutions, evaluates and selects a preferred solution, and proposes its implementation. Whereas traditional management assumes both objectives and environmental parameters as fixed, resilience management relaxes those assumptions by explicitly considering changing objectives and changing or even unknown environmental parameters (McKinney et al. 2015, Saleh et al. 2003). This triggeres novel design approaches, in particular, "design for robustness" and "design for flexibility." Robust design aims to satisfy a fixed set of requirements, despite changes in the environment or within the system after the system has entered service (Saleh et al. 2003). Design for flexibility further relaxes those assumptions, aiming to perform well in the face of changes in initial system requirements and objectives and any changes in the environment or within the system after the system has entered service (Saleh et al. 2003). Whereas design for robustness and flexibility entered with aerospace engineering, they have rarely been used in infrastructure engineering. Changing design philosophies for infrastructure systems will take at least a decade, and it is urgently needed. Operations research is providing a whole set of approaches to identify an optimal solution under specific constraints. Whereas traditional optimisation approaches are based on certainty assumptions, robust optimisation techniques provide methods for optimisations under uncertainty, based on said-based uncertainty models (Bertsimas et al. 2011). In this sense, robustness means to identify a solution that is feasible for any realisation of uncertainty in a given set. At present, we are just beginning to explore the use of robust optimisation techniques for infrastructure extension and protection problems. The field offers many opportunities for future developments.

5.3.5 *Communication*

Communication (Fig. 5.4, centre) brings together the four risk governance processes (pre-assessment, appraisal, evaluation, management). Resilience assessment and management usually happens in ad hoc workgroups, consisting of people who have never worked together, which is always the case when stakeholders are involved. Communication within workgroups serves to balance the power between stakeholders by reducing information asymmetries and building trust, which is an important regulator of behaviours and responses of team members (Altschuller and Benbunan-Fich 2010). Assuming that the workgroup is a system of distributed cognition, communication can raise awareness and support joint sense making, which is crucial for taking weak signals and ambiguities serious and integrating them into the work group process (Weick et al. 1999).

Once the resilience appraisal process is finished, the role of communication becomes that of ex-plaining the results of the resilience appraisal process to public and private actors responsible for building infrastructure resilience. The communication process should enable them to make in-formed decisions (IRGC 2012) and to build resilience within their organisational units. In Sect. 5.4, we explore in greater detail the resilience building process.

5.4 Building Resilience with Knowledge Transfer and Continuous Learning

Building resilience requires adaptation, which means changing a systems' architecture and changing mental models of individuals, teams, organisations, governance, and government systems. As mentioned earlier, mental models are "small-scale models" of reality (Johnson-Laird 2001) that people, teams, and even organizations use to assess the state of the system and its environment, to understand its significance and meaning, to retrieve or develop one or more courses of action, and to select and execute the most meaningful course. In Sect. 5.2.3, we characterised those capabilities with four cognitive resilience functions: keeping awareness, anticipating critical events, remembering actions that were successful in the past, and adapting the behaviour of individuals and organisations.

Knowledge transfer is a concept that fosters and supports collaborations between research organisations, business entities, and public sectors. There are other concepts, such as "implementation science," "knowledge for action," or "knowledge management" that embrace similar purposes. Whereas "knowledge transfer" and "implementation science" are clearly devoted to the transfer activities between research, public, and private entities, "knowledge management" concerns an organisation's use and management of knowledge and information. The concept of "Knowledge transfer" first appeared in the 1990s, and it assumes research organisations push messages to research users (Mitton et al. 2007). The effectiveness of "knowledge transfer" mechanisms seems to be modest, and "best practice" approaches seem to be based at best on anecdotal evidence or even rhetoric rather than rigorous evidence (Mitton et al. 2007). One major reason might be the underlying notion of knowledge, which is often assumed to be transferable, like transferring files from one computer to another. This perception of what is transferable has its shortcomings, which motivated Collins (1993) to propose four types of knowledge/ability/skills: encoded, embodied, embrained, and encultured. While symbols – text, mathematical formula, computer code, data, etc. – represent encoded knowledge that is easily transferred, the three remaining types of knowledge have different representations, which require encoded knowledge to be transformed. Embodied knowledge/abilities/skills, such as playing tennis, dancing, etc. has to be acquired by training, and it is not sufficient to study books and manuals alone. In building resilience, it will be necessary to change "embrained" and "encultured" knowledge/abilities/skills, which is much more than transferring knowledge. Embrained knowledge is – a simplistic way – the set of concepts stored in an individual brain that is constantly updated based on experiences. Encultured knowledge is an element of social embeddedness (Collins 1993) that encapsulates social rules and habits. These concepts suggest building resilience will require the use of different channels of knowledge transfer and transformation that

- support policymaking with the best available knowledge,
- lead to the development of best professional practices based on the latest scientific knowledge,
- provide for updating mental models of individuals for crucial behavioural issues.

Below, we will sketch a vision of building resilience with knowledge transfer and transformation in the three fields of action. We are aware that this is one point of view and that there are others that should be considered.

5.4.1 Knowledge Transformation for Policy Making

Policymaking is a process of deciding on principles and courses of actions for a certain subject domain, such as infrastructure systems. This process is always a compromise between balancing different values against the best possible rational solutions. In a seminal paper, Lasswell (1951) argued that science-based policymaking relies on a systematic policy process on the one hand and a policy intelligence process on the other. Whereas our understanding of policy process has increased considerably over the last 60 years, policy intelligence continues to be poorly developed. A recent review (Rutter et al. 2013) of the health science domain concluded that two key processes– policymaking and scientific knowledge generation – are represented by two different communities of policy practitioners and researchers, who are for the most part disconnected which makes informed policy making rather difficult to achieve. Therefore, the key question is how to design an interface between the two communities. From an information perspective, there are four requirements. The transferred knowledge/information has to be (1) accessible, (2) relevant, (3) understandable, and (4) on time (Mitton et al. 2007, Rutter et al. 2013). Knowledge published in peer-reviewed journals are not a good way to reach policymakers (Rutter et al. 2013) and might not fit the specific requirements of the issue to be solved. Most scientific publications are written in scientific jargon, which non-scientists find difficult to understand, and which calls for different writing style or even a narrative.

Above, we took a process-oriented view on policy and knowledge generation. More recently, actor-oriented approaches became popular, looking at the key actors in the process. A knowledge broker is a person who facilitates the use of scientific knowledge in a policy process, replacing knowledge producers (researchers) who were originally thought to take this role (Rutter et al. 2013). A recent study found knowledge brokers embedded in the policy process but performed by members of the policy community rather than by external researchers or academics (Oliver et al. 2013). Another study concluded that members of the public administration and independent "evidence institutions," such as the National Institute for Health and Care Excellence (UK), played an important role in knowledge brokering. There is also informal evidence that research groups supported former members in their efforts to secure jobs in ministries close to political and administrative operations. Such hires create an opportunity to augment the policymaking processes with scientific information. Here, we hypothesise that knowledge brokering is one of the preferred alternatives to imbuing policy intelligence, as proposed by Lasswell in his seminal paper on policy science (Lasswell 1951). To the best of our knowledge, the means by which information and knowledge are best used to support the policy making is not well-established. Science-policy intelligence could imitate the intelligence cycle that has been used in the intelligence community, which consists of the following steps: direction, collection, processing, analysis, dissemination, and feedback. The "direction" step is crucial because it defines the knowledge-intelligence requirements of key actors in the policy process. Whereas scientists often have the "push" model in mind, the "direction" step initiates a "pull" process that makes certain that the knowledge/information supplied is relevant and timely. Brokerage specialists, which have to know both the relevant fields of science and the policy process, have to handle the steps for "collection," "processing," and "analysis.". The "dissemination" step is – besides the "direction" step – the most crucial one. Whereas scientists think in terms of papers and reports, brokerage specialists – in analogy to the intelligence business – consider different dissemination formats. Memos are fast knowledge containers; whereas, bulletins are regular updates that highlight new developments and trends. Finally, assessment reports are less focused and more in-depth, usually compiling the essence of state-of-the-art in certain discipline. The effective use of "memos," "Bulletins", and "Assessment Reports" with key actors within the policy process is more an art than a science.

5.4.2 Building Best Practices

Best practices are approaches, methods, and techniques that are expected to produce superior results when adopted by public and private organisations. Viewed in the context of infrastructure-operating organisations, the first question is: "What best practices produce resilient organisational performance?" Whereas, follow-up questions are: "How can such an organisation be built?" and "How is knowledge transfer used to expedite critical transformations that foster resilience within an organisation?"

In the early 1990s various disciplines explored the nature of organisations that demonstrated resilience as they underwent transformations or directly suffered internal or external threats to their existence. Weick investigated aircraft carrier organisations that function under expectations of error-free performance (Weick and Roberts 1993). He concluded that highly-reliable organisations are fundamentally different from highly-efficient organisations, and that "a well-developed collective mind" makes the difference. The "collective mind" concept builds upon the theory of distributed cognition that also emerged at the beginning of the 1990s (Hutchins 1995). Traditional views assume cognition is encapsulated in individuals; whereas, distributed cognition assumes cognitive processes are socially distributed across members of a group (Hollan et al. 2000). Consequently, a social organisation is a form of cognitive architecture, bound together by information transmission and transformation trajectories, some of which are newly created, while other are continuously reconfiguring and adapting. Hence, organisational performance is to a high degree a reflection of the topology of cognitive processes distributed across members of the social group. Those findings produced a shift in perceptions, whereby organisations were then viewed as a collective of cognitive processes. Weick emphasised that highly-reliable organisations struggle to be alert and to act through mindful conduct. Investigating a 1949 firefighting disaster in the US, Weick concluded (Weick 1993) that a disintegration of role structure and sense making played a critical role in the disaster. Situational assessment and sense making – trying to make things rationality accountable – turn out to be key factors for failure, which supports the argument that highly-reliable organizations should shift their focus from decision-making to sense making and meaning. Weick also postulated what makes an organisation more resilient and offered the following factors: (1) improvisation and bricolage, (2) virtual role systems, (3) the attitude of wisdom, and (4) respectful interaction. All of these factors are counter-agents of highly rigid, hierarchical organisations, promoting distributed cognition and intelligence. In 1999, Weick published a contribution, which provides a framework for highly-reliable, mindful organisations (Weick et al. 1999). He emphasised that a highly-reliable organisation is more grounded in adaptive human cognition and action than is the engineering view of "reliable outcomes resulting from repetition cognition and action". His mindfulness concept is based on the assumption that stable and interactive cognitive processes aim to discover and correct errors, but also attend to the quality of attention and the conservation of attention. Weick defined three classes of processes that enable mindful, resilient organizational behaviour: (1) discovery processes, that is, when such processes are absent, there is a preoccupation with failure, a reluctance to simplify interpretations, or an extant sensitivity to operations; (2) correction, to be specific, commitments to resilience; (3) organizational structure, in particular "on the specification of structures". Highly-reliable organisations overcome the tendency to make assumptions and socialise people to ignore contextual change, which are both obstacles to error detection. They have the capacity to cope with anticipated threats after they have become manifest and to bounce back. Their organisation is based on adaptive structures, enacting "moments of organised anarchy", being aware that quarterly hierarchies can amplify errors, and in particular for those cases in which miscues happen near the top (Weick et al. 1999).

While Weick's work materialized from research focused understanding the behaviour of organiza-tions, a similar stream of research emerged from the safety and risk management community. Scholars in industrial safety and system safety research initiated a paradigm shift from "human-centered" to "organizational-centered" approaches that they termed "resilience engineering" (Hollnagel 2006). It was a new way of thinking about safety and looking for ways to enhance the ability of organisations to create processes that are robust as well as flexible, to monitor and revise risk models, and to target resources proactively. NASA became convinced that a new approach to safety issues was required after a series of major incidents, in particular, the Mars exploration fail-ures of1999, and the Columbia space shuttle accident in 2003 (Woods 2006). In 2004, the first symposium on resilience engineering was held in Sweden, out of which resulted a book "Resilience Engineering: Concepts and Precepts" (Hollnagel 2006). The symposium series has since continued, and in parallel, the Resilience Engineering Association has launched the professional society for the resilience engineering specialist.

The two research directions – resilience engineering and organizational mindfulness – are only loosely coupled, but predicated on similar foundations, such as (1) focus on organizations rather than the individual or specific human decisions or actions, (2) focus on processes instead of structures, and (3) work from principles of collective cognition. Hollnagel's four cornerstones of resilience (Hollnagel 2009) – learn, anticipate, monitor, respond – are a subset of all cognitive processes, while Weick's five processes (Weick et al. 1999) of mindfulness are enablers for collective cognitive processes, in particular, sensing, anticipating, appraising (sense-making), retrieving action patterns, and responding. Both research directions emerged in the last 20 years, and their approaches and tools are still in the early stages of development, offering a considerable potential for further improvement.

If we want to build organisational resilience, experts have to make informed judgments regarding "how resilient" or to what extent an organisation should become resilient. A recent, comprehensive review on mindfulness in organisations reported that quantitative assessments of collective mindfulness and its constituent processes are just beginning to occur (Sutcliffe et al. 2016). There are several psychometric constructs, for which there is however only limited validated testing, such as the eight-item scale for organisation mindfulness, and a five-factor, 38-item measure of organizational mindfulness processes, etc. (see (Sutcliffe et al. 2016)). Resilience engineering's four basic abilities are seen as a natural starting point for understanding how an organisation functions: how it responds, how it monitors, how it learns, and how it anticipates (Hollnagel 2009). It has been characterised as "resilience analysis grid", which is a kind of scorecard to assess organisational resilience. The same author proposed the function resonance analysis model (FRAM) which aims to capture the dynamics of complex social-technical systems, their non-linear dependencies, and their emerging behaviours (Hollnagel 2012). The SCALES framework is based on the same intellectual tradition and aims to link resilience engineering and enterprise architecture principles into a frame-work that enables context-driven analysis of organisational resilience (Herrera 2016). As suggested above, resilience engineering approaches are in the early phases of development and offer many opportunities for further improvement (Herrera 2016).

Following the concept of "mindfulness organisations", building resilience means purposeful organizational development. Sutcliffe et al. reported that organizational practices, such as (1) organizational audits, (2) and boundary-spanning facilitators, (3) active socialization through vivid stories, (4) simulation of rare events, (5) pro-active interpersonal monitoring practices, etc. foster more in-tense, continual attention to weak signals and overall attention to quality (Sutcliffe et al. 2016). IT-based support tools have proven to enhance and inhibit collective mindfulness, careful and collaborative analysis of issues, and the enrichment of response repertoires. The Resilience Engineering Association documented a case study of how to introduce resilience thinking into an aviation industry firm. A three-step approach, executed over several years, was designed and implemented. The approach consisted of first introducing a theoretical framework; next, working on an aircraft accident case study; and last, Integrating resilience thinking into the line oriented flight training in simulator sessions (FRANK and STEINHARDT, n.d). The effectiveness of this more classical, training-oriented approach is not yet known, but there is evidence in the mindfulness literature that overtraining and "over-experience" negatively affect mindfulness within the task domain (Sutcliffe et al. 2016). This contradicts the common statement that "the more training and the more experience, the better", and thus, has to be rethought carefully.

5.4.3 Adapting Individual Behaviour

Infrastructure disruptions affect the living space and the people that live in that space. The behaviour of those people immediately before, during, and after a disruptive event contributes considerably to the overall resilient behaviour of infrastructure systems. Their capabilities to contribute to the biophysical resilience functions – (1) to stop system degradation, (2) to re-establish the critical system functions, and (3) to rebuild system functions back to normalcy – depends heavily on their system of distributed cognition. As Hutchins pointed out, any system of labour division requires distributed cognition to coordinate the activities of its members (Hutchins 1995). Therefore, distributed cognition is the foundational concept on which the resilience behaviour of infrastructure users and inhabitants is constructed. The key cognitive functions, on which resilience behaviour is built, are (1) awareness, (2) anticipation, (3) remembering and retrieving patterns of useful actions, and (4) adaptation and learning.

How to raise awareness – the first cognitive function - and to continually keep it on at an elevated-level, is the first issue. In the past, cultures that were often exposed to strategies that enhanced risk awareness such as memorising disruptive events using cultural witnesses, such as historical flood gauges, that were and are visible marks within the living spaces of communities (Pfister 2011). A flood gauge is a visual cue to an event memorised and its associated risks that make all extreme events visible to all the inhabitants and contributes to increasing risk awareness. For each event, chroniclers wrote up a narrative to amplify this type of "collective memory" and to contribute to sense-making and understanding (Pfister 2011). Smokey Bear is an example of a personalised risk witness and advocate that was created in 1944 and is successfully running since then. Smokey Bear is an advertising mascot created to frame public behaviour in facing the risk of forest fires, and in turn becoming a part of the American popular culture in the 1950s. So successful was the Smokey campaign that it led to a level of fire suppression far beyond what ecologists recommend for healthy forests. This is because it produced larger stores of unburnt residual wood, which are now fuelling the larger and more frequent forest fires seen today. However, what we can learn from this successful campaign is that a personalised mascot figure can gain the power to raise public awareness and to shape public behaviour.

A meaningful, purposeful response is a compound action of the operating organisation, government agencies, system users, and inhabitants. Following the analogy of the immune system, the inhabitants and system users must acquire a collective memory, out of which they can retrieve action patterns that proved useful in the past. What this collective memory looks like and how it may be updated and maintained over the long term are again critical issues that must be resolved. Historians and anthropologists showed that stories (narratives) have been powerful to store cultural knowledge and to transfer it from generation to generation. Historically, chroniclers wrote narratives about disruptive events, thus making them memorable and transferable to society (Pfister 2011). However, the human mind has a natural mechanism for forgetting that knowledge or those memories if not recalled frequently or updated periodically. Pfister documented a loss of collective disaster memory for Central Europe in the twentieth century. In this study, he found the initial loss of collective memory promulgated a "disaster gap", but eventually further memory loss resulted in the disappearance of cultural disaster memories, such as flood gauge marks in flat-prone regions (Pfister 2011). The 2014 Korean ferry accident is perhaps a more recent example illustrating the importance of narratives (Jeon et al. 2016). On April 16, 2014, a Korean ferry capsized, resulting in 304 causalities. To overcome this catastrophe, some Korean public institutions investigated and reported on the case. Resulting documents were quite technical, voluminous, and their style adequate to report the accident to high-level officials. Nevertheless, they failed to explain to the public how this accident happened, which is not atypical for such reports. A group of authors published a book entitled "the Sewol-records of the day" with the intention of producing a comprehensive narrative of how the ship capsized and why the rescue operation failed and why official reports failed to inculcate a larger public understanding of the event (Jeon et al. 2016). The main lesson learned is that overcoming of disasters requires that the public understands what happened and that the disaster narrative creates meaning. The official Korean reports failed to do both.

Acknowledgement Several colleagues contributed to this chapter. We would like to thank Raymond Nyer and Igor Kozine for their valuable inputs. Many thanks to Sarah Thorne and Cate Fox-Lent for their editing work, which contributed to the improvement of the understandability and readability of the text. Finally, our thank goes to the members of the infrastructure working group of the NATO Workshop "Resilience-Based Approaches to Critical Infrastructure Safeguarding" for their contributions.

Further Suggested Readings

Ahuja RK, Magnanti TL, Orlin JB (1993) Network flows theory, algorithms, and applications. Pearson Education Limited, Harlow

Altschuller S, Benbunan-Fish R (2010) Trust, performance, and the communication process in Ad Hoc decision-making virtual teams. J Comput-Mediat Commun 16(1):27–47

Amin M (2000) Toward self-healing infrastructure systems. Computer 33(8):44–53

Amin M (2001) Toward self-healing energy infrastructure systems. Comput Appl Power IEEE 14(1):20–28

Amin M (2002) Toward secure and resilient interdependent infrastructures. J Infrastruct Syst 8(3):67–75

Amin SM, Giacomoni AM (2012) Smart grid—safe, secure, self-healing: challenges and opportunities in power system security, resiliency, and privacy. IEEE Power and Energy Magazine

Amsler LB (2016) Collaborative governance: integrating management, politics, and law. Public Adm Rev 76(5):700–711

Backhaus T, Faust M (2012) Predictive environmental risk assessment of chemical mixtures: a conceptual framework. Environ Sci Technol 46(5):2564–2573

Barabási AL (2009) Scale-free networks: a decade and beyond. Science 325(5939):412–413

Barabási AL, Albert R (1999) Emergence of scaling in random networks. Science 286(5439):509–512

Becker T, Nagel C, Kolbe TH (2011) Integrated 3D modeling of multi-utility networks and their interdependencies for critical infrastructure analysis. In: Kolbe TH, König G, Nagel C (eds) Advances in 3D geo-information sciences. Springer, Berlin/Heidelberg, pp 1–20

Bertsimas D, Brown DB, Caramanis C (2011) Theory and applications of robust optimization. SIAM Rev 53(3):464–501

Callaway DS, Newman MEJ, Strogatz SH, Watts DJ (2000) Network robustness and fragility: percolation on random graphs. Phys Rev Lett 85(25):5468–5471

Cannon-Bowers JA, Salas E (2001) Reflections on shared cognition. J Organ Behav 22(2):195–202

Cardellini V, Casalicchio E, Galli E (2007) Agent-based modeling of interdependencies in critical infrastructures through UML. In: Proceedings of the 2007 spring simulation multiconference-Volume 2, 119–126. Society for Computer Simulation International

Casalicchio E, Galli E, Tucci S (2007) Federated agent-based modeling and simulation approach to study interdependencies in IT critical infrastructures. In: Distributed simulation and real-time applications, 2007. DS-RT 2007. 11th IEEE International Symposium, 182–189. 22–26 Oct. 2007

Cimellaro GP, Reinhorn AM, Bruneau M (2010) Framework for analytical quantification of disaster resilience. Eng Struct 32(11):3639–3649
Collins HM (1993) The structure of knowledge. Soc Res 60(1):95–116
Craik KJW (1943) The nature of explanation. University Press, Cambridge. 123 p
Day JC, Ingham MD, Murray RM, Reder LJ, Williams BC (2015) Engineering resilient space systems. INCOSE Insight 18(1):23–25
Deiotte R (2016) An architechtural patterns library to aid systems of systems engineering. INCOSE Insight 18(3):31–34
Dove R, Ring J, Tenorio T (2012) Systems of the third kind: distinctions, principles, and examples. INCOSE Insight 15(2):6–8
Dulac N, Owens B, Leveson N (2007) Demonstration of a new dynamic approach to risk analysis for NASA's constellation program. MIT CSRL final report to the NASA ESMD associate administrator. Complex Systems Research Laboratory, MIT. Accessed on 04 Feb 2013. http://sunnyday.mit.edu/ESMD-Final-Report.pdf
Espinoza S, Panteli M, Mancarella P, Rudnick H (2016) Multi-phase assessment and adaptation of power systems resilience to natural hazards. Electr Power Syst Res 136:352–361
Eusgeld I, Nan C, Dietz S (2011) "System-of-systems" approach for interdependent critical infrastructures. Reliab Eng Syst Saf 96(6):679–686
Farmer F (1967) Siting criteria–a new approach. In Proceedings of the IAEA symposium on nuclear siting, 303–29
Fitzgerald J (2016) Comprehensive model-based engineering for systems of systems. INCOSE Insight 19(3):59–62
Francis R, Bekera B (2014) A metric and frameworks for resilience analysis of engineered and infrastructure systems. Reliab Eng Syst Saf 121:90–103
Frank M, Steninhardt G (n.d.) Small details - big difference. Can resilience development hange the future of aviation? Resilience Engineering Association. Accessed on 02 Nov 2016. http://www.resilience-engineering-association.org/wp-content/uploads/2016/09/Introduction-note-about-resilience-V4.pdf
Friedenthal S, Burkhart R (2015) Evolving SysML and the system modeling environment to support MBSE. INCOSE Insight 18(2):39–41
Fuchs J (2015) Model based systems engineering. INCOSE Insight 18(2):9
Gao J, Buldyrev SV, Stanley HE, Havlin S (2011) Networks formed from interdependent networks. Nat Phys 8(1):40–48
Gao J, Barzel B, Barabási A-L (2016) Universal resilience patterns in complex networks. Nature 530(7590):307–312
Gorod A, Sauser B, Boardman J (2008) System-of-systems engineering management: a review of modern history and a path forward. IEEE Syst J 2(4):484–499
Haimes YY (2009) On the complex definition of risk: a systems-based approach. Risk Anal 29(12):1647–1654
Haimes Y (2016) Risk modeling, assessment, and management, 4th, Wiley series in systems engineering and management. Wiley, Hoboken. 690 S. p
Haimes YY, Crowther K, Horowitz BM (2008) Homeland security preparedness: balancing protection with resilience in emergent systems. Syst Eng 11(4):287–308
Harris RL (1999) Information graphics a comprehensive illustrated reference. Oxford University Press, New York. 448 p
Harvey T (1980) Disruptive event uncertainties in a perturbation approach to nuclear waste repository risk analysis. Lawrence Livermore Laboratory, Livermore. 52 p
Henry D, Ramirez-Marquez JE (2016) On the impacts of power outages during Hurricane Sandy—a resilience-based analysis. Syst Eng 19(1):59–75
Herrera I (2016) Resilience engineering and indicators of resilience. In: Resource guide on resilience. IRGC, Editor. EPFL International Risk Governance Center, Lausanne. Accessed on 02 Nov 2016. https://www.irgc.org/wp-content/uploads/2016/04/Herrera-Resilience-Engineering-and-Indicators-of-Resilience.pdf

Hollan J, Hutchins E, Kirsh D (2000) Distributed cognition: toward a new foundation for human-computer interaction research. ACM Transact Comput Hum Interact (TOCHI) 7(2):174–196
Holling CS (2001) Understanding the complexity of economic, ecological, and social systems. Ecosystems 4(5):390–405
Hollnagel E (2006) Preface. In: Hollnagel E, Rigaud E (eds) Ssecond resilience engineering symposium. Presses Mines ParisTech, Antibes - Juan les Pins, pp v–vi
Hollnagel E (2009) The four cornerstones of resilience engineering. In: Nemeth CP, Hollnagel E, Dekker S (eds) Resilience engineering perspectives, Volume 2: preparation and restoration. Ashgate, Aldershot/Burlington, pp 117–134
Hollnagel E (2012) FRAM, the functional resonance analysis method: modelling complex socio-technical systems. CRC Press, Boca Raton
Hu Y, Ksherim B, Cohen R, Havlin S (2011) Percolation in interdependent and interconnected networks: abrupt change from second- to first-order transitions. Phys Rev E 84(6):066116
Hutchins E (1995) Cognition in the wild. MIT Press, Cambridge, MA. 381 p
IRGC (2012) An introduction to the IRGC Risk Governance Framework. International Risk Governance Council. Accessed on 29 Sept 2016. http://www.irgc.org/wp-content/uploads/2015/04/An_introduction_to_the_IRGC_Risk_Governance_Framework_final_v2012.pdf
ISO (2009) ISO 31000 - risk management - principles and guidelines. International Standard Organization, Geneva. 27 p
Jeon C, Choi H, Kim S (2016) How not to learn from disasters: disaster reports and sociotechnical resilience in South Korea. In: Amir S, Kant V (eds) The sociotechnical constitution of resilience - structures, practices and epistemologies. Nanyang Technological University, School of Humanities and Social Sciences. Future Resilient Systems, Singapore, p 323
Johnson-Laird PN (2001) Mental models and deduction. Trends Cogn Sci 5(10):434–442
Kaplan S, Garrick BJ (1981) On the quantitative definition of risk. Risk Anal 1(1):11–27
Keller W, Modarres M (2005) A historical overview of probabilistic risk assessment development and its use in the nuclear power industry: a tribute to the late Professor Norman Carl Rasmussen. Reliab Eng Syst Saf 89(3):271–285
Laprie J, Kanoun K, Kaaniche M (2008) Modelling interdependencies between the electricity and information infrastructures. Lect Notes Comput Sci 4680:54–67
Lasswell HD (1951) The policy orientation. In: Lerner D, Lasswell HD (eds) The policy sciences. Recent developments in scope and method. Stanford University Press, Stanford, pp 3–15
Leu G, Abbass H (2016) A multi-disciplinary review of knowledge acquisition methods: from human to autonomous eliciting agents. Knowl-Based Syst 105:1–22
Leveson N (2004) A new accident model for engineering safer systems. Saf Sci 42(4):237–270
Leveson N (2008) Technical and managerial factors in the NASA challenger and columbia losses: looking forward to the future. Mary Ann Liebert, Inc., Controversies in Science & Technology - Volume 2 From Climate to Chromosomes
Lewis HW, Budnitz RJ, Rowe WD, Kouts HJC, von Hippel F, Loewenstein WB, Zacharsiasen F (1978) Risk assessment review group report to the U.S. Nuclear Regulatory Commission. National Technical Information Service, Springfield. 66 p
Linkov I, Eisenberg DA, Bates ME, Chang D, Convertino M, Allen JH, Flynn SE, Seager TP (2013a) Measurable resilience for actionable policy. Environ Sci Technol 47(18):10108–10110
Linkov I, Eisenberg DA, Plourde K, Seager TP, Allen J, Kott A (2013b) Resilience metrics for cyber systems. Environ Syst Decis 33(4):471–476
Macal C, North M (2002) Simulating energy markets and infrastructure interdependencies with agent based models. In Social agents: ecology, exchange, and evolution conference, 195–214. University of Chicago, Chicago, IL US. Decision and Information Sciences, Division Office, Argonne National Laboratory, 9700 South Cass Avenue, Argonne, Illinois 60439–4832
Macal CM, North MJ (2010) Tutorial on agent-based modelling and simulation. J Simul 4(3):151–162

McKinney D, Iyengar N, White D (2015) Situational awaremess: a transformational perspective. INCOSE Insight 18(3):44–48
McNeil AJ, Frey R, Embrechts P (2015) Quantitative risk management: concepts, techniques and tools. Princeton university press, Princeton
Mitton C, Adair CE, McKenzie E, Patten SB, Perry BW (2007) Knowledge transfer and exchange: review and synthesis of the literature. Milbank Q 85(4):729–768
Mohammed S, Ferzandi L, Hamilton K (2010) Metaphor no more: a 15-year review of the team mental model construct. J Manag 36(4):876–910
Moors A, Ellsworth PC, Scherer KR, Frijda NH (2013) Appraisal theories of emotion: state of the art and future development. Emot Rev 5(2):119–124
Murray RM, Day JC, Ingham MD, Reder LJ, Williams BC (2013) Engineering resilient space systems. Keck Institute for Space Studies, Pasadena. Accessed on 22 Aug 2016. http://www.kiss.caltech.edu/study/systems/, 82 p
Nan C, Sansavini G (2017) A quantitative method for assessing resilience of interdependent infrastructures. Reliab Eng Syst Saf 157:35–53
Newman MEJ (2010) Networks an introduction. Oxford University Press, Oxford. 772 p
Oliver K, Money A, de Vocht F (2013) Knowledge brokers or policy entrepreneurs? Strategies to influence the policy process. J Epidemiol Community Health 67(Suppl 1):A76
Olsen JR (2015) Adapting infrastructure and civil engineering practice to a changing climate. American Society of Civil Engineers, Reston. Accessed on 18 Dec 2016. http://ascelibrary.org/doi/pdf/10.1061/9780784479193, 93 p
Osorio I, Frei MG, Sornette D, Milton J, Lai Y-C (2010) Epileptic seizures: quakes of the brain? Phys Rev E 82(2):021919
Panzieri S, Setola R, Ulivi G (2004) An agent based simulator for critical interdependent infrastructures. In Securing critical infrastructures, Grenoble
Park J, Seager TP, Rao PSC, Convertino M, Linkov I (2013) Integrating risk and resilience approaches to catastrophe management in engineering systems. Risk Anal 33(3):356–367
Parshani R, Buldyrev SV, Havlin S (2010) Interdependent networks: reducing the coupling strength leads to a change from a first to second order percolation transition. Phys Rev Lett 105(4):48701
Parshani R, Rozenblat C, Ietri D, Ducruet C, Havlin S (2011) Inter-similarity between coupled networks. EPL (Europhys Lett) 92(6):68002
Pederson P, Dudenhoeffer D, Hartley S, Permann M (2006) Critical infrastructure interdependency modeling: a survey of US and international research. 1–20 pp.
Pennock MJ, Wade JP (2015) The top 10 illusions of systems engineering: a eesearch agenda. Proc Comput Sci 44:147–154
Pfister C (2011) "The Monster Swallows You": disaster memory and risk culture in Western Europe, 1500–2000. Rachel Carlson Center for Environment and Society, Munich. 23 p
Renn O, Klinke A (2004) Systemic risks: a new challenge for risk management. EMBO Rep 5(1S):S41–S46
Roege PE, Collier ZA, Mancillas J, McDonagh JA, Linkov I (2014) Metrics for energy resilience. Energ Policy 72:249–256
Rutter A, Hawkins B, Parkhurst J (2013) Knowledge transfer and exchange: a look at the literature in relation to research and policy. Working Paper. **#4**. London School of Hygiene and Tropical Medicine. Accessed 0n 10 Oct 2016. http://www.lshtm.ac.uk/groups/griphealth/resources/grip_health_working_paper_3.pdf
Saadat H (2010) Power system analysis, 3rd edn. PSA Publishing LLC, Alexandria. 752 p
Saleh JH, Hastings DE, Newman DJ (2003) Flexibility in system design and implications for aerospace systems. Acta Astronaut 53(12):927–944
Schlapfer M, Kessler T, Kroger W (2012) Reliability analysis of electric power systems using an object-oriented hybrid modeling approach. arXiv preprint arXiv:1201.0552
Schneider CM, Moreira AA, Andrade JSJ, Havlin S, Herrmann HJ (2011) Mitigation of malicious attacks on networks. Proc Natl Acad Sci 108(10):3838–3841

Simon HA (1960) The new science of management decision. Harper & Row, New York
Simon HA, Dantzig GB, Hogarth R, Plott CR, Raiffa H, Schelling TC, Shepsle KA, Thaler R, Tversky A, Winter S (1986) Report of the research briefing panel on decision making and problem solving. In: E. Committee on Science, and Public Policy (ed) Research briefings 1986. National Academy Press, Washington, DC, pp 17–36
Slovic P, Fischhoff B, Lichtenstein S, Roe F (1981) Perceived risk: psychological factors and social implications [and discussion]. In: Proceedings of the Royal Society of London A: mathematical, physical and engineering sciences. The Royal Society, London, pp 17–34
Sornette D (2009) Dragon-kings, black swans and the prediction of crises. Swiss Finance Institute Research Paper, (09–36)
Sornette D, Ouillon G (2012) Dragon-kings: mechanisms, statistical methods and empirical evidence. Eur Phys J Spec Top 205(1):1–26
Sutcliffe KM, Vogus TJ, Dane E (2016) Mindfulness in organizations: a cross-level review. Ann Rev Organ Psychol Organ Behav 3:55–81
Swyngedouw E (2006) Circulations and metabolisms:(hybrid) natures and (cyborg) cities. Sci Cult 15(2):105–121
Taleb NN (2007) The black swan the impact of the highly improbable. Random House, New York. 366 p
UN (2014) World urbanization prospects: 2014 revision, highlights. United Nations, New York. Accessed on 18 Dec 2016. https://esa.un.org/unpd/wup/publications/files/wup2014-highlights. Pdf, 27 p
Walden DD, Roedler G, Forsberg K, Hamelin R, Shortell T (2015) INCOSE systems engineering handbook: a guide for system life cycle processes and activities, 4th edn. Wiley, Hoboken. 290 p
Watts DJ, Strogatz SH (1998) Collective dynamics of 'small-world'networks. Nature 393(6684):440–442
Weick KE (1993) The collapse of sensemaking in organizations: the Mann Gulch disaster. Adm Sci Q:628–652
Weick KE, Roberts KH (1993) Collective mind in organizations: heedful interrelating on flight decks. Adm Sci Q:357–381
Weick KE, Sutcliffe K, Obstfeld D (1999) Organizing for high reliability: processes of collective mindfulness. Res Organ Behav 21:81–123
Weick KE, Sutcliffe KM, Obstfeld D (2005) Organizing and the process of sensemaking. Organ Sci 16(4):409–421
Wildberger AM (1997) Complex adaptive systems: concepts and power industry applications. IEEE Control Syst 17(6):77–88
Williams T, Shepherd D (2016) Building resilience or providing sustenance: different paths of emergent ventures in the aftermath of the Haiti Earthquake. Acad Manag J 2015(0682)
Woods DD (2006) Resilience engineering: redefining the culture of safety and risk management. Hum Fact Ergonom Soc Bull 49(12):1–3
Yu G, Qi X (2004) Disruption management: framework, models and applications. World Scientific, Singapore/River Edge. 294 p

Chapter 6
Engineering Resilience in Critical Infrastructures

Giovanni Sansavini

Abstract This short paper is a result of several intense days of discussion following a talk at the NATO Advanced Research Workshop "Resilience-Based Approaches to Critical Infrastructure Safeguarding", which took place in Ponta Delgada, Portugal on June 26–29, 2016. This piece elaborates on the definition of resilience, the need for resilience in critical interdependent infrastructures, and on resilience quantification. An integrated metric for measuring resilience is discussed and strategies to build resilience in critical infrastructures are reviewed. These strategies are presented in the context of the research work carried out at the Reliability and Risk Engineering Laboratory, ETH Zurich, namely, (a) planning ahead for resilience during the design phase, (b) carrying out effective system restoration, (c) quickly recovering from the minimum performance level, (d) self-healing, adaptation and control, and (e) exploiting interdependencies among infrastructures. This paper embraces a fundamentally engineering perspective and is by no means an exhaustive examination of the matter. It particularly focusing on technical aspects and does not touch upon the rich work on community resilience and the possible measures to strengthen the response of communities to disasters.

Keywords Critical infrastructures • Cascading failures • Self-healing • Adaptation • Recovery • Restoration • Robust optimization • Resilience

6.1 Defining Resilience

Resilience has emerged in the last decade as a concept for better understanding the performance of infrastructures, especially their behavior during and after the occurrence of disturbances, e.g. natural hazards or technical failures. Recently, resilience has grown as a proactive approach to enhance the ability of infrastructures to

---

G. Sansavini (✉)
Reliability and Risk Engineering Laboratory, Institute of Energy Technology,
Department of Mechanical and Process Engineering, ETH Zurich, Zürich, Switzerland
e-mail: sansavig@ethz.ch

© Springer Science+Business Media B.V. 2017

I. Linkov, J.M. Palma-Oliveira (eds.), *Resilience and Risk*, NATO Science for
Peace and Security Series C: Environmental Security,
DOI 10.1007/978-94-024-1123-2_6

prevent damage before disturbance events, mitigate losses during the events and improve the recovery capability after the events, beyond the concept of pure prevention and hardening (Woods 2015).

The concept of resilience is still evolving and has been developing in various fields (Hosseini et al. 2016). The first definition described resilience as "a measure of the persistence of systems and of their ability to absorb change and disturbance and still maintain the same relationships between populations or state variables" (Holling 1973). Several domain-specific resilience definitions have been proposed (Ouyang et al. 2012) (Adger 2000) (Pant et al. 2014) (Francis and Bekera 2014). Further developments of this concept should include endogenous and exogenous events and recovery efforts. To include these factors, resilience is broadly defined as "the ability of a system to resist the effects of disruptive forces and to reduce performance deviations" (Nan et al. 2016). Recently, the AR^6A a resilience framework has been proposed based on eight generic system functions, i.e. attentiveness, robustness, resistance, re-stabilization, rebuilding, reconfiguration, remembering, and adaptiveness (Heinimann 2016).

Assessing and engineering systems resilience is emerging as a fundamental concern in risk research (Woods and Hollnagel 2006) (Haimes 2009) (McCarthy et al. 2007) (McDaniels et al. 2008) (Panteli and Mancarella 2015). Resilience adds a dynamical and proactive perspective into risk governance by focusing (i) on the evolution of system performance during undesired system conditions, and (ii) on surprises ("known unknowns" or "unknown unknowns"), i.e. disruptive events and operating regimes which were not considered likely design conditions. Resilience encompasses the concept of vulnerability (Johansson and Hassel 2010) (Kröger and Zio 2011) as a strategy to strengthen the system response and foster graceful degradation against a wide spectrum of known and unknown hazards. Moreover, it expands vulnerability in the direction of system reaction/adaptation and capability of recovering an adequate level of performance following the performance transient.

6.2 Need for Resilience in Critical Interdependent Infrastructures

Resilience calls for developing a strategy rather than performing an assessment. If on the one hand it is important to quantify and measure resilience in the context of risk management, it is even more important that the quantification effort enables the engineering of resilience into critical infrastructures (Guikema et al. 2015). Especially for emerging, not-well-understood hazards and "surprises" (Paté-Cornell 2012), resilience integrates very smoothly into risk management, and expediently focuses the perspective on the ex-ante system design process. Following this perspective, risk thinking becomes increasingly embedded into the system design process.

The application of resilience-building strategies look particularly promising for critical interdependent infrastructures, also called systems-of-systems, because of its dynamical perspective in which the system responds to the shock event, adapting and self-healing, and eventually recovers to a suitable level of performance. Such perspective well suits the characteristics of these complex systems, i.e. (i) the coexistence of multiple time scales, from infrastructure evolution to real-time contingencies; (ii) multiple levels of interdependencies and lack of fixed boundaries, i.e. they are made of multiple layers (management, information & control, energy, physical infrastructure); (iii) broad spectrum of hazards and threats; (iv) different types of physical flows, i.e. mass, information, power, vehicles; (v) presence of organizational and human factors, which play a major role in severe accidents, highlighting the importance of assessing the performance of the social system together with the technical systems.

As a key system of interdependent infrastructures, the energy infrastructure is well suited to resilience engineering. In the context of security of supply and security of the operations, resilience encompasses the concept of flexibility in energy systems. Flexibility providers, i.e. hydro and gas-fired plants, cross-border exchanges, storage technologies, demand management, decentralized generation, ensure enough coping capacity, redundancy and diversity during supply shortages, uncertain fluctuating operating conditions and unforeseen contingencies (Roege et al. 2014) (Skea et al. 2011).

6.3 Quantifying Resilience

Resilience is defined and measured based on system performance. The selection of the appropriate MOP depends on the specific service provided by the system under analysis.

The resilience definition can be further interpreted as the ability of the system to withstand a change or a disruptive event by reducing the initial negative impacts (absorptive capability), by adapting itself to them (adaptive capability) and by recovering from them (restorative capability). Enhancing any of these features will enhance system resilience. It is important to understand and quantify these capabilities that contribute to the characterization of system resilience (Fiksel 2003). Absorptive capability refers to an endogenous ability of the system to reduce the negative impacts caused by disruptive events and minimize consequences. In order to quantify this capability, robustness can be used, which is defined as the strength of the system to resist disruption. This capability can be enhanced by improving system redundancy, which provides an alternative way for the system to operate. Adaptive capability refers to an endogenous ability of the system to adapt to disruptive events through self-organization in order to minimize consequences. Emergency systems can be used to enhance adaptive capability. Restorative capability refers to an ability of the system to be repaired. The effects of adaptive and restorative

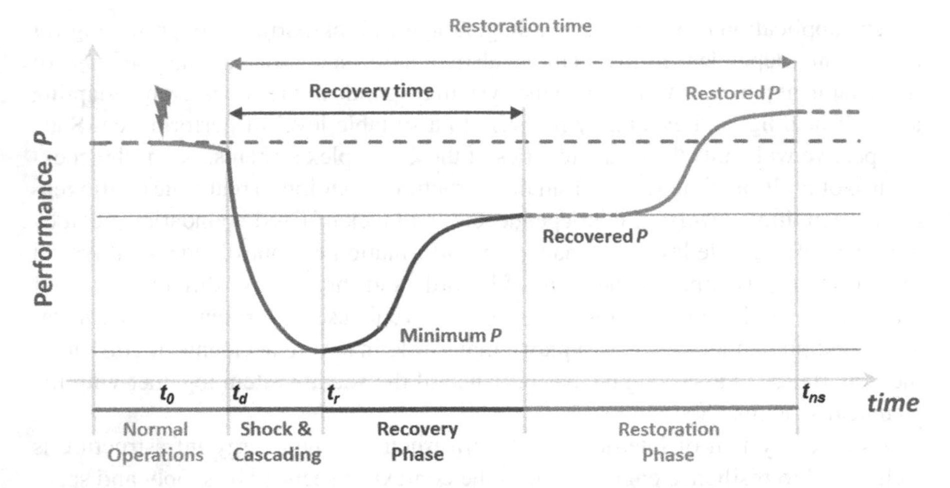

Fig. 6.1 The "resilience curve", i.e. the performance transient after disturbance, and its phases

capacities overlap and therefore, their combined effects on the system performance are quantified by rapidity and performance loss.

Resilience can be quantified though computational experiments in which disruptions are triggered, the system performance is analyzed (Fig. 6.1), and integrated resilience metrics are computed (Nan and Sansavini 2017). By repeating this process, different system design solutions can be ranked with respect to resilience. By the same token, resilience against various disruptions can be assessed, and resilience-improving strategies compared. The selection of the appropriate *MOP* depends on the specific service provided by the infrastructure under analysis. For generality, it is assumed that the value of *MOP* is normalized between 0 and 1 where 0 is total loss of operation and 1 is the target *MOP* value in the steady phase. As illustrated in Fig. 6.1, the first phase is the original steady phase ($t < t_d$), in which the system performance assumes its target value. The second phase is the disruptive phase ($t_d \leq t < t_r$), in which the system performance starts dropping until reaching the lowest level at time t_r. During this phase, the system absorptive capability can be assessed by identifying appropriate measures. *Robustness* (or *Resistance*) (*R*) is a measure to assess this capability, which quantifies the minimum *MOP* value between t_d and t_{ns}:

$$R = \min\{MOP\ (t)\}\ (for\ t_d \leq t \leq t_{ns}) \tag{6.1}$$

where t_d represents the time when the system is in disruptive phase and t_{ns} represents the time when the system reaches the new steady phase. This measure is able to identify the maximum impact of disruptive events; however, it is not sufficient to reflect the ability of the system to absorb the impact. Two additional complementary measures are further developed: *Rapidity* ($RAPI_{DP}$) and *Performance Loss* (PL_{DP})

during disruptive phase. The measure *Rapidity* can be approximated by the average slope of the *MOP* function.

$$RAPI_{DP} = \frac{MOP(t_d) - MOP(t_r)}{t_r - t_d} \quad (6.2)$$

To improve the accuracy of the estimation of *RAPI*, ramp detection is applied to quantify the average slope (Ferreira et al. 2013). According to (Kamath 2010) and (Zheng and Kusiak 2009), a ramp is assumed to occur if the difference between the measured value at the initial and final points of a time interval Δt is greater than a predefined ramping threshold value:

$$\frac{MOP(t + \Delta t) - MOP(t)}{\Delta t} > \Delta X_{ramp} \quad (6.3)$$

where ΔX_{ramp} represents the predefined ramping threshold value. The system rapidity can then be calculated as the average of slope of each ramp:

$$\text{RAPI} = \frac{\left| \sum_{i=1}^{K} \frac{MOP(t_i) - MOP(t_i - \Delta t)}{\Delta t} \right|}{K} \quad (6.4)$$

where K represents the number of detected ramps and $MOP(t_i)$ represents the *MOP* value at the *i*-th detected ramp. Compared to (2), this method better captures the speed of change in the system performance during disruption and recovery phases. According to this approach, the rapidity during disruptive phase can be calculated as:

$$\text{RAPI}_{DP} = \frac{\left| \sum_{i=1}^{K_{DP}} \frac{MOP(t_i) - MOP(t_i - \Delta t)}{\Delta t} \right|}{K_{Dp}} \left(for\ t_d \leq t_i < t_r \right) \quad (6.5)$$

where K_{DP} represents number of detected ramps during the disruptive phase.

The *performance loss* in the disruptive phase (PL_{DP}), using the system illustrated in Fig. 6.1 as an example, can be quantified as the area of the region bounded by the *MOP* curve before and after occurrence of the negative effects caused by the disruptive events, i.e. between t_d and t_r which is referred to as the system impact area:

$$PL_{DP} = \int_{t_d}^{t_r} \left(MOP(t_o) - MOP(t) \right) dt \quad (6.6)$$

Where t_o represents the time when the system is in original steady phase. A new measure, i.e. the *time averaged performance loss* (*TAPL*), is introduced. Compared to *PL*, *TAPL* considers the time of appearance of negative effects due to disruptive events up to full system recovery and provides a time-independent indication of both adaptive and restorative capabilities as responses to the disruptive events. $TAPL_{DP}$ in the disruptive phase ($t_d \leq t < t_r$) can be calculated as:

$$\mathrm{TAPL}_{DP} = \frac{\int_{t_d}^{t_r} \left(MOP(\mathrm{t_o}) - MOP(t) \right) dt}{t_r - t_d} \tag{6.7}$$

The third phase is the recovery phase ($t_r \leq t < t_{ns}$), in which the system performance increases until the new steady level. During this phase, the system adaptive and restorative capability can be assessed by developing appropriate measures: *rapidity* ($RAPI_{RP}$), *performance loss* (PL_{RP}) and *time average performance loss* ($TAPL_{RP}$).

$$\mathrm{RAPI}_{RP} = \frac{\left| \sum_{i=1}^{K_{RP}} \frac{MOP(t_i) - MOP(t - \Delta t)}{\Delta t} \right|}{K_{Rp}} \quad \left(for\ t_r \leq \mathrm{t_i} < t_{ns} \right) \tag{6.8}$$

where K_{RP} represents the number of detected ramps in recovery phase.

$$PL_{RP} = \int_{t_r}^{t_{ns}} \left(MOP(t_0) - MOP(t) \right) dt \tag{6.9}$$

$$\mathrm{TAPL}_{RP} = \frac{\int_{t_r}^{t_{ns}} \left(MOP(t_0) - MOP(t) \right) dt}{t_{ns} - t_r} \tag{6.10}$$

The fourth phase is the new steady state ($t \geq t_{ns}$), in which system performance reaches and maintains a new steady level. As seen in Fig. 6.1, the newly attained steady level may equal the previous steady level or reach a lower level. It should be noted that the new steady state may even be at a higher level than the original one. In order to take this situation into consideration, a simple quantitative measure *Recovery Ability* (*RA*) is developed:

$$RA = \left| \frac{MOP(t_{ns}) - MOP(t_r)}{MOP(t_o) - MOP(t_r)} \right| \tag{6.11}$$

Different system phases and related system capabilities are summarized in Table 6.1.

Table 6.1 Summary of different resilience phases

Phases	Time scope	Transition point	Capabilities (features)	Measurements
Original steady phase	$t < t_D$		Susceptibility	Susceptibility
Disruptive phase	$t_D \leq t < t_R$	TRNS(D)	Absorptive capability	R
				$RAPI_{DP}$
				PL_{DP}
Recovery phase	$t_R \leq t < t_{NS}$	TRNS(R)	Adaptive capability	$RAPI_{RP}$
			Restorative capability	PL_{RP}
New steady phase	$t \geq t_{NS}$	TRNS(NS)	Recovery capability	RA

R Robustness, $RAPI_{DP}$ Rapidity in disruptive phase, PL_{DP} Performance Loss in disruptive phase, $RAPI_{RP}$ Rapidity in recovery phase, PL_{RP} Performance Loss in recovery phase, *RA* Recovery ability

6.3.1 The Integrated Resilience Metric

Although the measurements introduced and discussed in Sect. 6.6.3 are useful in assessing the system behavior during and after disruptive events, an integrated metric with the ability of combining these capabilities is needed in order to assess system resilience with an overall perspective and to allow comparisons among different systems and system configurations. The basic idea of incorporating various resilience capacities into one metric has been proposed by Francis and Bekera to develop resilience factor (Francis and Bekera 2014). The idea is also supported by (McDaniels et al. 2008). Therefore, the resilience metric (*GR*) is proposed, which integrates the previous measures:

$$\begin{aligned} GR &= f\left(R, RAPI_{DP}, RAPI_{RP}, TAPL, RA\right) \\ &= R \times \left(\frac{RAPI_{RP}}{RAPI_{DP}}\right) \times \left(TAPL\right)^{-1} \times RA \end{aligned} \tag{6.12}$$

where $TAPL_{DP}$ and $TAPL_{RP}$ have been combined into one *TAPL* measure ($\frac{\int_{t_d}^{t_{ns}} \left[MOP(t_0) - MOP(t)\right] dt}{t_{ns} - t_d}$) in order to incorporate effects of total performance loss during disruptive and recovery phases.

The functional form of the proposed resilience metric assumes that robustness *R*, recovery speed $RAPI_{RP}$ and recovery ability *RA* have a positive effect on resilience, and, conversely, performance loss *TAPL* and loss speed $RAPI_{DP}$ have a negative effect. To compile the integrated metric (12), no weighting factor is assigned to the measures so that no bias is introduced, i.e. they contribute equally to resilience. *GR* is consistent with the definition proposed in Sect. 6.6.1:

1. If the system is more capable of resisting a disruptive event or force (large *R*, small $RAPI_{DP}$), the system is more resilient (large *GR*).

2. If the system is more capable of reducing the magnitude and duration of deviation of its performance level between original state and new steady state (small *TAPL*, large $RAPI_{RP}$), the system is more resilient (large *GR*).
3. Additionally GR also incorporate the possibility of improvement of the system performance after the occurrence of the disruptive event. If the new performance level is larger than the original (large *RA*), the system is more resilient (large *GR*).

GR is a non-negative metric and its value equals zero in the following relevant cases:

1. System performance level drops to zero after the disturbance ($R = 0$).
2. After the disturbance, system performance immediately drops to its lowest level ($RAPI_{DP} \rightarrow \infty$, i.e. no absorptive capability).
3. System performance never increases past the lower level, *R*, which is the new steady phase ($RAPI_{DP} = 0$, i.e. no adaptive and restorative capability).

GR is dimensionless and is most useful in a comparative manner, i.e. to compare the resilience of various systems to the same disruptive event, or to compare resilience of same system under different disruptive events. This approach of measuring system resilience is neither model nor domain specific. For instance, historical data can also be used for the resilience analysis. It only requires the time series that represents system output during whole time period. In this respect, the selection of the *MOP* is very important.

During the last decade, researchers have proposed different methods for quantifying resilience. In 2003, the first conceptual framework was proposed to measure the seismic resilience of a community (Bruneau et al. 2003), by introducing the concept of Resilience Loss, later also referred to as "resilience triangle".

In recent years, the importance of improving the resilience of interdependent critical infrastructures has been recognized, and research works have developed. Historically, knowledge-based approaches have been applied to improve the understanding of infrastructures resilience (McDaniels et al. 2008). Lately, model-based approaches have been developed to overcome the limitations of data-driven approaches, such as System Dynamics (Bueno 2012), Complex Network Theory (Gao et al. 2016), and hybrid approaches (Nan et al. 2016).

Approaches to quantify system resilience should be able to

- capture the complex behavior of interdependent infrastructures
- cover all phases of the transient performance following the disruption, and to include all resilience capabilities
- clarify the overlap with other concepts such as robustness, vulnerability and fragility.

Resilience quantification of interdependent infrastructures is still at an early stage. Currently, a comprehensive method aiming at improving our understanding of the system resilience and at analyzing the resilience by performing in-depth experiments is still missing.

6.4 Building Resilience in Critical Infrastructures

In the context of critical infrastructures, resilience can be developed by focusing on the different phases of the transient performance following a disturbance (also called resilience curve), and devising strategies and improvements which strengthen the system response.

Focusing mainly on the technical aspects, these strategies can be summarized as:

6.4.1 Planning Ahead During the Design Phase

Robust or stochastic optimization against uncertain future scenarios, i.e. attacks or uncertain future demand in the energy infrastructure, can be used in the system planning or expansion process; uncertain scenarios provide the basis to design resilient systems.

In (Fang and Sansavini 2017), the combination of capacity expansion and switch installation in electric systems that ensures optimum performance under nominal operations and attacks is studied. The planner-attacker-defender model is adopted to develop decisions that minimize investment and operating costs, and functionality loss after attacks. As such, the model bridges long-term system planning for transmission expansion and short-term switching operations in reaction to attacks. The mixed-integer optimization is solved by decomposition via two-layer cutting plane algorithm. Numerical results shows that small investments in transmission line switching enhance resilience by responding to disruptions via system reconfiguration (Fig. 6.2). Sensitivity analyses show that transmission planning under the assumption of small-scale attacks provides the most robust strategy, i.e. the minimum-regret planning, if many constraints and limited investment budget affect

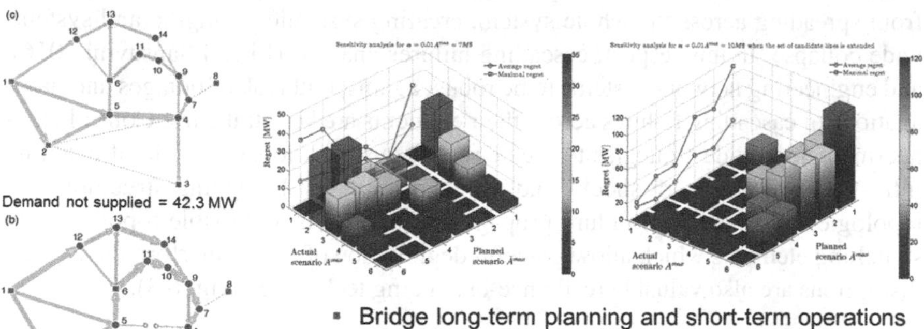
(c)
Demand not supplied = 42.3 MW
(b)
Demand not supplied = 26.7 MW
Regret [MW]
Actual scenario
Planned scenario
▪ Bridge long-term planning and short-term operations
▪ Switching devices reduce demand not supplied
▪ Constrained expansion -> Plan for small attacks
▪ Flexible expansion -> Plan for large attacks

Fig. 6.2 Integrated planning of system expansion and recovery devices against uncertain attack scenarios (Fang and Sansavini 2017)

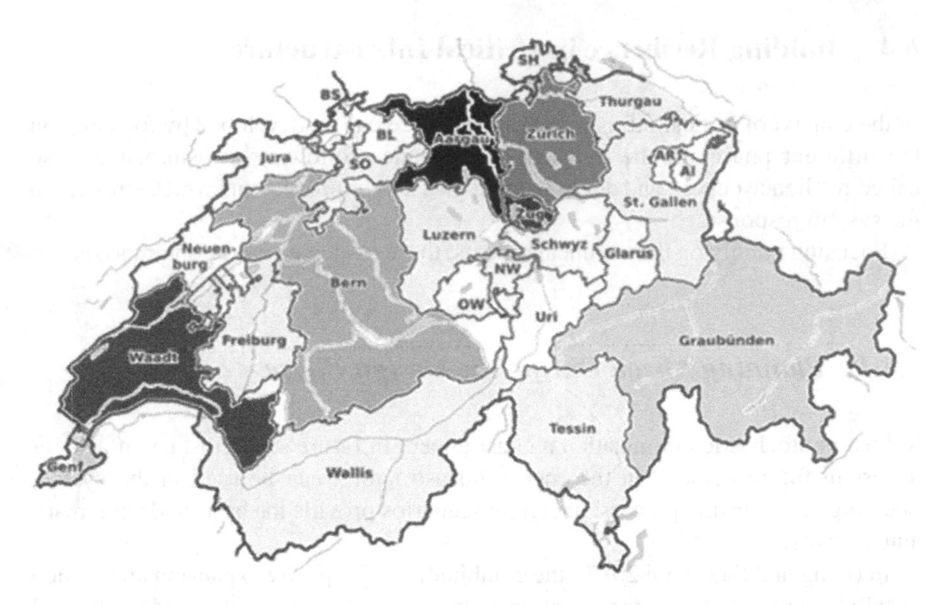

Fig. 6.3 The heat map of the cumulative economic losses at each canton of Switzerland due to propagation of cascading failures in the electric power system (Li et al. 2015)

the planning. On the other hand, the assumption of large-scale attacks provides the most robust strategy if the planning process involves large flexibility and budget.

6.4.2 Self-Healing, Adaptation and Control

Graceful degradation: the system cannot be designed with respect to every uncertain scenario, therefore a resilient design should consider how to prevent the disturbance from spreading across the whole system, creating systemic contagion and system-wide collapse. In this respect, cascading failures analysis (Li and Sansavini 2016), and engineering network systems to be robust against outbreak of outages and propagations of cascading failures across their elements are key strategies. Control engineering can provide strategies to create robust feedback loops capable of enabling infrastructures to absorb shocks and avoid instabilities. Designing structures and topologies which prevent failure propagation, and devising flexible topologies by switching elements which allow graceful degradation of system performances after disruptions are also valuable resilience-enhancing techniques (Fig. 6.3).

6.4.3 *Recovering Quickly from the Minimum Performance Level*

Robust or stochastic optimization of the recovery and restoration process in the face of uncertainties in the repair process or in the disruption scenarios.

System restoration and its contribution to the resilience of infrastructure networks following disruptions have attracted attention in recent years. Optimization approaches usually guide the identification and scheduling of restoration strategies for rapid system functionality reestablishment under limited resources. Most of the related studies rely on deterministic assumptions such as complete information of resource usage and deterministic duration of the repair tasks. However, restoration activities are subject to considerable uncertainty stemming from subjective expert judgment and imprecise forecasts that may render the scheduling solution obtained by a deterministic approach suboptimal or even infeasible under some uncertainty realizations. Restoration planning and scheduling under uncertainty can be investigated within a credibility-based fuzzy mixed integer programming (PMIP) approach, in which the imprecise parameters are modelled by fuzzy numbers (Fang and Sansavini 2016). To solve the proposed fuzzy optimization problem, an interactive fuzzy solution technique is utilized which provides the decision maker (DM) the flexibility to consider two significant factors when making decision: the degree of achievement of his/her aspiration level and the risk of violation of the constraints. A computational experiment involving the Swiss high voltage electric power transmission network demonstrates the significance and applicability of the developed approach for DM to determine efficient restoration actions aimed to enhancing system resilience. Generally, the system restoration curves, i.e. the system performance levels evolving over time, show that decreasing the degree of feasibility of the constraints results a faster system restoration (Fig. 6.4).

6.4.4 *Effective System Restoration*

Through the combination of restoration strategies, e.g. repairing the failed elements and building new elements, the infrastructure can achieve a higher performance with respect to the pre-disruption conditions, and display the anti-fragility property (Taleb 2012; Aven 2015).

A system is anti-fragile if its performance improves as the result of exposure to stressors, shocks or disruptions. This behavior is typical of complex systems and it is not usually exhibited by engineered technical systems. In fact, technical systems can display anti-fragility when new investments are allocated, e.g. after disasters. In post-disaster restoration planning of infrastructure networks, the possibility of combining the construction of new components and the repair of failed ones can lead to anti-fragile behavior. The strategic goal is to determine the optimal target system structure so that the performance of the target system is maximized under the

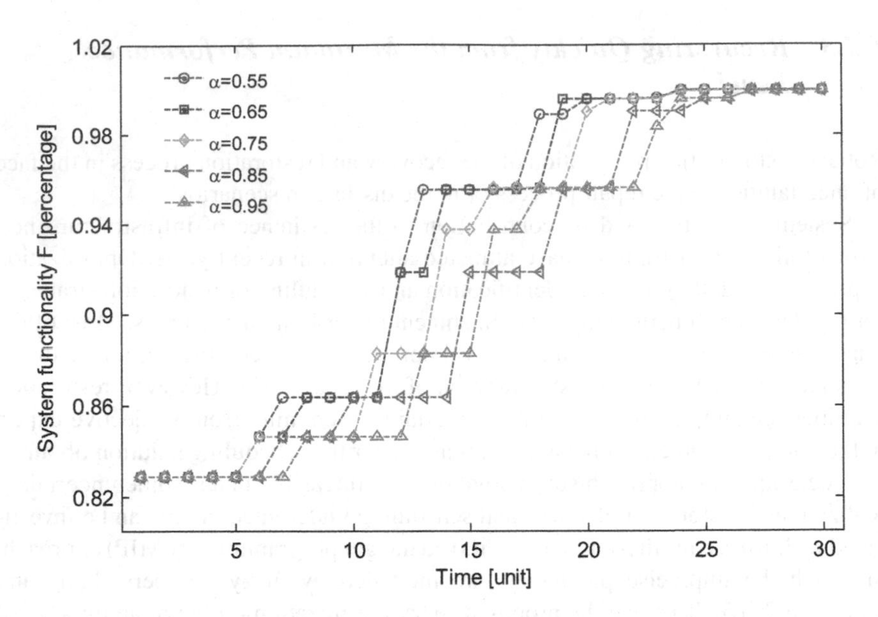

Fig. 6.4 System restoration curves for five different feasibility levels of the solution vector, i.e. the set of decision variables concerning the restoration process

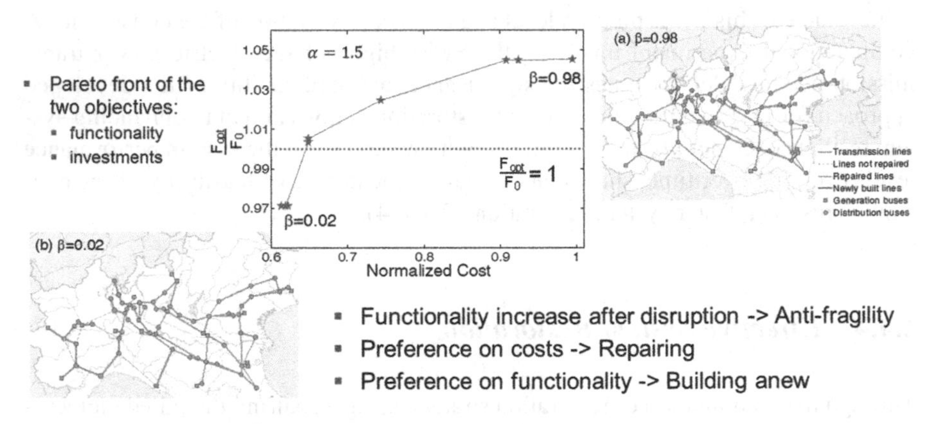

Fig. 6.5 Optimum restoration by repairing and building anew

constraints of investment cost and network connectivity. The problem can be formulated as a mixed-integer binary linear programming (MILP). The preliminary results (Fig. 6.5) show that the restored network can achieve an improved functionality as compared to the original network if new components are constructed and some failed components are not repaired, even when the former is much more expensive than the latter. Therefore, different investment allocations schemes define whether an infrastructure network is fragile or anti-fragile. In particular, the tested infrastructure exhibits anti-fragile behavior even for restoration investments that amount at 62% the cost of complete repair. Furthermore, antifragility provides an opportunity for the system to meet future service demand increase, and a perspective under which disruptions can be seen as chances for system performance improvements.

6.4.5 *Exploiting Interdependencies Among Infrastructures*

Interdependencies and couplings in systems operations can foster the propagations of failure across coupled system; on the other hands, interdependencies might also provide additional flexibility in disrupted conditions and additional resources that can facilitate achieving stable conditions of the coupled system.

Cyber interdependencies are pervasive in critical infrastructures (CIs) and particularly in electric power networks, which are dependent on information and communications technology (ICT), e.g., supervisory control and data acquisition (SCADA) systems, to transmit measurements signals to control centers and to dispatch control signals to actuators. The requirements towards ICT to transmit these signals with tolerable communication delays for timely balancing of power demand and supply have increased due to changes in the operating conditions of electric power networks. On the one hand, its operating conditions are pushed closer to its stability limits due to amplified loading conditions. On the other hand, the increasing share of distributed inverter-connected renewable energy, e.g., wind and PV, on the distribution level has led to a decrease in the inertia and an increase in the volatility in the power grid further reducing its stability margins. Under these conditions, severe consequences, e.g. system-wide blackouts, can be caused by disturbances in the electric grid. In the face of these challenges, ICT is expected to turn the current electric grid into a "smart grid" in order to assure reliable, efficient and secure operations of the electric grid. An application that benefits from the ICT in power systems is grid splitting, also referred to as controlled islanding, which relies on real-time system-wide measurements to enable the detection and recovery from failures in real time, i.e., by applying system topology changes. Grid splitting is a special protection scheme that separates a power system into synchronized islands in a controlled manner in response to an impending instability, i.e., generator rotation desynchronization triggered by a component fault. By appropriately disconnecting transmission lines, severe consequences, e.g., system-wide blackouts, are mitigated through the formation of stable islands. The successful application of grid splitting depends on the communication infrastructure to collect system-wide synchronized measurements and to relay the command to open line switches. Grid splitting may be ineffective if communication is degraded and its outcome may also depend on the system loading conditions. The effects of degraded communication and load variability on grid splitting are investigated in (Tian and Sansavini 2016). To this aim, a communication delay model is coupled with a transient electrical model and applied to the IEEE 39-Bus and the IEEE 118-Bus Test System. Case studies show that the loss of generator synchronism following a fault is mitigated by timely splitting the network into islands. On the other hand, the results show that communication delays and increased network flows can degrade the performance of grid splitting. The developed framework enables the identification of the requirements of the dedicated communication infrastructure for a successful grid-splitting procedure.

Acknowledgments The author acknowledges the CTI – Commission for Technology and Innovation (CH), and the SCCER-FURIES – Swiss Competence Center for Energy Research – Future Swiss Electrical Infrastructure, for their financial and technical support to the research activity presented in this paper.

Further Suggested Readings

Adger W (2000) Social and ecological resilience: are they related? Prog Hum Geogr 24:347–364

Aven T (2015) The concept of antifragility and its implications for the practice of risk analysis. Risk Anal 35(3):476–483

Bruneau M, Chang SE, Eguchi RT, Lee GC, O'Rourke TD, Reinhorn AM (2003) A Framework to Quantitatively Assess and Enhance the Seismic Resilience of Communities. Earthquake Spectra 19:733–752

Bueno NP (2012) Assessing the resilience of small socio-ecological systems based on the dominant polarity of their feedback structure. Syst Dyn Rev 28:351–360

Fang Y, Sansavini G (2016) Optimum post-disruption restoration for enhanced infrastructure network resilience: a possibilistic programming approach. In: Proceedings of ESREL 2016 European Safety and Reliability Association annual conference. Glasgow, UK, pp 25–29

Fang YP, Sansavini G (2017) Optimizing power system investments and resilience against attacks. Reliab Eng Syst Saf 159:161–173

Ferreira C, Gama J, Miranda V, Botterud A (2013) Probabilistic ramp detection and forecasting for wind power prediction. In: Billinton R, Karki R, Verma AK (eds) Reliability and risk evaluation of wind integrated power systems. Springer India, India, pp 29–44

Fiksel J (2003) Designing Resilient, Sustainable Systems. Environ Sci TechnolEnviron Sci Technol 37:5330–5339

Francis R, Bekera B (2014) A metric and frameworks for resilience analysis of engineered and infrastructure systems. Reliab Eng Syst Saf 121:90–103

Gao J, Barzel B, Barabási A-L (2016) Universal resilience patterns in complex networks. Nature 18:307–312

Guikema S, McLay L, Lambert JH (2015) Infrastructure Systems, Risk Analysis, and Resilience—Research Gaps and Opportunities. Risk Anal 35(4):560–561

Haimes YY (2009) On the Definition of Resilience in Systems. Risk Anal 29:498–501

Heinimann HR (2016) A generic framework for resilience assessment. In IRGC (2016). Resource guide on resilience. Lausanne: EPFL International Risk Governance Center. v29-07-2016

Holling CS (1973) Resilience and stability of ecological systems. Annu Rev Ecol Syst 4:1–23

Hosseini S, Barker K, Ramirez-Marquez JE (2016) A review of definitions and measures of system resilience. Reliab Eng Syst Saf 145:47–61

Johansson J, Hassel H (2010) An approach for modelling interdependent infrastructures in the context of vulnerability analysis. Reliab Eng Syst Saf 95:1335–1344

Kamath C (2010) Understanding wind ramp events through analysis of historical data. Transmission and distribution conference and exposition, IEEE PES 2010;1–6.

Kröger W, Zio E (2011) Vulnerable Systems. Springer-Verlag, London

Li B, Sansavini G (2016) Effective multi-objective selection of inter-subnetwork power shifts to mitigate cascading failures. Electr Power Syst Res 134:114–125

Li B, Barker K, Sansavini G (2015) Measuring the societal and multi-industry impact of cascading failures in power systems, Proceedings of ESREL 2015 European Safety and Reliability Association Annual Conference, 7–10 September 2015. Zurich, Switzerland, pp 4445–4453

McCarthy J, Pommering C, Perelman L, Scalingi P, Garbin D, Shortle, J (2007) Critical Thinking: Moving from Infrastructure Protection to Infrastructure Resilience. George Mason University, Fairfax, pp 97–109

McDaniels T, Chang S, Cole D, Mikawoz J, Longstaff H (2008) Fostering resilience to extreme events within infrastructure systems: Characterizing decision contexts for mitigation and adaptation. Glob Environ Chang 18:310–318

Nan C, Sansavini G (2017) A quantitative method for assessing resilience of interdependent infrastructures. Reliab Eng Syst Saf 157:35–53

Nan C, Sansavini G, Kröger W (2016) Building an integrated metric for quantifying the resilience of interdependent infrastructure systems. In: Panayiotou C, Ellinas G, Kyriakides E, Polycarpou M (eds) Critical information infrastructures security. CRITIS 2014. Lecture Notes in Computer Science, vol 8985. Springer, Cham

Ouyang M, Wang Z (2015) Resilience assessment of interdependent infrastructure systems With a focus on joint restoration modeling and analysis. Reliab Eng Syst Saf 141:74–82

Ouyang M, Dueñas-Osorio L, Min X (2012) A three-stage resilience analysis framework for urban infrastructure systems. Struct Saf 36–37:23–31

Pant R, Barker K, Zobel CW (2014) Static and dynamic metrics of economic resilience for interdependent infrastructure and industry sectors. Reliab Eng Syst Saf 125:92–102

Panteli M, Mancarella P (2015) Modeling and evaluating the resilience of critical electrical power infrastructure to extreme weather events. IEEE Syst J 99:1–10

Paté-Cornell E (2012) On "Black Swans" and "Perfect Storms": risk analysis and management when statistics are not enough. Risk Anal 32:1823–1833

Roege PE, Collier ZA, Mancillas J, McDonagh JA, Linkov I (2014) Metrics for energy resilience. Energy Policy 72:249–256

Skea J, Chaudry M, Ekins P, Ramachandran K, Shakoor A, Wang X (2011) A resilient energy system. In: Paul Ekins JS (ed) Energy 2050: making the transition to a secure low-carbon energy system. Taylor & Francis, Hoboken, pp 145–186

Taleb NN (2012) Anti-Fragile. Penguin, London

Tian D, Sansavini G (2016) Impact of degraded communication on interdependent power systems: the application of grid splitting. Electron Spec Issue Smart Grid Cyber Secur 5(3):49

Woods DD (2015) Four concepts for resilience and the implications for the future of resilience engineering. Reliab Eng Syst Saf 141:5–9

Woods DD, Hollnagel E (2006) Resilience Engineering: Concepts and Precepts. CRC Press, Boca Raton

Zheng H, Kusiak A (2009) Prediction of wind farm power ramp rates: a data-mining approach. J Sol Energy Eng 131:31011

Chapter 7
Seaport Climate Vulnerability Assessment at the Multi-port Scale: A Review of Approaches

R. Duncan McIntosh and Austin Becker

Abstract In the face of climate change impacts projected over the coming century, seaport decision makers have the responsibility to manage risks for a diverse array of stakeholders and enhance seaport resilience against climate and weather impacts. At the single port scale, decision makers such as port managers may consider the uninterrupted functioning of their port the number one priority. But, at the multi-port (regional or national) scale, policy-makers will need to prioritize competing port climate-adaptation needs in order to maximize the efficiency of limited physical and financial resources and maximize the resilience of the marine transportation system as a whole. This chapter provides an overview of a variety of approaches that set out to quantify various aspects of seaport vulnerability. It begins with discussion of the importance of a "multi-port" approach to complement the single case study approach more commonly applied to port assessments. It then addresses the components of climate vulnerability assessments and provides examples of a variety of approaches. Finally, it concludes with recommendations for next steps.

Keywords Seaport • Port • Shipping • Climate assessment • CIAV • CCVA • Resilience • Climate change vulnerability assessment • Comparative assessment • Multi-port assessment • Indicator-based assessment • Regional scale assessment

7.1 Seaports Are Critical, Constrained, and Exposed

Seaports represent an example of spatially defined, large scale, coast-dependent infrastructure with high exposure to projected impacts of global climate change (Becker et al. 2013, Hanson et al. 2010, Melillo et al. 2014). Seaports play a critical role in the global economy, as more than 90% of global trade is carried by sea (IMO

R. Duncan McIntosh (✉) • A. Becker
Department of Marine Affairs, University of Rhode Island, Kingston, RI 02881, USA
e-mail: mcintosh@uri.edu

© Springer Science+Business Media B.V. 2017

I. Linkov, J.M. Palma-Oliveira (eds.), *Resilience and Risk*, NATO Science for Peace and Security Series C: Environmental Security,
DOI 10.1007/978-94-024-1123-2_7

2012). A disruption to port activities can interrupt supply chains, which can have far reaching consequences (Becker et al. 2011b, 2013, IPCC 2014a). Seaports are inextricably linked with land based sectors of transport and trade, and serve both the public and private good. Globally, climate change adaptation is still in the planning stages for most seaports (Becker et al. 2011a), yet the inevitable imperative for climate resiliency looms, as atmospheric concentrations of greenhouse gasses, the primary driver of climate change (IPCC 2013), continue to accumulate (WMO 2015). Indeed, most aspects of climate change will persist for centuries even if anthropogenic emissions of carbon dioxide were halted today (IPCC 2013).

Functionally restricted to the water's edge, seaports will face impacts driven by changes in water-related parameters like mean sea level, wave height, salinity and acidity, tidal regime, and sedimentation rates, yet they can also be affected directly by changes in temperature, precipitation, wind, and storm frequency and intensity (Koppe et al. 2012). The third U.S. National Climate Assessment (NCA) (Melillo et al. 2014) of the U.S. Global Change Research Program notes that impacts from sea level rise (SLR), storm surge, extreme weather events, higher temperatures and heat waves, precipitation changes, and other climatic conditions are already affecting the reliability and capacity of the U.S. transportation system. While the U.S. NCA predicts that climate change impacts will increase the total costs to the nation's transportation systems, the report also finds that adaptive actions can reduce these impacts.

In the face of these challenges, port decision makers have the responsibility to manage risks for a diverse array of stakeholders and enhance seaport resilience against climate and weather impacts. At the single port scale, decision makers such as port managers may consider the uninterrupted functioning of their port the number one priority. But, at the multi-port (regional or national) scale, policy-makers will need to prioritize competing port climate-adaptation needs in order to maximize the efficiency of limited physical and financial resources and maximize the resilience of the marine transportation system as a whole.

Recognizing a regional or national set of ports and waterways as part of an interconnected marine transportation system (MTS),[1] how should responsible decision makers prioritize the climate adaptation decisions for systems that involve multiple ports? This chapter provides an overview of a variety of approaches that set out to quantify various aspects of seaport vulnerability. It begins with discussion of the importance of a "multi-port" approach to complement the single case study approach more commonly applied to port assessments. It then addresses the components of climate vulnerability assessments and provides examples of a variety of approaches. Finally, it concludes with recommendations for next steps.

[1]The marine transportation system, or MTS, consists of waterways, ports, and inter-modal landside connections that allow the various modes of transportation to move people and goods to, from, and on the water. (MARAD 2016)

7.2 Impediments to Multi-port Adaptation

A 2016 study which quantified the resources, time and cost of engineering minimum-criteria "hard" protections against sea level rise for 223 of the world's most economically important seaports, suggested insufficient global capacity for constructing the proposed protective structures within 50–60 years (Becker et al. 2016). As individual actors and governments consider climate-adaptation solutions for seaports, a global uncoordinated response involving heavy civil infrastructure construction may be unsustainable simply from a resource availability perspective (Becker et al. 2011b, 2016, Peduzzi 2014). Given limited financial and construction resources for the implementation of engineered protection across many ports, some form of prioritization for national and regional-scale climate-adaptation will likely be necessary. Port authorities have expressed that although general concern for climate change exists, awareness of sea level rise is limited and the planning for adaptation is lacking (Becker et al. 2010).

The implementation of strategic adaptation on a multi-port scale is further challenged by complex and dynamic regional differences defined by varying landscapes and geographies that are far from uniform in their climate change vulnerability. Some ports, for example, may by surrounded by lowlands at risk to inundation from sea level rise. For these ports, the ground transportation systems may by more threatened than the port itself (e.g., Port of Gulfport, MS). In other areas, storm surge might be amplified by the geomorphology of an estuarine system (e.g., Providence, RI).

At the single port scale, the design of engineering protection during a port's expansion can benefit by estimating how long the infrastructure will last and withstand future impacts (Becker et al. 2015). However, justifying major investments is challenged by the uncertainty involved in projecting the extent to which ports will be impacted this century (Becker and Caldwell 2015). In the following section, we first discuss the concept of measuring vulnerability, risk, and resilience, then describe assessment methods employed by individual ports. Following, we discuss the need for multi-port assessment approaches and work in this area to date.

7.3 Assessing Climate Vulnerabilities to Facilitate Far-Sighted Resilience Planning

Vulnerability and resilience are two theoretical concepts, sometimes defined complementarily, other times described as opposite sides of the same coin, (Gallopín 2006, Linkov et al. 2014) that have gained increasing attention in the climate change adaptation and hazard risk reduction literature. As theoretical notions, resilience and vulnerability are not directly measurable, and some researchers (Barnett et al. 2008, Eriksen and Kelly 2007, Hinkel 2011, Klein 2009, Gudmundsson 2003) have criticized attempts to assess them as unscientific and or biased. However, policymakers are increasingly calling for the development of methods measure relative risk, vulnerability, and resilience (Cutter et al. 2010, Hinkel 2011, Rosati 2015).

The International Association of Ports and Harbors (IAPH) defines seaport vulnerability using three components: *exposure*, *sensitivity*, and *adaptation capacity* (Koppe et al. 2012). Measuring a port's *exposure* requires downscaled regional climate projections which may not yet be available for some port regions, and where they are available, necessarily contain uncertainty. A port on the west coast of the U.S., for example, may be considered less exposed to hurricanes than a port on the east coast. Port exposure, then, may be analyzed using a multiple scenario approach, with a range of values for the applicable climate variables. Measuring port *sensitivity* and *adaptation capacity* generally requires site-specific analyses. By analyzing the impacts of projected changes in regional or even local climate variables and evaluating a port's design criteria in light of those impacts, the sensitivity to those changes can be determined for a port and its assets. Recently constructed infrastructure designed for higher intensity storms, for example, may be considered as less sensitive to a given storm event than infrastructure that is in a state of disrepair already. An assessment of a port's adaptive capacity, taking into account the port system's planning parameters, management flexibility and existing stresses, can reveal obstacles to a port system's ability to cope with climate change impacts. A port with robust planning procedures and more wealth, for example, may be considered to have a higher adaptive capacity than a port that has lesser planning and resources. In 2011, Becker and collaborators made a first attempt at quantifying international seaport adaptive capacity by developing a scoring system based on port authority responses regarding climate adaptation policies currently in place (Becker et al. 2011a).

Because exposure and vulnerability are dynamic (IPCC 2012), varying across spatial and temporal scales, and individual ports are differentially vulnerable and exposed, assessments should be iterative with multiple feedbacks, shaped by people and knowledge (IPCC 2014a), and take a "bottom up" approach by including input from a diverse stakeholder cluster to ensure that the variables representing exposure, sensitivity and adaptive capacity are empirically identified by and important to the stakeholders, rather than presupposed by the researchers or available data (Smit and Wandel 2006).

A concept related to vulnerability, *risk* is a measure of the potential for consequences where something of value is at stake and where the outcome is uncertain (IPCC 2014b). Risk can be quantitatively modeled as $Risk = p(L)$, where L is potential loss and p the probability of occurrence, however, both can be speculative and difficult to measure in the climate-risk context. Risk, in the context of climate change, is often defined similarly to vulnerability (Preston 2012, IPCC 2014a), but with the added component of *probability*, thus making *vulnerability* a component of *risk*.

Resilience, another closely related term with a more positive connotation than *vulnerability*, is defined by the IPCC as "the capacity of social, economic and environmental systems to cope with a hazardous event or trend or disturbance, responding or reorganizing in ways that maintain their essential function, identity and structure, while also maintaining the capacity for adaptation, learning and transformation" (IPCC 2014b). The National Academy of Science (The National Academies 2012) and the President of the United States (Obama 2013) define critical infrastructure resilience as, "the ability to prepare, resist, recover, and more successfully adapt to the impacts of adverse events." With *resilience* defined in terms of ability, and *vulnerability* defined in terms of susceptibility, it is tempting to consider them polar opposites (Gallopín 2006), however, resilience can also be considered a broader concept than vulnerability. Most working definitions of resilience involve a process that begins before a hazardous impact, but also includes temporal periods during and after the impact. Resilience, like vulnerability, can also encompass coping with adverse effects from a multitude of hazards in addition to climate change. By increasing our understanding of the distribution of seaport climate vulnerabilities, the overall *resilience* of the MTS may be enhanced.

7.4 CIAV Decision-Support for the Seaport Sector

As port decision makers face climate impact, adaptation, and vulnerability (CIAV)[2] decisions, climate change vulnerability assessments (CCVA), including risk and resilience assessments support those decisions by addressing the "adapt to what" question (IPCC 2014a). The process enables a dialog among stakeholders and practitioners on planning and implementation of adaptation measures to enhance resilience. The Intergovernmental Panel on Climate Change (IPCC) describes vulnerability and risk assessment as "the first step for risk reduction, prevention, and transfer, as well as climate adaptation in the context of extremes." [p. 90] (IPCC 2012) The U.S. NCA considers vulnerability and risk assessment an "especially important" [p. 137] (Melillo et al. 2014) area in consideration of adaptation strategies in the transportation sector. Such assessments can be made at the single-port scale or at the multi-port scale, with each approach having benefits for different types of decision makers.

7.4.1 Single-Port Scale

Among climate change vulnerability, resilience, and risk assessment methods applied to seaports, most efforts to date have been limited in scope to exposure-only assessments (Hanson et al. 2010, Nicholls et al. 2008), or limited in scale to a single port; either as case studies (Koppe et al. 2012, Cox et al. 2013, USDOT 2014, Messner et al. 2013, Chhetri et al. 2014) or as self-assessment tools (NOAA OCM 2015, Semppier et al. 2010, Morris and Sempier 2016).

[2] Climate impact, adaptation, and vulnerability (CIAV) decisions are choices, the results of which are expected to affect or be affected by the interactions of the changing climate with ecological, economic, and social systems.

While single-port scale CCVA inform CIAV decisions within the domain of one port (e.g., Which specific adaptations are recommended for my port?), a CCVA approach that objectively compares the relative vulnerabilities of multiple ports in a region could support CIAV decisions at the multi-port scale (e.g., Which ports in a region are the *most* vulnerable and urgently in need of adaptation?). The hitherto focus on individual port scale assessments presents a challenge for how to describe the *distribution* of climate-vulnerabilities across multiple ports.

7.4.2 Multi-port Scale

At the multi-port scale, an evaluation of *relative* climate-vulnerabilities or the *distribution* of those vulnerabilities among a regional or national set of ports requires standard measures (e.g. indicators, or metrics). Directly immeasurable, concepts such as resilience and vulnerability are instead made operational by mapping them to functions of observable variables called indicators. *Indicators* are measurable, observable quantities that serve as proxies for an aspect of a system that cannot itself be directly, adequately measured (Gallopin 1997, Hinkel 2011). Indicator-based assessment methods, therefore, are generally applied to assess or 'measure' features of a system that are described by theoretical concepts. The indicator-based assessment process of operationalizing immeasurable aspects of a system consists (Hinkel 2011) of two or sometimes three steps: (1) defining the response to be indicated, (2) selecting the indicators, (3) aggregating the indicators (this step is sometimes omitted but necessary to yield a numerical 'score' or create a comparative index). In this section, we investigate examples of indicator-based assessment methods applied to multi-port systems to aid the further development of such methods for the port sector, which can yield benefits including the ability to not only 'measure' immeasurable concepts like vulnerability and resilience, but also to index and compare them across entities.

7.4.3 Factors Considered in Port Resilience Evaluation

The US National Oceanic and Atmospheric Administration (NOAA) Office for Coastal Management (OCM) along with the federal interagency Committee on the Marine Transportation System (CMTS) produced a port resilience planning web-based tool (NOAA OCM 2015), tailored towards communities undergoing a port expansion or reconstruction, that assembles resilience indicators and their datasets. This web-based prototype tool came online in 2015 with the stated purpose of assisting transportation planners, port infrastructure planners, community planners, and hazard planners to explore resilience considerations and options in developing marine transportation projects. Inspired by and aligned with broader resilience objectives called for in the CMTS's strategic action plan (USCMTS 2011), this tool shows port communities what to look for in resilient freight transportation infrastructure. While the Port Tomorrow resilience planning tool assembles seaport resilience indicators, provides links to their potential data sources, and organizes them with categories and subcategories into a framework for assessing port resilience, the tool stops short of providing a method to normalize and aggregate the indicators into a comparative score.

7.4.4 Assessing Global Port City Exposure

One of the few CCVA to comparatively assess multiple ports, the 2010 work by Hanson, Nichols, et al. (Hanson et al. 2010) made some of the first progress towards comparative seaport CCVA by focusing on assessing the *exposure* component of seaport climate-vulnerability. Part of a larger project on Cities and Climate Change that was sponsored by the Organization for Economic Cooperation and Development (OECD), this global screening study assesses the exposure[3] of all 136 international port cities with over one million inhabitants in 2005 to coastal flooding. The analysis considers exposure to present-day extreme water levels (represented by a 100-year flood) as well as six future scenarios (represented by the decade 2070–2080) that include projected changes in sea level and population. The researchers base the methods used on determining the numbers of people who would be exposed to the water level of interest and then using that number to estimate the potential assets exposed within each city. The researchers then rank the cities by number of people exposed and by 2005 U.S. dollar value of assets exposed. These two response variables, i.e. people and dollar value of assets, are semi-empirical quantities rather than theoretical concepts, and as such, the methods involved in this study are not directly analogous to other indicator-based assessment methods. Instead of using indicators to serve as proxies for some immeasurable concept, this study uses indicators to approximate concrete numbers that, due to scale, are difficult to measure.

This study took the form of a Geographic Information System (GIS) elevation-based analysis, after authors (McGranahan et al. 2007). The researchers used 100-year historic flood levels taken from the Dynamic Interactive Vulnerability Assessment (DIVA) database as current extreme water levels to be modeled in GIS for each city. For the future water levels, the researchers calculate two different scenarios, one that considers only natural factors (i.e. a calculated "storm enhancement factor," historic subsidence rates, and sea level rise (SLR)), and another that adds to those factors one representing anthropogenic subsidence.

For current population, the study takes the ambient population distribution estimates from LandScan 2002 (Bright and Coleman 2003) for each city, delimited by city extents from post code data. The postcodes are taken from geocoding data and, for cities in the USA, from Metropolitan Statistical Areas (MSAs) from Census

[3] Exposure refers to the nature and extent to which a system is subjected to a source of harm, taking no account of any defenses or other adaptation.

data. The authors resample the 1 km LandScan 2002 data to 30 m for all cities in the US and UK and resampled to 100 m for the remaining cities. To determine population distribution by elevation, the authors use 90 m resolution topographic data from the Shuttle Radar Topography Mission (SRTM) for most cities, 30 m SRTM data for the US, and a 10 m Digital Elevation Model (DEM) provided by Infoterra for the UK. The authors then overlay each LandScan population distribution over the relevant Digital Terrain Model (DTM), yielding for each city a map of geographical cells with defined population and elevation. From these maps, the authors are able to isolate total population within 1 m vertical bands of elevation. To represent future population, the authors start with baseline population projections from the OECD ENV-Linkages model, which itself is based on United Nations (UN) medium variant projections to 2050. To bring these projections to 2070, the authors extrapolate them forward using national growth rates and UN projected rates of urbanization.

To indicate the dollar value of assets, the researchers use what they describe as a "widely used assumption in the insurance industry" (Hanson et al. 2010, 92) (p 92) that as urban areas are typically more affluent than rural areas, each person in a city has assets that are 5 times the national Gross Domestic Product (GDP). This simple calculation is based on the national per capita GDP Purchasing Power Parity (PPP) values for 2005 from the International Monetary Fund (IMF) database. To indicate future GDP, the study uses OECD baseline projections to 2075. To find the total value of assets exposed then, the researchers take the number of people exposed (from the GIS maps described above) and multiply that number by a country's GDP PPP times five.

Using the indicators described above, and organized in Table 7.1, this study is ultimately able to produce rankings of port cities exposed to coastal flooding by number of people and by dollar value of assets exposed to extreme water levels in 2005 and for projected extreme water levels in 2075.

7.4.5 Assessing Regional Port **Interdependency** *Vulnerabilities*

Another example of CCVA that extends beyond the single-port scale is the 2013 work by Hsieh et al. that examines the vulnerability of port failures from an interdependency perspective using four commercial ports in Taiwan as empirical case studies (Hsieh et al. 2013). The method determines factors vulnerable to disasters by reviewing literature and conducting an in-depth interview process with port experts; in this way, the researchers developed 14 'vulnerable factors' that can be considered similar to our described indicators (Berle et al. 2011).

To develop the 14 indicators, the authors held a series of discussions in open participatory meetings. Eleven experts participated, including port officials, government officials, planners, and scholars. The discussions classified the indicators into four categories: accessibility, capability, operational efficiency, and industrial cluster/energy supply, as shown in Table 7.2. The process to determine weights for the indicators followed the analytic network process (ANP) of Jharkharia and

Table 7.1 Indicators, categories and data sources used in (Hanson et al. 2010)

Indicator categories	Indicator sub-categories	Indicators	Data source
Elevation	Elevation	elevation	Shuttle Radar Topography Mission (SRTM)
Population	Population	population distribution	Landscan 2002
Future Population	Future Population	Projected Population in 2075	OECD ENV-Linkages Model
	Projected Urbanization Rate (assumed uniform within country)	2005–2030 trends extrapolated to 2075, assuming that urbanization rates will saturate at 90%, except where it is already larger than this value (e.g. in special cases like Hong Kong)	UN projected urbanization rates 2005–2030 (are then extrapolated to 2075)
Current Water Level	Current Water Level	100 yr. storm surge	DIVA
Future Water Level	SLR	assumes a homogenous global rise of 0.5 m by 2070	assumed from lit.
	Anthropogenic Subsidence	assumes uniform 0.5 m decline in land level (from 2005 to 2070) in port cities located in deltas	assumed
	Natural Subsidence	Annual Rate of subsidence extrapolated to 2070	used annual sub. Rate from DIVA
	Storm Enhancement Factor	10% increase in extreme water level assumed for cities exposed to TC, 10% increase assumed for cities bet. 45 and 70 deg. latitude which are assumed exposed to Extra-TC	CHRR (Columbia), historical TC tracks, Munich Re
Value of Assets	Value of Assets	national per capita GDP PPP (assuming each person in a city has assets 5 x annual GDP per capita)	www.imf.org
Future Value of Assets	Future Value of Assets	Projected GDP per capita	OECD Baseline projections to 2075

Shankar (2007) (Jharkharia and Shankar 2007), and involved constructing an impact matrix via fuzzy cognitive maps (FCMs) developed and evaluated during these participatory meetings. The impact matrix represents magnitudes of causal effects of each indicator compared to every other indicator.

To standardize the indicators, the experts completed a questionnaire that had them identify threshold values for each indicator. The researchers provided a scale from 0–4, with 0 indicating that the port can operate normally, and 1–4 indicating that the port would experience slight, average, significant effects, and complete port

Table 7.2 Indicators, categories, and data sources used in (Hsieh et al. 2013)

Indicator categories	Indicators	Data source
Accessibility	Ground access system (%)	GIS maps
	Travel time (minute)	GIS maps
	Shipping route density (lines)	port annual statistics overviews
Capability	Gantry crane capacity (TEUs)	Ministry of Transportation and Communications
	Facility supportability (%)	port annual statistics overviews
	Wharf productivity (10^3 tons/meter)	Ministry of Transportation and Communications
Operational Efficiency	EDI connectivity (%)	Ministry of Transportation and Communications
	Turnaround time (hr)	Ministry of Transportation and Communications
	Labor productivity (tons/person)	port annual statistics overviews
	Berth occupancy rate (%)	port annual statistics overviews
Industrial Cluster/Energy Supply	Investment growth (10^9 NTD[a])	national industry, commerce, and service census
	FTZ business volume (10^9 NTD)	national industry, commerce, and service census
	Electric power supply (%)	GIS maps
	Gas supply (%)	GIS maps

[a]*NTD* New Taiwan Dollars

failure, respectively. Using this scale, the experts identified a threshold value (i.*E. minimum* or maximum value, depending upon whether the indicator indicates vulnerability or competitiveness) for each indicator that would lead the port to each of the five results described in the scale 0–4. The researchers used the Delphi method during three rounds, allowing the experts to revise their earlier answers in light of the replies of other members of their panel and achieve consensus. Table 7.3 shows the standardized indicators (called "Vulnerable factors"), their units, and their threshold values.

The data for the indicators come from published statistics, literature, and GIS maps. Table 7.2 shows the specific data source for each of the 14 indicators. To score a port's vulnerability, the researchers standardize a port's raw indicator data using Table 7.3, then sum the standardized indicators multiplied by their weights to produce a total vulnerability score. The results for the 4 Taiwanese case study ports are shown in Table 7.4.

In addition to the vulnerability assessment method herein described, Hsieh et al. also conducted an interdependency analysis to determine how strongly each indicator affects and is affected by the other indicators of the port system. This analysis uses groups of experts who fill out a matrix form during an iterative Delphi-style process, similar to that used during the first stages of this project.

Table 7.3 Standardized indicators showing threshold values from (Hsieh et al. 2013)

Vulnerable factors		Rating				
		0	1	2	3	4
(1)	Ground access system (%)	>90	90–80	80–50	50–20	<20
(2)	Travel time (minute)	<90	90–120	120–150	150–180	>180
(3)	Shipping route density (lines)	<15	15–100	100–200	200–300	>300
(4)	Gantry crane capacity (TEUs*)	>90	90–70	70–50	50–35	<35
(5)	Facility supportability (%)	>80	80–70	70–50	50–40	<40
(6)	Wharf productivity (103 tons/ meter)	>5	5–4	4–2	2–1.5	<1.5
(7)	EDI connectivity (%)	>90	90–80	80–50	50–20	<20
(8)	Turnaround time (hr)	<24	24–36	36–48	48–72	>72
(9)	Labor productivity (tons/person)	>350	350–250	250–150	150–100	<100
(10)	Berth occupancy rate (%)	>70	70–50	50–30	30–10	<10
(11)	Investment growth (109 NTD**)	>10	10–8	8–4	4–2	<2
(12)	FTZ business volume (109 NTD**)	>10	10–8	8–4	4–2	<2
(13)	Electric power supply (%)	>90	90–80	80–50	50–20	<20
(14)	Gas supply (%)	>50	50–30	30–20	20–5	<5

Table 7.4 Results of port vulnerability analysis from (Hsieh et al. 2013)

	Score of vulnerable factors	Keelung	Taipei	Taichung	Kaohsiung
(1)	Ground access system	3	2	2	1
(2)	Travel time	2	1	0	0
(3)	Shipping route density	1	1	1	4
(4)	Gantry crane capacity	3	3	1	0
(5)	Facility supportability	0	3	2	0
(6)	Wharf productivity	0	2	0	1
(7)	EDI connectivity	1	1	1	1
(8)	Turnaround time	0	1	1	1
(9)	Labor productivity	0	0	1	1
(10)	Berth occupancy rate	3	1	2	2
(11)	Investment growth	4	2	0	0
(12)	FTZ business volume	4	1	0	0
(13)	Electric power supply	2	0	1	0
(14)	Gas supply	1	0	0	0
Port	vulnerability	1.6131	1.8063	0.8746	0.7724

7.4.6 *Assessing Relative Port* Performance

At the multi-port, MTS scale, CCVA have been sparse. Indicator-based multi-port assessments to date have tended to focus on port *performance* rather than *vulnerabilities* or *resilience.* Here, we investigate some of the methods used to assess relative port *performance* in an effort to inform new CCVA methods at the multi-port scale.

7.4.7 *Port Performance Indicators: Selection and Measurement (PPRISM)*

Carried out from 2010 to 2011 by the European Seaports Organization (ESPO) and co-funded by the European Commission, the Port Performance Indicators: Selection and Measurement (PPRISM) program was designed to take a first step towards establishing a culture of performance measurement in European ports by identifying a set of relevant and feasible performance indicators for the European port system. The aim of this project was to develop indicators that allow the port industry to measure, assess, and communicate the impact of the European port system on society, the environment, and the economy. Although PPRISM does document equations (ESPO 2011) used to aggregate numbers used for individual indicators, this study does not aggregate the indicators themselves into a total performance score. The future plans for PPRISM include the establishment of a Port Sector Performance Dashboard (as part of a European Port Observatory website) that will not publish or compare interport performance, but illustrate the performance of the whole European system of ports.

The indicator selection process began with input from five European Universities: University of the Aegean, Institute of Transport and Maritime Management Antwerp, Eindhoven University of Technology, Vrije Universiteit Brussel, and Cardiff University. These academic partners came up with 159 port performance indicators based on a literature review and industry current practices and organized them under the following five categories: Market Trends, Logistic Chain and Operations, Environmental Indicators, Socio-economic Indicators, and Governance Indicators. The academic partners excluded indicators that did not fulfill one of the following criteria (ESPO 2010):

P: Policy relevance – Monitor the key outcomes of strategies, policies and legislation and measure progress towards policy goals. Provides information to a level appropriate for policy decision – making.

I: Informative – Supplies relevant information with respect to the port's activities.

M: Measurable – Is readily available or made available at a response cost/benefit ratio. Updated at regular intervals in accordance with reliable procedures.

R: Representative – Gives clear information and is simple to interpret. Accessible, publicly appealing and therefore likely to meet acceptance.
F: Feasible / Practical – Requires limited numbers of parameters to be established. Uses existing data and information wherever possible. Simple to monitor.

Following the academic pre-selection process, the 159 indicators were assessed by ESPO members. ESPO organized four special workshop sessions for this purpose in combination with its Technical Committee meetings. During these workshops, ESPO members screened the pre-selected indicators and discussed their proposed definitions and calculation methods with the academic partners. ESPO members considered and provided qualitative feedback on the data availability and relevance of the proposed indictors. Additionally, ESPO members provided quantitative feedback on the feasibility and acceptability of each indicator by using a five point Linkert-style scale during two rounds, following the Delphi methodology.[4] The first round of this Delphi-style assessment process by ESPO members narrowed the 159 indicators down to 39. The second round with the modified indicators resulted in additional indicators, adjustments to indicator definitions and calculation formulas, renamed indicators, and produced a new list of 45 indicators.

The four rounds involved in the Delphi-style indicator assessment included only internal stakeholders (i.e. representatives of the European port authorities). In an effort to increase the validity and reliability of the work, the scope was then expanded to include external stakeholders, targeting a "representative external stakeholder response panel" (ESPO 2011) to include port users, government, and academics. This external stakeholder assessment made use of an online survey that was freely available without restrictions on who was invited to participate. The survey was advertised in social media, specialized presses, and personal networks and remained open for 4 months (February–May 2011). This external stakeholder assessment helped to narrow the list of indicators further to 42.

The results of the internal and external stakeholder assessments guided the final choice of 14 indicators that were then tested in a pilot phase. The 42 indicators were narrowed down to 14 (Table 7.5) through a process of weighing stakeholders' acceptance vs the feasibility of implementation of each indicator.

The pilot consisted of an EU-wide project to test the feasibility of the 14 selected indicators, with the intent to uncover the real-world availability of data and the willingness of port authorities to provide data. For the pilot study, the PPRISM group sent an electronic form to all port authorities associated with ESPO accompanied by an explanatory letter from ESPO Secretary General Patrick Verhoeven and received back a total of 58 forms fully or partially filled out. The pilot revealed problems with data availability, unclear data requests, and port participation. Given that data provision is voluntary, and hence, the number of ports submitting could fluctuate from year to year, the pilot study recommended that, at least for the initial stages of any

[4] The Delphi method is an iterative, multistage response process designed to generate expert consensus.

Table 7.5 Findings and conclusions for each piloted indicator (ESPO 2012)

Indicators	Pilot result	Next steps
1. Maritime traffic	Relevant and feasible	Building a "time series" mainly focusing on the relative changes in traffic volumes over time. A three dimensional approach is suggested with respect to the dimension of 'time', (quarterly figures), of 'commodity'[total throughput plus 5 categories of cargoes plus passenger traffic (7 in total)] and 'geography'(all European ports)
2. Call size	Relevant and feasible	Building a "time series" mainly focusing on the relative changes in traffic volumes over time. A three dimensional approach is suggested with respect to the dimension of 'time', (yearly figures), of 'commodity'[total throughput plus 5 categories of cargoes plus passenger traffic (7 in total)] and 'geography'(all European ports)
3. Employment (Direct)	Relevant and feasible	Getting data from a larger number of ports
4. Added value (Direct)	Relevant and feasible	Getting data from a larger number of ports
5. Carbon footprint	Relevant and feasible	Make Tool available to port associations and authorities. Provide training support where requested.
6. Total water consumption	Relevant and feasible	
7. Amount of waste	Relevant and feasible	
8. Environmental management	Relevant and feasible	Promote using Tool (see above) and populate from SDM and PERS responses.
9. Maritime connectivity	Relevant and feasible	Building a 'time series' to monitor maritime connectivity over time.
10. Intermodal connectivity	Relevant and feasible	Getting data from a larger number of European ports.
11. Quality of customs procedures	Relevant and feasible	This indicator can be substituted by something more detailed in the medium run. Until then, this is the best available indicator.
12. Integration of port cluster	Relevant and feasible	Revision of criteria used. The need to reduce the number of criteria is already anticipated. More detailed info for each criteria will be asked. Efforts to standardize and collect quantitative data as well. In the long run the objective is to measure the efficiency of a PAs initiatives related to the respective indicators.
13. Reporting Corporate and Social Responsibility	Relevant and feasible	
14. Autonomous management	Relevant and feasible	

port performance dashboard, reporting data in the form of trends rather than single values is the best approach. The results of the pilot study are shown in Table 7.5.

Upon conclusion of the pilot study, the PPRISM project group published its executive report (ESPO 2012), with the recommendation that the development of European Ports Observatory be phased in over time, starting small. Though a printed version of a Dashboard was presented at the 2012 ESPO Conference in Sopot, Poland, the current status of the dashboard remains unclear.

7.4.8 *USCMTS Marine Transportation System Performance Measures*

The World Association for Waterborne Transport Infrastructure (PIANC) report, *Performance Measures for Inland Waterways Transport (*PIANC Inland Navigation Commission 2010*)*, identifies three general purposes for performance measures (operational, informational, referential) and nine thematic areas (infrastructure, ports, environment, fleet and vehicles, cargo and passengers, information and communication, economic development, safety, and security). Building upon the PIANC report and aiming to create an initial picture of the overall state of the U.S. MTS using authoritative data, the United States Committee on the Marine Transportation System (USCMTS) Research and Development Integrated Action Team in 2015 published a compilation of MTS *performance measures* (USCMTS 2015) developed from publicly available data sources. Serving as standard metrics, such indicators allow standardized comparison of the components of port *performance* including; Economic Benefits to the Nation, Capacity and Reliability, Safety and Security, Environmental Stewardship, and Resilience.

While the USCMTS study suggests two "Resilience Performance Measures," (i.e., *Age of Federally Owned and Operated Navigation Locks*, and *Physical Condition Rating of Critical Coastal Navigation Infrastructure owned by USACE*[5]), these measures do not consider private, state, or locally owned container terminals or port facilities, and the authors conclude that more work is needed to capture the concept of port or MTS resilience using standard metrics. Table 7.6 compares the indicator selection and aggregation methods of the aforementioned indicator-based seaport assessments.

7.5 Discussion

To date, there are relatively few examples of multi-port assessments. The approaches discussed in this chapter, and summarized in Table 7.6, tend to lean heavily on expert judgement in the selection and evaluation for indicators of climate vulnerability or focus exclusively on the "exposure" aspect of vulnerability.

Worth note is the use of indicators to develop a score or rating of climate vulnerability (or resilience). Such assessment may be welcome or rejected, depending on the goals and objectives of the audience. For example, a high "vulnerability" score

[5] United States Army Corps of Engineers.

Table 7.6 Examples of multi-port, indicator-based assessments

Study	Response Indicated	Indicator Selection Method	Indicator Aggregation Method
PPRISM	Port performance	(i) Academic pre-selection	Not aggregated
		(ii) Delphi Method with internal stakeholders	
		(iii) Delphi Method with external stakeholders	
USCMTS Performance Measures	Port performance	Internal review: An ideal MTS performance measure would be collected locally, using the same method across all areas of responsibility, so that state, regional, and national summaries could be easily compiled for comparison.	Not aggregated
Nichols and Hanson et al.	Coastal flood exposure measured in number of people and dollar value of assets	Response variables are semi-empirical quantities rather than theoretical concepts.	Does not involve selecting and aggregating indicators; rather it involves a more straightforward calculation of the responses.
Hsieh et al.	Port interdependency vulnerability	(i) Participatory discussion process with experts	(i) Experts develop weights via analytic network process (ANP)
		(ii) Delphi method with experts	(ii) Raw indicator data is standardized, weighted, and summed to yield a vulnerability score
NOAA Port Tomorrow	Port resilience	Indicator selection is led by a guiding question for each indicator subcategory	Not aggregated

may help a port petition a funding agent to build a case for needed resilience investments. On the other hand, a high score could also leave a port at a competitive disadvantage if tenants perceive higher levels of storm risk. Thus, while aggregations, scores, and rankings may be desired by regional or national-level decision makers, creating multi-port assessment tools is not without controversy.

That said, such tools *can* help inform the decision-making process. And, as demand for climate-critical resources (both funding and materials) increases, the need to better understand relative vulnerability of coastal systems, such as ports, will also increase. Our review of the literature suggests a need for better tools that can be used to gain an objective understanding of various aspects of port vulnerability. Although expert judgement will likely be necessary to a certain extent, due to the inherent difficulty of measuring and quantifying fuzzy concepts such as "adaptive capacity," publicly available data (e.g., historical storm tracks, types of cargo handled, throughput) can also be leveraged to help decision makers gain a better sense of which areas are more vulnerable, in what ways, and how this vulnerability might be reduced.

7.6 Conclusion

Seaports are critical to global trade and national security, yet sit on the front-line for extreme coastal weather and climate impacts, and such impacts are projected to worsen globally. As port decision-makers wrestle with the myriad of climate adaptation options (including the option of making no adaptations at all), their CIAV decisions can and should be supported with data. For CIAV decision-support, the first step often involves assessing vulnerabilities. For an individual seaport, this process tends to take the shape of CCVA, either as a participatory self-assessment, or as a site-specific case study. For multiple port systems, however, we suggest an opportunity exists for further research and development of standardized, comparative CCVA methods for seaports and the marine transportation system, with the objective of supporting CIAV decisions with information products that allow decision makers to compare mechanisms and drivers of climate change across multiple ports.

Further Suggested Readings

Barnett J, Lambert S, Fry I (2008) The hazards of indicators: insights from the environmental vulnerability index. Ann Assoc Am Geogr 98(1):102–119

Becker A, Caldwell MR (2015) Stakeholder perceptions of seaport resilience strategies: a case study of Gulfport (Mississippi) and Providence (Rhode Island). Coast Manag 43(1):1–34

Becker A, Wilson A, Bannon R, McCann J, Robadue D (2010) Rhode Island ports & commercial harbors: a GIS-based inventory of current uses and infrastructure, edited by Susan Kennedy. Rhode Island Statewide Planning Program, US Department of Transportation, Federal Highway Administration

Becker A, Inoue S, Fischer M, Schwegler B (2011a) Climate change impacts on international seaports: knowledge, perceptions, and planning efforts among port administrators. Clim Chang 110(1–2):5–29. doi:10.1007/s10584-011-0043-7

Becker A, Newell D, Fischer M, Schwegler B (2011b) Will ports become forts? Climate change impacts, opportunities and challenges. Terra et Aqua 122:11–17

Becker A, Acciaro M, Asariotis R, Cabrera E, Cretegny L, Crist P, Esteban M, Mather A, Messner S, Naruse S, Ng AKY, Rahmstorf S, Savonis M, Song DW, Stenek V, Velegrakis AF (2013) A note on climate change adaptation for seaports: a challenge for global ports, a challenge for global society. Clim Chang 120(4):683–695. doi: DOI 10.1007/s10584-013-0843-z

Becker A, Toilliez J, Mitchell T (2015) Considering sea level change when designing marine civil works: recommendations for best practices. In: Esteban M, Takagi H, Shibayama T (eds) Handbook of coastal disaster mitigation for engineers and planners. Elsevier, Waltham
Becker A, Chase NTL, Fischer M, Schwegler B, Mosher K (2016) A method to estimate climate-critical construction materials applied to seaport protection. Glob Environ Chang 40:125–136
Berle Ø, Asbjørnslett BE, Rice JB (2011) Formal vulnerability assessment of a maritime transportation system. Reliab Eng Syst Saf 96(6):696–705
Bright EA, Coleman PR (2003) LandScan 2002. Oak Ridge National Laboratory, Oak Ridge
Chhetri P, Corcoran J, Gekara V, Maddox C, McEvoy D (2014) Seaport resilience to climate change: mapping vulnerability to sea-level rise. J Spat Sci 60:1–14. doi:10.1080/14498596.2014.943311
Cox RJ, Panayotou K, Cornwell RM (2013) Climate risk assessment for Avatiu Port and connected infrastructure. Water Research Lab, University of New South Wales
Cutter SL, Burton CG, Emrich CT (2010) Disaster resilience indicators for benchmarking baseline conditions. J Homeland Secur Emergen Manag 7(1). doi:10.2202/1547-7355.1732
Eriksen SH, Mick Kelly P (2007) Developing credible vulnerability indicators for climate adaptation policy assessment. Mitig Adapt Strateg Glob Chang 12(4):495–524
ESPO (2010) Work Package 1 (WP1): pre-selection of an initial set of indicators. In: PPRISM: Port PeRformance Indicators: Selection and Measurement. European Sea Ports Organization (ESPO)
ESPO (2011) Work Package 2 (WP2): stakeholders' dialogue to evaluate and select a shortlist of indicators. In: PPRISM: Port PeRformance Indicators: Selection and Measurement. European Sea Ports Organization (ESPO)
ESPO (2012) Project Executive report (PPRISM WP4 D4.2). In: PPRISM: Port PeRformance Indicators: Selection and Measurement. European Sea Ports Organization (ESPO)
Gallopin GC (1997) Indicators and their use: information for decision-making. In: Boldan B, Bilharz S (eds) Sustainability indicators. A report on the project on indicators of sustainable development. SCOPE, pp 13–27
Gallopín GC (2006) Linkages between vulnerability, resilience, and adaptive capacity. Glob Environ Chang 16(3):293–303
Gudmundsson H (2003) The policy use of environmental indicators—learning from evaluation research. J Transdiscipl Environ Stud 2(2):1–12
Hanson S, Nicholls R, Ranger N, Hallegatte S, Corfee-Morlot J, Herweijer C, Chateau J (2010) A global ranking of port cities with high exposure to climate extremes. Clim Chang 104(1):89–111. doi:10.1007/s10584-010-9977-4
Hinkel J (2011) "Indicators of vulnerability and adaptive capacity": towards a clarification of the science-policy interface. Glob Environ Chang Hum Policy Dimens 21(1):198–208. doi: DOI 10.1016/j.gloenvcha.2010.08.002
Hsieh C-H, Tai H-H, Lee Y-N (2013) Port vulnerability assessment from the perspective of critical infrastructure interdependency. Marit Policy Manag 41(6):589–606. doi:10.1080/03088839.2013.856523
IMO (2012) International shipping facts and figures – information resources on trade, safety, security, environment. Maritime Knowledge Centre, International Maritime Organization
IPCC (2012) Managing the risks of extreme events and disasters to advance climate change adaptation: Special Report of the Intergovernmental Panel on Climate Change (SREX). Cambridge University Press, Cambridge
IPCC (2013) Climate change 2013: the physical science basis: working group I contribution to the fifth assessment report of the intergovernmental panel on climate change. In: Stocker T, Qin D, Plattner G-K, Tignor M, Allen SK, Boschung J, Nauels A, Xia Y, Bex V, Midgley PM (eds) Fifth assessment report of the intergovernmental panel on climate change. Intergovernmental Panel on Climate Change, Cambridge/New York
IPCC (2014a) Climate change 2014: impacts, adaptation, and vulnerability. Part A: global and sectoral aspects. Contribution of working group II to the fifth assessment report of the intergovernmental panel on climate change. In: Fifth assessment report of the intergovernmental panel on climate change. Intergovernmental Panel on Climate Change, Cambridge/New York

IPCC (2014b) WGII AR5 glossary. In: Fifth assessment report of the intergovernmental panel on climate change. IPCC, Geneva

Jharkharia S, Shankar R (2007) Selection of logistics service provider: an analytic network process (ANP) approach. OMEGA Int J Manag Sci 35(3):274–289. doi: DOI 10.1016/j.omega.2005.06.005

Klein RJT (2009) Identifying countries that are particularly vulnerable to the adverse effects of climate change: an academic or political challenge. Carbon Climate L Rev,284

Koppe B, Schmidt M, Strotmann T (2012) IAPH-report on seaports and climate change and implementation case study for the port of Hamburg

Linkov I, Bridges T, Creutzig F, Decker J, Fox-Lent C, Kröger W, Lambert JH, Levermann A, Montreuil B, Nathwani J, Nyer R, Renn O, Scharte B, Scheffler A, Schreurs M, Thiel-Clemen T (2014) Changing the resilience paradigm. Nat Clim Chang 4(6):407–409. doi:10.1038/nclimate2227

MARAD (2016) Marine Transportation System (MTS). Maritime Administration. Accessed On 25 May 2016. https://www.marad.dot.gov/ports/marine-transportation-system-mts/

McGranahan G, Balk D, Anderson B (2007) The rising tide: assessing the risks of climate change and human settlements in low elevation coastal zones. Environ Urban 19(1):17–37

Melillo JM, Richmond TTC, Yohe GW (2014) Climate change impacts in the United States. In The third national climate assessment, US Global Change Research Program

Messner S, Moran L, Reub G, Campbell J (2013) Climate change and sea level rise impacts at ports and a consistent methodology to evaluate vulnerability and risk. ENVIRON International Corp

Morris LL, Sempier T (2016) Ports resilience index: a port management self-assessment. U.S. Department of Commerce, Gulf of Mexico Alliance

Nicholls RJ, Hanson S, Herweijer C, Patmore N, Hallegatte S, Corfee-Morlot J, Château J, Muir-Wood R (2008) Ranking port cities with high exposure and vulnerability to climate extremes: exposure Estimates. In: OECD environment working papers. Organization for Economic Cooperation and Development, Paris

NOAA OCM (2015) Port tomorrow: port resilience planning tool [Prototype]. NOAA Office for Coastal Management. Accessed on 3 Apr 2015. http://www.coast.noaa.gov/port/

Obama B (2013) Presidential policy directive 21: critical infrastructure security and resilience. Washington, DC

Peduzzi P (2014) Sand, rarer than one thinks. Environ Dev 11:208–218

PIANC Inland Navigation Commission (2010) Performance indicators for inland waterways transport: user guideline, PIANC Report No. 111. The World Association for Waterborne Transport Infrastructure (PIANC), Bruxelles

Preston BL (2012) Climate change vulnerability assessment: from conceptual frameworks to practical heuristics. CSIRO Climate Adaptation Flagship Working paper

Rosati JD (2015) PhD, PE, D.CE, Coastal & Hydraulics Laboratory, Engineer Research & Development Center, U.S. Army Corps of Engineers. Personal Communication. Accessed on 01 Oct 2015

Semppier TT, Swann DL, Emmer R, Sempier SH, Schneider M (2010) Coastal community resilience index: a community self-assessment. Mississippi-Alabama Sea Grant Consortium

Smit B, Wandel J (2006) Adaptation, adaptive capacity and vulnerability. Glob Environ Chang 16(3):282–292

The National Academies (2012) Disaster resilience: a national imperative. edited by Committee on Increasing National Resilience to Hazards and Disasters; Committee and Engineering on Science, and Public Policy. Washington, DC

USCMTS (2011) Strategic action plan for research and development in the marine transportation system. US Committee on the Marine Transportation System, Washington, DC

USCMTS (2015) Marine transportation system performance measures: executive summary. 1200 New Jersey Ave SE. Washington, DC. 20590: U.S. Committee on the Marine Transportation System, Research and Development Integrated Action Team

USDOT (2014) Impacts of climate change and variability on transportation systems and infrastructure the gulf coast study, phase 2 screening for vulnerability final report, Task 3.1. edited by ICF International. US Department of Transportation, Washington, DC
WMO (2015) The state of greenhouse gases in the atmosphere based on global observations through 2014. In: Greenhouse Gas Bulletin. World Meteorological Organization, Geneva

Chapter 8
Resilience of Critical Infrastructures: Benefits and Challenges from Emerging Practices and Programmes at Local Level

Paolo Trucco and Boris Petrenj

Abstract Since the beginning of 2010 there has been a boom of Public-Private Partnerships (PPPs) with a goal of Critical Infrastructure Protection and Resilience (CIP-R) and Emergency Management (EM) in North America and partly in Europe and Australia as well. Currently having PPPs as one of the main ways to cope with CI interdependencies through engaging all stakeholders in order to build 'full-spectrum' resilience, it is important to look up to the best practices. Previous research has set the theoretical base of PPPs and claimed their high potential for enhancing CIP-R that is vastly unexploited due to challenges in their establishment and management. It is now necessary to move forward to studying partnerships' practical side – common issues they face, ways to overcome them and concrete benefits they are able to bring. Through studying seven cases, this work compares different PPP approaches and their contribution to CIP-R. The study demonstrates how challenges are faced and solved in an innovative way and how the benefits are reached. It also shows approaches and joint activities that support information sharing and trust building as the main ingredients that hold partners together and enable progress in other aspects, from which both public and private parties may benefit. Starting from the findings and a subsequent analysis within and between the seven cases, the study proposes a framework for the development of regional CIP-R. programmes in the context of a PPP.

Keywords Critical infrastructures • Public-private partnership • Resilience • Protection • Case study

P. Trucco (✉) • B. Petrenj
School of Management, Politecnico di Milano,
Piazza Leonardo da Vinci 32, 20133 Milan, Italy
e-mail: paolo.trucco@polimi.it

© Springer Science+Business Media B.V. 2017

I. Linkov, J.M. Palma-Oliveira (eds.), *Resilience and Risk*, NATO Science for Peace and Security Series C: Environmental Security,
DOI 10.1007/978-94-024-1123-2_8

8.1 Introduction

An infrastructure is a set of basic facilities, services, and installations that are necessary for the functioning of a community (American Heritage Dictionary of the English Language 1996) or society, such as electricity, gas and oil production, transport and distribution; communication and transportation systems; water supply; public health; financial and security services, etc. Contemporary societies are increasingly dependent on availability, reliability, correctness, safety and security of many technological infrastructures, commonly referred to as Critical Infrastructure (European Commission 2005). A Critical Infrastructure (CI) is an array of assets and systems that, if disrupted, would threaten national security, economy, public health and safety, and way of life (McNally et al. 2007; Hilton 2007). Concurrently, the importance of infrastructures has skyrocketed as modern societies increasingly rely on their functioning (Ouyang 2014).

Despite all protection measures, including physical protection of the facilities, surveillance, cyber protection of information and control (SCADA) systems, screening people entering the site, etc., it is impossible to reach risk '0' level. Since the preventive effort itself is not sufficient (cannot be completely reliable or otherwise costs would be unsustainable), more effort is put in enhancing resilience, in order to cope with inevitable events. Counting both high prices of highly reliable preventive efforts and private sector reluctance to invest more in preventing very-low-probability events, despite their expected high-impact, the advantages of resilience-based approaches are reduction of expenses of protection amelioration for certain risk scenarios (which may or may not occur) and improvement of response and recovery activities that cover all hazards (Pursiainen 2009; De Bruijne and Van Eeten 2007).

CI resilience is emerging as one of the utmost critical issues of this decade. Resilience generally means the ability to recover from shock, insult, or disturbance, and the quality or state of being flexible, and it is used quite differently in different fields (Bouchon 2006). In the disaster management domain, it is generally defined as *"the capacity of a system, community or society potentially exposed to hazards to adapt, by resisting or changing in order to reach and maintain an acceptable level of functioning and structure. This is determined by the degree to which the social system is capable of organizing itself to increase this capacity for learning from past disasters for better future protection* and *to improve risk reduction measures"* (United Nations 2005). **Technical resilience** consists of improving the level of resilience of infrastructures (e.g. adding redundancy, geographical isolation, backups, etc.). In its further development, resilience moved towards the '*full spectrum resilience*' (Boone 2012) by adopting broader approach including **organizational resilience** (covering strategic, operational, and tactical levels of intra- and inter-organisational coordination and collaboration, addressed across a range of potential impacts) and **societal resilience** (including e.g. preparation of the authority, population and economical world – emergency plans, business continuity plans, evacuation plans, alternative resources).

The US Department of Homeland Security (DHS) in its National Infrastructure Protection Plan (NIPP) defined resilience as "*the ability to resist, absorb, recover from, or successfully adapt to adversity or a change in conditions*". More specifically, **infrastructure resilience** is "*the ability to reduce the magnitude and/or duration of disruptive events. The effectiveness of a resilient infrastructure or enterprise depends upon its ability to anticipate, absorb, adapt to, and/or rapidly recover from a potentially disruptive event*" (National Infrastructure Advisory Council (NIAC) 2009). The NIPP (Department of Homeland Security (DHS) 2013) aims to unify Critical Infrastructure and Key Resource (CIKR) protection efforts across the US. It outlines how government and private sector participants in the critical infrastructure community work together to manage risks and achieve security and resilience outcomes. It has evolved from concepts introduced in the initial NIPP in 2006 (Department of Homeland Security (DHS 2006) and revised version in 2009 (Department of Homeland Security (DHS) 2009), until the latest version in 2013 (Department of Homeland Security (DHS) 2013) that focused on partnering for CIP-R. NIPP is supported through supplements in form of tools and resources that can be used for the implementation of specific aspects (such as sector-specific plans, training courses). The Federal Emergency Management Agency (FEMA) made significant efforts to increase the level of private sector collaboration at all levels. FEMA offers a variety of tools to help organizations interested in starting PPPs, such as courses, stories and models of successful partnerships, funding, etc.

The concept of resilience as European strategy had not been mentioned at all either in the 'Green Paper on a European Programme for Critical Infrastructure Protection' (European Commission 2005) in 2005, the Directive Proposal (COM 2006) in 2006 or the final Council Directive (Pursiainen 2009; European Council 2008). The Stockholm Programme from 2009 (Conclusions of the European Council 2009) invited the Council, the Commission, the European Parliament, and the Member States to draw up and implement policies to improve measures for the protection, security preparedness and resilience of critical infrastructure. It also called for Directive 2008/114/EC (European Council 2008) to be analysed and reviewed in order to consider including additional policy sectors. Ultimately, the review of the EPCIP Programme (European Commission 2012) in 2012 called for improved resilience of Critical Infrastructures as a part of comprehensive EU Internal Security Strategy. Most of the EU nations have addressed the issue by developing national CIP-R plans and initiating actions.

On the other side, since interconnected infrastructures largely have a regional scope, their interdependencies and service restoration need to be addressed regionally as well. Local level is where the CIP-R issues are first tackled. Depending on the organization of a country, its population and infrastructure density, 'local' ranges from a big city metropolitan/urban area, parish, region, a few regions acting as one when dealing with CIP-R, all the way to a (small) country. As FEMA Administrator Craig Fugate explained "*We have realized that a federal-centric approach will not yield success and that instead we must collaborate and engage with partners at every level of government as well as the non-profit and private sector.*" (FEMA 2011) CI systems are not limited or designed to fit geographical borders. CIP-R resilience is largely cross-border issue in many areas worldwide. Considering diverse and complex aspects and challenges of protection and resilience of CIs including distributed networks, varied organisational structures and operating models, interdependent functions and systems, multi-level authorities, partners, responsibilities, and regulations (The White House 2013), it is clear that it would not be efficient to tackle CIP-R only from national or regional level. Protecting CIs is a shared responsibility requiring cooperation among all levels of government (national, regional, local) and the involvement of the private sector (Department of Homeland Security (DHS) 2009).

In the face of many CI breakdowns current CIP-R approaches have often proved inadequate and with major limitations (Boin and McConnell 2007; Kröger 2008). Recent years have brought major governmental initiatives and rapidly increasing number and spectrum of activities all over the world addressing the issues regarding CIP-R. There are pervasive efforts to improve protection and resilience of CIs and ensure their operational continuity in wake of broadened range of hazards and treats. Effective CIP-R depends on numerous stakeholders collaborating at different institutional and operational levels and exchanging information by means of a variety of channels. In this regard, regional initiatives have emerged worldwide as one of the key strategies to deal with CIP-R issues in the context of Emergency Management (EM) and Community Resilience policies. Since the beginning of 2010 there has been a boom of Public-Private Partnerships (PPPs) in North America and partly in Europe and Australia as well, as the main approach for today's practitioners around the world to deal with CIP-R issues. Strong steps are being taken in all the CI sectors to bolster coordination and information sharing across the government-business border, and even more attention should be placed on growing and nurturing PPPs in CIP-R.

PPPs hold great promise to provide resounding value for both government and businesses, but also face significant obstacles that will need to be overcome. Indeed, PPPs come with challenges in their establishment and management so they sometimes fail to perform and bring benefits as expected, a phenomenon that may lead to a fracture between the appearance and the reality of PPPs on CIP-R. This is why the characteristics of the PPP that runs a specific Regional CIP-R Programme have strong influence on the scope, objectives, activities, and also on the quality of achievements of the programme itself. Recent research has set the theoretical base of Public-Private Partnerships (PPP) and claimed their high potential for enhancing CIP-R that is vastly unexploited due to challenges in their establishment and management. We move forward by studying partnerships' practical side. Through exploratory case study analysis, we try to understand the role and contribution of regional programmes in shaping the contents and results of CIP-R efforts. We identify and consider all the relevant aspect when it comes to these partnerships, such as PPP models, common issues they face, ways to overcome them (good practices in use), alignment with higher level programmes, contribution to information sharing, collaboration and efficiency of crisis response, ability to bring benefits and sustain CI system resilience in general. We sum up all the findings into a framework for the development of regional CIP-R programmes.

The rest of the chapter is organised as follows. Section 8.2 gives the theoretical background on the topic, related aspects and current developments. In Sect. 8.3 we explain the aim of the present study and its methodology. The main findings in form of case studies description and their analysis are presented in Sect. 8.4. The cases are summarised, emphasising their common and distinct features and specific activities. Section 8.5 introduces the framework for the development of regional CIP-R programmes and explains its main parts. The final conclusions are drawn in Sect. 8.6.

8.2 Theoretical Background

8.2.1 Governance Issues and Approaches to Support CIP-R

After the process of privatization and market liberalization during 1980s and 1990s, significant amount of infrastructures passed under ownership of private enterprises. At the same time some public services were being outsourced from the state to private companies. Government's interest, and also obligation, is to ensure providing of essential services that are vital for national security and the well-being of population. On the other hand, the focus of private organizations is on running their business (business continuity) and the security issue is not at the top of their priorities, so there is '*a different sense of urgency in concerning the problem*' among two partnering sides (Cavelty and Suter 2009). Private sector doesn't have funds earmarked for this purpose or is just unwilling to invest more in security. There are exceptions, but in many cases costs of improving security measures or vulnerabilities mitigation outweigh the benefit of reduced risk (Auerswald et al. 2005).

On the other side, every infrastructure disruption, with an outcome of temporary reduction or loss of services, causes significant economical loses and damage to prosperity of the nation. Therefore passing the responsibility for security issues to the private sector is an extremely delicate matter for the government (Percy 2007). For example the role of the US government during the Deepwater Horizon Oil Spill in Gulf of Mexico (national issue) has been perceived unsatisfactory and criticized by BP Commission for failing to assume leadership and effectively coordinate public and private sector (Heineman 2011). Government oversight, necessarily accompanied with industry's internal revisions, is needed to adequately reduce risks and effectively prepare to respond in emergencies (National Commission on the BP Deepwater Horizon Oil Spill and Offshore Drilling 2011).

In situation where control commenced to slowly slip away from the state's hands, a new role for the government presented itself as a possible more effective strategy. In 'meta-governance' approach governments serve as coordinators and stimulators of operators networks (Cavelty and Suter 2009). Another method of resiliency development at both strategic and operational level is through the implementation of Public-Private Partnerships (PPPs). PPPs '*serve as the medium through which that infrastructure functions and protects itself* (Barnes and Newbold 2005). Protecting and ensuring the resilience of critical infrastructure became a shared responsibility among government and the private sector (U.S. President's Commission on Critical Infrastructure Protection 1997). In fact, no single organization has all the necessary resources, relevant information and competence to cope with complex inbound and outbound interdependencies under different accident scenarios (Petrenj et al. 2012), or as US Congress stated: "*Disaster preparedness, mitigation, response, and recovery are efforts that particularly lend themselves to public and private partnerships. In order to effectively respond and recover from an event, the two sectors must work together to protect citizens during a disaster, and help communities rebuild after*" (U.S. DHS 2012). Through its grant program in 2012, DHS has provided supplemental resources to support Public-Private collaboration in order to enhance regional disaster resilience and emergency management.

There is a wide range of PPP forms, characterized by their objectives, models, organization, relationships, leadership, contracts, size, type of actors, etc. While original concept of PPP is projectbased and aims to add value and increased efficiency to the specific service, compared to other options such as concluding a more traditional contract (COM 2005), PPPs with a purpose of collaborative efforts for CI protection and resilience (in scope of this work) are more programmeoriented (i.e., not limited by time periods) and aimed not at enhancing operational efficiency, but at increasing security and vital service continuity (Cavelty and Suter 2009). Main goals of this kind of partnerships should be quite clear and common – protecting property and lives and ensuring continuity of essential services in the face of a turbulent environment where different types of hazards are present. However, in specific incidents primary objectives can become mismatched. Conflicts can appear about selecting priorities, followed by prioritizing actions and resources.

8.2.2 Hierarchical vs PPP Approaches in EM

During the last decades public policy and Emergency Management theorists have increasingly recognised the need for a different approach, rather than traditional hierarchical framework used in normal operating conditions (Comfort 2007). Hierarchy model works very well under relatively stable and fairly predictable conditions (routine emergencies), with time to plan. On the other hand, when coping with dynamic, complex and largely uncertain events hierarchies tend to break down. Information gets lost due to compression, has to cross many levels which takes too much time and non-functioning link stops information completely (Helbing et al. 2006). Obstructed information flow up and down the hierarchy undermines the flexibility, improvisation and urgency expected from crisis responders (Boin 2005). It is impossible for authorities to control each and every move of first responders, and furthermore, organizational diversity makes it impossible to establish an uppermost hierarchy. Blurred boundaries between public and private sectors also make traditional top-down approach inappropriate.

Ability to handle unanticipated and non-routine events is critical and information processing plays a crucial role for the effectiveness of organizations' response to crisis. As complexity and uncertainty rise, transition to flatter organizational structures is a quick way to increase information processing and keep up to the challenge ahead. Command-and-Control (C2) becomes unreliable and flatter structures become more appropriate. An effective response is flexible and networked, recombining the joint potential of the response network (Boin 2005). Several tests showed that network teams were overall faster and more accurate in difficult scenarios than hierarchical teams (Boin and McConnell 2007). Network teams also shared more knowledge in the difficult scenarios, compared with the easier scenarios (Schraagen et al. 2010). More horizontal and networked organizational structures turned out to be more appropriate to crisis management than classic C2. 'Edge organizations' (Roby and Alberts 2012) empower the first lines in situations when plans don't work, and authorities should limit themselves to making only critical decisions – decisions only they can make (Boin 2005). There is no single 'best' approach for each incident, but organizations have to adapt according to the emergency management stage, complexity of the event that they are encountering and environmental characteristics (Lemyre et al. 2011).

Sharing power/authority and even resources is still far from what is the situation in practice and might eventually come up in future as partnerships develop and mature. Even the most of information sharing still occurs through informal channels, relying on acquaintances, personal contacts and connections. Information sharing and coordination of operations is the first step in this direction and basis for establishment higher levels of collaboration, including e.g. pooling of resources, mutual support, and joint decision-making. Beaton et al. (2010) have developed a list of 13 essential collaboration capabilities needed to support actors in their crisis response information sharing (Fig. 8.1).

Networks have become prevalent form of multi-organizational governance since they are seen as superior way to deal with malefic problems. Networks consist of legally autonomous organizations that work together to achieve not only their own but collective goals as well. Networks offer enhanced learning and planning, and enough resources and knowledge available to deal with complex problems. However "*some form of governance is necessary to ensure that participants engage in collective and mutually supportive action, that conflict is addressed, and that network resources are acquired and utilized efficiently and effectively*" (Provan and Kenis 2008). Research carried out by Provan and Kenis (2008) presented three ways to govern a network: self-governance, governance by a lead organization and governance by a network administrative organization (NAO). They argue that the successful adoption of a particular form of governance will be based on four key structural and relational contingencies: trust, size (number of participants), goal consensus and the nature of the task (need for network level competencies) – Table 8.1. Approaches to Inter-Organizational network governance when it comes to CIP-R are defined as (CRN Report 2009):

- Meta-governance of identities: Defining Priorities and Strategies
- Hands-on Meta-governance: Network Participation
- Hands-off Meta-governance: Indirect Steering of Networks

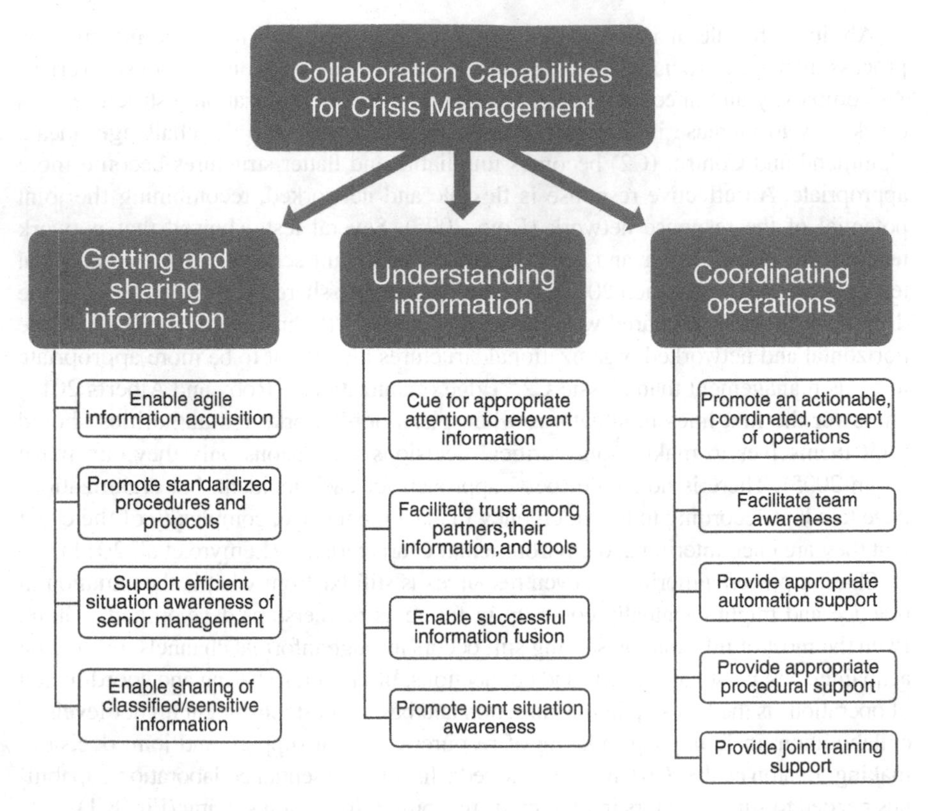

Fig. 8.1 Collaboration capabilities required for crisis management (Adapted from Beaton et al. 2010)

Table 8.1 Key predictors of effectiveness of network governance forms (Provan and Kenis 2008)

Governance forms	Trust	Number of participants	Goal consensus	Need for network-level competencies
Shared governance	High density	Few	High	Low
Lead organization	Low density, highly centralized	Moderate number	Moderately low	Moderate
Network Administrative Organisation (NAO)	Moderate density, NAO monitored by members	Moderate to many	Moderately high	High

Each of the approaches has its advantages and drawbacks. Scholars are aware of the governance form impact to the network functioning and effectiveness as well as on crisis response (Moynihan 2009), but further analysis should be conducted for a better understanding and assessment of the impact on the information sharing and collaboration forms within CIP-R PPPs.

8.2.3 The Key Role of Information Sharing

Effective Critical Infrastructure Protection and Resilience (CIP-R) is dependable on numerous actors collaborating at different institutional and operational levels and exchanging information by means of a variety of channels. In this regard Public-Private Partnerships (PPPs) have emerged as the most important governance model all around the world to deal with CIP-R issues (Cavelty and Suter 2009). Indeed, PPPs present themselves as a comprehensive way for enhancing proactive risk management through an all-hazard approach, as well as for increasing the effectiveness of responsiveness and recovery by matching complementary skills, expertise and resources from public and private sectors. Arguably, PPPs improve both protection and resilience of interdependent CI systems and enhance all phases of the emergency management cycle and thus are emerging as the new and most promising governance model to develop effective CIP-R strategies (US Department of Homeland Security (US DHS) 2013).

In particular information sharing is nowadays generally recognised as the key element of government and private sector efforts to protect CI (Sue 2005). Timely, trusted information sharing and collaboration among stakeholders are crucial within the CIP-R mission (US Department of Homeland Security (US DHS) 2013). NIAC's extensive analysis (National Infrastructure Advisory Council (NIAC) 2012) concluded that "*information sharing is perhaps the most important factor in the protection and resilience of critical infrastructure*" and that trust is the 'essential glue' to make public-private system work. The US Presidential Policy Directive (PPD-21) (The White House 2013) on Critical Infrastructure Security and Resilience aims to enhance coordination, collaboration and information sharing, as well as to encourage and strengthen PPPs.

The European Programme for Critical Infrastructure Protection (EPCIP) sets the overall framework for CIP-R activities in Europe – across all EU States and in all relevant sectors of economic activity.EU is also aiming to strengthen information-sharing on CIP-R between member states by establishing a Critical Infrastructure Warning Information Network (CIWIN), running since mid-January 2013. Information exchange tool should contribute to increasing security in the EU, building trust among relevant stakeholders, standardizing and better integrating national CIP-R programs (European Commission (EC) 2013).

Partnerships and information sharing are perhaps the most important concepts within the CIP-R mission, according to several authors. However, it remains difficult and complicated to establish trusted relationships and implement information sharing mechanisms effectively (Cavelty and Suter 2009; Sue 2005; Natarajan 2013). From this point of view it is worth investigating PPPs ability to improve information sharing and collaboration and to raise the level of CIP-R. Increased attention should be placed on growing and nurturing CIP-R PPPs and concrete steps are required to bolster these partnerships in order to realize their great promise (Givens and Busch 2013). Hence it is relevant to examine the characteristics of PPP themselves and assess different factors that could increase benefits of this kind of approach as a whole.

8.3 Study Methodology

The goal of the CIP-R PPPs is to bring stakeholders together. We can say that the sense of industry–government collaboration (PPPs) activities, in a nutshell is:

- Knowledge and best-practices sharing (information and techniques related to risk management and identification of vulnerabilities/weak-spots, technology to prevent attacks and disruptions, etc.) (Pursiainen 2009);
- Collaborative risk assessment (vulnerabilities identification, interdependencies mapping and analysis, incident consequence estimation);
- Collaborative crisis/emergency management (collaborative preparation and response to the emergency situations).

We argue that through these collaborative activities, each of which requires building trust and specific type of information shared, resilience and protection of CIs could be enhanced. Here again issues may occur – such as unwillingness to share information, lack of interest for partnering, lack of trust to partners, etc. – so the effort is also directed to overcoming existing barriers.

The overall aim of this study is to conduct a case study analysis to understand the role of different PPP models in shaping the contents and results of regional CIP-R programmes. More specifically, the main purpose is to determine whether well-established PPPs are able to improve crisis response and sustain CI system resilience in general. To this end the paper analyses PPP approach to CIP-R, its strengths, possible weaknesses and contribution to CIP-R at higher levels. It seeks to understand the organization and functioning of PPPs with a goal of CIP-R in different settings, the challenges and issues they are facing for efficient functioning, their contribution to enhanced information sharing and collaboration, as well as to higher level resilience of Critical Infrastructures.

In this study, a *region* is understood as an area that is recognised as such by its stakeholders. A region can be a single or multi-jurisdiction area, portion of a state (or province), or may span national borders. Regions have established cultural characteristics, and are cemented by common social and economic activities; as such, they are restricted by geographic boundaries and tend to coincide with the service area of the infrastructures that serve them.

With a focus on emerging PPPs at regional level to address CIP-R issues, the questions this study aims to answer are:

- What are the characteristics and the added value of regional Critical Infrastructure Protection and Resilience (CIP-R) strategies and programmes?
- What are successful practices/approaches to support implementation of regional CIP-R programmes?

- What are the expected and perceived benefits of PPP establishment – results achieved? What are the advancements over time, experience and lessons learned?
- How regional CIP-R strategies and programmes are promoted and supported?

Focusing on the main questions, the analysis does not cover merely the basics of partnership but all the aspects that emerged as relevant in practice. We consider each side's (public and private) position, perspective and concerns towards PPP, as well as tools that have been developed in order to satisfy emerging needs and support spectrum of partnership activities.

As the prior research into practical aspects of PPPs with a goal of CIP-R is quite limited, the case method is well suited to the research questions at hand (Benbasat et al. 1987; Walsham 1995). Case research allows a relatively full understanding of the nature and complexity of phenomena and lends itself to exploratory investigations when phenomena are still insufficiently understood (Meredith 1998; Voss et al. 2002; Yin 2003; Seuring 2008; Eisenhardt 1989). Case studies are suitable for exploring issues that are too complex for empirical survey or experimental research.

Therefore, we decided to adopt an *exploratory-explenatory multiple-case* study research strategy (Yin 2003) as the most suitable choice, focusing on local PPPs with a goal of CIP-R as the unit of analysis. This approach is suitable for understanding of CIs as one of the biggest and the most complex socio-technical systems in combination with PPPs that are concurrently coping with issues of different nature. The cases were selected for the analysis due to the fact that they are among the leaders in the field (regarded as best practices among practitioners) and at the same time diverse in characteristics and with different focuses (Table 8.2). We do not use 'extreme cases' but major and representative ones and in this way we partly deal with the issue of generisability. Seven PPPs have been studied, one in Canada (CRP), one in the US (LA BEOC), one operating across the border and covering both Canadian territories and American states (PNWER), and four in Europe (Lombardy region – Italy; Kennemerland Safety Region – The Netherlands; Scottish Government – UK; Øresund cross-border region – Denmark and Sweden). In this way, the diversity of the cases has been assured by means of location, size and main focus. Each individual case presents a complete study where facts are gathered and conclusions drawn. In the further step, using cross-case analysis and being able to look from a broader perspective, we capture some common and distinctive features and thus generalise beyond the influence of location specific factors (e.g. cultural, political characteristics).

In order to better analyse and confirm the validity of the findings, multiple sources of data have been used (data source triangulation – (Denzin 1984)). Source materials for the analysis of the cases included (1) a set of semi-structured interviews with people engaged in PPPs and some partnering organisations (CEOs, Managers, Private Sector Coordinators, Civil protection representatives, etc.); (2) documents, reports, action plans, websites and other publications; (3) participation in meetings, roundtables, focus groups and tabletop exercises; (4) contributions by involved personnel. An overview by cases is given in Table 8.3.

Table 8.2 Cases general features

	Location	Focus	Size/Level	Cross-Border
Copenhagen	Denmark (Europe)	Emergency management	Trans-national region	Yes
Kennemerland (VRK)	The Netherlands (Europe)	Safety and emergency management	Safety region	No
Lombardy	Italy (Europe)	Emergency management	Administrative region	No
Louisiana (LA BEOC)	USA	Business continuity and community resilience	State	No
Montreal	Canada	CI interdependencies identification, assessment and mitigation	Big city – metropolitan area	No
Pacific NorthWest Economic Region (PNWER)	USA/Canada	Disaster resilience and cross-border emergency management	Multi-state economic region	Yes
Scottish Government	UK (Europe)	Critical national infrastructure protection and resilience	Country with separate jurisdiction	No

Table 8.3 Data sources used

	Interviews	Documentation, reports, action plans, other pub.	Focus groups	Table-top exercise	Website	Contribution to the case description by involved personnel	On site visits
Copenhagen		X			X		X
Kennemerland		X	X			X	X
Lombardy		X	X	X		X	X
Louisiana	X	X			X		X
Montreal	X	X				X	X
PNWER	X	X		X	X		X
Scottish Gov.		X	X		X	X	X

Semi-structured interviews, being flexible, allow new questions to be raised during the interviews based on the response of the interviewees. Interviews were typically of 30–60 min duration and notes were taken during all of them. Besides being a source of data they helped to refine our research questions and led to further rounds of interviews. The rigour and validity of the findings were further ensured (Yin 2003; Eisenhardt 1989) through the follow-up interviews with several respondents; reviewing of the case summaries by the interviewees; discussion of the analysis of the cases and research findings with members of some of the studied PPPs. This has been done in order to collect possible missing details, get more comments, clarifications as well as to remove possible misunderstandings and ambiguities.

8.4 Findings

In this section all the seven case studies are described in full, followed by the summary of their goals and objectives (Table 8.4) and their main practices (Table 8.5).

8.4.1 Copenhagen Capital Region

The Copenhagen case study has a specific focus on the Øresund (or Öresund) Region – a transnational region in northern Europe. The region was created after the construction of the Øresund Link that connects Copenhagen (Denmark) and Malmö (Sweden) comprising of a motorway route and a railway route. It was opened in 2000 and is jointly owned by the Danish and Swedish governments. The link is approximately 16 km long comprises a 4 km immersed tunnel, an artificial island, Peberholm, which is 4 km long, and an 8 km cable-stayed bridge (Fig. 8.2). The Oresund Region consists of Southern Sweden (Skania) and Eastern Denmark (Zealand). The region's two centres, Copenhagen on the Danish side and Malmö-Lund-Trelleborg on the Swedish side, both border Øresund. The Øresund link has created one physically connected region of 3.6 million people with interlinked transport systems for Skåne and Zealand, thus turning Copenhagen and Malmö into a new European metropolis.

The Oresund bridge (7845 m) between Peberholm and Lernacken, which forms the eastern section of the fixed link between Denmark and Sweden, is divided into three main sections: a 3014 m western approach bridge leading from the artificial island to the high bridge, a 1092 m long high bridge and a 3739 m eastern approach bridge between the high bridge and Lernacken on the Swedish coast. The bridge comprises a cable-stayed bridge with a main span of 490 m (world's longest cable-stayed bridge for both road and railway), two side spans of 160 m each and two approach bridges with 141 m spans between the piers.

The Oresund Tunnel is 4050 m long and consists of a 3510 m immersed tunnel under Drogden and two portal buildings of 270 m each. Together, these make up the western section of the fixed link between Denmark and Sweden.

Rail traffic is operated by the rail authority, Banedanmark (Rail Net Denmark) and Banverket (the Swedish National Rail Administration), and is monitored by the train stations in Malmo and Copenhagen – Copenhagen Central Station (RFC) and Train Traffic Management in Malmö (DLC).

Table 8.4 Summary of goals and objectives of the cases

Local CIP-R programme	Mission and goals	Objectives
Copenhagen capital region	Own, operate and maintain the railway and the entire motorway and the land works on both sides of the bridge	Provide an efficient, safe and accessible traffic facility with minimum impact on the environment
	Maintain a high level of accessibility and safety on the link	Provide fast, safe and reliable passage across Oresund at competitive prices
	Repay the loans raised to the construction of the link within a reasonable time frame	Develop, implement and update a joint Danish-Swedish contingency plan
		Repay the bridge's loans within 30 years after its opening (1991) with most of the revenue deriving from road traffic
		Achieving financial stability in a long-term perspective
Kennemerland safety region	To deliver CIP-R through:	
	Assurance of conformance with legal instruments,	To ensure legal conformance of the regional disaster plan so as to maximize public safety and security.
	Maintenance of a public and private partnership for planning and crisis management,	Preparation of measures concerning prevention of, and response to disasters and serious accidents in the municipality, and so the partnership aims to ensure CIP-R conforms to and benefits from regional safety planning.
	Assessment and updating of plans,	Establishing an emergency plan in compliance with the legal provisions of combating accidents and disasters to maximize safety and security.
	Conduct of exercises to prove the practical viability and value of such plans.	Assessment and updating of the crisis plan.
		Conduct exercises at least once every 2 years to demonstrate and test the emergency planning
Lombardy region	Evolution of the governance processes, decision-making and operational resilience of regional CIs;	Characterisation of the critical nodes of major regional transport and energy infrastructures; globally more than 200 regional nodes have been identified and documented;

(continued)

Table 8.4 (continued)

Local CIP-R programme	Mission and goals	Objectives
	Maintaining a continuous process and shared identification and monitoring of threats, vulnerabilities and consequent risk analysis;	Analysis of the accidents influencing regional CIs and creating a series of historical cases;
	Definition of procedures and protocols for the exchange of information and operational interaction between all the actors involved;	Development of vulnerability and resilience studies based on specific quantitative simulation tool;
	Studying the most appropriate technologies, enabling the operating model of reference and able to guarantee security of access and protection of information	Design, validation and implementation of collaborative emergency plans;
		Standardization of communication among the actors – mapping information relevant and communication channels, dealing with interoperability and security of IS.
Louisiana	To create a disaster resilient business community by building from current preparedness efforts, thereby helping Louisiana businesses to become more disaster resistant and able to support the various response and recovery efforts of the State and local community.	Facilitating bi-directional communication of critical information between the State and private sector and promote the resumption of normal business operations;
	To improve disaster preparedness, response and self-sufficiency, reduce reliance on FEMA, and maximize business, industry and economic stabilization.	Enhancing participation by businesses and non-profit organizations in disaster management efforts
	To Provide support in any major disaster – focus on providing situational awareness and resource support, supporting community recovery, mitigation, and economic stabilization.	Joint trainings and exercises with the public and private sectors;
		Economic assessment of events impact to major State economic drivers and the resulting impacts to regional, State, and national economies;

(continued)

Table 8.4 (continued)

Local CIP-R programme	Mission and goals	Objectives
		Maximizing the use of Louisiana businesses and national private sector resources and distribution capabilities to provide needed emergency response products and services;
		Supporting the coordination of voluntary donations from businesses through the Voluntary Organizations Active in Disaster (VOADs) and individuals.
Montreal metropolitan community	The Centre Risque & Performance (CRP) is dedicated to the study of interdependencies between critical infrastructures. In concert with partners from the public and private sectors, its mission is to integrate risk and resilience evaluation into the management mechanisms of industrial and governmental systems.	Developing a methodology of interdependency modelling and evaluation
		Creating operational planning tools of emergency measures.
		Validating and integrate the CRP tools into day-to-day professional activities of network administrators.
		Training highly qualified personnel in the risk management and analysis field, in organizational resiliency and interdependency evaluation.
Pacific NorthWest Economic Region (PNWER)	To improve the Pacific Northwest's ability to withstand and recover and to protect its critical infrastructures from all-hazards disasters	Developing and conducting regional infrastructure interdependencies initiatives focused on various threat scenarios that include regional cross-sector/cross discipline workshops and exercises;
	To coordinate regional Sector Councils, public and private critical infrastructures and key businesses stakeholders to examine interdependencies and cascading impacts resulting from different disasters.	Seeking funding and other resources to support regional pilot projects and other activities and to enable State and local agencies to address regional preparedness needs;
	To develop regional public-private partnerships.	Overseeing the implementation of priority projects and activities in a cost-effective, timely and ethical manner;

(continued)

Table 8.4 (continued)

Local CIP-R programme	Mission and goals	Objectives
	To provide training, education and developing tools, technologies, and approaches to secure interdependent infrastructures and improve all-hazards disastr preparedness and resilience.	Conducting outreach and develop and facilitate seminars, workshops, and targeted exercises to raise awareness and test the level of preparedness.
		Communicating stakeholder validated regional disaster resilience recommendations to State and provincial governments and policymakers.
Scottish Government	Lead the way in reducing the vulnerability of CNI in the Devolved Sectors in Scotland by ensuring that appropriate protective security and resilience arrangements are in place.	*Pursue*
		Enhance local intelligence gathering opportunities and capability in the vicinity of CNI sites;
	Support UK Government in their efforts to reduce the vulnerability of the CNI in Reserved Sectors and sub-sectors (e.g. Energy, Finance) through enhanced protective security & resilience.	Increase awareness and enhance quality of intelligence submissions.
	Minimize disruption to the Scottish public and business community by ensuring that relevant Consequence Management response plans are in place and Scotland is able to deal with a civil emergency (e.g. Preparing Scotland hub and spoke model).	*Prevent*
	Develop a Scotland CNI partnership framework to ensure shared understanding and ownership of CNI issues in Scotland.	Develop Community Engagement strategies,16 where appropriate and agreed (subject to further consultation) which are relevant to the needs of the communities living in the vicinity of certain CNI sites;
	Adopt a robust, proactive approach to all aspects of CNI planning and protection in Scotland, in line with the UK Government National Security Strategy mindful of the distinction between devolved and reserved areas.	Develop Community Impact Assessments, where appropriate and agreed (subject to further consultation) which will assist in the implementation of new protective security and resilience projects.

(continued)

Table 8.4 (continued)

Local CIP-R programme	Mission and goals	Objectives
		Protect
		Lead the way to reduce the vulnerability of the CNI in Devolved Sectors in Scotland by ensuring that appropriate protective security and resilience arrangements are in place;
		Support UK Government in reducing the vulnerability of CNI sites in the Reserved Sectors in Scotland;
		Work in partnership with CPNI, SSDs and others to develop protective security arrangements on the approach to CNI sites where appropriate, which are realistic and proportionate based on current threat and risk assessments;
		Monitor the development of site specific incident response plans;
		Monitor the development of Generic Counter Terrorist incident response Plans;
		Where appropriate, and in Reserved Sectors in consultation with the SSD, encourage infrastructure sectors to protect critical assets to avoid disruption to services from natural hazards.
		Prepare
		Develop a detailed understanding of the interdependencies and impact of loss issues for Scotland as a whole and for each of the SCG areas;
		Develop local planning arrangements, which seek to integrate emergency planning and counter terrorist planning teams, with the aim of providing realistic and effective contingency plans for all CNI sites;

(continued)

Table 8.4 (continued)

Local CIP-R programme	Mission and goals	Objectives
		Policies to adapt to increasing threat from climate change;
		Support information sharing on infrastructure to improve emergency planning and response arrangements for natural hazards;
		Promote policies to ensure location, layout and design of new infrastructure considers risks from natural hazards;
		Work with infrastructure sectors to improve the resilience of networks and systems providing essential services.

The Oresund Bridge is owned by the Oresundsbro Konsortiet. Oresundsbro Konsortiet is a client company that was set up on the basis of the agreement of 1991 between the Governments of Denmark and Sweden, jointly owned by the two companies, A/S Oresund and Svensk-Danska Bro-förbindelsen SVEDAB AB. The collaboration between the two companies is laid down in a consortium agreement approved by the two governments. Oresundsbro Konsortiet's primary task is to operate the fixed link across Oresund, including to maintain a high level of accessibility and safety on the link, and to repay the loans raised to construct the Oresund Bridge within a reasonable time frame. Each side is also responsible for the ownership and operations of the land works on their respective sides of the Oresund Bridge. The full organizational structure, as shown in Fig. 8.3 is complex, with the stock of Oresundsbro Konsortiet being equally owned by the Danish holding company A/S Oresund and the Swedish holding company SVEDAB AB, which in turn are controlled by the Danish and Swedish transportation ministries.

"Vägverket" and "Banverket" are the Swedish road and rail authorities, respectively, while "Sund & Bælt" is the Danish authority which oversees the major Danish island linkages. A/S Storebælt acts as a holding company for the Great Belt Fixed Link, much as A/S Oresund does for the Oresund Bridge.

The partnership arrangement (Fig. 8.4) is essentially a public-public partnership between two nations, which assumed full traffic and revenue risk for the project. In order to ensure the safety of the link, the Oresundsbro Konsortiet Company is in partnership with 9 Danish and 6 Swedish agencies, including police, fire, rescue, medical, alarm units and the traffic and rail control agencies. Oresundsbro Konsortiet does not have its own fire brigade or police; it depends on the local authorities for these services. Therefore, it has established a partnership with several agencies from both Swedish and Danish sides to ensure the safety of the link. Involved parties

Table 8.5 Summary of the main practices

Local CIP-R programme	Main practices
Copenhagen capital region	Regulations for transport of hazardous goods.
	A holistic risk analysis (2010) to identify and prioritise the company's risks. Once a year, the Board of Directors presents a report that sets out the company's key risks and specific proposals for handling them.
	Joint contingency plans including an internal crisis response, to handle accidents on the link. The contingency plans are set as a part of the national safety plans of both Denmark and Sweden, and are tested regularly through exercises. Implementation of the Joint contingency plans.
	Continuous exercises and training s including full-scale exercises every 4 years, table-top exercises, small-scale exercises (scenarios) and weekly alerting exercises.
	An e-learning platform for involved parties to learn safety issues and get prepared for accidents.
	Two control rooms located on the Danish and Swedish sides.
	Partnership with 9 Danish and 6 Swedish agencies, including local police, fire, rescue, medical, alarm units and the traffic and rail control agencies.
	Tetra – RAKEL/SINE Gateway System: SINE is a digital radio network based on the Tetra standard and is used by all Danish emergency services dealing with public order, safety and health.
	Other communication tools:
	Radio
	Dark Fibre Link
	COMputer-Based Alarm System Oresundsbron (COMBAS O): A computer-based alarm system for the Oresund Fixed Link to ensure efficient and rapid alarms to relevant parties and immediately accessible action plans.
Kennemerland safety region	Emergency response plans and their implementation:
	Joint Local Emergency Response Action Plan (LERAP) implemented at Schiphol as local crisis management plan driven by Airline and local First Responder
	2014 updated Crisis Management Plan Schiphol issued by VRK including roles for Private Partners (Rail, Cargo Companies, etc.). The plan will be updated on a yearly basis.
	The Emergency Plan of AAS
	Emergency plan for the Schiphol tunnel of ProRail
	Operational plan of the Royal Military Constabulary
	Plans of the Fire Department and Health Department (GHOR)
	Municipal plans Shelter & Care, CRIB and Communication
	Procedures/plans LVNL, KLM, and other private organisations

(continued)

Table 8.5 (continued)

Local CIP-R programme	Main practices
	Regular exercises of the safety region and those held by the Amsterdam Airport Schiphol.
	Exchange of relevant documents, as well as discussion in meetings regarding safety and response topics.
	The cooperation between Amsterdam Airport Schiphol and the Safety Region Kennemerland is based on activities at specific levels:
	Dispatch Centers
	Executive-operational level partnership
	Crisis response
	Steering and administrative groups Schiphol
	LCMS (National Emergency Management System): a net-centric, web-based shared data system that ensures the information used in the organisation is at all times the same, known and verified, and this applies to neighbouring safety regions and national agencies.
Lombardy region	Mapping of emergency management processes and vital node analysis:
	More than 200 regional critical nodes have been identified and documented as a ranking list of most critical nodes and clusters of nodes
	Analysis of the accidents influencing regional CIs and creating a series of historical cases
	Mapping the organizational models and operational processes of emergency management of the main CI operators active in the region
	Thematic Task-Forces (TTF): 3 TTF have been established, one focused on mapping of the information flows and communication channels among actors, another focused on developing collaborative procedures for coping with major meteorological events and the third one to set up collaborative activities in case of large blackout events.
Louisiana (LABEOC)	CI/KR interdependencies and risk analyses including CI Consequence Analysis and Infrastructure Surveillance and Risk Assessment.
	"Big business-small business" Emergency Management Mentorship program: engaging big businesses (willing and able to mentor), with the small ones helping them to strengthen their disaster preparedness and reduce recovery time. The program also improves response capabilities and results in business recovery plans.
	Web portal for the LA BEOC where businesses are asked to register with the state before a disaster and identify any products or services they might provide to assist communities in the state that have been affected by a disaster.
	Functioning of the NIMSAT web portal during emergencies in providing products and services listed by businesses.
	Improving business resilience and survivability through temporary finding alternative ways of providing essential services until the infrastructure functioning has been recovered.

(continued)

Table 8.5 (continued)

Local CIP-R programme	Main practices
Montreal metropolitan community	Centre Risque & Performance (CRP) of the École Polytechnique de Montréal supports the programme by consolidating the theory of organisational resilience, establishing a common set of terms and developing a method to evaluate resilience.
	DOMINO: a modelling, mapping, decision and planning assistance tool which is a system for managing interdependencies and analysing domino effects.
	The *Civil Security Center* of The Organisation of Civil Protection of Montreal metropolitan area (OSCAM) has made special arrangements with external suppliers/stakeholders in the event of a disaster. OSCAM is activated when a disturbing situation represents a significant risk to the life and health security of the population.
	Table-top exercises with an emphasis on the importance to work together before, during and after a disruption event.
	2008 Quebec government's initiative to increase the resilience of its essential systems coordinated by the Civil security of Quebec (OSCQ), focused on maintaining or restoring the functioning of essential systems to an acceptable level despite any failures that might occur.
Pacific NorthWest Economic Region (PNWER)	Center for Regional Disaster Resilience (CRDR): coordinates public and private critical infrastructures and key businesses stakeholders to examine interdependencies and cascading impacts resulting from different disasters. It also coordinates several regional 'sector councils' including cyber security, energy, fusion center info sharing, etc.
	CIP Task Force – initiated coordination of regional Critical Infrastructure Protection (CIP) managers from the states and provinces as well as federal partners to build relationships with one another, share information and best practices on a regular basis, and thus increase infrastructure and community resilience, leading to many states and provinces sharing CIP plans and training and exercise opportunities.
	'Northwest Warning, Alert and Response Network' NWWARN information sharing platform: a regional alert and warning system to encourage cross-sector information sharing, which is now the communication backbone of the Washington State Fusion Center (WSFC), routinely used for two-way communications with around 3000 CI/KR stakeholders.
	Blue Cascades Exercise Series: developed to explore infrastructure interdependencies, at the same time building relationships and trust – supporting NWWARN use, including 6 exercises addressing variety of topics (e.g. cyber security, earthquake recovery, pandemics, supply chain resilience). Exercises have resulted in lessons learned and a lot of jurisdiction emergency and recovery plans.
	Regional Supply Chain Resilience Project: developing a supply chain resilience public-private sector working group that is able to provide input and advice on issues related to regional supply chain resilience to strengthen the region's ability to withstand and rapidly recover from disasters.

(continued)

Table 8.5 (continued)

Local CIP-R programme	Main practices
	Pacific Northwest Emergency Management Arrangement: a bi-national plan for recovering from a disaster in a cross-border area (CRDR).
	Private-Sector-Led Exercises: The CRDR participates in private-sector-led exercises, with trusted relationships as a primary benefit.
	Region 6 Critical Infrastructure Protection Work Group (CIP WG): made up of the region's key agencies and voluntary private-sector representation from some of the county's largest employers and owners and operators. The Region 6 CIP WG and the DHS Protective Security Advisor (PSA) have worked with the CRDR to facilitate interdependency workshops, tabletop exercises, and other partnership-building activities.
	Supply Chain Resilience Task Force: During an emergency, the CRDR may activate this task force to communicate directly with an EOC about what the critical elements and decisions are that would affect the region three to 6 months from the time of the incident.
Scottish Government	'Secure and Resilient' supports the all-risks approach outlined in the UK National Security Strategy by addressing UK Government strategies in tackling the priority risks outlined in the UK Government National Security Risk Assessment (NSRA) and National Risk Assessment (NRA), as well as other identified risks specific to Scotland.
	Critical Infrastructure Resilience Unit (CIRU) as the main unit devoted to CIP-R related activities in Scotland aimed to ensure that effective and appropriate resilience arrangements are in place across the devolved sectors in Scotland.
	'Preparing Scotland' – set out as a 'hub and spokes' model – established as the guidance to responders assisting them in planning, response and recovery and aimed to establish good practice based on professional expertise, legislation and lessons learned from planning for and dealing with major emergencies at regional and local levels. Available in full to relevant responders and organisations involved in the operation, protection and resilience of Cis.
	Operation Estrela – infrastructure resilience exercise programme to threat from insider attack.

from both Denmark and Sweden include organisations as Police, Fire Brigades, Train and Traffic Control Centres, Hospitals, Alarm Centres, etc. In collaboration with the relevant authorities in Denmark and Sweden, Oresundsbro Konsortiet maintains a comprehensive contingency plan, including an internal crisis response, to handle accidents on the link. The contingency plans are set as a part of the national safety plans of both Denmark and Sweden, and are tested regularly through exercises.

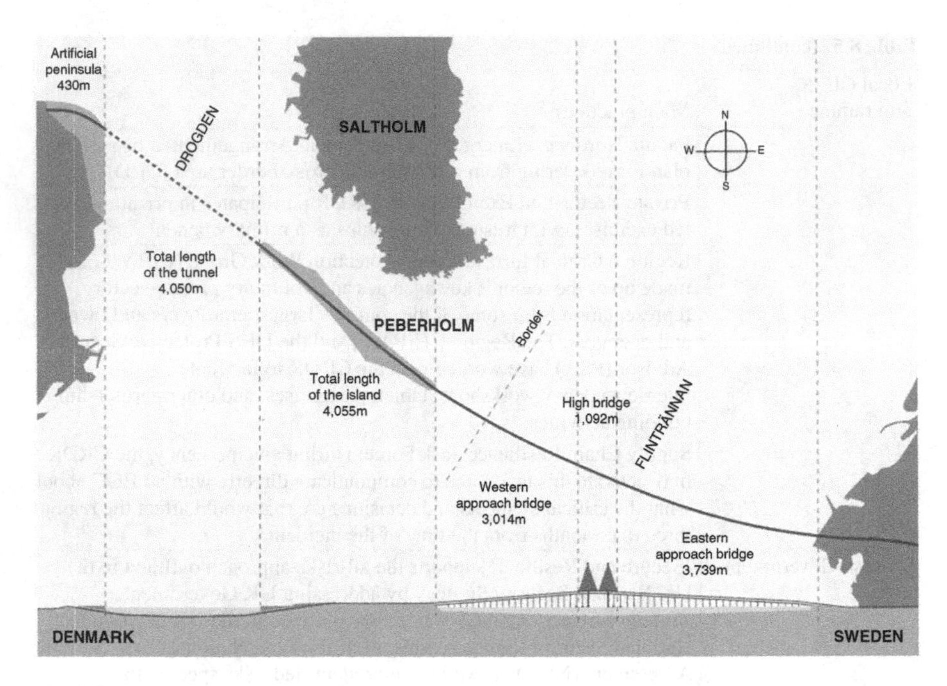

Fig. 8.2 An overview of the Oresund link

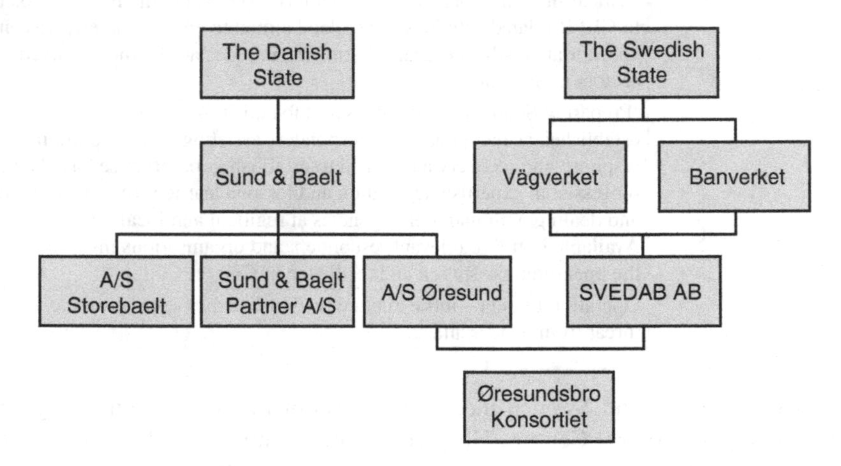

Fig. 8.3 Oresund Bridge organizational structure

8.4.1.1 Information Sharing

The main system for sharing information and communication between partners is the Tetra – RAKEL/SINE Gateway System (Fig. 8.5). RAKEL (acronym for radio communications for effective management) is the Swedish national digital communications system (mobile system) used by the emergency services and others in the

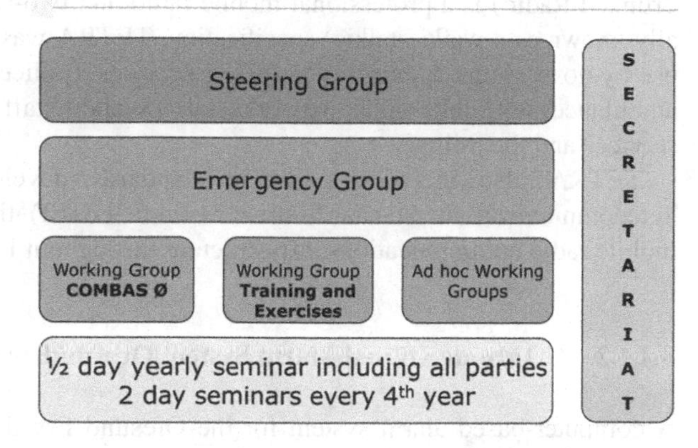

Fig. 8.4 Structure of the partnership

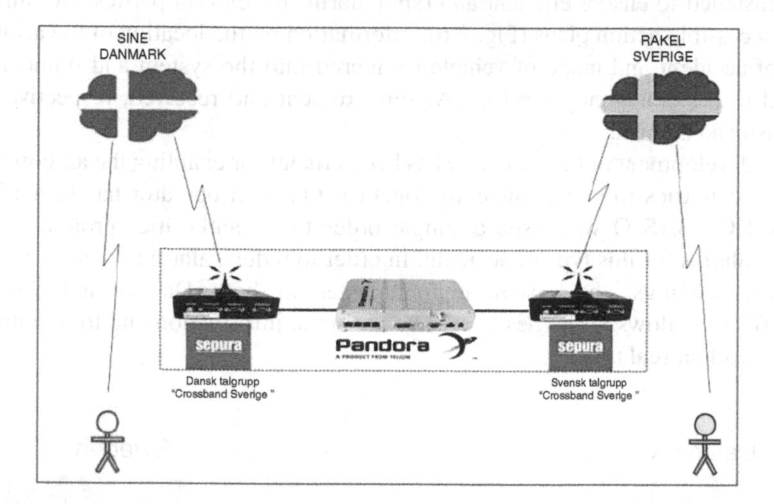

Fig. 8.5 Tetra – RAKEL/SINE gateway system

fields of civil protection, public safety and security, emergency medical services and healthcare. It is used mainly by police, military police, rescue, ambulance services, emergency alarming (RAPS) and local/state emergency management. RAKEL also helps increase societal preparedness. The system streamlines everyday communications, and enables new ways of working, which increases readiness and with it, ultimately the ability to manage an emergency.

RAKEL is meant to merge all civil protection agencies and organisations into one common forum, increasing information exchanges across organisational and sector boundaries. During the recovery phase of an emergency the system can be used as a tool for monitoring and evaluation, where communications routines and operations can be easily analysed.

Terrestrial Trunked Radio (TETRA) (formerly known as Trans-European Trunked Radio) is a professional mobile radio and two-way transceiver (colloquially known as a walkie-talkie) specification. TETRA was specifically designed for use by government agencies, emergency services, (police forces, fire departments, ambulance) for public safety networks, rail transport staff for train radios, transport services and the military.

TETRA also includes a set of standards developed by the European Telecommunications Standardisation Institute (ETSI) that describes a common mobile radio communications infrastructure throughout Europe.

8.4.1.2 COMputer-Based Alarm System Oresundsbron (COMBAS O)

A computer-based alarm system for the Oresund Fixed Link (COMBAS O) has been installed to ensure efficient and rapid alarms to relevant parties and immediately accessible action plans (Fig. 8.6). Information on the location of the accident, type of accident and make of vehicle is entered into the system and immediately passed to the emergency services. Alarms are sent and received, respectively, in Swedish and Danish.

The development of COMBAS O has been crucial for enabling the authorities in the two countries to work efficiently together. Once an operator has keyed in an alarm, COMBAS O will issue a simple order to all authorities programmed to receive alarms for this type of accident. In order to reduce alarm time and language misunderstandings, the system communicates in both Danish and Swedish. COMBAS O allows all parties to receive the same information and to monitor the rescue work in real time.

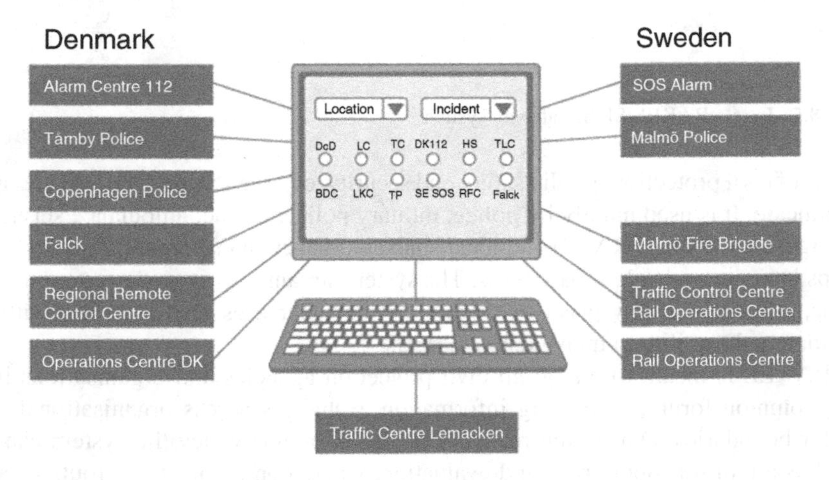

Fig. 8.6 COMBAS O alarm system

8.4.1.3 Risk Assessment and Emergency Management

Once a year, the Board of Directors presents a report that sets out the company's key risks and specific proposals for handling them. This was done for the first time in 2010 and is updated on an annual basis.

Generally, the emergency response in Denmark has 2 levels: *Municipal (The municipal fire and rescue services) and State (The national fire and rescue services)*. The Danish crisis management organisation is presented in Fig. 8.7.

Local contingency planning for the Oresund Bridge started 3 years before the commissioning of the fixed link. The task began with the preparation of a contingency concept which set out existing plans, parties involved in both countries and outlined the framework for a joint contingency plan which could overcome the differences in the two countries (Fig. 8.8). Once the authorities had accepted the concept, detailed planning of the contingency measures could begin.

In collaboration with the relevant authorities in Denmark and Sweden, Oresundsbro Konsortiet maintains a comprehensive contingency plan, including an internal crisis response, to handle accidents on the link. The contingency plans are tested regularly through exercises. The emergency and response plans contain incident level classifications and geographical dimension considerations.

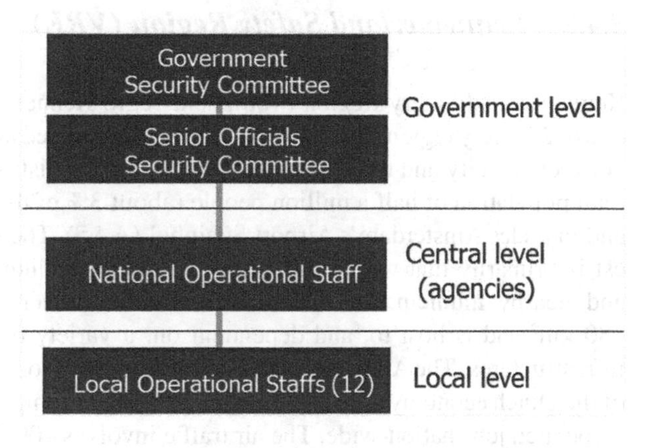

Fig. 8.7 Danish crisis management organisation

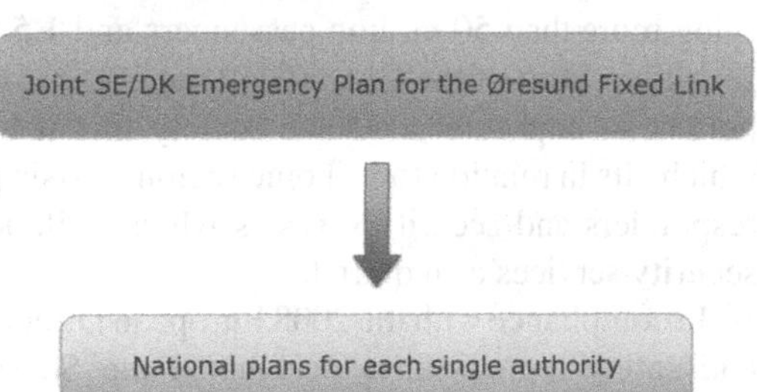

Fig. 8.8 Emergency plans

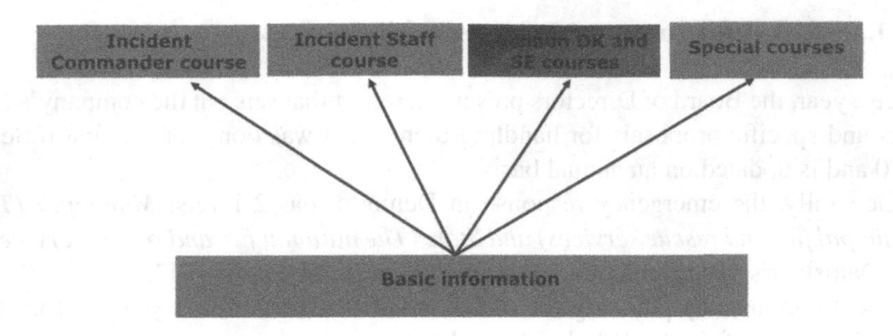

Fig. 8.9 Going towards e-learning

8.4.1.4 Education

To achieve the contingency objectives, joint training of staff from all relevant authorities and at all levels is required. Oresundsbro Konsortiet has developed an e-learning platform for involved parties (Fig. 8.9), along with trainings and exercises, such as full-scale exercises every 4 year, table-top exercises, small-scale exercises (scenarios) and weekly alerting exercises.

8.4.2 Kennemerland Safety Region (VRK)

Kennemerland Safery Region (Veiligheidsregio Kennemerland, or shortly VRK) is one of 25 safety regions in The Netherlands. It is situated northwest from Amsterdam, between the city and the North Sea. The region consists of ten communities with a total population of half a million people (about 3% of the Netherlands' population) and includes Amsterdam's Airport Schiphol (AAS). The geographical area of interest is primarily that within the airport boundary including the city of Amsterdam and nearby municipalities such as Haarlemmermeer. The immediate region is 180 km^2 and is host to, and dependent on, a variety of sophisticated and critical infrastructures. The AAS, is a major gateway to and from Europe; it is a key element of the Dutch economy and employment with 64,000 employees on site, plus 290,000 associated jobs nation-wide. The air traffic involves 400,000 flights per annum, carrying more than 50 million passengers and 1.5 million tonnes of cargo, generating 26 billion Euro of the Dutch GNP via 500 companies located at Schiphol. The airport is so important to Dutch society that it has its own crisis planning activity, which sits in relation to its home region's crisis planning activity. It also has its own responders and security services which collaborate with regional responders and security services as required.

In compliance with the 2008 European Directive, the CI sectors in the Netherlands and safety regions have designated the Security Liaison Officers (SLOs), and addressed the obligations regarding Operator Security Plan (OSP) for each potential European Critical Infrastructure (ECI). In addition, the Dutch CIP Contact Point has been set up as a priority, so that the actual implementation of the directive can be channelled through the CIP contact point.

The partnership within VRK is not formed as a single entity or organisation, and in fact operates through a series of PPPs whereby some of the partners are involved for different purposes. Within this complex mix of partners there are 13 public and 6 private organisations of several kinds (medical services, fire services, AAS, National Rail Company, Air Traffic Control and KLM Airlines). Specific plans exist for several Critical Infrastructures in each safety region – the following case of Schiphol Airport CIP-R is only one of the localized partnership, used as illustration. The specific goals and objectives of the partnership are determined by the Statute. In summary, the objectives aim to deliver CIP-R through assurance of conformance with legal instruments, maintenance of the PPP for planning and crisis management, assessment and updating of plans, and conduct of exercises to prove the practical viability and value of such plans.

The public-private partnership between the Safety Region of Kennemerland and Amsterdam Airport Schiphol was formalized in 2007, and builds upon the pre-existing partnerships between AAS and parts of the present Safety Region. Until 2010, the responsibility for responding to crises and disasters were within the local governments. In 2010, according to the law regarding the safety regions, these responsibilities were transferred from the local level to the regional level of government called *Safety Regions*. Therefore, Safety Regions are responsible for CIP-R issues and crisis response at the regional level, while Mayors are responsible for public order and for crisis response at the local (municipality) level. All levels of government work together with critical infrastructure owners/operators to ensure a sufficient level of protection at their respective levels.

8.4.2.1 History

In 2007 the so-called territorial congruence (territoriale congruentie) took place. The Ministry of Internal Affairs (Binnenlandse Zaken en Koninkrijksrelaties – BZK) which was then responsible for disaster management and fire safety, decided it was more efficient for the police, fire and ambulance services to work together when they covered the same working areas.

Amsterdam Airport Schiphol is in the municipality of Haarlemmermeer, and the regional fire services and disaster management then the responsibility of the fire services of Amsterdam. The police services, however, was the responsibility of Kennemerland, while the medical services were provided by the health organization (GGD) that worked not only for Haarlemmermeer, but also for other municipalities in the area.

Because the regional police of Kennemerland had the same working area as the prosecution district, the BZK decided that it was more efficient if the regional fire services of Kennemerland took over fire control and disaster management. The medical services would then be provided by a new health organization (GHOR) for the whole region of Kennemerland.

The present Safety Region Kennemerland (Veiligheidsregio Kennemerland – VRK) is now formed out of the regional fire service, plus the medical/health service.

The first task of the VRK was to organize the disaster management and crisis response of the municipality of Haarlemmermeer, including Amsterdam Airport Schiphol. VRK also began to organize that for other municipalities in the region, as well as addressing the large/complex risks such as those related to Tata Steel, sea-worthy cargo and cruise ships over the North Sea Channel to and from the harbours of Amsterdam, large public events and a critical/vital infrastructure with a number of important roadways and railways including multiple tunnels.

VRK deployed its new Safety Bureau whose primary task is the preparation and support of the multidisciplinary crisis response tasks of the Safety Region. The Safety Bureau provides the planning, facilities support, training, exercise and evaluation of the main crisis response structure of the Safety Region.

To ensure alignment between the stakeholders, liaison and support was seconded to the Safety Bureau.

Another important step was the arrangement of a joint co-located dispatch centre for the police, ambulance and fire services in the region of Kennemerland.

In 2010, new legislation revised arrangements between fire services, medical services and disaster management. The mayors remained responsible, but now as one board. The new legislation is more focussed on modern crisis response instead of classical disaster management.

Within the above background, things changed in the public-private partnership between Amsterdam Airport Schiphol and the public emergency services, but the basis remains.

8.4.2.2 Public-Private Partnership – Operational Levels

The cooperation between Amsterdam Airport Schiphol and the Safety Region Kennemerland is based on activities at specific levels:

- Dispatch centres
- Executive/operational level (regular/daily incident response)
- Crisis response

8.4.2.3 Dispatch Centres

Amsterdam Airport Schiphol has its own coordination centre where all the business processes of the airport are coordinated, supported and aligned. Its own dispatch centre for the airport fire and the medical service is part of that centre. In case of small incidents this centre can deploy the airport fire and the medical service on its own. The joint dispatch centre of the safety region monitors these deployments. The dispatch systems are connected, as are the alert (P2000) and communication systems (C2000).

8.4.2.4 Operational Partnerships

In case of escalation or need for support, the joint dispatch centre of the safety region will deploy additional units. For example, the airport medical service can provide first aid but is by legislation not permitted to transport patients to a hospital. An ambulance of the safety region has to take it over. In the case of escalation of a fire or accident, a duty officer of the safety region and additional regional units will be deployed to the scene. The preparation and execution of the fire and medical services at the airport are organized in close cooperation between the private services of the airport and the public services of the safety region. This ensures alignment between planning and procedures, equipment of vehicles, materials, training and exercise.

8.4.2.5 Crisis Response

In case of an incident that disrupts the business processes of the airport, the Operations Manager of Amsterdam Airport Schiphol can take over the coordination of that incident and will assemble a management committee with representatives of the involved business partners. Amsterdam Airport Schiphol has prepared this in its own incident response plan. An example is a major disruption of the luggage handling system that will lead to delays of incoming and outgoing flights. But in the case of a major/complex fire or accident, the coordination is the responsibility of the safety region. The safety region ensures systematic crisis response through its regional crisis response plan, and a subset of that plan addresses incidents at the Schiphol Airport through formulation of a specific crisis response plan for the Schiphol area. The main scenarios addressed are:

- airplane crash (at or nearby the airport),
- hazardous materials incident at Aircraft Fuel Supply (large storage tanks) or at KLM Engineering & Maintenance (large storage of chemicals),
- incident in the railway underpass (underground platforms with switch lanes).

8.4.2.6 Steering and Administrative Groups Schiphol

The administrative management of the specific crisis response plan for the Schiphol area has links with activities such as training and exercises, management of the facilities (crisis response centre with systems) and judgement of evaluations (as a PDCA-circle).

To align this, a steering and a management group are instituted. The *management group* comprises tactical representatives of involved partners, both public and private. The *steering group* comprises strategic representatives under chairmanship of the mayor of Haarlemmermeer.

8.4.2.7 Most Recent Public Lesson

The main task of the Safety Region Kennemerland is to organize the crisis response in the municipality of Haarlemmermeer and Schiphol airport. A major test was seen in February 2009 when a large passenger plane crashed in farmland just before the landing strip (early touchdown). The cooperation between private services of Amsterdam Airport Schiphol and public services of the Safety Region Kennemerland was very successful, as confirmed by evaluations and investigations.

Points for improvement emphasised after-care of passengers and relatives. Before the municipality could get responders to the site, public care was organized by citizens, supported by motorists of the nearby motorway and farmers of the nearby farms. This form of self-reliance was a signal to all municipalities in the Netherlands to change its public care in case of crisis to facilitate the needs of the public rather than control it.

8.4.3 Lombardy Region

Lombardy (*Lombardia* in Italian) is one of the 20 Italian regions, located in the north. A sixth of Italy's population lives in Lombardy (around ten million citizens) and it accounts for around 20% of Italy's GDP, making it the most populous and richest region in the country and one of the richest in Europe. It has a constant population growth, a highly developed infrastructure system and hosted the Expo 2015.

To establish a risk-informed policy making process, the Regional Administration launched in 2007 a 4-year research programme named "PRIM -Integrated Regional Program for the mitigation of major risks" (Lombardy Region 2007). The aim of the programme was the identification of the most critical areas, following an all-hazard approach, the expected impacts on population and economic activities, and the related prevention and mitigation actions. The programme allowed developing a multi-risk assessment methodology that integrates information with different degree of accuracy into a limited set of leading indicators.

The continuous development of high-value services characterizing the Lombardy region society, one of the most industrialized in Europe, deeply relies on complex infrastructure systems. Considering the results of its first study in 2007, it became evident that hazards identified over the territory, not only can threat the citizen life, but can also cause severe disruptions of infrastructure service continuity inducing wide cascading effects. As a consequence, following the release of the EC Directive 114/EC (2008), the Lombardy Region Administration decided to set up a preliminary study to investigate CI vulnerability and to assess current emergency practices in the sector.

It emerged that there is a great potential for an increase in the flow of shared information regarding criticality and accidents which can increase efficiency of the invested resources and also bring an improvement in the security level. The objective of the Lombardy region policy in CIP/R is therefore not to add new mechanisms or control processes, but to **promote and advance collaborative processes**. In light of this logic, from 2010 Lombardy Region has launched a program of activities aimed at defining a model of integrated and shared management, capable of supporting a higher level of collaboration within the processes of prevention, risk monitoring and emergency management related to regional CIs. The program was named "Programma Regionale per la Collaborazione ed il Coordinamento nella Sicurezza delle Infrastrutture Critiche (PReSIC)". In December 2010 a Memorandum of Understanding was signed by 18 operators of energy and transport CIs operating in the Lombardy region.

The key elements that define the scope of the PPP in Lombardy are (Fig. 8.10):

- evolution of the governance processes, decision-making and operational resilience of regional CIs;
- maintaining a continuous process and shared identification and monitoring of threats, vulnerabilities and consequent risk analysis;
- definition of procedures and protocols for the exchange of information and operational interaction between all the actors involved;
- studying the most appropriate technologies, enabling the operating model of reference and able to guarantee security of access and protection of information.

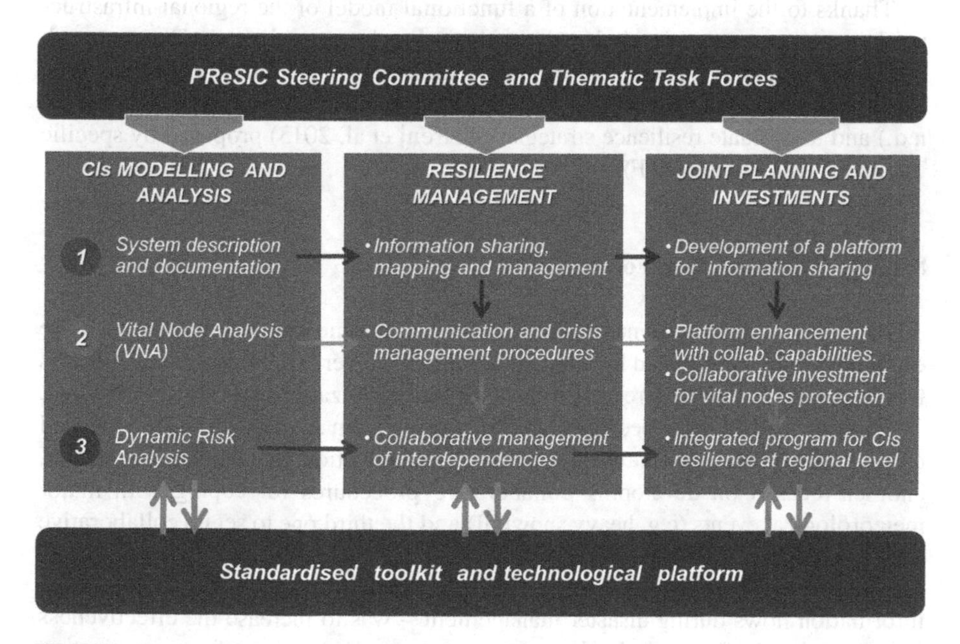

Fig. 8.10 Roadmap for the development and evolution of PReSIC

PReSIC strategy and objectives call for a deep involvement of public and private CI operators. Since this is clearly the most challenging point of the programme, several resources and means of collaboration has been mobilized.

8.4.3.1 Mapping of Emergency Management Processes and Vital Node Analysis

The preliminary study, carried out by a team of academics and consultants, provided a complete picture of the actual status of the vulnerability of regional infrastructural nodes and the corresponding emergency management processes adopted by the most important CI operators. More specifically the study focused on:

- Carrying out a census of the critical nodes of major regional transport (road, rail, air and underground) and energy (electricity, gas and fuels) infrastructures; globally more than 200 regional nodes have been identified and documented;
- Analysis of the accidents influencing regional CIs and creating a series of historical cases;
- Mapping the organizational models and operational processes of emergency management of the main CI operators active in the region.

The scientific and technical team of PReSIC offered a constant support to operators in preparing and gathering useful information, mainly by means of document analysis, FMECA-like (Failure Mode Effect and Criticality Analysis) questionnaires, direct interviews and process mapping tools.

Thanks to the implementation of a functional model of the regional infrastructural system a systematic vital node analysis has been carried out (Trucco et al. 2012) and returned a ranking list of most critical nodes and clusters of nodes. The functional model is also normally used to support scenario analysis (Cagno et al. n.d.) and to evaluate resilience strategies (Petrenj et al. 2013) proposed by specific Thematic Task Forces (TTF).

8.4.3.2 Thematic Task-Forces (TTF)

TTFs represent the backbone of the PReSIC programme implementation; they are established and coordinated by a higher level PPP Governance Committee which is formed by the managing directors from all of the organizations that signed the MoU.

So far three TTF have been established starting from January 2011, one focused on mapping of the information flows and communication channels among actors, another focused on developing collaborative procedures for coping with major meteorological events (e.g. heavy snowfall) and the third one to set up collaborative activities in case of large blackout events.

The primary objective of the first TTF – focused on the mapping of multi actor information flows during disaster management – was to increase the effectiveness and operational efficiency thanks to a greater standardization of communication flows and channels among actors in the regional system (Fig. 8.11). The first analy-

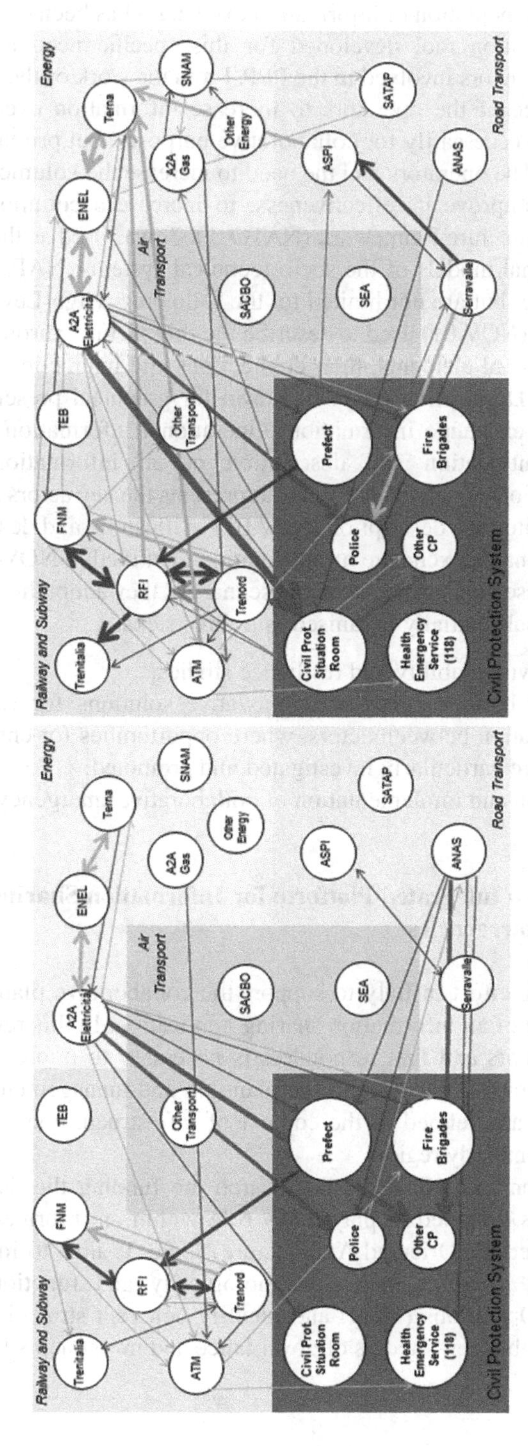

Fig. 8.11 Information flows before (*left*) and after (*right*) PPP establishment (Operation context: service interruption of a generic CI)

sis and the final documentation of information exchanges has been carried out using a web-based application tool developed for this specific need and constantly accessible by all the actors involved in the PPP. From the work of the roundtables it is evident preference of the operators to increase information exchanges in the future, although not necessarily for collaborative purposes, but primarily for informational purposes. The operators feel the need to increase the volume of communication, or at least improve its effectiveness, to increase a common operational picture. NATO Architecture Framework (NATO 2007) was used as the standard for presenting operational models of the socio-technical systems. NAF views used in this research include, but are not limited to, the following: High-Level operational concept description (NOV-01) used to describe the 'big picture' through geographical location, operational elements, their connections and interactions; Operational Node Connectivity Description (NOV-02) used for graphical presentation of the nodes that need to exchange information; Operational information requirements (NOV-03) for identification and description of all information exchanges; Organizational relationship chart (NOV-04) to presents the key actors and their relationships; System interface description (NSV-01) to illustrate and describe systems and interfaces that enable exchange of information identified in NOV-03.

As for TTFs focused on specific accident scenarios, they adopt the same methodological approach, substantially organised into three steps:

- Development of vulnerability and resilience studies;
- Identification of best practices and innovative solutions for risk mitigation through collaboration between actors, where opportunities for enhancing information sharing are particularly investigated and promoted;
- Design, validation and implementation of collaborative emergency plans.

8.4.3.3 Towards an Integrated Platform for Information Sharing During Emergencies

There is an ongoing effort in Italy to support the collaborative plans between CI operators by release of an information sharing application. In this regard, the integration of CI operators and first responders is necessary to improve information sharing and collaborative processes in the planning and management of emergencies. Requirements are defined in the context of infrastructure systems and civil protection of the Lombardy region.

Lombardy Region and Ministry of Research are funding the development of application modules designed to play a key role within an information platform, realized in SOA (Service Oriented Architecture) logic. It aims to improve operational management of emergencies, technologically and functionally support Network Enabled Operations (NEO) and identify coherent strategies in terms of PPPs that would enable new models of governance and investments for CIP/R.

Innovative solutions are being developed at different levels:

- Standardization of information content based on: (i) extension/adaptation of standard protocols already existing in the field of Civil Protection, such as Tactical Situation Object (TSO) (Henriques and Rego 2008); (ii) automatic translators to ensure the specificity of glossaries adopted by each operator.
- Development of shared ontology and algorithms for semi-automatic generation of operational information from the data available in the IT systems of each CI operators
- Prediction of vulnerability and domino effects through Pattern Recognition Algorithms, applied to the information exchange process, and discrete event simulation, both powered by real-time operational data;
- Adoption of technological and architectural features that ensure interoperability, easy customization and reconfiguration, access security and resilience to emergency

8.4.4 Louisiana

Louisiana is a state located in the southern region of the United States, by the Gulf of Mexico, with a surface area of about 135,000 km^2 (1.35% of all US territory) and a population of around 4.65 million (1.45% of total US population). According to the US Department of Commerce, the Gross State Product of Louisiana in 2013 was about 253.6 billion dollars that accounted for around 1.5% of US total GDP. The main cities are Baton Rouge (the capital) and New Orleans – the largest city and a major US port. Louisiana is the only state in the U.S. with political subdivisions (local governments) named 'parishes', which are equivalent to counties.

Louisiana is often affected by tropical cyclones, thunderstorms, and is very vulnerable to strikes by major hurricanes, particularly the lowlands around and in the New Orleans area. New Orleans was catastrophically affected when the Federal levee system failed during Hurricane Katrina in 2005. It was the costliest natural disaster, as well as one of the five deadliest hurricanes, in the history of the US.

Another major event was the Deepwater Horizon Oil Spill in Gulf of Mexico (2010). It was a national issue that is considered the largest accidental marine oil spill in the history of the petroleum industry and the worst environmental disaster America has ever faced.

The Louisiana Business Emergency Operations Center (LA BEOC) is a joint partnership between Louisiana Economic Development (LED), the Governor's Office of Homeland Security and Emergency Preparedness (GOHSEP), the National Incident Management Systems & Advanced Technologies (NIMSAT) Institute at the University of Louisiana at Lafayette and the Stephenson Disaster Management Institute (SDMI) at Louisiana State University. The LA BEOC has been recognized by FEMA as a best practice model for PPPs. It was launched in 2010 to support the coordination of activities and resources of businesses and volunteer organizations in

Louisiana and across the nation. The four institutions are equipped with an IT system that enables them communicate between themselves. It is operated as a state-of-the-art facility on the LSU (Louisiana State University) campus, the development of which was supported with in-kind donations of technology and software and cash donation by major national and Louisiana based businesses. LA BEOC doesn't own any resources to give or land to private sector, nor is there a lot of decision making inside LA BEOC – it is getting the information and forwarding to who needs it. There are 30 seats at LA BEOC for representatives of business associations, each of whom have outreach to all of their members.

The mission of the LA BEOC in support of any major disaster is to focus on providing situational awareness and resource support, supporting community recovery, mitigation, and economic stabilization. Its goal is to improve response and self-sufficiency, reduce reliance on FEMA, and maximize business, industry and economic stabilization. It is operated as a state-of-the-art facility on the LSU (Louisiana State University) campus, the development of which was supported with in-kind donations of technology and software and cash donation by major national and Louisiana based businesses. LA BEOC doesn't own any resources to give or land to private sector, nor is there a lot of decision making inside LA BEOC – it is getting the information and forwarding to who needs it. There are 30 seats at LA BEOC for representatives of business associations, each of whom have outreach to all of their members.

Loss of one or a few critical infrastructure services significantly affects functioning of private businesses causing multiple ripple effects. Establishment of the LABEOC had a goal of mitigating disaster effects and consequences supporting state private businesses continuity. It consists of temporary finding alternative ways of providing essential services until the infrastructure functioning has been recovered. Besides improving business resilience and survivability, it is also important since:

- **Incentivizes new companies to enter the state market** – if the state is willing to help businesses during an emergency and make them safer, it is a good image and motivation for other companies to enter the market.
- **Brings economic benefits in two ways**
  - Through money saving – local goods and services are significantly cheaper than the ones requested from federal level
  - Every local purchase supports the state economy through tax income
- **Citizens are more satisfied** using local products and services that they are accustomed to.

LABEOC model offers an improved crisis communication with state EOC:

- **Private businesses have who to contact and request help** – LA BEOC is handling requests that are not going to be considered if asked directly to state.
- **Communication B2B** – many needs are satisfied locally by making bond between different business, matching ones needs and others resources or ser-

vices, and thus making benefits for both sides without engagement of the public authorities
- **Serves as filter for information between businesses and state government** – State EOC was getting overwhelmed by phone calls and requests from individual businesses. LA BEOC liaison at the State EOC is able to receive the request and needs that are not fulfilled on local level and address them in an appropriate way.
- Information on the state of infrastructures collected by government office (reliable) is wrapped as 'situational awareness report' and is sent to LA BEOC for use.
- The private sector participants with positions in LA BEOC support the activities of the state EOC – utilize their relationships to source goods and services needed, and capture damage assessment critical to assisting the state in developing accurate situation awareness and economic impact assessments.

During emergencies everything starts local – city or parish. In many cases business need something that cannot be supplied locally. Businesses are registered from all over the state, so in case of an incident in one area businesses from other parts are able to help. The NIMSAT Institute has developed a web portal for the LA BEOC where businesses are asked to register with the state before a disaster and identify any products or services they might provide to assist communities in the state that have been affected by a disaster. Communication with neighbouring states is on a higher level and in charge of the state.

8.4.4.1 "Big Business-Small Business" Emergency Management Mentorship Program

In January 2012, FEMA announced a new campaign "Small Business is Big" and made an effort to help small businesses, often lacking the resources and knowledge, to be better prepared for all-hazards disasters. The need for improvement of businesses resilience is strongly supported by the statistics from the Institute of Business and Home Safety (*25% of all businesses do not reopen after a major disaster*) and the U.S. Chamber of Commerce (*when a business does not have a formal emergency plan in place the figure rises to 43%*) (National Incident Management Systems and Advanced Technologies (NIMSAT) Institute 2012a).

"Big Business – Small Business" is an innovative effort in the area of PPPs that engages big businesses, willing and able to mentor, with the small ones helping them to strengthen their disaster preparedness and reduce recovery time. Private-private partnership model is voluntary based and promotes proactive (whole-community) emergency management approach. ***Why is this programme important and what are the mutual benefits?*** Big businesses benefit from strengthening their supply chains (where small businesses are often located), raising reputation and positive branding. Small businesses get an opportunity to learn about resilience/business continuity, get missing resources and adopt best practices from experienced leaders who have been through disasters and know what it takes to survive.

Considering the social and economic importance of SMEs it creates a great contribution to community resilience. Businesses also build beneficial long-term relationships that round this win-win environment. "Big business-small business" platform has been launched by NIMSAT institute in June 2012.

8.4.4.2 CI/KR Interdependencies and Risk Analyses

The NIMSAT Institute seeks to advance the understanding of risk faced as a nation due to the interdependencies between various Critical Infrastructure/Key Resources (CI/KR) assets, the dependency of various public and private sector supply chains on these assets, and the consequences of disruptions to the way of life regardless of the cause or location of disruption. The main activities in this direction include:

- **Critical Infrastructure Consequence Analysis** – The NIMSAT Institute, the National Infrastructure Simulation and Analysis Center (NISAC) of the US DHS, Sandia National Labs, and the LA-1 Coalition collaborated on the assessment of the national consequences of disruptions to Louisiana's energy corridor (Port Fourchon/Louisiana Offshore Oil Port/Grand Isle/Louisiana Highway 1).
- **Infrastructure Surveillance and Risk Assessment** – The NIMSAT Institute is working with the Louisiana Office of Coastal Protection and Restoration (OCPR), in the development of a state-of-the art Intelligent Flood Protection Monitoring, Warning and Response System (IFPRMWRS) at strategic locations within levee systems in the New Orleans region. This system will include the ability to monitor and warn of undesirable performance that could lead to catastrophic consequences.

8.4.5 Montreal Metropolitan Community

Montreal is the largest city in province of Quebec and the second largest city of Canada with 4 million citizens (11% of total population of Canada) that covers a small portion of the country (about 4250 km^2). According to the Quebec Institute of Statistics, Montreal Metropolitan Community's GDP in 2013 was around 161 billion dollars which accounts for about 9% of the total GDP of Canada.

The Great Ice Storm in 1998 (strongly hit eastern Ontario, southern Quebec and parts of the US) brought into focus the need for all stakeholders to work together, form partnerships and toil spirit of full collaboration. It also raised awareness of the possible consequences of damaged infrastructure in Canada. At this point Federal and Provincial Acts stated that (Lecomte et al. 1998):

- emergency operations are most effective when managed at the lowest level of government
- the response structure should be built upon permanent organizations

- coordinated support from government (federal and provincial) should come from their external partners
- intervention must respect the responsibilities of the participants
- the response and recovery structure must be flexible enough to accommodate all circumstances

In the period after the storm a few of the regional organisations in Quebec decided to give money for the university research on interdependencies. Subsequently, in 2004, a grant from the Natural Sciences and Engineering Research Council of Canada and Public Safety and Emergency Preparedness Canada (now Public Safety Canada) was given to 6 universities/teams across Canada for a Joint Infrastructure Interdependencies Research Program (JIIRP), where Centre risque & performance (CRP) of École Polytechnique de Montréal was assigned to study interdependencies and domino effects.

At the provincial level, in 2008 Quebec launched a government initiative to increase the resilience of its essential systems. Coordinated by the Civil security of Quebec (Organisation de la sécurité civile du Québec – OSCQ), initiative focused primarily on maintaining or restoring the functioning of essential systems to an acceptable level despite any failures that might occur. OSCQ resilience subcommittee's mandate was to mobilize the owner and operators of CIs, whether private or public, to build partnerships, and to ensure the coherence and complementarity of the preventive and preparatory measures envisaged by the stakeholders. CRP of the École Polytechnique de Montréal was asked to give support by consolidating the theory of organisational resilience, establishing a common set of terms and developing a method to evaluate resilience.

8.4.5.1 Interdependencies and Domino-Effects Study

The ***preventive approach*** (Robert et al. 2007) adopted by the CRP implies the proactive risk management. It emphasizes the anticipation of harmful consequences and establishing a bilateral communication of risk among CIs that interact within a single socioeconomic environment. In order to anticipate the consequences caused by potential failure, and take into account the changing status of the CIs, *coordinative space* must be set up, where it could be possible to share information relevant for planning efficient, effective and realistic protective measures. The preventive approach deliberately focuses on anticipation and effective, targeted communication of the relevant information in order to protect populations by reducing the domino effects generated by interdependencies. Advantages of the preventive approach include cooperation, communication, anticipation, planning and continuous risk management.

Consideration of the consequences rather than the causes of failures (***consequence-based risk management approach/All-hazard approach***) leads to the vulnerability assessment of the entities making up an environment. At the same time, it allows ranking of the employment of emergency measures based on the acceptability of the potential consequences. It leads people in charge to better prepare for the risks related to interdependencies among CIs, but calls for initial evaluation of interdependencies in order to estimate a) possible domino effect in case of a disruption, and b) users that have to be informed, so the protective measures could be put in place on time.

As CRP experienced, there were four main barriers for information sharing at the point of interdependency identification and analysis:

- **Confidentiality –** Dissemination of information may represent an additional vulnerability for a network. While security reasons are concern for every organization when it comes to sharing confidential information, competition was problem only in certain sectors. This was not an issue for water and gas operators since they are unique in the region. On the other side, in telecommunication sector situation was significantly different since more enterprises were competing over the market.
- **Interpretation –** Managers of a system are the only ones able to interpret correctly information regarding their system. Receiving a basic level ('raw') information/data makes it prone to misinterpretation by the managers, leads them to analysis that is not good (since they are not experts), to come up with a wrong conclusion and make errors when taking action. (e.g. creating the maps without the key to read them, or without a clear idea how to use them.)
- **Value and property –** Acquisition and management of information is costly. Organizations are not ready to share their information if they do not receive something in return. A lot of the infrastructural systems had been laid underground many years ago and they exact position/location as well as their structural condition (status) is not always precisely known (sometimes even unknown). These data have an intrinsic market value. The acquisition of information requires human and technological resources, and after, there are costs of managing and updating the data on the systems.
- **Update –** The data of an organization are numerous. The update is complex and must be done continuously. Only the organization itself can perform this task efficiently.

How did CRP cope with these issues? Since geographical data are essential dimension in order to properly target and coordinate actions in the field, CRP has developed an innovative flexible cartography approach (Robert and Morabito 2010) in which, rather than representing infrastructures, represents location sectors in which the consequences of the resource failures are synthesized. Approach with flexible representation allows for a targeted intervention while preserving the confidentiality of information. The size of the sectors used may vary based on needed analysis detail level, geographical zone studied, and the level of confidentiality CI managers wish to maintain. In this specific case, where the methodology has been applied in downtown Montreal, the study zone has been divided into 1 km^2 sectors.

Subsequently, a modelling, mapping, decision and planning assistance tool, DOMINO, was developed. It is a prototype of a system for managing interdepen-

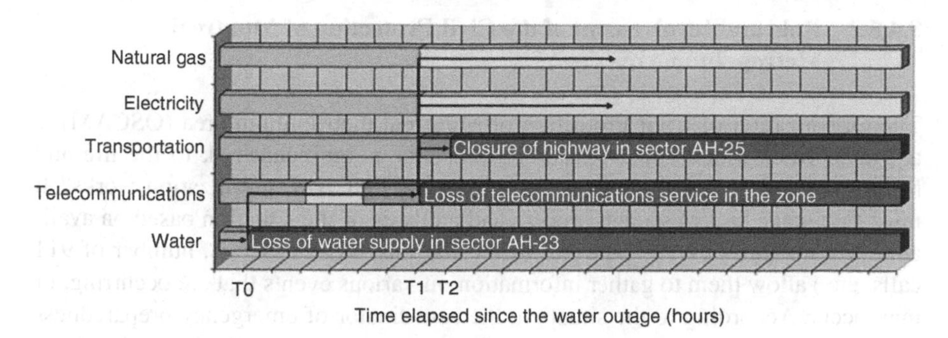

Fig. 8.12 An example of DOMINO simulation (Robert et al. 2008)

dencies and analyzing domino effects (Fig. 8.12). DOMINO uses a flexible cartography approach to locate system infrastructures and simulate domino effects, ensuring at the same time data confidentiality (agreements had been signed with partners). The online database is organized in that way that each organization has a password protected access to its own private section of the database where they can manage the information they are sharing, used for domino effects analysis. Module that contains the results of the simulations (analysis of domino effects) is available for all systems including the Civil Security Center of the City of Montréal.

In cases of high sensitivity, confidential geographical information needed for identifying anticipated impacts of geographic interdependencies in some sectors is exchanged in the interaction only between system owners, without unnecessary sharing it with other members. Once the meeting is over, each participant takes away the strategic and confidential information related to its system. Thus, this is only a temporary pooling of information, though a vital one to enable the subsequent analysis. This approach for confidential information protection can be also used during the actions aimed at mitigation of vulnerability. Where points of high vulnerability have been identified through functional and/or geographical interdependencies analysis, involved organizations are left to work together to find a possible improvement. Their activities can include technical or organizational changes, changes in flow and use of primary and alternative resources, etc. After mutual activities are finished operators can come back to partners, so the information about the interdependencies can be updated and used for simulation. The presented tool works in the manner of Early Warning System (EWS). EWSs are generally composed of four inter-related key elements: risk knowledge, monitoring and warning service, dissemination and communication and response capability (United Nations 2006), and since it addresses only the first three key elements it is not a real EWS but more system able to make a good mobilization of the resources – so can be defined as *Early Mobilization and Cooperation System.* The future development should include utilization in the real-time environment – during the response phase of EM.

8.4.5.2 Role and Involvement of the Civil Protection of Montreal Metropolitan Area

The Organization of Civil Protection of Montreal metropolitan area (OSCAM) is activated when a disturbing situation represents a significant risk to the life and health security of the population. How does the OSCAM mobilization works? It must first make an assessment process and analysis of the situation based on available information. Several tools (telemetry stations, weather alerts, number of 911 calls, etc.) allow them to gather information on various events that are occurring, or may occur. According to the situation, the coordinator of emergency preparedness will determine if one goes to standby, alert or intervention mode. Each alert level corresponds to a different level of mobilization (used to determine who will be mobilized) that are also different from one risk from another. Different indicators, are established by the people who are directly involved in the risk – experts in the domain. The indicators are constantly followed and when the threshold is reached (defined for every risk) mobilization starts. If there are no specific indicators the coordinator will always have the final say.

OSCAM is able to reach each people who run (are responsible for) each major infrastructure. They can get in touch with anybody who is involved in municipality at any kind of level. Automatic phone system can call each stakeholder or its replacement very quickly. Message will reach to every phone number and email until somebody answers and acknowledges that he will report on duty. System is automatic so it sends very short situation update, and tells what actions OSCAM requires – to come to work, or to get ready to be able to come to work in a few hours. Every municipal stakeholder has pre-defined missions, so there are standard pre-planned procedures (who does what) that people would have to follow in the event of a disaster. If there is a risk that has no specific plan, then it will be the emergency responders on the scene that will determine if they are overwhelmed or if they need emergency measures to give them special powers.

Coordination centre – half of the room are people who are in touch with the people in the field (fire department, police department, ambulance, representative from public health, representative of public transport) – on the other side there are people in charge of gathering information, people in charge of financial aspects, logistics, elected people, people in charge of communications – each of members is just in touch with his entire team in a different room. Representatives of each infrastructure operator have their own centre and communication with representative – liaison agent who has power to make decision. Collecting information that would facilitate strategic decisions is the responsibility at the center. Collected situational awareness information is transferred to the coordinator who then decides who he wants at the table. Decisions are made based on the impact on the population. Not how to fix a damaged infrastructure but how to minimise the impact on the surrounding population. At the emergency coordination center the site is handled but also the consequences on the rest of the population. It's easier to make a decision when persons from very different backgrounds/or different organizations are together, having a multi angle on things to consider (e.g. Doctor, toxicologist, CBR specialist, surveillance – all talking to each other and making wiser decisions). The fire department is responsible for rescue operations.

The role of *Civil Security Center* is to coordinate among all the stakeholders in the city region. One of its responsibilities is to make special arrangements with external suppliers/stakeholders in the event of a disaster. The provincial level has very similar missions that could provide support if needed.

Sometimes, during the planning phase, it takes a lot of time to get information at that time but when they get into intervention there are never problems for getting any kind of information. It always remains a challenge when new players/personnel (due to promotions/retirements) come to play, but once they get to know people from OSCAM and why the information is needed – it gets easier. They're always afraid that OSCAM is going to ask some technical aspects/information, which it doesn't, only if they have something going on in the sector that OSCAM needs to know about.

In the tabletop exercises emphasis was made on the **importance to work together before, during and after a disruption event**. During exercises it easy for an organization to say something that they might not be able to deliver in real life. "*We get to know people; we get to make them think about what they just said; we get to make them realize what they would be responsible for delivering if that would really happened*", said Michel Bonin (Civil Security Center– City of Montréal) about the exercise benefits. "*We have established very close network of people – strategic intelligence – who talk a few times a week on any kind of subject, usually by a conference call. We've been working together so often and so long that now we know exactly what we can expect in a real emergency.*"

There are two basic ways to measure success (evaluate improvements):

- In preplanning every year report card is given for every person responsible for a mission – to evaluate his level of preparedness;
- After every kind of intervention debriefing is always made – out of the debriefing come recommendations – one person will be responsible to make follow ups to those recommendations. There are not many interventions but we still they get better every time – lessons are learned.

8.4.6 Pacific North-West Economic Region

Pacific NorthWest Economic Region (PNWER) is a statutory public/private non-profit created in 1991 by five US states (Idaho, Montana, Oregon, Washington and Alaska) and five Canadian jurisdictions (British Columbia, Alberta, Yukon, Saskatchewan and Northwest Territories) focused on issues impacting the economy of the Pacific NorthWest. State/jurisdiction governments understood that there are regional impacts that don't stop at borders but impact everyone, and realized as well that each of the governments had influence only within their own borders. By establishing PNWER as a statutory non-profit they are able to cross the borders, get all the people together and have a collective approach to tough issues. It is also much easier to make consistent government decisions. Nothing will adversely impact the economic vitality of the region – that is the essence of what is PNWER all about.

The first initiative to address regional infrastructure security issues was the creation of The Partnership for Regional Infrastructure Security in November 2001 and launch of the Regional Disaster Resilience and Homeland Security Program with the goal of improving the Pacific Northwest's ability to withstand and recover and to protect its CIs from all-hazards disasters. PNWER has, through its Center for Regional Disaster Resilience (CRDR), coordinated public and private critical infrastructures and key businesses stakeholders to examine interdependencies and cascading impacts resulting from different disasters. It also coordinates several regional 'sector councils' including cyber security, banking and finance, livestock health, energy, fusion center info sharing, etc. CRDR is committed to working with states, provinces, territories, and communities to develop regional public-private partnerships, develop action plans, and undertake pilot projects and activities to further this important mission. PNWER also provides training, education and developing tools, technologies, and approaches that build on existing capabilities, in order to secure interdependent infrastructures and improve all-hazards disaster preparedness and resilience. PNWER was listed as a best practice for working with other states and provinces to address critical infrastructure security issues in the NGA's Governors Guide to Homeland Security (in March 2007) and also referenced in the National Infrastructure Protection Plan (NIPP) as the model for bringing the public and private sectors together to address critical infrastructure protection issues (in July 2009).

The Washington State Fusion Center (WSFC) is a unified counterterrorism, "all crimes," fusion center, incorporating agencies with intelligence, critical infrastructure, public safety and preparedness, resiliency, response and recovery missions. The WSFC is Washington State's single fusion center and concurrently supports federal, state, tribal agencies and private sector entities, regional and local law enforcement, public safety and homeland security by providing timely, relevant and high quality information and intelligence services.

8.4.6.1 NWWARN Information Sharing Platform

One of the major achievements was the development of a regional alert and warning system named 'Northwest Warning, Alert and Response Network' (NWWARN), to encourage cross-sector information sharing. NWWARN project started in 2004 as a joint project between Federal Bureau of Investigation (FBI), DHS and PNWER, with assistance of regional CI operators as well as key business and government managers with responsibilities for security, preparedness, strategic planning, emergency management, response and recovery from all disasters and terrorism threats. DHS planned to use it for its own needs but never completed its implementation, so it was finally built as a notification platform adjusted to PNWER needs by MyStateUSA (Idaho). It is now the communication backbone of the Washington

State Fusion Center (WSFC), routinely used for two-way communications with around 3000 CI/KR stakeholders.

Inside NWWARN platform information is shared through gatekeepers – experts in a particular infrastructure (water, electric utilities, shipping, defence industries). Gatekeepers are the trusted sources of information within an infrastructure, designed primarily to approve members within their infrastructure to be added to the system. Any of the gatekeepers could inquire with another gatekeeper for information on something that they need to know. Proprietary business information that can be very confidential is not needed in this kind of exchange, but mainly information on facilities and interdependencies with other systems.

Suspicious activity report had been identified as a gap and this capability was added to the platform afterwards. Social media integration enables to directly push information to Twitter or Facebook, while capability to draw information in (integrate e.g. *Google crises/alerts*) relies on crowdsourcing mechanisms to collect information. Ability to see in real time what kind of information is being posted online gives better situational awareness picture. Next big step would be to create a portion of NWWARN as a business operation center tool – in order to have a single source of information for business community to get and request information during crises. Businesses want accurate information from one place – informative to make decisions about their businesses.

In a nutshell, the goal of information sharing to help protect regional/national infrastructures, communities and the public has been achieved by:

- Maximizing near real-time, two-way sharing of situational information without delay
- Providing immediate distribution of critical information to the members who need to act on it
- Providing a place for members and non-members to submit suspicious activity reports to the FBI and Washington State Fusion Center
- Using commonly used, popular mediums for disseminating messages (phone calls, emails, text messages, etc.)

8.4.6.2 CIP Task Force and Blue Cascades Exercise Series

Information sharing and collaboration are a matter of relationships and trust – virtually never works, but physically – meeting people and building trust.

PNWER established the ***CIP Task Force*** – initiated coordination of regional Critical Infrastructure Protection (CIP) managers from the states and provinces as well as federal partners (Department of Homeland Security, Department of Defense, the Department of Energy, the U.S. Army Corps of Engineers, etc.) to build relationships with one another, share information and best practices on a regular basis, and thus increase infrastructure and community resilience. This coordination has led to many states and provinces sharing CIP plans and training and exercise opportunities and has helped build regional trust.

Blue Cascades Exercise Series have been developed to explore infrastructure interdependencies, at the same time building relationships and trust – supporting NWWARN use. Since 2002, PNWER has held six exercises addressing variety of topics (e.g. cyber security, earthquake recovery, pandemics, supply chain resilience), each designed by stakeholders and reflecting regional concerns. Blue Cascades has become a model for bringing together public and private sector stakeholders to discuss cascading impacts across the region. It has been mostly about "who to talk to and about what, when something happens". Recovery and mitigation activities are often topics that don't get enough attention in other kind of venues, so having the opportunity to get into the recovery and restoration sidc of it (in a Blue Cascades type of exercise) is important to move everybody forward. Blue Cascades offer an opportunity to discuss about emergency plans with various types of jurisdictions and companies (like Boeing, Microsoft), decide on the best practices from each of the type of approach, implement best practices and modify own plans. One of the main outcomes of the exercises is that everyone ended up with much more comprehensive plans than individual departments or jurisdictions could create on their own. After each exercise, stakeholders assist in developing an action plan to address the issues uncovered during the exercise. Results of the exercises are kind of a roadmap – identify key areas to think about in planning and sometimes have specific topics that are necessary to make the region more resilient. Exercises have resulted in lessons learned and a lot of jurisdiction recovery plans that have not even had a thought in the past.

8.4.7 Scottish Government

Scotland is a country and a part of the United Kingdom that covers the northern third of the Great Britain. With a population of around 5.3 million (8.3% of UK population) and a total surface area of 78,800 km^2 (one third of UK area), Scotland is a rather small country. According to the Scottish Government website, the Scotland's GDP in 2013 was about 245.3 billion dollars that is about a tenth of the UK total GDP.

The flooding in England and Wales during the summer of 2007 was a timely point when it was acknowledged that Critical National Infrastructure (CNI) in Scotland is both critical to Scotland and the wider UK, and therefore, appropriate plans and strategies must be developed, involving all levels of the Government to protect the UK CNIs. Towards this aim, *Preparing Scotland* was established as the guidance to responders assisting them in planning, response and recovery. It is aimed to establish good practice based on professional expertise, legislation and lessons learned from planning for and dealing with major emergencies at regional and local levels. It is set out as a hub and spokes model; the hub, including philosophy, principles, governance structures and regulatory guidance and the spokes a range of detailed guidance on specific matters such as caring for people, mass fatalities and communicating with the public. '*Secure and Resilient*' is one of the spokes of Preparing Scotland, available publically in summary form and, reflecting its security status, available in full to relevant responders and organisations involved in the operation, protection and resilience of CIs. Secure and Resilient seeks to implement the UK National Security in relation to Critical National Infrastructure in Scotland. It sits under and meshes with the UK National Security Strategy, the UK CONTEST Strategy and the UK Critical National Infrastructure Protection Framework. It is intended to describe in more detail the Scottish Government contribution to these UK strategies including aims, responsibilities and delivery arrangements. It also clarifies areas where Scottish Government leads (on devolved matters) and areas which are reserved where Scottish Government aims to work closely in support of Whitehall departments.

The main unit devoted to CIP-R in Scotland is the Critical Infrastructure Resilience Unit (CIRU). CIRU works closely with relevant colleagues from within Scottish Government, the Cabinet Office and CNI Site Operators, in order to ensure that effective and appropriate resilience arrangements are in place across the devolved sectors in Scotland.

8.5 Discussion

Despite all the cases refer to a local dimension, they differ a lot in terms of institutional context and size of the Public-Private Partnership. It may also be noted that goals and objectives are largely heterogeneous, ranging from enhancing Emergency Management coordination to the development of a fully integrated regional resilience strategy.

Additional relevant findings that emerged by the in-depth analysis of the case studies are:

- The continuous improvement strategy implemented in most of the cases is based on a sequence of small but touchable win-win achievements; there are no cases of large programmes fully financed over a long time horizon;
- Activities and implemented technological or organisational solutions are largely focused on EM cycle; resilience functions are not emphasised as core dimensions to develop thc contents of the programmes;
- Collaborative and qualitative approaches to solution design are dominant;
- Understanding/modelling and documenting interdependencies are issues addressed by almost all the programmes as part of the key prevention activities;
- Enhancing information sharing among all the public and private stakeholders is regarded as one of the key success factors and is deserved of specific support platforms and reference agreements;
- Exercises are the most common practice used to enhance awareness, trust and to build inter-organisational collaboration culture between public and private stakeholders.

The framework for the Development of Regional CIP-R Programmes provides the list and the relationships between the key elements that are needed for a successful and sustainable programme design and implementation (Fig. 8.13).

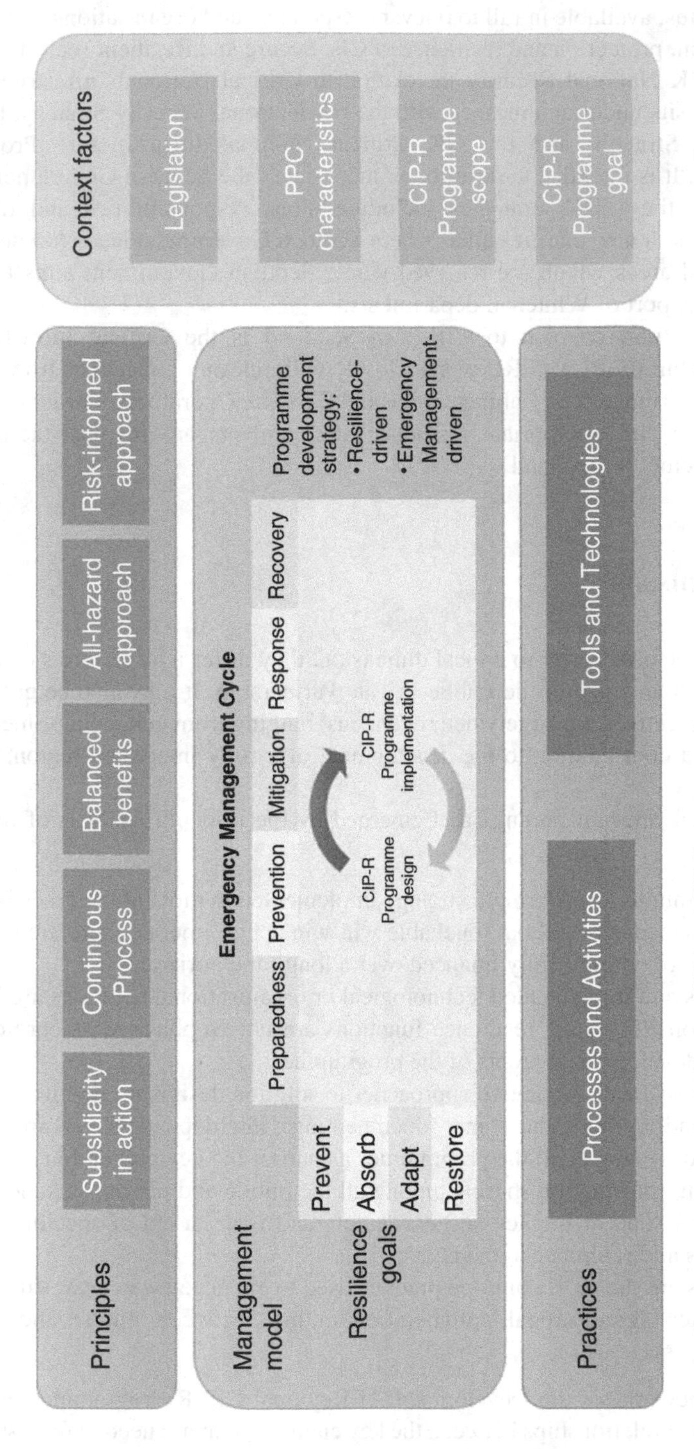

Fig. 8.13 Framework for the development of Regional CIP-R programmes

The framework comprises four types of elements: the fundamental principles for a successful CIP-R programme development, the components of the management model, the set of practices that can be used to implement the programme, and the relevant context factors that influence the programme over its entire life cycle.

8.5.1 The Fundamental Principles

The development of regional CIP-R programmes and their CIP-R planning approach are grounded on the following five fundamental principles.

1. Subsidiarity "in action"

Subsidiarity and complementarity principles are at the roots of the EPCIP Programme. Regional strategies and programmes for CIP-R represent the best existing examples of a bottom-up approach to the subject and, as such, the most promising opportunity for a deeper and more effective deployment of the subsidiarity principle in the CIP-R domain.

Involving stakeholders at regional level does not mean involving 'regional' CI operators only, but more precisely establish collaborative processes with relevant/national CI operators at a level that is closer to the field, thus closer to the implementation arena. A bottom up approach can leverage on existing local experiences to design and implement an effective CIP-R programme. This subsidiarity 'in action' is made of recognition, support, involvement, harmonisation and sometimes devolution to single stakeholders or group of stakeholders that brings distinctive capabilities for building the programme. Thus, subsidiarity "in action" can ensure that regional knowledge and expertise can fully address CIP-R through PPP where the bottom-up approach is regionally focused yet responsive to, and connected with, relevant national concerns.

2. Continuous improvement process

Existing successful experiences demonstrate that many local CIP-R Programmes rapidly evolved over the time, thanks to the virtuous cycle of:

- gaining commitment of some key stakeholders on relevant disruption scenarios;
- fixing achievable and win-win objectives in the short term;
- communicating tangible results and achievements to all stakeholders to involve them in the programme and expand the PPP;
- revising and enhancing scope, goals and objectives of the CIP-R programme thanks to the new entries.

3. Balanced benefits

A Regional CIP-Programme led by a local PPP is generally driven by a mix of different interests and needs: public authorities, responders' organisations, CI operators and owners, businesses … Normally, it also covers a mid-long term planning horizon. Hence, the prioritisation of different types of achievable results is a key issue. It emerges that the strategy of pursuing balanced benefits – government vs business needs; short vs long term – is the most effective to assure long term sustainability of the programme and the achievement of tangible results.

4. All-hazard approach

The term 'All-hazard approach' denotes a way of CIP-R development able to comprise all conditions, environmental or manmade, either accidental or intentional, that have the potential to cause injury, illness, death, or loss of assets, service delivery, or other intangibles; or alternatively causing functional social, economic, or environmental harm.

Three closely related factors necessitate the development of a holistic, all-hazards approach to regional CIP-R:

- infrastructure vulnerabilities and interdependencies
- information sharing processes and solutions
- public-private collaboration

5. Risk-informed approach.

Risk-informed decision-making (RIDM) is a deliberative process that uses a set of performance measures, together with other considerations, to "inform" decision-making. The RIDM process acknowledges that human judgment has a relevant role in decisions, and that technical information cannot be the unique basis for decision-making. This is because of inevitable gaps in the technical information, and also because decision-making is an intrinsically subjective, value-based task. In tackling complex decision-making problems involving multiple, competing objectives, the cumulative knowledge provided by experienced personnel is essential for integrating technical and nontechnical elements to produce dependable decisions.

8.5.2 Management Model

The key components of the management model of a Regional CIP-R Programme are:

1. Contents development matrix (Emergency Management cycle vs Resilience core functions);
2. CIP-R Programme design and implementation process;
3. CIP-R Programme long term development strategy.

A long lasting CIP-R Programme is expected to pass through different phases in its life cycle. Its evolution is strongly influenced by the origin and goal set at the beginning of the Regional CIP-R Programme life. Sometimes the goal may change due to changes in political priorities (e.g. security and protection issues vs safety and resilience issues) or dramatic evolutions in the most relevant threats a certain region is exposed to (e.g. due to Climate Change). Another evolutionary dimension is the increase in size of the PPP, thanks to new members, or of the geographical extension of the programme.

The temporal evolution of the CIP-R programme, managed through its design-implementation cycles, can be driven, time by time, by different priorities or by a different mix of perspectives. Here we highlight two main development strategies, two perspectives (or logic) are equally important and complementary:

- ***Emergency Management-driven development strategy*** emphasises the operational integration of the CIP-R Programme with the management of real events. It is preferable when the PPP is led by public authorities, with security roles, or by responders; in these cases, the goal of the CIP-Programme is generally more focused on EM improvement (e.g. Kennemerland Safety Region, LABEOC, PNWER).
- ***Resilience-driven development strategy*** emphasises the supportive role of the PPP and its Regional CIP-R Programme. The programme is developed to build protection and resilience capabilities into the regional system, that are exploited by different organisations (e.g. EM, Civil Protection agencies, or Police forces) through decision making and operational processes that are largely out of the scope of the programme (e.g. Lombardy Region, Scotland). This is typical when the PPP is led by private stakeholders (Montreal Metropolitan Community).

One of the key issues in developing successful CIP-R programmes is that these two perspectives are not completely overlapped, even though they share some common traits. Hence, a way must be found to assume both and harmonise them into a unique and consistent programme. Having a Regional CIP-R Programme aligned with the Emergency (or Disaster) Management cycle is also vital to check and assure the compatibility and full integration of the programme with the EM or Civil Protection system already in place in the region. An approach which facilitates emergency services and CI operators to collaborate in addressing resilience improvement measures, while planning to cope with CI disruptions (Fig. 8.14), was proposed by Kozine and Andersen (2015).

This framework for integrating the resilience capacities of CI into the EM cycle reflects the main characteristics of such emergencies (e.g. interdependent, multi-sectoral, multi-stakeholder) and supports the identification, assessment and development of specific resilience (technical and organizational) capabilities. The Capability Building Cycle (Fig. 8.15) presents an operational approach for continuous process of programme design and implementation (as in the framework – Fig. 8.13) of a Regional CIP-R Programme. A pilot case in Lombardy Region (Italy) fostered collaborative EM in the context of public-private collaboration for CI resilience

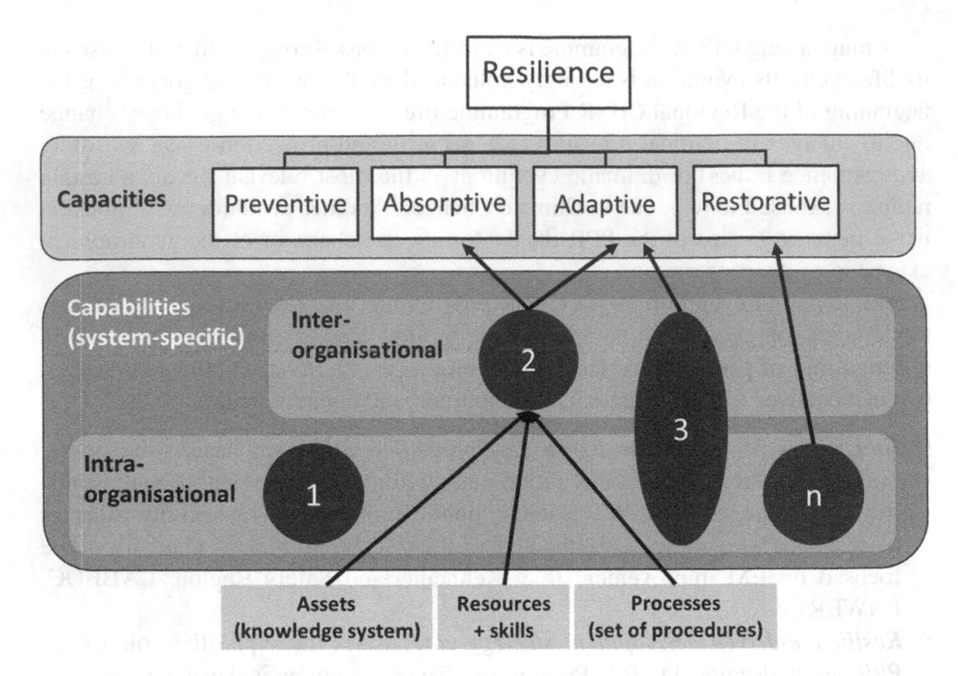

Fig. 8.14 Building system resilience (Adapted from Kozine and Andersen 2015)

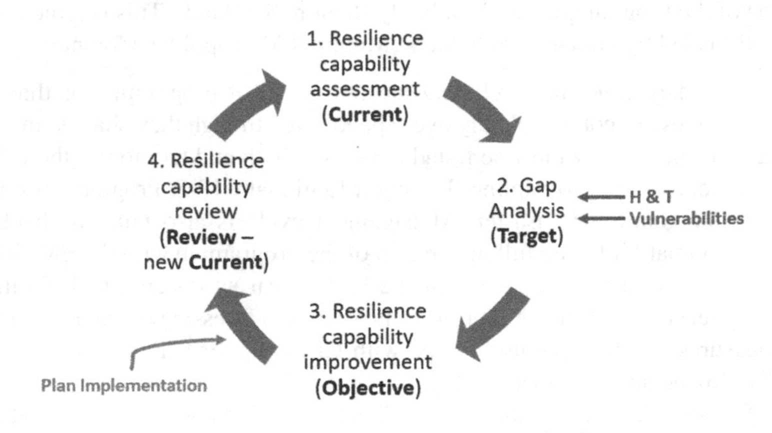

Fig. 8.15 Capability building cycle (Adapted from Trucco et al. 2016)

(by supporting the preparedness and collaborative planning activities) and demonstrated the applicability of the approach (Trucco et al. 2016).

8.5.3 Good Practices

Good practices (GPs) are generally defined as 'Commercial or professional procedures that are accepted or prescribed as being correct or most effective'. It is any collection of specific methods that when applied solve an existing problem, produce expected results and bring benefits. Within the context of these guidelines, the concept applies to available knowledge to addressing:

- Establishment and management of regional PPPs for Critical Infrastructure Protection and Resilience;
- The implementation of a CIP-R programme in an efficient and effective way, thus assuring the achievement of its main objectives and goal.

The rationale is to disseminate and promote a set of practices that have been effective in addressing key issues in regional CIP-R initiatives and, as such, can be deemed as a reference or source of expertise to set-up and develop regional CIP-R programmes.

The practices can be classified into two main categories:

- Activities and processes;
- Tools and technologies.

The collection of the identified GPs (Trucco et al. 2015) had also been evaluated along three main parameters (implementation effort, transferability potential, type and relevance of expected benefits) through engagement of international experts, professionals and researchers the GPs.

8.5.4 Context Factors

Finally, different context factors influence the contents and the management model of a Regional CIP-R Programme in its start-up and evolutionary process phases.

1. **Legislation** (national and local) – In addition to EC legislation on CIP (EC Directive 2008/114/EC) and the related European Programme for Critical Infrastructure Protection (EPCIP), national legislation in single Member States plays a crucial role in shaping regional initiatives on CIP-R.

   In several EU countries, opportunities for engaging stakeholders in the development of a local CIP-R programme might be found directly in extant legislation:

- thanks to awareness and concern of CI operators induced by responsibilities and enforced requirements for action (e.g. in Scotland and Italy);
- well established culture and standards at national level thanks to a coherent regulatory framework (e.g. The Netherlands);
- resources – financial, technological, skills and knowledge – made available at nation-al level (e.g. in the Netherlands and Denmark);
- full or partial devolution of CIP-R responsibilities with clear interfaces with national and EC levels (e.g. Scotland).

2. **PPP Characteristics** – characteristics of the PPP that leads a specific Regional CIP-R Programme have strong influence on the scope, objectives, activities, and also on the quality of achievements of the programme itself.

   Governance by a Lead Organisation is the most common form of governance for PPPs addressing CIP-R issues, where the leading organisation is a public body with specific responsibilities for CIP-R at national or regional level. There are some examples of shared governance where private CI operators are more directly engaged in defining scope and goals of the collaborative network, such as the case of Montreal Metropolitan Community (Canada). In this case, the PPP is collectively led by CI operators and technically sup-ported by the Ecole Polytechnique de Montreal, scope and goal of the partnerships is however limited to the understanding and assessment of interdependencies among CI in the area. Public authorities receive the results of the assessment and are only involved in the evaluation of potential win-win solutions for the removal of most critical dependencies or their better management. These types of PPPs are characterised by a few number of member organisation, a narrow and well-focused goal but with a high level of consensus.

   When the PPP is a direct result of a public policy for involving the private sector in the development and/or implementation of CIP-R programmes, such as in The Netherlands where "Security Regions" are mandated to do so, the size is generally larger, and the scope and objectives are predominantly set by public authorities. This highest potential for impact on the region is balanced by the challenges brought by a relatively lower level of trust and goal consensus. As an example, it may happen that in practice not all the involved CI operators are willing to commit themselves to implementing collaborative plans, containing additional responsibilities or the mobilisation of additional resources, put in place under the strong leadership of public authorities.
3. **CIP-R Programme Scope and Goal** – Characteristics of a Regional CIP-R programme strongly depend on the scope of the programme and on the goals that the leading Organisation or PPP want to achieve through the programme. The most important scoping factor is the policy/strategy background set for the programme, that is: protection-centred vs resilience-centred programmes. Other elements that influence the overall scope: The set of CI sectors covered, the type of regional dimension of the programme, the Phases of the EM cycle covered by the programme...

Different goals or priorities are also possible for Regional CIP-Programmes, such as:

- Improvement of Emergency and Disaster Management (also cross-border);
- Identification and assessment of CI interdependencies;
- Contribute to a better protection and resilience of National Critical Infrastructure;
- Business continuity of different private sectors, CI operators, SMEs;
- Community resilience.

8.6 Conclusions

PPP's effectiveness and contribution to CIP-R largely depends on the way it has been implemented, its main focus, and the achieved maturity level its inherent relationships. We argue that PPPs present a good way to tackle CIP-R issues on low (regional) level, but there is still a long way to go before reaching their full potential. Things are constantly changing so it makes it a never-ending process to make sure that you have a partnership and system that can respond effectively and efficiently when you need it. The emergence of CIP-R focused PPPs opens to the need of (re) defining missions, mutual relationships and governance models of multilevel CIP-R programmes (continental, national, local, ...) as key factors for effective and efficient policies. Another challenge is the important role of universities and research centres in developing and incorporating appropriate and relevant scientific knowledge into the activities and efforts in protection and resilience of CI.

Regional level is the operational level where the CIP-R concerns are first tackled, the first one to carry the burden of incident response, and the level where the major portion of resilience capabilities are built. The implementation process has to be mainly developed according to a bottom-up approach, since the top-down approach (which policy-making follows) does not ensure the success of CIP-R measures (FEMA 2011). In order to properly address all the challenges, it is important and necessary to align programmes on different levels – i.e. in which way are regional CIP-R strategies able to address CIP-R issues at higher levels, and vice versa, how can CIP-R policies and strategies support a bottom-up approach in the form of regional programmes. Such an alignment needs to consider many aspects and dimensions in the context of CIP-R, such as principles, goals and objectives of these programmes and the multi-level partners, responsibilities, and activities.

What is even more important, is to avoid fragmentation within a regional CIP-R programme, from strategic to operational level. Regional programmes are based on the willingness of each partner to assume responsibility for a share of the effort. The focus is on the process of building (collaborative) capabilities to anticipate, absorb, adapt to, and/or rapidly recover from a potentially disruptive event. The development of those capabilities (whether technical, organisational, social or economic), needs to be aligned with the programme strategy. Well-developed programmes following this approach are able to form cross-regional (and cross-border) relationships and possess ability to easily scale up, if needed.

This study can be used by both public and private entities to better understand the distinctive features of CIP-R related PPPs, their establishment, functioning and managements, possible strengths and weaknesses and different ways of achieving practical objectives.

Acknowledgment The authors would like to express their gratitude to all the people involved in the four PPPs and collaborating organizations – namely, Centre Risque & Performance (CRP), Department of Mathematics and Industrial Engineering at École Polytechnique de Montréal (Québec, Canada), Organization of Civil Protection of Montreal (L'Organisation de sécurité civile de l'agglomération de Montréal – OSCAM), National Incident Management Systems and Advanced Technologies (NIMSAT) Institute at University of Louisiana, Lafayette (LA, USA), Pacific NorthWest Economic region, Seattle (WA, USA), GAP-Santé research unit at University of Ottawa (Ontario, Canada), Directorate General on Security and Civil Protection, Lombardy Region (Milan, Italy), Scottish Government (UK), Kennemerland Safety Region (The Netherlands), Oresundsbron (Denmark-Sweden), Technical University of Denmark (DTU), Danish Emergency Management Agency (DEMA) and Copenhagen Capital Region Pre-Hospital Services (Pre-Hosp) – for their help, time, collaborative efforts, availability for interviews, documentation and kind hospitality.

The present study has been funded with support from European Commission, under CIPS/ISEC Work Programme. The financial support is gratefully acknowledged. This publication reflects the views only of the author, and the European Commission cannot be held responsible for any use which may be made of the information contained therein.

Further Suggested Readings

American Heritage Dictionary of the English Language (1996) 3rd edn

Anderson JJ, Malm A (2006), Public-private partnerships and the challenge of critical infrastructure protection. In: Dunn M, Mauer V (eds) International critical information infrastructure protection handbook 2006, vol II. Center for Security Studies, ETH Zurich, pp 139–167

Auerswald P, Lewis MB, La Porte TM, Michel-Kerjan E (2005) The challenge of protecting critical infrastructure. Issue Sci Technol XXII(1):77–83

Barnes J, Newbold K (2005) Humans as a critical infrastructure: public-private partnerships essential to resiliency and response. First IEEE international workshop on critical infrastructure protection, November 3–4, 2005 – Darmstadt, Germany

Beaton EK, Boiney LG, Drury JL, GreenPope RA, Henriques RD, Howland M, Klein GL (2010) Elements needed to support a crisis management collaboration framework. Integrated Communications Navigation and Surveillance Conference (ICNS), 11–13 May 2010, Herndon, VA

Benbasat I, Goldstein DK, Mead M (1987) The case research strategy in studies of information systems. MIS Q:369–386

Boin A (2005) Designing effective response structures: a discussion of established pitfalls, best practices and critical design parameters. A background paper prepared for the Swedish Tsunami Commission

Boin A, McConnell A (2007) Preparing for critical infrastructure breakdowns: the limits of crisis management and the need for resilience. J Conting Crisis Manag 15(1):50–59

Boone W (2012) Full spectrum resilience: an executive summary, CIP report June 2012. Center for Infrastructure Protection and Homeland Security, George Mason University, Fairfax, VA (USA)

Bouchon S (2006) The vulnerability of interdependent. Critical infrastructures systems: epistemological and conceptual state-of-the-art. Institute for the Protection and Security of the Citizen, EC JRC

Cagno E, De Ambroggi M, and Trucco P (n.d.) Interdependency analysis of CIs in real scenarios. In: Bérenguer G, Soares G (eds) Proceedings of ESREL 2011 – advances in safety, reliability and risk management. Taylor & Francis Group, London, pp 2508–2514, ISBN 978-0-415-68379-1

Cavelty MD, Suter M (2008) Early warning for critical infrastructure protection and the road to public-private information sharing. Inteligencia Y Seguridad 4

Cavelty MD, Suter M (2009) Public-private partnerships are no silver bullet.... Int J Crit Infrastruct Protect 2(4):179–187

COM (2005) European Commission's communication to the European Parliament, the Council, the European Economic and Social Committee and The Committee of the Regions on Public-Private Partnerships and Community Law on Public Procurement and Concessions, COM(2005) 569 final

COM (2006) Communication from the Commission of 12 December 2006 on a European programme for critical infrastructure protection, COM(2006) 786 final

Comfort LK (2007) Crisis management in hindsight: cognition, communication, coordination, and control. Public Adm Rev 67(Supplement s1):189–197

Conclusions of the European Council of 10/11 December 2009 on 'The Stockholm programme – an open and secure Europe serving and protecting citizens (2010–2014)'; 17024/09

CRN Report (2009, March) Focal report 2: critical infrastructure protection. Zurich

CRN Roundtable Report (2009, November 27) 6th Zurich roundtable on comprehensive risk analysis and management: network governance and the role of public- private partnerships in New Risks

De Bruijne M, Van Eeten M (2007) Systems that should have failed: critical infrastructure protection in an institutionally fragmented environment. J Contingen Crisis Manag 15(1):18–29

Denzin N (1984) The research act. Prentice Hall, Englewood Cliffs

Department of Homeland Security (DHS) (2006) National infrastructure protection plan (NIPP). Washington, DC

Department of Homeland Security (DHS) (2009) National infrastructure protection plan (NIPP): partnering to enhance protection and resiliency, Washington, DC

Department of Homeland Security (DHS) (2013) National infrastructure protection plan (NIPP): partnering for critical infrastructure security and resilience. Washington, DC

Eisenhardt K (1989) Building theories from case research. Acad Manag Rev 14:532–550

European Commission (2005) Green paper on a European programme for critical infrastructure protection, COM(2005) 576 final

European Commission (2012) Staff working document on the review of the European Programme for Critical Infrastructure Protection (EPCIP), SWD(2012) 190 final

European Commission (EC) (2013) DG Home Affairs – Critical Infrastructures. Available at: http://ec.europa.eu/dgs/home-affairs/what-we-do/policies/crisis-and-terrorism/critical-infrastructure/index_en.htm

European Council (2008) Directive 2008/114/EC of 8 December 2008 on the identification and designation of European critical infrastructures and the assessment of the need to improve their protection. Official Journal of the European Union

FEMA (2011) Five years later: an assessment of the post Katrina Emergency Management Reform Act, Written Statement of Craig Fugate. FEMA Administrator

FEMA (2012) After action report of the first national conference on "Building Resilience through Public-Private Partnerships", August 3–4, 2011, Washington, DC. Progress published January 2012

Givens AD, Busch NE (2013) Realizing the promise of public-private partnerships in U.S. critical infrastructure protection. Int J Crit Infrastruct Prot 6(1):39–50
Heineman BW Jr (2011) Crisis management failures in Japan's reactors and the BP spill. Harvard Business Review
Helbing D, Ammoser H, Kuhnert C (2006) Information flows in hierarchical networks and the capability of organizations to successfully respond to failures, crises, and disasters. Phys A Stat Mech Appl 363(1):141–150
Henriques F, Rego D (2008) OASIS tactical situation object: a route to interoperability. In: Proceedings of the 26th annual ACM international conference on Design of communication (SIGDOC 2008). ACM, New York, pp 269–270
Hilton BN (2007) Emerging spatial information systems and applications. Idea Group Publishing, Hershey
Kozine I, Andersen HB (2015) Integration of resilience capabilities for critical infrastructures into the emergency management set-up. In: Proceedings of the annual European safety and reliability conference – ESREL 2015 conference, September 2015, Zurich, Switzerland
Kröger W (2008) Critical infrastructures at risk: a need for a new conceptual approach and extended analytical tools. Reliab Eng Syst Saf 93(12):1781–1787
Lecomte EL, Pang AW, Russell JW (1998) Ice storm 98, The Institute for Catastrophic Loss Reduction (ICLR). Diane Pub Co, Toronto
Lemyre L, Pinsent C, Boutette P, Corneil W, Riding J, Riding D, Johnson C, Lalande-Markon M, Gibson S, Lemus C (2011) Research using in vivo simulation of meta-rganizational Shared Decision Making (SDM). Task 3: testing the shared decision making framework in vivo. Defence Research and Development Canada Ottawa (Ontario), Centre For Security Science
Lombardy Region (2007) Regione Lombardia: PRIM 2007–2010, Programma Regionale Integrato di Mitigazione dei Rischi, Studi Preparatori – Incidenti ad elevata rilevanza sociale in Lombardia, Regione Lombardia – Protezione civile, Prevenzione e Polizia Locale, 2007 (in Italian)
McEvily B, Perrone V, Zaheer A (2003) Trust as an organizing principle. Organ Sci 14:91–103
McNally RK, Lee S-W, Yavagal D, Xiang W-N (2007) Learning the critical infrastructure interdependencies through an ontology-based information system. Environ Plan B Plan Des 34(6):1103–1124
Meredith J (1998) Building operations management theory through case and field research. J Oper Manag 16:441–454
Moteff JD, Stevens GM (2003, January 29) Congressional research service, critical infrastructure information disclosure and homeland security, RL31547
Moynihan DP (2009) The network governance of crisis response: case studies of incident command systems. J Public Adm Res Theory 19(Issue 4):895–915
Natarajan N (2013) Partnerships and information sharing: The administration's efforts to enhance critical infrastructure security and resilience. CIP report April 2013, Center for Infrastructure Protection and Homeland Security, George Mason University, VA (USA)
National Commission on the BP Deepwater Horizon Oil Spill and Offshore Drilling (2011) Final report: BP deepwater horizon oil spill and offshore drilling, released January 11th, 2011
National Incident Management Systems and Advanced Technologies (NIMSAT) Institute (2012a), Big business – small business mentorship program launches for enhanced disaster resiliency, press release on June 1st, 2012
National Incident Management Systems and Advanced Technologies (NIMSAT) Institute (2012b) Compendium of public-private partnerships for emergency management
National Infrastructure Advisory Council (NIAC) (2006) Public-private sector intelligence coordination – final report and recommendations by the council
National Infrastructure Advisory Council (NIAC) (2009) Critical infrastructure resilience – final report and recommendations, September 8, 2009
National Infrastructure Advisory Council (NIAC) (2012) Intelligence information sharing, Final Report and Recommendations

NATO (2007) Architecture Framework v3
Ouyang M (2014) Review on modeling and simulation of interdependent critical infrastructure systems. Reliab Eng Syst Saf 121:43–60
Percy S (2007) Mercenaries: strong norm, weak law. Int Organ 61:367–397
Petrenj B, Lettieri E, Trucco P (2012) Towards enhanced collaboration and information sharing for critical infrastructure resilience: current barriers and emerging capabilities. Int J Critic Infrastruct Spec Issue Next Gen Critic Infrastruct Syst Challeng Sol Res 8(2/3):107–120
Petrenj B, De Ambroggi M, Trucco P (2013) Simulation-based characterisation of critical infrastructure system resilience: application to a snowfall scenario. In: Proceedings of ESREL 2013
Prieto DB (2006) Information sharing with the private sector: history, challenges, innovation, and prospects. In: Auerswald PE, Branscomb LM, La Porte TM, Michel- Kerjan EO (eds) Seeds of disaster, roots of response: How private action can reduce public vulnerability. Cambridge University Press, Cambridge, pp 404–428
Provan K, Kenis P (2008) Modes of network governance: structure, management, and effectiveness. J Public Adm Res Theory 18(Issue 2):229–252
Pursiainen C (2009) The challenges for European critical infrastructure protection. J Eur Integr 31(6):721–739
Reason P, Bradbury H (eds) (2008) The sage handbook of action research: participative inquiry and practice. Sage, Thousand Oaks
Robert B, Morabito L (2010) An approach to identifying geographic interdependencies among critical infrastructures. Int J Crit Infrastruct 6(1):17–30
Robert B, Morabito L, Quenneville O (2007) The preventive approach to risks related to interdependent infrastructures. Int J Emerg Manag 4(2)
Robert B, de Calan R, Morabito L (2008) Modelling interdependencies among critical infrastructures. Int J Crit Infrastruct 4(4):392–408
Roby CJ, Alberts DS (2012) NATO NEC C2 maturity model. DoD command and control research program. Washington, DC. Available at: www.dodccrp.org
Schraagen JM, Veld MH, De Koning L (2010) Information sharing during crisis management in hierarchical vs. network teams. J Conting Crisis Manag 18(2):117–127
Seuring S (2008) The rigor of case study research in supply chain management. Supply Chain Manag Int J 13(2):128–137
Shani AB, Pasmore WA (1985) Organization inquiry: towards a new model of the action research process. In: Warrick D (ed) Contemporary organization development. Scott, Foresman, Glenview, pp 438–448
Shani AB, David A, Willson C (2004) Collaborative research: alternative roadmaps. In: Adler N, Shani AB, Styhre A (eds) Collaborative research in organisations, foundations for learning, change and theoretical development. Sage Publications, Thousand Oaks, pp 83–100
Sue E (2005) Eckert, protecting critical infrastructure: the role of the private sector in guns and butter. In: Dombrowski P (ed) The political economy of international security. Lynne Rienner Publishers, Boulder
Susman GI, Evered RD (1978) An assessment of the scientific merit of action research. Adm Sci Q 23:583–603
The White House (2013) Presidential Policy Directive – Critical Infrastructure Security and Resilience (PPD-21). Available at: http://www.whitehouse.gov/the-press-office/2013/02/12/presidential-policy-directive-critical-infrastructure-security-and-resil
Trucco P, Cagno E, De Ambroggi M (2012) Dynamic functional modelling of vulnerability and interoperability of critical infrastructures. Reliab Eng Syst Saf 105:51–63. doi:10.1016/j.ress.2011.12.003
Trucco P, Petrenj B, Bouchon S, Di Mauro C (2015) The rise of regional programmes on critical infrastructure resilience: identification and assessment of current good practices. WIT Transac Built Environ 150:233–245
Trucco P, Petrenj B, Kozine I, Andersen BH (2016) Emergency management involving critical infrastructure disruptions: operationalizing the deployment of resilience capabilities. In: Walls

R, Bedford (eds) (2017) Risk, reliability and safety: innovating theory and practice, ESREL 2016.Taylor & Francis Group, London, pp 548–555, ISBN 978-1-138-02997-2
U.S. DHS (2012) Homeland security grant program – supplemental resource: support for public-private collaboration
U.S. President's Commission on Critical Infrastructure Protection (1997)
United Nations (2005) Report of the world conference on disaster prevention, Kobe (Hyogo, Japan), 18–22, January 2005
United Nations (2006) International Strategy for Disaster Reduction (UN/ISDR, 2006) – platform for the promotion of early warning. Available at: http://www.unisdr.org/2006/ppew/whats-ew/basics-ew.htm
United States General Accounting Office (US GAO) (2004) Critical infrastructure protection. Establishing Effective Information Sharing with Infrastructure Sectors
US Department of Homeland Security (US DHS) (2013) US Department of Homeland Security (US DHS) website. Critical infrastructure protection partnerships and information Sharing (http://www.dhs.gov/critical-infrastructure-protection-partnerships-and-information-sharing), visited on 10/04/2013
Voss C, Tsikriktsis N, Frohlich M (2002) Case research in operations management. Int J Oper Prod Manag 22(2):195–219
Walsham G (1995) The emergence of interpretivism in IS research. Inf Syst Res 6(4):376–394
Whyte WF, Greenwood D, Lazes P (1989) Participatory action research. Am Behav Sci 32:513–551
Willis HH, Lester G, Treverton GF (2009) Information sharing for infrastructure risk management: barriers and solutions. Intellig Nat Secur 24(3):339–365
Yin R (2003) Case study research: design and methods. Sage Publications, Thousand Oaks

Part IV
Social

Chapter 9
Social Resilience and Critical Infrastructure Systems

Benjamin D. Trump, Kelsey Poinsatte-Jones, Meir Elran, Craig Allen, Bojan Srdjevic, Myriam Merad, Dejan M. Vasovic, and José Manuel Palma-Oliveira

Abstract Resilience analysis and thinking serve as emerging conceptual frameworks relevant for applications assessing risk. Connections between the domains of resilience and risk assessment include vulnerability. Infrastructure, social, economic, and ecological systems (and combined social-ecological systems) are vulnerable to exogenous global change, and other disturbances, both natural and anthropologically derived. Resilience analysis fundamentally seeks to provide the groundwork for a 'soft landing', or an efficient and robust restoration following disturbance as well as the ability to reduce harms while helping the targeted system rebound to full functionality as quickly and efficiently where possible. Such applications are consistent

B.D. Trump (✉)
US Army Corps of Engineers, Engineer Research and Development Center, Vicksburg, MS, USA

University of Lisbon, Lisbon, Portugal
e-mail: Benjamin.D.Trump@usace.army.mil

K. Poinsatte-Jones
US Army Corps of Engineers, Engineer Research and Development Center, Vicksburg, MS, USA

M. Elran
Tel Aviv University, Tel Aviv-Yafo, Israel

C. Allen
University of Nebraska, Lincoln, NE, USA

B. Srdjevic
University of Novi Sad, Novi Sad, Serbia

M. Merad
National Center for Scientific Research, Paris, France

D.M. Vasovic
University of Niš, Niš, Serbia

J.M. Palma-Oliveira
University of Lisbon, Lisbon, Portugal

© Springer Science+Business Media B.V. 2017

I. Linkov, J.M. Palma-Oliveira (eds.), *Resilience and Risk*, NATO Science for Peace and Security Series C: Environmental Security,
DOI 10.1007/978-94-024-1123-2_9

with The National Academy of Sciences (NAS) definition of resilience, which more broadly denotes the field as "the ability to plan and prepare for, absorb, recover from, and adapt to adverse events" (Larkin S, Fox-Lent C, Eisenberg DA, Trump BD, Wallace S, Chadderton C, Linkov I (2015) Benchmarking agency and organizational practices in resilience decision making. Environ Sys Decisions 35(2):185–195). Given this definition, we seek to describe how resilience analysis and resilience thinking might be applied to social considerations for critical infrastructure systems. Specifically, we indicate how resilience might better coordinate societal elements of such infrastructure to identify, mitigate, and efficiently recover from systemic shocks and stresses that threaten system performance and service capacity.

9.1 Introduction

Resilience analysis and thinking serve as emerging conceptual frameworks relevant for applications assessing risk. Connections between the domains of resilience and risk assessment include vulnerability. Infrastructure, social, economic and ecological systems (and combined social-ecological systems) are vulnerable to exogenous global change, and other disturbances, both natural and anthropologically derived. Here, resilience analysis fundamentally seeks to provide the groundwork for a 'soft landing', or an efficient and robust restoration following disturbance as well as the ability to reduce harms while helping the targeted system rebound to full functionality as quickly and efficiently where possible. Such applications are consistent with The National Academy of Sciences (NAS) definition of resilience, which more broadly denotes the field as "the ability to plan and prepare for, absorb, recover from, and adapt to adverse events" (Larkin et al. 2015).

However, resilience is also being applied to the context of individual and collective social behaviors – i.e. social resilience (Berkes et al. 2008; Pelling 2012; Cacioppo et al. 2011). Specifically, Cacioppo et al. (2011) note that "social resilience [...] is inherently a multilevel construct, revealed by capacities of individuals and groups to respectively foster, engage in, and sustain positive social relationships and to endure and recover from stressors and social isolation." In other words, social resilience serves as a metaphor for the ability of individuals and communities to adapt, reorganize, and improve in the midst of external shocks and stresses (Norris et al. 2008; Pelling 2012). However, regardless of an individual or collective focus relative to social resilience, phenomena systems focus is essential where even individual resilience is strongly influence by social and physical factors "outside" of the particular individual (Sippel et al. 2015).

Further, in almost all levels of systems resilience, resilience is directly affected by the involvement and participation of key stakeholders before, during, and after a disruption to a system. Specifically, the actions of such stakeholders may mitigate or exacerbate such disruptions based upon the public's awareness of the challenge and to its relevant response strategies, which find expression in (i) the resources and readiness committed to promoting resilience for a given system and community, (ii) the priorities defined and the risks evaluated and (iii) the efforts at crafting effective involvement and risk communication to align community actions, reactive or preventive. Based upon the system and community in question, such stakeholders may range from the communities at large to local, regional, and national as well as policymakers and decision makers, where specific focus must be levied upon engaging the correct and relevant stakeholders to prepare for and recover from adverse events. In this way, stakeholder engagement is an essential element of promoting social resilience in response to various adverse events and system stresses, and serves as a key task in aligning incentives and unifying members of government, industry, and the lay public to address potential or ongoing challenges that may otherwise generate lasting harms to given communities.

This chapter reviews discussion on social resilience held at a NATO Conference in the Azores, Portugal, from June 26–29, 2016. Specifically, this chapter includes the perspectives of various participants of the Social Resilience Working Group, which was tasked with the goal of addressing (i) the purpose and definition of social resilience, (ii) how social resilience is fostered, and (iii) challenges that complicate social resilience or otherwise must be considered for future work in the field. In this respect, the Working Group discussed social resilience as comprising both 'resilient societies' as well as 'the actions and involvement of key social groups to improve resilience of systems from engineering applications'. As such, this chapter begins with a general review of social resilience and resilient societies, and then reviews how social mechanisms such as stakeholder engagement are critical to promote system resilience for ecological or infrastructural applications.

9.2 Resilience in the Social Domain

Societal resilience has come to include several different activities and areas of study, such as with community resilience amid gradual stresses (Adger et al. 2002; House 2007; Simich and Andermann 2014) and more acute events such as climatological and/or ecological shocks (Pendall et al. 2009; Leichenko 2011; Smith and Stirling 2010). Unpacking such discussion, 'shocks' include those events that generate unexpected and fast-acting effects upon individuals or communities within a particular area (Mitchell and Harris 2012). Typically referred to as 'one-off events', or in situations of high uncertainty and surprise 'black swans', shocks are characterized by the relatively rapid manner in which they appear and subside (although their resulting impact can cause long-lasting damage). From a climatological perspective, examples of such shocks include Hurricanes Katrina and Sandy in the United States, which made landfall on certain coastal communities and dramatically damaged local economies and public health over an extended period.

On the other hand, 'stresses' include those influences or factors that, over a more extended period of time, challenge and potentially compromise individual or group resilience. Rather than overwhelm existing infrastructural and systemic resources within a brief period, such stresses act slowly to reduce such system's efficiencies and abilities to perform at a high level. Examples here include challenges from climate change, which can hinder agricultural output and community wellbeing over time without proper protections and countermeasures. Other examples may include the effect of mass migration upon society, where gradual effects can stress infrastructural systems in the absence of a plan to accommodate a large influx of people. While stresses generally may not possess dramatic adverse effects within a short period, the consequences of such gradual effects may be equally challenging to societal resilience and the ability of respective communities to absorb and recover from potential threats and problems over time.

Social resilience within this context may apply to societies and communities of various size, ranging from local neighborhoods and towns to more regional or national governments. For smaller communities, organizations, and businesses, discussions of resilience may center on the ability of local governments and set communities to address long-term concerns such as with the impact of climate change (Berkes and Jolly 2002), ecological disasters (Adger et al. 2005; Cross 2001), earthquakes (Bruneau et al. 2003), and cybersecurity concerns (Williams and Manheke 2010), as well as other manmade hazards such as transnational wars, civil wars, terrorism, migration, and industrial hazards. For larger communities and governments, such concerns are similar yet often more complex and varied in nature, where they involve hundreds to potentially thousands of stakeholders and include the interaction of various infrastructural systems.

Regardless of the size and characteristics of the community observed, an important consideration for any social resilience exercise includes the notion of panarchy, or the ability of differing systems and sub-systems to affect and potentially harm other systems and sub-systems during various shocks and stresses (Walker et al. 2004; Garmestani et al. 2008). Within the concept of panarchy, a systemic shock or stress may generate cascading effects and feedback loops that overwhelm system capacities to absorb and recover from adverse events. Panarchy serves as a framework for a complex series of interaction effects that, without a contextually rich and thorough understanding of how a given system interacts and operates with other infrastructural and societal elements, can greatly exacerbate the damages wrought by shocks and stresses over time. Palma-Oliveira and Trump (2016) argue that "understanding the consequences and magnitude of such cascade effects is crucial to identify areas where systems may be brittle or resilient."

Panarchic effects are particularly troublesome for large governments due to the many stakeholders and interconnected infrastructural systems that must be accounted for on a grand scale (Angeler et al. 2016; Cross 2001). For social considerations, such cascading effects from various shocks and stresses may overtax the ability of societies to absorb adversity and maintain normatively beneficial growth and development. DeWitte et al. (2016) note epidemic disease as one example of such complications throughout history, where the arrival bubonic plague of the fourteenth century-onward often shattered commerce and daily life in Europe. Linkov et al. (2014) applied similar lessons to modern epidemics as with the ebolavirus, which Ali et al. (2016) describe as overtaxing local public health authorities and drastically limiting economic activity within ebola-endemic areas.

Further, Walker et al. (2004), Magis (2010), and Briske et al. (2010) argue that systems generally have two outcomes in the face of an external shock event – either (i) they absorb the shock and any temporary losses in system optimality in order to return to full function at a later time, or (ii) collapse and reorganize under the strain of the shock. For the former, the system seeks to preserve itself by adapting to adverse events and recovering to near or total efficiency over a period of time (Walker et al. 2004; Briske et al. 2010). Further, such systems are considered resilient to a varying degree to those particular shocks or stresses (predictable or unpredictable), where they possess the ability to weather such challenges without completely collapsing and inducing permanent damage to social function. While often discussed as being an inherently positive trait within published literature, this is not necessarily always true, where a resilient system may be harmful or a reinforcing social trap in nature (Palma-Oliveira and Trump 2016).

For the latter, shocks that overwhelm system capacity to absorb challenges and operate normally can cause the system to fail outright (Briske et al. 2010). Upon such failure, the system might be reorganized in a manner that differs from its original state, either in the form of more beneficial and robust action in response to similar shocks in the future, or as a more brittle and/or negatively reinforcing set of actions and behaviors. History is crowded with countless examples of societal collapse on the micro and macro scales – some of which were able to rebuild and prosper, while others struggled under social traps like with recurring environmental damage, economic weakness, poor public health, and many others (Redman and Kinzig 2003; Dai et al. 2012; Schwartz and Nichols 2010).

With respect to improving social resilience, Linkov et al. (2014), Larsen et al. (2011), Bosher et al. (2009), and Djalante (2012) state that a dedicated response by key stakeholders and government decision makers is required to mitigate and manage shocks to societal resilience, where such actors foster 'safe-to-fail' and recovery options when an adverse shock arises. In this vein, stakeholders and decision makers that are actively involved with promoting social resilience in anticipation of various shocks and stresses (and take appropriate steps to fund and create relevant systems to shore up social resilience) may improve the resilience capacity of such systems before, during, and after the imposition of a shock or stress (Djalante 2012; Bosher et al. 2009).

Looking at an example of multi-purpose water resource systems, surface reservoirs are often located near important urban or industrial infrastructures. Breakage in water containment capacities as with dams, levees, and other infrastructure can contribute to sudden flood events without proper controls put into place beforehand. A critical component of such controls (and thereby bolstering water system resilience) centers on the willingness of decision makers and key stakeholders to invest resources into promoting such helpful countermeasures – where a failure to do so may result in widespread and potentially lasting damages to infrastructure, the local economy and public health.

Over extended time horizons, the critical infrastructural resilience for applications like water resource management (dams, pumping facilities, evacuation structures, spills etc.) should be modeled as an attempt to reduce recovery time post-shock as much as feasible (Hashimoto 1980; Hashimoto et al. 1982). In turn, this requires, based on the magnitude of damage caused by the disruption: (i) the existence of precise plan of recovery, (ii) pre-defined responsibilities, and (iii) sets of relevant operational actions to ensure control and reduction of damages. In each of these considerations, there exists a clear need for systems analysis techniques in various planning and construction phases to tackle aforementioned problems (Moy et al. 1986). For instance, systems analysis may help to simulate possible hydrologic conditions and various operational scenarios to control water flows into and from reservoirs (McMahon et al. 2006). With such analysis, social impacts may be weighted to be of high importance, including consideration of time intervals to determine shifting social priorities and/or infrastructural functions to meet the demands of local economic, health, and social needs (Srdjevic and Srdjevic 2016).

9.3 The Various Levels of Social Groups Relevant to Infrastructural Resilience

An essential consideration of promoting resilience in infrastructural systems includes an understanding of the differing levels of stakeholder involvement within the process of developing resilience. In essence, risk analysis and governance for uncertain technologies and infrastructures such as with critical infrastructural systems do not exist in a vacuum (Berkes and Jolly 2002; Pelling 2012; Larkin et al. 2015; Trump et al. 2017; Trump 2016), and often depend heavily on the actions, inaction, and interactions of various individuals and groups that operate and use such services. In this way, policymakers and stakeholders concerned with critical infrastructure resilience must be mindful of the social factors and drivers that may facilitate or hinder response and recovery from adverse shocks and stresses (Chapin et al. 2004; Palma-Oliveira et al. 2017).

A key requirement of reviewing critical infrastructure resilience is to consider what it is that must be protected. For many cases in ecological, cyber, medical, and energy security, a primary concern is the continued safety and delivery of services to the local population – not simply the protection of a singular infrastructure project or service (Pelling 2003). As such, to promote social wellbeing, safety, and economic action, key stakeholders and decision makers within various communities and governments must account for social interaction and response to adverse shocks to various systems.

For example, amid pandemic disease, steps must be taken to reassure the public and promote continued economic activity while working to reduce disease incidence. Simply focusing on combating disease may help stave off a larger epidemic and reduce the rate of incidence, but ignoring social factors and not communicating clearly with the public may still generate disastrous economic, social, and medical harms (Davtyan et al. 2014). One example of this includes the 2014 outbreak of the ebolavirus, where poor communication and networking by governments with the public not only caused significant damage to the local economy, but also instilled mistrust by locals and health workers of the government's effort to combat the virus (Torabi-Parizi et al. 2015). Another example includes HIV/AIDS, where limited public engagement, risk communication, or consideration of social factors directly contributed to the increased stigma of HIV positive individuals and discouraged many from seeking treatment at the end of the twentieth century (Vega 2016). Internationally, similar observations were noted with the SARS epidemic in China, where a mistrust of local government caused many to refuse to seek medical care and contributed to a more lasting outbreak of the disease (Chan et al. 2016).

In this way, policymakers and key stakeholders must be mindful of the need to strengthen and protect infrastructural systems to fend off, as much as possible, ecological, medical, and anthropologically-derived shocks, and recover as fast as possible and return to normal functionality. An equally important consideration includes how individuals and groups will behave and interpret information during and after such a shock. Such considerations may include public awareness and risk communication drives, the allocation of financial and physical resources to be activated before and in the event of an emergency, and well-established and distributed safety guidelines and best practices, among others. With such preparation, local and national governments may be able to limit the negative social impacts that inevitably arise from consequential shocks to energy, medical, environmental, or economic infrastructure, and may help reduce recovery time in the aftermath of the event in question.

We review stakeholder involvement in the promotion of resilience as a function of three critical stakeholder groups, which are noted below:

1. Stakeholder groups that deal with infrastructural resilience
2. Stakeholder groups and key parties that deal with infrastructural resilience in disasters
3. Stakeholder groups that evaluate the importance of infrastructural systems and assess risk

For the first group, these stakeholders typically comprise those individuals tasked with the daily management and preservation of the critical infrastructural system at hand. These stakeholders work to build critical infrastructure resilience agnostic of any particular threat, and instead work to promote overall system health and its ability to prevent, protect, mitigate, absorb, and recover from a diverse array of threats. Such stakeholders will often have permanent positions managing such systems, and serve a role of 'maintenance and preparation' as opposed to higher-level resource allocation and system evaluation.

The second group focuses more upon promoting resilience and rebounding system function and efficiency in the midst of disasters and shocks. While such stakeholders are not typically involved in the daily management of a given infrastructural system, their services are engaged in events categorized as disasters or shocks that require abnormal or extraordinary involvement from higher level decision makers and policymakers. More scenario driven, these stakeholders respond to specific threats to a system and its nested sub-systems, and generally maintain a level of expertise on their specific subject (i.e. expertise in cybersecurity, in ecosystem health, in disease control, etc.). Such individuals will interface with the first group of stakeholders, yet may have additional authority to make decisions and report findings should a threat within their area of expertise surface.

Lastly, the third group serves as those stakeholders that assess system risk and evaluate the importance of a given infrastructural system (or group of systems). Given their ability to strategize and distribute resources based upon their decisions, such stakeholders may be more senior level policymakers and decision makers. As such, the crucial role with stakeholder engagement for resilience projects at this stage is to demonstrate the value and necessity of a given system to maintain function and quickly rebound from adverse shocks. Without such a valuation, these stakeholders may not provide the necessary level of resources and manpower needed to plan for, respond to, and recover from such adverse events.

Interfacing with each level of stakeholder is of crucial importance to acquire resources, manpower, and political willpower needed to shore up system resilience to various threats. The first two stakeholder groups are required to engage in 'on-the-ground' tasks to assess risk and work to promote resilience, while the last group must be engaged to ensure that such a system's resiliency remains a priority both now and in the future.

9.4 Future Developments

The growing sophistication and interconnectedness of critical infrastructure ranging from energy to cybersecurity to transportation allow for an improved coordination and delivery of services over time. However, such system interconnectedness and complexity also generates the potential for more consequential and lasting damage to accrue should those systems fail. As noted above, these failures may arise suddenly as shocks or gradually as stresses, and could yield lasting harmful effects upon local societies if not adequately prepared for and recovered from before, during, and after the shock or stress occurs.

Though more substantial and consequential events are relatively rare, recent history has demonstrated how social issues may improve or detract from an infrastructural system's resilience in the face of adversity. One such example includes the impacts of Hurricanes Katrina and Rita, which devastated New Orleans and surrounding areas in Louisiana and resulted in lasting damage to the local economy and public health (Goodman and West-Olatunji 2008; Colten et al. 2008).

Emerging challenges such as with climate change, mass migration, economic instability, pandemic disease, technology innovation, cybersecurity risk, terrorism, and many others all yield potential threats to social resilience and stability (DeWitte et al. 2016; Maguire and Hagan 2007; Seager et al. 2017; Keck and Sakdapolrak 2013). As critical systems such as with energy, medical care, communication, defense, and others continue to centralize and grow in interdependency, external shocks and stresses may cause substantial and cascading system failure that may cause lasting damage to social strength and wellbeing. As such, methods and strategies are required that adequately assess system and nested sub-system resilience across society and inform decision makers of the actions under high uncertainty that must be taken before, during, and in the aftermath of an adverse event.

A further concern includes the future development of quantitative and qualitative methods to correctly assess and measure resilience both for social systems and from a general perspective. As resilience continues to mature and enter the lexicon of micro and macro-scale risk management stakeholders, a necessary development in the field includes the use of practical and user-friendly risk and decision models that illustrate and compute system and nested sub-system resilience. Currently, such methods and regulatory approaches are limited in scope (Linkov et al. 2015). Similarly, Larkin et al. (2015) noted that as of 2015, few methodological approaches of resilience analysis were formally used to facilitate risk assessment within United States government agencies. However, Larkin et al. (2015) did state that many agencies have begun experimenting with prototypical resilience models for local and regional governments and communities, where such methods may become more standardized and mandatory as their use is proven beneficial for bolstering various elements of social and infrastructural resilience.

Given the complexity of the systems and the different levels at present, one further question includes how to integrate and distinguish between the various levels of a system and, within such systems, the different actions that could generate cascading action. The panarchy framework (Cutter et al. 2008) jointly with Linkov et al.'s systematic measurement of indices could help advance our understanding of such complex systems and their resilience. Within such a focus, the involvement and training of stakeholders is a central element of improving social resilience.

Further Suggested Readings

Adger WN, Hughes TP, Folke C, Carpenter SR, Rockström J (2005) Social-ecological resilience to coastal disasters. Science 309(5737):1036–1039

Adger WN, Kelly PM, Winkels A, Huy LQ, Locke C (2002) Migration, remittances, livelihood trajectories, and social resilience. AMBIO J Hum Environ 31(4):358–366

Ali H, Dumbuya B, Hynie M, Idahosa P, Keil R, Perkins P (2016) The social and political dimensions of the Ebola response: global inequality, climate change, and infectious disease. In: Climate change and health. Springer International Publishing, Berlin/Heidelberg, pp 151–169

Angeler DG, Allen CR, Garmestani AS, Gunderson LH, Linkov I (2016) Panarchy use in environmental science for risk and resilience planning. Environment Systems and Decisions 36(3):225–228

Berkes F, Jolly D (2002) Adapting to climate change: social-ecological resilience in a Canadian western Arctic community. Conserv Ecol 5(2):18

Berkes F, Colding J, Folke C (eds) (2008) Navigating social-ecological systems: building resilience for complexity and change. Cambridge University Press, Cambdrige/New York

Bosher L, Dainty A, Carrillo P, Glass J, Price A (2009) Attaining improved resilience to floods: a proactive multi-stakeholder approach. Disaster Prevention and Management: An International Journal 18(1):9–22

Briske DD, Washington-Allen RA, Johnson CR, Lockwood JA, Lockwood DR, Stringham TK, Shugart HH (2010) Catastrophic thresholds: a synthesis of concepts, perspectives, and applications. Ecol Soc 15(3):37

Bruneau M, Chang SE, Eguchi RT, Lee GC, O'Rourke TD, Reinhorn AM, von Winterfeldt D (2003) A framework to quantitatively assess and enhance the seismic resilience of communities. Earthquake Spectra 19(4):733–752

Cacioppo JT, Reis HT, Zautra AJ (2011) Social resilience: the value of social fitness with an application to the military. Am Psychol 66(1):43

Chan KL, Chau WW, Kuriansky J, Dow EA, Zinsou JC, Leung J, Kim S (2016) The psychosocial and interpersonal impact of the SARS epidemic on Chinese health professionals: implications for epidemics including Ebola. In: The psychosocial aspects of a deadly epidemic: what Ebola has taught us about holistic healing, vol 287. Praeger, Santa Barbara

Chapin FS III, Peterson G, Berkes F, Callaghan TV, Angelstam P, Apps M et al (2004) Resilience and vulnerability of northern regions to social and environmental change. AMBIO J Hum Environ 33(6):344–349

Colten CE, Kates RW, Laska SB (2008) Three years after Katrina: lessons for community resilience. Environment: science and policy for sustainable development 50(5):36–47

Cross JA (2001) Megacities and small towns: different perspectives on hazard vulnerability. Global Environ Change B Environ Hazard 3(2):63–80

Cutter S, Barnes L, Berry CB, Evans E, Tate E, Webb J (2008) A place-based model for understanding community resilience to natural disasters. Glob Environ Chang 18(4):598–606

Dai L, Vorselen D, Korolev KS, Gore J (2012) Generic indicators for loss of resilience before a tipping point leading to population collapse. Science 336(6085):1175–1177

Davtyan M, Brown B, Folayan M (2014) Addressing Ebola-related stigma: lessons learned from HIV/AIDS. Glob Health Action 7:26058

DeWitte SN, Kurth MH, Allen CR, Linkov I (2016) Disease epidemics: lessons for resilience in an increasingly connected world. J Public Health

Djalante, R (2012) Adaptive governance and resilience: the role of multi-stakeholder platforms in disaster risk reduction

Garmestani AS, Allen CR, Cabezas H (2008) Panarchy, adaptive management and governance: policy options for building resilience. Neb L Rev 87:1036

Goodman RD, West-Olatunji CA (2008) Transgenerational trauma and resilience: improving mental health counseling for survivors of hurricane Katrina. J Ment Health Couns 30(2):121

Hashimoto, T (1980) Robustness, reliability, resilience and vulnerability criteria for water resources planning, Ph.D. dissertation, Cornel Univ., Ithaca, N.Y.

Hashimoto T, Stedinger JR, Loucks DP (1982) Reliability, resiliency and vulnerability criteria for water resource system performance evaluation. Water Resour Res 18(1):14–20

House QE (2007) Environmentally displaced people: understanding the linkages between environmental change, livelihoods and forced migration

Keck M, Sakdapolrak P (2013) What is social resilience? Lessons learned and ways forward. *Erdkunde* 67(1):5–19

Larkin S, Fox-Lent C, Eisenberg DA, Trump BD, Wallace S, Chadderton C, Linkov I (2015) Benchmarking agency and organizational practices in resilience decision making. Environ Sys Decisions 35(2):185–195

Larsen RK, Calgaro E, Thomalla F (2011) Governing resilience building in Thailand's tourism-dependent coastal communities: Conceptualising stakeholder agency in social–ecological systems. Glob Environ Chang 21(2):481–491

Leichenko R (2011) Climate change and urban resilience. Curr Opin Environ Sustain 3(3):164–168

Linkov I, Fox-Lent C, Keisler J, Della Sala S, Sieweke J (2014) Risk and resilience lessons from Venice. Environ Sys Decisions 34(3):378–382

Linkov I, Larkin S, Lambert JH (2015) Concepts and approaches to resilience in a variety of governance and regulatory domains. Environ Sys Decisions 35(2):183

Magis K (2010) Community resilience: an indicator of social sustainability. Soc Nat Resour 23(5):401–416

Maguire B, Hagan P (2007) Disasters and communities: understanding social resilience. Aust J Emer Manag 22(2):16

McMahon TA, Adeloye AJ, Sen-Lin Z (2006) Understanding performance measures of reservoirs. J Hydrol 324:359–382

Mitchell T, Harris K (2012) Resilience: a risk management approach. ODI Background Note. London, Overseas Development Institute
Moy WS, Cohon JL, Revelle CS (1986) A programming model for analysis of reliability, resilience and vulnerability of a water supply reservoir. Water Resour Res 22(4):489–498
Norris FH, Stevens SP, Pfefferbaum B, Wyche KF, Pfefferbaum RL (2008) Community resilience as a metaphor, theory, set of capacities, and strategy for disaster readiness. Am J Community Psychol 41(1–2):127–150
Palma-Oliveira J, Trump B (2016) Modern resilience: moving without movement. In: International risk governance council handbook on resilience
Palma-Oliveira J, Trump B, Wood M, Linkov I (2017) The tragedy of the anticommons: a solutions for a "NIMBY" post-industrial world. Risk analysis
Pelling M (2003) The vulnerability of cities: natural disasters and social resilience. Earthscan, London
Pelling M (2012) The vulnerability of cities: natural disasters and social resilience. Earthscan, London
Pendall R, Foster KA, Cowell M (2009) Resilience and regions: building understanding of the metaphor. Cambridge J Reg Econ Soc
Redman CL, Kinzig AP (2003) Resilience of past landscapes: resilience theory, society, and the longue durée. Conserv Ecol 7(1):14
Schwartz GM, Nichols, J J (Eds) (2010) After collapse: the regeneration of complex societies. University of Arizona press
Seager T, Trump BD, Poinsatte-Jones K, Linkov I (2017) Why life cycle assessment does not work for synthetic biology. Environ Sci Technol 51(11):5861–5862
Simich L, Andermann L (eds) (2014) Refuge and resilience: promoting resilience and mental health among resettled refugees and forced migrants, vol 7. Springer, New York
Sippel LM, Pietrzak RH, Charney DS, Mayes LC, Southwick SM (2015) How does social support enhance resilience in the trauma-exposed individual? Ecol Soc 20(4):10. http://dx.doi.org/10.5751/ES-07832-200410
Smith A, Stirling A (2010) The politics of social-ecological resilience and sustainable socio-technical transitions. Ecol Soc 15(1):11
Srdjevic B, Srdjevic Z (2016) Multicriteria analysis of the water resources system performance. J Water Resour Vodoprivreda in Press (in Serbian)
Torabi-Parizi P, Davey RT, Suffredini AF, Chertow DS (2015) Ethical and practical considerations in providing critical care to patients with ebola virus disease. Chest J 147(6):1460–1466
Trump, BD (2016) A comparative analysis of variations in synthetic biology regulation. Doctoral dissertation, University of Michigan
Trump B, Cummings C, Kuzma J, Linkov I (2017) A decision analytic model to guide early-stage government regulatory action: applications for synthetic biology. Regul Gov. doi:10.1111/rego.12142
Vega MY (2016) Combating stigma and fear: Applying psychosocial lessons learned from the HIV epidemic and SARS to the current Ebola crisis. The psychosocial aspects of a deadly epidemic: What Ebola has taught us about holistic healing:271
Walker B, Holling CS, Carpenter SR, Kinzig A (2004) Resilience, adaptability and transformability in social–ecological systems. Ecol Soc 9(2):5
Williams PA, Manheke RJ (2010) Small business-a cyber resilience vulnerability

Chapter 10
Societal Resilience: From Theory to Policy and Practice

Meir Elran

Abstract Societal resilience is defined in this essay as the capacity of communities to flexibly contain major disruptions and to rapidly bounce back and forward following the unavoidable decline of their core functionalities. The article examines the ways to translate the theory to a clear and determined policy and hence implementation in the field, before the disruption occurs and henceforth. While the theory is necessarily universal in nature, its practical ramifications are presented within the Israeli context. It is suggested that the Israeli scene, being challenged for decades by protracted terror, as a man made recurrent hazard, can be perceived as a national laboratory for practicing societal resilience.

Keywords Resilience • Societal resilience • Terrorism • Response strategies • Israel

Resilience has become, in recent years, a frequently used "buzzword" in public discourse. Different people interpret resilience in a variety of ways, and not always with reference to mass disasters. In fact, the notion of non-individual, non-psychologically-related resilience is often challenged in the professional realm (Bonanno et al. 2015), if not altogether refuted (Aguirre 2007). Those who have accepted the notion of societal or community resilience tend to interpret it in different ways (Norris et al. 2008). In the present article, the term "resilience" will be used to characterize the capacity of any system, whether it be social, infrastructure, economic, or any combination of these at the local or national level, to withstand the consequences of a major disruption, and to expeditiously recover to its initial level of core functionality.

The purpose of this article is to discuss the ways in which the universal theory of societal resilience is applied through policy and implemented in the field, as well as its practical applications within the Israeli context. Israel has been challenged for decades by protracted terror as a manmade recurrent hazard and the primary model

M. Elran (✉)
The Institute for National Security Studies (INSS), Tel Aviv University, Tel Aviv, Israel
e-mail: meiryelran@gmail.com

© Springer Science+Business Media B.V. 2017

I. Linkov, J.M. Palma-Oliveira (eds.), *Resilience and Risk*, NATO Science for Peace and Security Series C: Environmental Security,
DOI 10.1007/978-94-024-1123-2_10

of societal resilience has been introduced and practiced in Israel long before it became a common nomenclature worldwide. While resilience may not have been optimally exercised in Israel, the mastery of resilience by Israelis can serve as valuable lessons for those who are willing to learn from predicaments and mistakes of experienced nations.

10.1 The Theoretical Level: Resistance Versus Resilience

The mandatory frame of reference for the discussion of resilience is disasters, or rather "disaster consequence management". The term 'disaster' usually connotes a momentous event, a catastrophe, which causes abundant fatalities, destruction and financial losses. Indeed, such an extraordinary calamity usually draws the attention of the media and the public at large. However, disasters do not reflect the whole picture, because even smaller disruptions might result in severe and often acute breaches of a system's routine functionality, which might cause further-reaching ramifications. Hence, it is suggested, that the level of casualties or damage, though unfortunate and significant by itself, represents only part of the scene, to which should be added the perception of the victims impacted by the disruption regarding its severity.

Disruptions are commonly categorized by their source, usually divided between natural hazards and manmade. This categorization by itself is somewhat dubious, as many so called natural hazards – mostly those generated by climate change - are initially manmade in essence. But beyond that, the important question is not so much what brought the event about, but what its consequences are. This is the basis for the emerging 'All Hazard' approach (Adini et al. 2012; Waugh 2005), which is less interested in the causes of disruptions and more in the operational aspects of preparedness and post disruption response.

This is the background for another distinction between two basic strategies in disaster management: The Resistance approach and the Resilience approach. The first aims at best to prevent the expected – or unexpected – disruption all together. If this is not attainable, resistance strives to protect the system from the consequences of the disruption, or to mitigate its adverse results. It can be safely suggested that the vast majority of public resources, in terms of budget, manpower and human ingenuity, have been universally invested in resistance efforts. This is true in the field of natural hazards, and perhaps even more so in the domain of manmade threats, particularly in counterterrorism, during recent decades.[1] This is connected to the often criticized notion of 'securitization' of disaster management, which raises the question why so much of the always limited resources are being spent on security challenges, rather than on natural hazards, that often cause much more damage in lives and property. Without dwelling upon this question, it can be suggested that all the

[1] See, for example, an ICT report on aviation counterterrorism https://www.ict.org.il/Article/1757/trends-in-aviation-terrorism

immense appropriations to provide resistance to natural and manmade hazards have not proven to be sufficient to curb the growth in the number and especially the cost of disruptions. Would that decrease the expenditure in resistance efforts? Probably not. Human thinking and culture is still mostly inclined, a-priori, to give preference to the 'cave syndrome', presently replaced by the 'fence mentality'. Sophisticated countries like the US and Israel incline to surround themselves with walls and other technological obstacles, contending that they will prevent hazards and bloc the intruders. This is as sound as the reliability of levies and other hurdles designed to defend the beaches from Tsunamis and alike.

The strategy of resilience, which has emerged in recent years, is designed to offset the insufficiencies of the commonly employed strategy of resistance. The main rationale behind adopting resilience is that resistance, at least in its preventive modes, portraits a false pretense of hermetic insulation. 'Resiliencers' would argue that even though there is room for investing in resistance measures, despite their high cost, it is necessary to acknowledge their deficiencies and to comprehend their questionable yield of cost / benefit. Hence, it is advised to balance the huge expenditures on resistance with appropriate investments in resilience measures. This admittedly will not decrease the hazards, but might distinctly enhance the chances of 'bouncing back' following disruptions, towards the crucial stage of recovery.

The generic resilience approach focuses on the vision of reducing the effects of disruptions following their occurrence. It contends that as the resistance measures cannot provide full proof prevention, they cannot be accepted as the exclusive strategy of disaster management. Accordingly, resistance should be supplemented by a strategy that embraces an antithetical angle: As hazards repeatedly happen and cause vast damage, they frequently bring about an inherent functional degradation of the impacted systems. In such a chaotic situation, the most compelling challenge is to enhance the prospects of swift 'bouncing back' of the systems' capacities to regain its core missions. This can be achieved by thorough and continuous preparedness, which needs to be built on the constructive assets of the systems in question. Building the resilience of a critical infrastructure will focus on other components than that of a community or an economic entity. The common denominator of all these diverse systems is to enhance their capacity to quickly transform the unavoidable downward trajectory following the acute disruption to an upward trajectory of growth. Meticulous preparedness of the system is proposed to bolster the chances of resurgence and facilitate the process of 'bouncing forward' (Plodinec 2009). Under enhanced preparedness an upward course can be expected, to enable the impacted system to rapidly start its journey of flourishing following the disruption, and eventually reach a higher level of its core functionality (Folke et al. 2002). This pattern would present resilience at its peak.

Societal resilience is the most challenging goal to achieve. It entails the engagement with sensitive interactions between human beings, and between people and their socio-political environments. It is argued that communities which enjoy a higher level of socio-economic capital will generally be more resilient than those who are positioned in poverty and social stagnation (Aldrich 2012).

The concluding submission of the theoretical chapter is that advanced disaster preparedness should carefully consider the pros and cons of both the resistance and the resilience strategies. A balanced joint approach will enable the endangered system to mitigate the threat, and at the same time will promote its chances to bounce back following the severe disruption and to expeditiously reach the difficult stage of recovery from a position of strength and growth. The strate of resilience is comprehensive in nature and should be adopted and implemented in a wide gamut of practical spheres. The two most important of those, in the counter-terror struggle, are the realms of critical infrastructure and in the community social domain. The precondition for the actual success of resilience lies in its appreciation as a viable realistic blueprint which – side by side with resistance measures – can indeed promote reasonable level of security. Resilience should not remain an intellectual exercise. It has to be officially embraced as a complementary strategy which compensates for the shortcomings of resistance, in both physical infrastructure and social threatened realms. Once this is accepted, and translated to mandatory action, than implementation remains to be the challenge. Applying the resilience model in the critical infrastructure sphere is easier than in the complex social environment of communities and the nation as a whole. The journey to achieve the goals is long and strenuous. But it is necessary if nations are committed to stand up against the modern epidemic of terrorism.

10.2 The Policy Level – The Israeli Case

The State of Israel represents a unique position in the field of societal resilience. Because of its history of harsh and repeated bloody conflicts with its neighboring adversaries, and particularly those 'low intensity conflicts' that have taken place in the last three decades, Israel is haunted with security issues that mold its external and internal politics. This is also the case regarding its approach to mass disruptions, which is overwhelmed with the security bias. Military considerations clearly take first priority, with the response to the threat of terrorism as a paramount issue. In comparison, in terms of awareness, preparedness and investments, this preference has overshadowed the limited attention granted to natural hazards, such as earthquake and Tsunami, which are estimated to cause severe damage, much beyond the ramifications of the terrorist peril.[2] Indeed there is a growing awareness in Israel that this pattern has to be balanced. Significant changes have been made in the enhancement of the capacities of the fire fighters, following the failures exposed in

[2] According to a special report on the activities of the steering committee on the deployment for earthquakes, 2012, submitted to the Israeli government on 29 March, 2012, the outcome of an earthquake in Israel, in a magnitude for occurrence is 5% within a period of fifty years are: 7000 fatalities, 9700 trapped in ruins, 170,000 homeless displaced, 28,000 buildings heavily impacted, 290,000 buildings medium to light impacted.

the major open fire in Haifa in December 2010[3] and 2017 has been marked by the Israeli National Emergency Management Authority as the 'earthquake preparedness' year.

The clear preference to security challenges has been reshaped in the last 15 years, in which the country has been engaged in five severe terrorist-centered conflicts: The Second Palestinian Intifada (2000–2004), The Second Lebanon War with Hezbollah, followed by three consecutive 'rounds' of hostilities with Hamas in the Gaza Strip. This dire series of hostilities has changed the pattern from 'traditional' terror, represented mostly by 130 suicide bombers' onslaughts during the Second Intifada, to those of rockets and missiles launched by non-state entities on civilian targets. The latter has become the prevailing risk pattern, when the Israeli civilian home front has been stormed once and again by average daily barrages of 120 high trajectory weapon systems of different types, range and warheads. This abnormal disruptive phenomenon has elevated terror to the highest level of threat, which compelled a stringent policy of resistance.

The clear-cut emphasis on the resistance approach is not new in Israeli strategic thinking. The updated terrorist threat produced yet a refined policy, which focuses on strict 'securitization' of the challenge and the response. This implies granting clear priority to military measures over civilian defensive and containment means. Consequently, the military has been placed in the leading position – together with the Security Agency (Shabak)[4] and the National Police, as secondary partners, to manage the threat. Within the military, the Home Front Command[5] (HFC), the chief army agency for the preparing and managing the civilian front in times of disruptions, has been granted a lesser status in terms of responsibility and resources, compared with the other military branches. Consequently, first priority has been granted to two pillars of resistance: The first and foremost – deterrence, aimed to postpone as much as possible the next round of hostilities, by impressing upon the non states adversaries the high price they will pay if they provoke the Israeli home front again.[6] Deterrence is primarily built by the IDF military might and the demonstration of offensive capacities, in the air and on the ground. In fact, Israel prides itself that deterrence against Hezbollah holds already for more than 10 years, as a result of the overwhelming strikes that caused heavy losses and grave damage in the last round of 2006 in Lebanon. Similar situation of deterrence persists now for more than 2 years in the Gaza front.

The second major resistance pillar, the one of physical protection, finds expression in the Israeli modern manifestation of the 'cave syndrome' presented above. The first layer of defense has been the construction of protective fences, during the Second Intifada (at the prohibitive cost of more than 3.5 billion Dollars), virtually along the entire stretch of the West Bank. Presently, the government is financing the

[3] In which 44 people perished http://www.ynetnews.com/articles/0,7340,L-4883161,00.html

[4] https://www.shabak.gov.il/english/Pages/default.aspx

[5] http://www.oref.org.il/894-en/Pakar.aspx

[6] As presented by the special report entitled "The IDF Strategy", 2015 https://www.idfblog.com/s/Desktop/IDF%20Strategy.pdf

construction of the renewed sophisticated barrier along the Gaza Strip border, which is also designed to answer to the offensive tunnels' threat, at the cost of 600 Million Dollars. To respond to the updated threat of high trajectory weapon systems, Israel has introduced a second layer of defense, constructed by the formidable active defense system, the 'Iron Dome',[7] developed by the local military industry 'Rafael', and financed generously by the US at close to 1 billion Dollars. It has already been operationally tested in the last round against Hamas in 2014, and was proven to be hugely effective. This success had a vast positive bearing on the Israeli civilians' confidence and sense of standing against the stressful peril. Still, despite these significant accomplishments, Israel has not been totally sealed and insulated from the dire affects of terror. These continue to be an ominous part of Israeli reality. The question is whether the policy-makers realize that resistance by itself cannot provide the ultimate answer, and that there is room for considering alternative strategies to augment it.

This raises the question on the role of resilience as a counter-terror strategy in Israel. Historically speaking, the first manifestations of institutional resilience were already present in Israel back in the 1980's, when local communities in the north of the country, facing protracted acts of (traditional) terror, adopted the notion of collective mental enhancement as a principle of countering the affects of the stressful hazard. Since these first attempts, not much has been added to augment the practice of resilience in government investments and action. Indeed, senior officials frequently use resilience in their public rhetoric; but in fact, the government has done little, throughout the years, to promote resilience to an actual strategic response to terror. The reasoning behind this inaction can be found in the assertion, frequently voiced by senior decision-makers, that resistance is the best way to enhance resilience. Namely, if terror is deterred, and if there are sufficient means of solid protection, than the affects of the threat are expectedly curtailed, and security is attainable. It can contrarily be argued that even though resistance measures undoubtedly do save lives and significantly contribute to the curtailment of terror, they have not achieved the needed possible level of security. Investing in the enhancement of resilience can yield better performance, primarily in the capacity of the civilian population to rapidly bounce back following disruptions. This is a key feature in achieving the required functional continuity.

10.3 Societal Resilience in Practice

As of now, in most countries that are threatened with severe terrorism, the 'top down' approach has not been sufficiently conducive to position the strategy of resilience highly and affectively on their national security agenda. This has been also the case in Israel, despite its high level of alertness. However, this does not entail a failure in the manifestations of societal resilience in communities which have been

[7] http://www.rafael.co.il/marketing/SIP_STORAGE/FILES/6/946.pdf

disrupted by protracted terror. The Israeli experience shows that the opposite can be the case. Time and again evidence has shown the high level of resilience of the Israeli society during and following major terror incidents, as has been expressed by the capacity to rapidly bounce back to normal functionality following recurrent episodes. The Israeli resilience scene has presented a clear 'bottom up' mindset: Whatever the national authorities refrain from doing, the grass roots and the local authorities, together with organizations of the civil society, have been active and successful in promoting and achieving.

Two main explanations can be suggested for the high level of societal resilience in Israel: The first is the long history of heinous disasters that the Jewish people have endured, particularly in the last century. In all past calamities the final outcome has been a clear victory of recovery and growth. Positive experience has been proven to be a major source of high level of resilience (Bonanno 2004; Windle 2011). Referring specifically to the challenge of manmade disruptions, caused by terrorist acts, Israel has a long and frequent history of such episodes, many of them deadly. Even in the most severe cases of terror, as with the suicide attacks of the Second Intifada, many of which caused the death of dozens of people, the bouncing back syndrome has been remarkably apparent and swift. Practically following all fatal cases, even during the worst days of the Second Intifada, the return to normal functioning was full and short. Similar outcomes were in all other numerous harmful terrorist campaigns against Israel.

The second attribute to the high rate of societal resilience in Israel has been the relatively limited magnitude of threat and damage which has characterized most of the terrorist acts in the country's history (Braun-Lewensohn and Mosseri Rubin 2014). Even though those acts varied greatly in their ferocity, as can be measured by the number of fatalities per case, not one of the numerous episodes that resulted in multiple fatalities could be connected to a slow or low trajectory of social return to normal conduct. Forecast of future terror incidents, particularly those that are expected to be carried out by rockets and precise missiles aimed at civilian centers, are liable to result in heavier damage. This could possibly lower the rate of high societal resiliency that was typical of the Israeli public as of now (Elran et al. 2016).

Beyond these general attributes, which characterize the macro scene of Israeli societal resilience, one can discern several revealing differences when observing the micro picture of individual communities. A more pointed analysis of the social circumstances in communities located in close proximity to the Hamas controlled Gaza Strip reveals an absorbing picture: In the last decade, in three consecutive 'rounds' of hostilities, the Jewish communities close to the border were shattered by excessively heavy and continuous onslaughts of diverse means – from mortars and rockets to offensive penetrating tunnels – which endangered their lives and grossly disrupted their routine. The findings of an in-depth research (Elran et al. 2015) disclose a high to a very high level of societal resilience in reference to these severe manmade disruptions. These were expressed in the actual conduct of the communities' residents, more than in surveys taken to judge their own perception of resilience. The pattern of communal behavior has been consistent and clear: There was always a trend of actual degradation of their functionalities right subsequent to the disruptions, commonly followed, rather expeditiously, by a distinct trajectory of bouncing back to the former level of routine. The best example of this repeated pattern of resilience has been found in the issue of evacuation in perilous circumstances. Under excessive fire, about 50% of the inhabitants – mostly mothers and children - living in communities within a range of 7 km. from the fenced border, decided to leave. There was no governmental decision on the issue, and the local residents made the agonizing move on their own resolve, mostly on their own or the communities' expense. "Our home, which is supposed to be the safest place, has become the most dangerous place", narrated later one of the evacuees, according to an account of the Israeli Trauma Coalition.[8] By taking this bold decision, which is greatly contrary to the nation's historic narrative of 'no evacuation under enemy fire',[9] the people expressed a sense of flexibility, which by itself is a major trait of resilience. While staying in their places of refuge, many of them decidedly declared that they are not going to return to the perilous place that has been their home. In fact, more than 90% of them returned within 24 hours following the cessation of hostilities. Most of the rest returned within a week. Very few never returned, but plenty of new residents joined these communities since the gloomy episodes.[10]

This behavioral sequence of severe disruption, followed by diminishing functionality, and then by a swift bouncing back to normal conduct, clearly represents high level of societal resilience. Furthermore, in some instances, elements of 'bouncing forward' could be detected, to represent a higher level of community resilience. For example, Observing demographic trends, it could be seen that the overall population in the threatened region, in front of the Gaza Strip, has expanded significantly in the last decade, despite the security challenges. In the regional council of Eshkol, the population growth has been remarkable at 6.2%. In one of the Kibbutzim,[11] which was relatively hit more severely than the other adjacent communities, clear evidence of decrease and then an upsurge of the membership has been observed following the terrorist attacks, and renewed flourishing social activities indicated distinct growth in the communities' morale. A more penetrating investigation shows that some communities in the disrupted area enjoy a higher level of social capital than the others. A clear connection between the rate of the social capital in these communities and their level of societal resilience has been discerned.

What are the social attributes that were found to contribute to the high level of measured resilience (Cutter et al. 2014)? It is important to note that many of the communities located in immediate proximity to the hostile border of Gaza are Kibbutzim. This connotes primarily that their average socio-economic level is higher than many rural communities in the country and clearly higher than their

[8] http://www.israeltraumacoalition.org/

[9] This is a rather false narrative, as the factual history, since Israel's War of Independence, witnessed plenty of civilian evacuations under conditions of hostilities.

[10] The regional council "Eshkol" enjoys an increase of the population since 2010 of 2.4%, much higher than the national average, which was 1.9% in 2014. According to forecasts of "Urbanix", the rate of growth might reach 6.4% in 2022

[11] Nahal Oz. http://eng.negev-net.org.il/htmls/article.aspx?C2004=12747

neighboring communities that are not collective or semi-collective. The historical cooperative basis of the Kibbutzim, though gradually diminishing subsequent to the process of privatization, produces a unique social structure of bonding, collaboration, sense of togetherness, active participation and mutual assistance. These are commonly recognized as promoters of social resilience (Aldrich 2012). Interestingly, two of the communities studied for their level of resilience are religious (Sa'ad and Alumim), and were found to have higher rate of resilience than their counterparts. This corresponds to the assertion that common ideology and faith (not only religious) represents a source of societal resilience (Kaplan et al. 2005).

Another facet of societal resilience in these communities was found to be the high rate of trust in the local leadership. In a survey conducted in 2014, following the last round of hostilities, it was found that the rate of confidence in the leadership of the regional council was higher than that of the individual communities, while the two enjoyed significantly higher trust of the local population than the national political leadership[12] (Goral 2015). The local authorities deserve this high level of trust, as they mostly serve as the mediator between the residents and the communities and the national agencies. An interesting manifestation of this important role could be found in their successful efforts to create new and improved channels of direct communication between the communities and the military, which was not smooth enough during the hostilities. Similar contribution on their part was to convince the government to increase the subsidies to the region's communities, following the last round of hostility,[13] which by itself helps to promote the well being of local population and hence their resilience.

Beyond these, it is important to state that adequate preparedness serves as a major contributor for high resilience of the communities (Paton and Johnston 2001). In fact, the region under discussion is perhaps the best prepared for manmade disruptions in the country, both in the household domain, as well as in the communities and the regional council level. Each of the localities holds a voluntary 'Emergency Response Team' that is responsible and ready to manage the community during crises situations. On the regional level, there are five 'Resilience Centers',[14] which provide psycho-social care to the 60,000 residents. These centers, operated by a not for profit organization – The Israeli Trauma Coalition[15] - are budgeted in matching by the government's social services, and provide individual clinical care, training to the population at large and assistance to the local professionals. They are also engaged, in partnership with the local councils, with the ongoing emergency preparedness enhancement. Their work in the decade in the area has been highly valued and acknowledged as a significant contribution to the societal resilience of the population and communities.

[12] "Barometer Sapir", http://.sapir.ac.il/content/4545

[13] Strategic five years program for the advance of resilience for the communities of the "Gaza Envelope", September, 2014

[14] http://www.israeltraumacoalition.org/?CategoryID=211

[15] http://www.israeltraumacoalition.org/?CategoryID=211

All these components together construct a solid structure of societal resilience that proved itself under the acute challenges during three major manmade disruptions. Unfortunately, this is not a typical picture in most other communities in Israel, which are less prepared for high risk episodes, not to mention novel disturbances. Still, what have been detailed above shows that there is an adequate model of 'bottom up' readiness that can provide the necessary level of resilience, especially if augmented by 'top down' engagement, which provides the additional necessary funds, together with the protection umbrella of resistance measures.

10.4 Conclusions

There is a growing interest, among professionals and architects of homeland security, in the role of implementing the resilience concept into practice in different hazard scenarios. It is suggested here that lessons of the Israeli practice, in the realm of implementing resilience in general and in constructing a viable response to counter protracted terrorism in particular, can be used globally, with the necessary modifications. As shown in this article, the 'top down' approach has not been optimal in the realization of the principles of resilience, despite the repeated rhetoric. Still, this shortcoming can be compensated for by the utilization of the theoretical model in the field, wherever there is understanding, awareness and will. It has been shown that in the fight against terror, communities, the civil sectors, and local authorities, have managed to construct, in a fruitful partnership, a satisfactory system that provides high prospects of societal resilience. Their joint conduct, the investment of local resources, with passion and perseverance, have clearly exhibited that the affordable expenditures in resilience measures have paid off lavishly, as a supplementary effort to that of resistance technological means. Indeed, it has been understood by the local stakeholders that these measures of resilience are not only critical for the sake of preparedness for emergencies. Rather, most of the steps taken contribute to the bolstering of social growth and prosperity, which by themselves constitute leverage for the needed high level of resiliency.

It is important to submit that societal resilience is not an existing trait that is found in all communities and cultures. It is a capacity that has to be developed continuously and relentlessly, in line with a deep awareness of the relevant needs, in accordance with the expected hazards. In Israel resilience is defined as "the capacity of a social system (individuals, family, community) to contain, react and adapt to extreme conditions of crises, to return to the optimal functionality in the shortest period of time following the end of the emergency period, and to enhance coping practices for future acute disruptions". The practical components of resilience are proposed to include: good governance, the acceptance of the local governments as the basic brick for the preparedness efforts, credible leadership, social solidarity and mutual interdependence, situational awareness, and the readiness of the individuals in the community.[16] This framework, if implemented fully by all stakeholders, can produce the necessary level of preparedness to enable impacted communities to rapidly bounce back following major disruptions.

Further Suggested Readings

Adini B, Goldberg A, Cohen R, Laor D, Bar-Dayan Y (2012) Evidence-based support for the all-hazards approach to emergency preparedness. Isr J Health Policy Res 1(1):1

Aguirre BE (2007) Dialectics of vulnerability and resilience. Georgetown J Poverty Law & Policy 14:39

Aldrich DP (2012) Building resilience: Social capital in post-disaster recovery. University of Chicago Press

Bonanno GA (2004) Loss, trauma, and human resilience: Have we underestimated the human capacity to thrive after extremely aversive events? Am Psychol 59(1):20

Bonanno GA, Romero SA, Klein SI (2015) The temporal elements of psychological resilience: An integrative framework for the study of individuals, families, and communities. Psychol Inq 26(2):139–169

Braun-Lewensohn O, Mosseri Rubin M (2014) Personal and communal resilience in communities exposed to missile attacks: does intensity of exposure matter? J Posit Psychol 9(2):175–182

Cutter SL (2016) The landscape of disaster resilience indicators in the USA. Nat Hazards 80(2):741–758

Cutter SL, Ash KD, Emrich CT (2014) The geographies of community disaster resilience. Glob Environ Chang 29:65–77

Elran M (2005) Israel's National Resilience; The Influence of the Second Intifada on Israeli Society, Jaffe Center for Strategic Studies, JCSS, Memorandum # 81 (Hebrew)

Elran M, Israeli Z, Padan C, Altshuler A (2015) Social Resilience in the Jewish Communities around the Gaza Strip Envelop during and after Operation Protective Edge, INSS, *Military and Strategic Affairs*, Vol. 7 No. 2

Elran M, Shaham Y, Altshuler A (2016) An expanded comprehensive threat scenario for the home front in Israel. *INSS Insight* No 828. http://www.inss.org.il/uploadImages/systemFiles/No.%20828%20-%20Meir,%20Yonatan,%20and%20Alex%20for%20web571603989.pdf

Folke C, Carpenter S, Elmqvist T, Gunderson L, Holling CS, Walker B (2002) Resilience and sustainable development: building adaptive capacity in a world of transformations. AMBIO J Hum Environ 31(5):437–440

Goral A (2015) The Societal resilience in the Communities of the Eshkol Regional Council, Center for Emergency Response Research

Kaplan Z, Matar MA, Kamin R, Sadan T, Cohen H (2005) Stress-related responses after 3 years of exposure to terror in Israel: are ideological-religious factors associated with resilience? J Clin Psychiatry 66(9):1146–1154

Norris FH, Stevens SP, Pfefferbaum B, Wyche KF, Pfefferbaum RL (2008) Community resilience as a metaphor, theory, set of capacities, and strategy for disaster readiness. Am J Community Psychol 41(1–2):127–150

Paton D, Johnston D (2001) Disasters and communities: vulnerability, resilience and preparedness. Disaster Prev Manag: Int J 10(4):270–277

Plodinec MJ (2009) Definitions of resilience: an analysis. Community and Regional Resilience Institute (CARRI), Oak Ridge

Waugh WL Jr (2005) Terrorism and the all-hazards model. J Emerg Manag 2(1):8–10

Windle G (2011) What is resilience? A review and concept analysis. Rev Clin Gerontol 21(02):152–169

[16] As defined by a 'Background Paper' for the enhancement of societal resilience in emergency situations, The Prime Minister's Office's round table, in cooperation with the Ministry of Defense, June 2016.

Chapter 11
Planning Resilient Communities

Alexander Hay, Antonio Gómez-Palacio, and Nicholas Martyn

Abstract Our world is changing. Extreme events are becoming less predictable with greater consequences. Infrastructure hardening alone is proving both inadequate and unaffordable. Resilience is not about preventing change – change is inevitable – rather, it is about managing change and adapting, responding, and recovering from disruptive events. How we manage change will be defined by how we manage the risk context, using urban planning to reduce the consequence of shocks and stimulate the collective ability to respond and recover. By focusing on people and the community operations that support their lives as the essential purpose of resilience, we can focus our actions more effectively. Infrastructure is built to support a purpose. That purpose does not disappear during a shock. Therefore, we should plan and design infrastructure and services to support the continued delivery of that purpose. The net result is an holistic view of community function and how it can manage both stresses and shocks to protect livelihoods, continued prosperity and quality of life. In effect we propose a framework for planning resilient communities that can support society in an increasingly unpredictable world.

Keywords Urban planning • Resilience • Protection • Municipalities • All-hazards

A. Hay (✉)
Southern Harbour Ltd, The Queensway, P.O.Box 57002, Toronto,
ON M8Y 3Y2, Canada
e-mail: ahay@southernharbour.net

A. Gómez-Palacio
Urban Planning Leader, DIALOG,
1100 – 2 Bloor Street East, Toronto, ON M4W 1A8, Canada
e-mail: agp@dialogdesign.ca

N. Martyn
RiskLogik, 80 Little Bridge Street, Almonte, ON K0A 1A0, Canada
e-mail: nick.martyn@risklogik.com

© Springer Science+Business Media B.V. 2017

I. Linkov, J.M. Palma-Oliveira (eds.), *Resilience and Risk*, NATO Science for Peace and Security Series C: Environmental Security,
DOI 10.1007/978-94-024-1123-2_11

11.1 A Shifting Risk Context

The hazards and risks we face today are different – and in most cases more extreme – than a generation ago. In fact, what constitutes a hazard is also evolving. For example, 20 years ago a cyber attack would not have been thought of as a threat to communities. Today it is one of the greatest threats to the Systems Control and Data Acquisition (SCADA) systems that manage municipal infrastructure. A successful SCADA attack today could be catastrophic. Our policies and practices have yet to adapt to this and many other realities of modern community life.

Extreme weather events have become more frequent and more severe in recent years. The Canadian Disaster Database, http://cdd.publicsafety.gc.ca/srchpg-eng.aspx maintained by Public Safety Canada, shows clear upward trends for flooding, wildfire and wind. When these trends are compared to the preceding 100 years, there is a clear disparity between what has been used to base our infrastructure designs on and what we are today experiencing. It can be no surprise that much of the protection incorporated into our infrastructure designs is proving inadequate, sometimes with catastrophic results. Our codes of practice were not written for this.

Technical advancement has greatly benefitted communications and the conduct of trade and business. Each advancement has also introduced a new range of hazards that we have yet to fully comprehend. However, the pace of technological change is so great that many of the vulnerabilities and incremental hazards are buried deep in the systems architecture and remain unidentified, or simply unrecognised. For example, the use of the 'Cloud' for data storage and access has greatly improved business transformation and opportunity, (PriceWaterhouseCoopers (US) 2014), but substantially increased vulnerability to power outage, as well as a host of privacy and cyber security risks (TEDtalks 2013). These secondary and tertiary vulnerabilities are increasing in direct proportion to the sophistication of the systems we rely upon, making a clear understanding of what precisely we rely upon and to what extent all the more important.

The consequence of loss is similarly escalating. Over the last generation, we have seen a steady concentration of value in infrastructure 'nodes' and clusters, such that the consequences of that node being lost are ever greater. This concentration of value can be due to increased operational activity and dependency, such as the increased number of financial transactions passing through a stock exchange or a SCADA system for a regional transportation network. The insurance industry records a far sharper increase in the financial consequence of catastrophic events, in large part because of this increased concentration of value (Insurance Bureau of Canada 2013). When we combine all these trends, the Risk Context for any operation or community is highly complex with little relevant guidance to refer back to. It is in effect a highly complex amalgam of ad hoc sub systems without a map.

To illustrate, the power station at East 14th Street and Avenue C, New York, NY, was storm surge protected to 12.50 ft; greater than the historic worst case of 10.02 ft recorded during Hurricane Donna in 1960. Superstorm Sandy delivered a 13.88 ft storm surge (Sharp 2012) that flooded the station with a consequent catastrophic loss of power to Lower Manhattan Island. The damage caused by the storm is estimated at $30Bn, yet the subsequent losses are still being counted and exceed $50Bn at the last estimate (Philips 2012). The true losses caused by these events far exceed the direct infrastructure and asset loss, they are the loss to operations and functions that the infrastructure was designed to enable. We must look beyond the headline $10Bn losses when the New York Stock Exchange was disrupted for 2 days. It is the interdiction of community operations that directly affects livelihoods and the very viability of communities.

Modern societies are enormously complex networks of highly connected, and highly valuable systems. The consequences of a shock or stress to one part of the system is felt throughout the network and often far beyond the spatial boundaries of the affected area. Often these consequences are amplified as the impact is propagated from one system to another. If we are to make our communities resilient to the stresses and shocks we anticipate, we must understand this risk context. We need to map the connections between critical infrastructure systems and model the impact of the stresses and shocks to understand the cascade of consequences to all of the systems in the network. In effect, we must create a blueprint of community functions and the infrastructure that supports them. Recognizing that we cannot protect everything, this identifies what we must protect to preserve the capacity for self-recovery and resilience.

11.2 Are We Managing the Risk?

Traditionally, human settlements were acutely responsive to natural and man-made hazards – recognizing that catastrophic events will happen from time to time, but the settlement must endure. Floods, fires, famines, wars, all required resilience strategies. For example, traditional settlements would clear the scrub around the outside of the settlement to starve wildfires of fuel so that they could not enter the settlement. Typically, this would be achieved through use of farmland as the buffer. Today, the National Building Code 2010 makes no provision for building proximity to forest for wildfire protection, despite mounting evidence that it is necessary (Cotter 2012). Alberta has issued guidance of 30–50 m set back between institutional buildings and forest, (Alberta Infrastructure 2013) recommending many of the same considerations that earlier generations considered normal practice. Suburban sprawl and municipal budget constraints have in many communities resulted in no scrub clearance and an associated increase in direct wildfire threat to the communities themselves, particularly with the risk of spot fires resulting from a wildfire in moderate to high winds (Alberta Infrastructure 2013). Recent fire events in Australia and California point out the folly of this approach. This does not mean that the street tree canopy is lost and the quality of living space sacrificed, but rather that in understanding the hazard, the density of fuel available to the wildfire threatening the settlement was managed. Similarly, conventional wisdom dictated that we should not build on flood plains and when it was necessary, for mills etc., specific design precautions and recovery strategies could be taken to ensure that the impact of a flood was limited.

Recent experience would suggest that we no longer recognise our relationship with the natural environment. Cost savings measures have resulted in reduced urban watercourse maintenance (City of Toronto and Genivar 2011) with the result that during extreme rainfall periods the culverts become clogged, potentially leading to major road blow-outs. The Finch Culvert in Toronto is one such example, requiring 15 months for all services to be fully reinstated after the last flood (City of Toronto Environment & Energy Office). The increasing development on flood plains is another prime example, despite regular warnings from the insurance industry (SwissRe 2010) and the natural environment that it is unwise (Carter and Willson 2006). This increased development of flood plains changes the water flow during a flood, resulting in worse flooding conditions and an increased canalisation into downtown cores. Calgary, AB, is a case in point, where despite the experience of serious flooding in 2005, flood plain development of the Bow and Oxbow Rivers continued apace contributing significantly to the extent and impact of the 2013 floods (Great Alberta Flood 2013). There appears to be a reluctance to address the questions that such events raise (Wingrove 2013). Our day-to-day relationship with the natural environment is much the same as the contextual influence of demographics and technological change, manifested as stresses to our community and business operations. When these stresses are not actively managed they draw in progressively more operational resources, thereby increasing the overall logistic burden and constraining the capacity for response to a shock or catastrophic event. We are not managing the risk and have not learned from our parents.

In parallel to this steady dis-association with the natural world and increased urbanisation, we have shifted our focus from the protection of livelihood to the protection of life. In protecting livelihood, we were concerned with our community's survival, its continued economic growth and development and the enhancement of civil society. This forced a longer perspective of time, meaning that continuity and recovery were every bit as important as mitigating the immediate effects of a disruptive event itself. In shifting the focus to the preservation of life as opposed to livelihood, we refined our codes accordingly. The fire provisions of the National Building Code (National Building Code of Canada 2010) are concerned with safe evacuation of occupants rather than structural stability, operational continuity during the response or any residual occupancy value of the structure following the fire. It is interesting to note that 18% of firefighter deaths are caused by structural collapse (Naum 2010) and that 44% of businesses fail to reopen after a fire (ClinicIT 2011). These statistics are consequent to the prevailing infrastructure/asset performance focus on the immediacy of the event (evacuating to save lives) rather than the continuity of operations and what follows after the disruptive event. This suggests that we are not managing the risk because we don't understand it; a situation exacerbated by the shift in strategic decision making from time-based vision to commitment-based decisions (Ghemawat 1991). To understand the risks, we must understand our [operating] purpose in context.

11.3 Shifting the Focus

11.3.1 Understanding Purpose

To be truly resilient, a community, a city, a business, must begin by understanding its purpose. Only then will it be able to comprehend what is critical to that purpose and be able to prioritize strategies for mitigating the impacts to it. While some communities will be able to identify one particular purpose centered on its industrial function, others, especially larger ones, will have many – sometimes in conflict with each other. Each purpose must be defined so that the operations that support it can be identified, mapped and prioritized in terms of their critical contribution to resilience of the whole.

In analysing an operation, there will be certain defining essential functions that must continue through and following a disaster to enable operational continuity and self-recovery. Each function is in turn analysed to determine what services and infrastructure it depends upon. These are in turn analysed for what they depend upon and gradually we grow a dependency chart or *Causal Chain*. Using a directed graph, the dependency relationships between the operations, functions and supporting infrastructures are defined by consequence of failure (mission critical, political and financial). This means that when we shock any part of the Causal Chain with a hazard effect, we can cascade the consequential effects in terms of direct cost of loss/compromise, operating performance and market position / brand / influence. The ability of any one node to withstand the effects of shock and not cascade them up or down the chain will be determined in part by the contextual stresses that the operation experiences.

If we now apply this construct to a community or to a City's civic functions, we see how our management of normal civic stresses can influence our ability to self-recover in a catastrophe. However, if left unmanaged these same stresses can superimpose and cause catastrophic effects in themselves. Detroit is a useful case study of unmanaged stresses accumulating to be every bit as devastating as a major shock. Balancing this management of stresses and planning for degrees of failure allows communities to become resilient.

11.3.2 Constructing Community Resilience

In applying this approach, we necessarily need to make the process collaborative, encouraging a contribution of informed solutions and 'work-around' that allow us to limit the effects of a failure and mitigate cascading failure through the causal chain. A community's ability to respond and recover from a shock is an intensely human one. The best designed infrastructure in the most resilient strategic framework will not deliver resilience to shocks and stresses unless the human component is actively engaged. The community must 'own' its fate. Communities (Hay 2013) that self-recovered and thrived following a catastrophe display common characteristics, absent in those that failed. This does not reflect any level of development, wealth or education. All resilient communities have an identity and a focus, (Jacobs 1997) infrastructure in balance with its needs, they exist within a strategic relationship with neighbouring communities and the city/region, and had confidence in their leadership (Boin et al. 2006) during an emergency. These characteristics are consistent with the approach outlined above, specifically the local nature of community resilience. The measures employed to manage the trending stresses in one neighbourhood will be completely inappropriate in another neighbourhood in the same city. We, therefore, need a framework that allows us to apply this approach; to capture the operation requirements and risk context with the influencing community dynamics. This allows us to stimulate community resilience and identify which infrastructure and services require protection, in priority.

11.4 Shifting Practice

Having identified that change is inevitable, that we cannot protect everything and that a focus on infrastructure rather than the continuity of operations is not producing resilience, we propose four key shifts in resilience practice.

11.4.1 Focus on People and Operations, Not Just Infrastructure

Shifting the focus to operational continuity does not mean that we reject the protection of infrastructure and other physical assets; quite the reverse. By understanding operational continuity needs as they apply to community resilience, we can focus infrastructure protection efforts on those projects that will produce the greatest resilience effects. We use protection to address the routine stresses that are expected through normal routine operation. This promotes operating efficiency. The level of protection required is determined by the tolerance for interruption of the essential function in question. Specifically, the protection investment to remove power spikes and short power interruptions might be entirely appropriate for a business, whereas the investment in complete standby power generation facilities might not, since in an emergency all generator fuel is requisitioned by the City for hospitals. This small fact has caught many unawares. In determining our level of protection, we must anticipate some degree of failure in a catastrophe. To operate through this failure, we need to understand completely both our own operation and what it depends upon, and the risk context in which it all exists.

11.4.2 Examine Both Shocks (the Extraordinary) and Stresses (the Ordinary), Not Just Shocks

Greater media attention is focused around the big catastrophic shocks; events, such as floods, explosions and ice storms. It is also easier, in the public eye, to establish a causal relationship between the event and its impact. Nevertheless, it is not only the shocks that disrupt an operation or a community, it is also the slower-brewing stresses such as economic downturns, shifts in demographics, congestion, and so on. Both can constitute a hazard and source of change to which a community must adapt and respond in order to sustain a larger vision of itself.

11.4.3 Redefine the Concept of Failure (Around People and Operations, Not Just Infrastructure)

If we are to be true to the purpose, we must approach resilience by (re)defining what constitutes 'failure' for any given operation, community, city, or region. Failure will be defined very differently by a hospital than by a regular office building. The tolerance for downtime or recovery time for different operations will vary and so too the necessary strategies to mitigate hazards. Determining what constitutes failure requires a critical reflection of purpose. What truly matters to us, and what does our purpose depend upon?

11.4.4 Go Beyond Prevention, into Adaptation, Response and Recovery

Typically resilience initiatives have focused on preventing change with bigger and stronger infrastructure. However, prevention is only half of the equation. Resilience is also about how we adapt to change, recover from and manage the impacts of change. By planning for the aftermath of a hazard, we are better able to adapt, react, respond, and recover. Resilience Plans provide people and decision-makers with the proper tools to manage change. Infrastructure should be designed to support efficient operations and facilitate incident reaction, response and recovery. The strategic development plans and infrastructure design briefs must include the multi-phase functional requirements for resilience if they are to support sustainable development (UNISDR 2013).

11.5 A Framework for Community Resilience Planning

Implementing the shift in both the focus and practice of community resilience planning suggested so far requires a replicable process. The following *framework* provides a broad-stroke method for undertaking this re-focused Community Resilience Planning (CRP) approach. Inevitably each community and business situation requires a nuanced and tailored approach.

11.5.1 Step 1: Scope Definition

The first step is to define the boundaries, scale, magnitude, and context of the community to be studied. Resilience Planning (RP) is a methodology that can focus on succinct operations (an institution, a facility, an emergency services operation) and/or broader scaled communities (a town, neighbourhood, city, or region). In all instances, Resilience Planning requires a full understanding of the operation/community and its dependencies, as well as the background context in which it exists. This includes the geospatial context of the subject (Fig. 11.1).

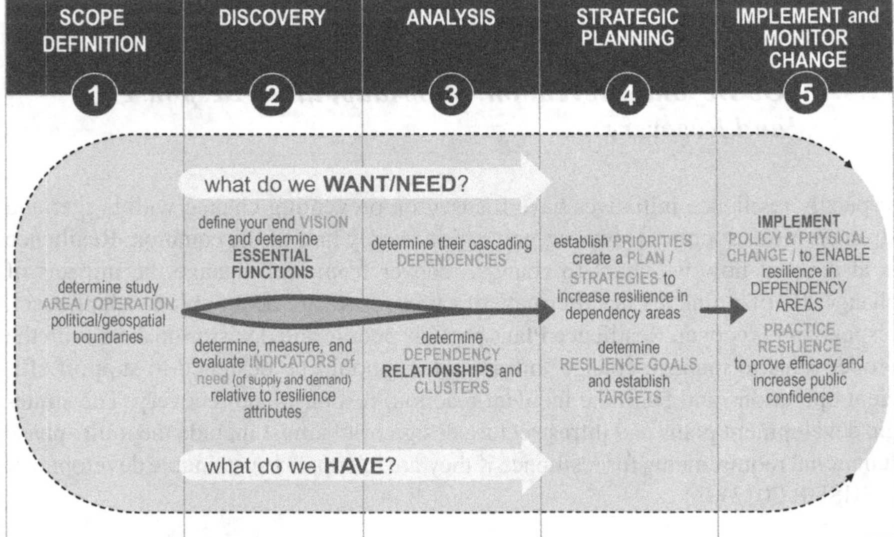

Fig. 11.1 The resilient community planning framework

11.5.2 Step 2: Discovery

This step in the process is twofold, and asks of the operation/community: what do we envision? and what do we have, today? It is the process of discovering the operations and functions and all the systems that support them. Establishing a vision for the operation/community is about determining the end goal, and elucidating the essential functions that support that Vision. For example, a community may have a goal of overcoming the booms and busts of the economy, and retaining its fleeting youth. A hospital may have a vision of continued operations through the most dramatic conditions. We need to understand the goal, in order to plan how we get there.

The second element in establishing a baseline is understanding the existing conditions. This is a mapping exercise that looks at a series of factors and indicators in order to establish the health or vulnerability of the operation/community, relative to broader resilience goals. Using a population density map as the base layer, we will employ a broad variety of indicators that will include:

1. People – the people for whom resilience is developed and upon whom it depends. The indicators will employ geo-referenced metrics of personal safety and health, access to shelter and sustenance, access to cash and freedom of movement.
2. Infrastructure and Ecosystems – the physical and socio-economic systems that sustain an operation/community. These encompass utilities infrastructure, distribution of goods and retail, access to areas of social gathering and coincidence (the public realm) and the natural ecosystems.
3. Organizational Capacity. The ability of people to manage and adapt infrastructure and ecosystems systems toward resilience goals. These will centre of the strategic vision and plan for the municipality or business and include both formal and informal systems of governance, social support structures and economic dependencies.

11.5.3 Step 3: Analysis

Once we have determined our end-state and situation, we need to understand how and where the vision and the existing conditions are at risk (Risk Context) and what the factors are upon which they depend (Dependency Mapping). The objective is to determine all the pathways of exposure to risk in the current state and analyse options to address those identified risks in order to deliver a more resilient operation/community. In order to understand the Risk Context, we need to evaluate the potential source of hazards, be they shocks or stresses.

From the Vision, we derive a chain of dependencies that are critical to realizing the vision. An operation comprises essential enabling functions, each of which will rely upon infrastructure systems, which in turn rely upon other systems and so on. As a minimum, we must be able to recognise the third order of dependency, though there is no specific limit. In mapping these operational dependencies, it must be very clear how the infrastructure systems inter-relate with the personnel and organisational dependencies. For example, the vision for an economically resilient downtown depends upon entrepreneurial start-ups, which depend upon attracting and retaining talent, which in turn depends upon a certain urban life-style.

The vision for continued operations at a hospital depends on power and water supplies, both of which depend on a power grid and when that fails a back-up power generator, which depends on a fuel supply and integrity of the fuel supply system. While the hospital controls its own generator for internal power, it does not control back-up power generation for the water supply upon which it has a critical upstream dependence. Therefore increasing auxiliary power generation for the hospital does not increase its resilience unless the critical dependence on an external water supply is also mitigated. In both cases, understanding the vision helps us map out the critical dependencies, identify third order vulnerabilities (and beyond) and mitigate them.

From the Mapping of Existing Conditions, we derive a clustering of highly dependent areas. In a community, it may be an area where people are particularly vulnerable given low income levels, non-availability of fresh food, and lack of transportation options. Within an operation, it may be that a multitude of processes are critically dependent on a single power supply outside of its control. Clusters of multiple sole-dependencies are particularly vulnerable and demand attention.

In both instances it is critical to understand the operation/community's Contextual Relationship. There are multiple operations within the Risk Context and so the operation that is the subject of the resilience analysis and planning should be considered a subset of that Context. It is therefore necessary to understand how the components of an operation and their respective dependencies link to and are influenced by the background context. For example, the engineer for the hospital back up power system mentioned earlier lives in a dormitory town 20 km away from the hospital. During a flood, public transport is unavailable and the back-up power system fails. His colleague, the engineer at the water purification and pumping station becomes stranded taking himself and his neighbour to work. The water plant doesn't have backup power but the ministry is sending a generator over and the engineer is required to hook it up. There are other stationary engineers within the municipality that in fact live closer to the hospital and the water plant but they cannot be accessed easily, even in an emergency. As a result all the emergency preparedness in the hospital is rendered moot and emergency hospital care compromised. This example is typical of anecdotal evidence from many emergency events and reinforces that resilience is as much about organisational and human factors as it is infrastructure protection. This does not mean that key personnel must reside where they work, but rather that for personnel who support critical operations or dependencies, the operation manager can access key capabilities in his immediate locale and allocate them to reduce vulnerability to disruptive events such as an ice storm or a flood. In this case, the hospital engineer might be assigned a more local emergency role or a workers bus schedule provided (as by the French national railway).

Once we understand the influences of the context on the essential functions and dependencies, we are able to recognise both the destabilising and stabilising properties of the whole. By being able to recognise what works and therefore how to reinforce / capitalise on that, we can add direct and significant value to the user. Furthermore, if we are to identify the stimuli for community resilience to any measure of detail, we must understand how the community is affected by and influences the operation and its components. Similarly, when looking at shocks and creeping stresses to the operation, we will need to recognise both the community tolerance for stress and the investment balance between hazard prevention and systems recovery.

11.5.4 Step 4: Strategic Planning

Every operation/community will have different Resilience Goals and priorities. These will depend on its Vision and on its operational requirements. Hospitals, for example, are usually a high priority in emergency situations and will require operational continuity throughout an extreme weather crisis. An office building, depending on its function, may tolerate a certain amount of down-time and the business or agency it houses can be mostly preoccupied with ensuring the wellbeing of its people and the protection of property. The converse would be true in a cyber attack where the office building housed the cyber defence team and their information technology infrastructure. In either scenario Resilience Goals and priorities will be established for the different milestones following a hazard (see Fig. 11.2):

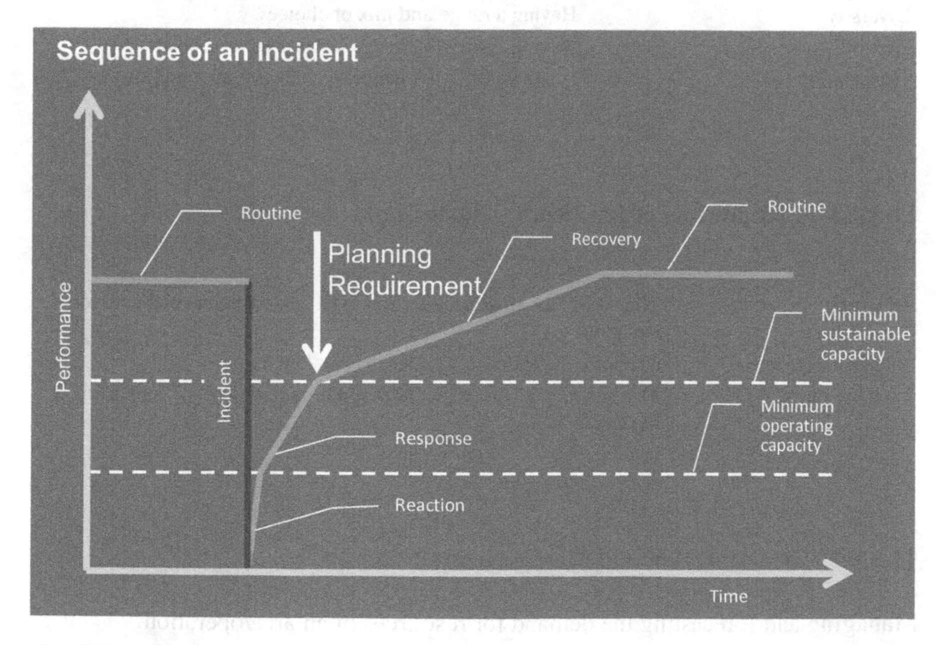

Fig. 11.2 Incident timeline showing the different stages of the incident

1. Routine: a base-line state of operational performance. Part of a routine operation should be implementing strategies to prepare for, and adapt to, future changes.
2. Reaction: the immediate period following a hazard, when systems 'react' automatically. These are the default, pre-established procedures that sustain life and minimum of operational capacity.
3. Response: the period following a hazard when people are able to orchestrate a 'response', ideally by following a Plan, but inevitably needing to make decisions and adapt to circumstances. These are the actions that enable the restoration of pre-determined business continuity.
4. Recovery: the longer, pre-defined and concerted effort, geared towards restoring the intended level of operational performance.
5. Routine (New Normal): a new base-line state of routine operational performance. Part of this new routine operation should be implementing strategies to prepare for the next changes.

As a general approach, consideration should be given to Dependency Management, Clustering and Demand Management.

11.5.4.1 Attributes of Resilience (Dependency Management)

Redundancy	Having viable alternatives for each critical system or resource.
Diversity	Having a range and mix of choices.
Flexibility	Having agility to change in purpose or dependencies.
Adaptability	Having agility to change to a new set of conditions.

11.5.4.2 Components of Resilience (Clustering)

Islanding	The ability for an area/operation to be self-sufficient and endure isolation.
Interconnectedness	The inter-dependencies between areas/operations.
Logistic burden	The critical path and expense (capital/human/time) of supporting a dependency

11.5.4.3 Demand (Demand Management)

Managing and forecasting the demand for resources of an area/operation.

From these it is possible to establish priorities and strategies for the subsequent creation of a Plan to deliver on the Resilience Goals. Strategically, a Plan will focus on the areas of high-dependency and on the critical functions of an operation/community, in balance with managing the demand and resources required. A Plan will outline:

Purpose	The vision, principles, goals, and objectives for Resilience.
Environment	Documenting a baseline along key indicators, and describing strengths and weaknesses (risk context, dependencies, and clusters) of both the operation/community and of the background context within which it exists.
Priorities	Focusing in from a comprehensive approach to determine the areas of both greatest concern and of greatest potential.
Strategies	Both pro-active and re-active strategies for managing dependencies and delivering on.
Resilience goals	These should include: strategies; actions; triggers, targets; responsibilities; and resources.
monitoring	Mechanisms to actively track progress and re-direct, as required.

11.5.5 Step 5: Implement and Monitor the Plan

Implementing the Plan is an ongoing exercise. Some of the strategies will be enacted prior to any occurrence, as part of adapting for change. Some of the strategies will unfold during the reaction, response, and recovery stages.

11.6 Conclusion

The accelerating pace of change in our world today makes any prediction of extreme event severity or frequency virtually impossible over the 25–100 year life of the infrastructure we build. Rather than resisting these changes through ever greater infrastructure hardening, we must manage it. In accepting that there will be failure, the focus switches to the continuity of essential functions through a catastrophe and into the recovery in order for communities and businesses to survive and prosper. This also means that we must address the contextual stresses that influence our ability to respond to and rapidly recover from a shock. The proposed Resilient Communities Framework relates the community operation and its dependencies to the community context to not only provide an holistic understanding of the risk context, but offer an indication of how managing the routine stresses can influence resilience of the whole. It affords a construct by which we can measure the value of infrastructure investment in terms of resilience and community survival.

Municipalities across the world are struggling with infrastructure investment to sustain economic development over the next generation amid the forecast migration of populations into cities. As we slowly emerge from the effects of the worst recession in history, we must make every investment count. While finding efficiencies in the procurement program will help, a clear strategic plan based upon a common vision of how a municipality/organisation sees itself in the future will ensure that each investment contributes to the common benefit. It is time to consider whether we will enable the sustainable development of the next generation or leave them beneath the Sword of Damocles.

Further Suggested Readings

Boin A et al. (2006) 'The politics of crisis management: public leadership under pressure' Cambridge University Press.

Carter K, Willson R (2006) 'Living with our rivers: flood 2005 lessons learned'. University of Calgary

City of Toronto Environment & Energy Office

City of Toronto and Genivar (2011) 'Climate change vulnerability assessment for culverts'

ClinicIT (2011) Also quoted by WikiBooks and by Wanja Eric Naef in Infocon Magazine October 2003 'Business continuity planning – a safety net for businsses'

Cotter J (2012) 'Canada wildfires: construction code change proposals to protect wildfire-prone areas rejected' The Canadian Press. http://www.huffingtonpost.ca/2012/12/02/canada-wildfires_n_2226393.html. Accessed 10 Mar 2014

Ghemawat P (1991) 'Commitment: the dynamic of strategy' 2nd Edition

Guideline for Wildfire Protection of Institutional Buildings in Forested Regions in Alberta' Alberta Infrastructure 2013

Hay A (2013) 'Surviving catastrophic events: stimulating community resilience' Institution of Engineering and Technology Infrastructure Risk and Resilience, pp 41–46

http://cdd.publicsafety.gc.ca/srchpg-eng.aspx

Insurance Bureau of Canada 'Facts of the Property & Casualty Insurance Industry in Canada 2013'

Jacobs 1997 'Death and life of great American cities' New Edition Random House.

National Building Code of Canada 2010

Naum CJ (2010) 'Structural anatomy: building construction, command risk management and fire-fighter safety' http://commandsafety.com/structural-anatomy/. Accessed 1 June 2011

Philips M (2012) 'After sandy's pain, there will be gain' Bloomberg Businessweek http://www.businessweek.com/articles/2012-10-31/sandy-after-the-pain-long-term-gain. Accessed 10 Mar 2014.

PriceWaterhouseCoopers (US) (2014) 'Drive business transformation through the cloud' http://www.pwc.com/us/en/issues/cloud-computing/index.jhtml. Accessed 10 Mar 2014.

Sharp T (2012) Superstorm sandy: facts about the Frankenstorm' LiveScience. http://www.livescience.com/24380-hurricane-sandy-status-data.html. Accessed 10 Mar 2014

SwissRe 2010 'Making flood insurable for Canadian homeowners' Institute of Catastrophoc Loss Reduction

TEDtalks: Hypponen, Mikko, 2013 'How the NSA betrayed the World's trust – time to act' http://www.ted.com/talks/mikko_hypponen_how_the_nsa_betrayed_the_world_s_trust_time_to_act accessed 10 March 2014.

'The 2013 Great Alberta flood: actions to mitigate, manage and control future floods' alberta watersmart, 2 August 2013.

UNISDR 'Towards the Post-2015 Framework for Disaster Risk Reduction Considerations on its possible elements and characteristics' 14 November 2013

Wingrove 2013 'Alberta shelved major flood report for six years' Globe&Mail 24 June 2013

Chapter 12
Integration of Risk and Resilience into Policy

The Case of the Seine River FLOODS in Paris and the Ile de France Basin

Raymond Nyer, Charles Baubion, Bérangère Basin, Adeline Damicis, Sonia Maraisse, Mathieu Morel, and Emmanuel Vullierme

Abstract France still remembers well the major flood which happened in 1910 and the dramatic damages it created at that time. Today with 17 million people and nine million jobs potentially impacted by the key rivers floods, France is very exposed to flooding hazards and particularly the Paris and the Ile de France Basin. Several EU Member States share similar significant flood risk and experienced, indeed, in UK,Germany and Central Europe very large floodings creating several Billion Euro of damages and a significant number of casualties. The recognition of this important flood risk across many EU Member countries led to the establishment of the October 2007, European Directive which provided a general framework for public management of floods. And Member States were required to incorporate this Directive into national law by 2010.To benefit from its International competences a Review of Risk Management Policies was conducted by the OECD Directoriate for Public Governance and Territorial Development.This review was supported by the French Ministry of Ecology, Sustainable Development and Energy and EU and led to a very comprehensive report published in 2014 providing a set of important recommendations on: governance of flood risk, on increasing resilience and on financing prevention. Based on the 2010 French law and helped by the recommendations of the OECD report the French government launched a National Flood Risk Management

R. Nyer (✉)
RNC Conseil, Paris, France
e-mail: raymond.nyer@centraliens.net

C. Baubion
OECD, Paris, France
e-mail: charles.baubion@oecd.org

B. Basin • M. Morel • E. Vullierme
French Ministry for Environment, Energy and Sea, General Directorate for Risk Prevention, Paris, France

A. Damicis • S. Maraisse
INHESJ National Institute of Advanced Studies on Security and Justice, Paris, France

© Springer Science+Business Media B.V. 2017

I. Linkov, J.M. Palma-Oliveira (eds.), *Resilience and Risk*, NATO Science for Peace and Security Series C: Environmental Security,
DOI 10.1007/978-94-024-1123-2_12

Strategy associating all flood management parties through the Joint Flood Commission (CMI). After a national public consultation the project of national strategy has been adopted on the 7th October 2014 by the Ministers of Environment, Interior, Agriculture and Housing. This document is the basis of the French Flood Management Policy at national, district, basin and local levels and actions plans have been devolopped accordingly over the whole country..To test the effectiveness of the Policies and actions put in place the Secretary General of the Paris Defense and Safety Zone (SGZDS) covering the whole Ile-de-France area organized an exercise simulating major Seine flooding in Ile-de-France basin.This exercise called EU SEQUANA lasted two whole weeks in 2016, March 7th–March 18th. More than 90 national and EU partners were involved in this project, most of them from vital importance sectors such as energy,water, transport, telecommunications….

Important recommandations to achieve better resilience of the flood risk management have been identified and will surely be integrated into future policies and processes.As an unexpected significant flood of the Seine River happened in May/June 2016, just after the EU SEQUANA exercise,some of the recommendations already implemented could be confirmed.

Keywords Paris and Seine basin floods • Risk resilience • OECD assessment and recommandations • EU and French policies • EU/French flood risk exercise Sequana

12.1 A Brief History of the Major Floods of the Seine River in Paris and Ile de France Basin

12.1.1 A Recurring Risk

The Ile-de-France, with Paris in the center, counts about 12 million inhabitants living in 1300 communes and is the main engine of the French economy.

This region crossed by the Seine River Basin, concentrates the most important political and economic powers of the country. It is also at the center of three affluent junctions the Seine/Yonne, the Seine/the Marne and the Seine/Oise, About 60 major floods occurred during the last century, and the last one just happened in May/June 2016. The floods are part of the natural behavior of the waterways in the Ile de France basin, but they can be of exceptional dimension in the event of intense and durable rains and generate dramatic impacts on the territory structured along the river (Fig. 12.1).

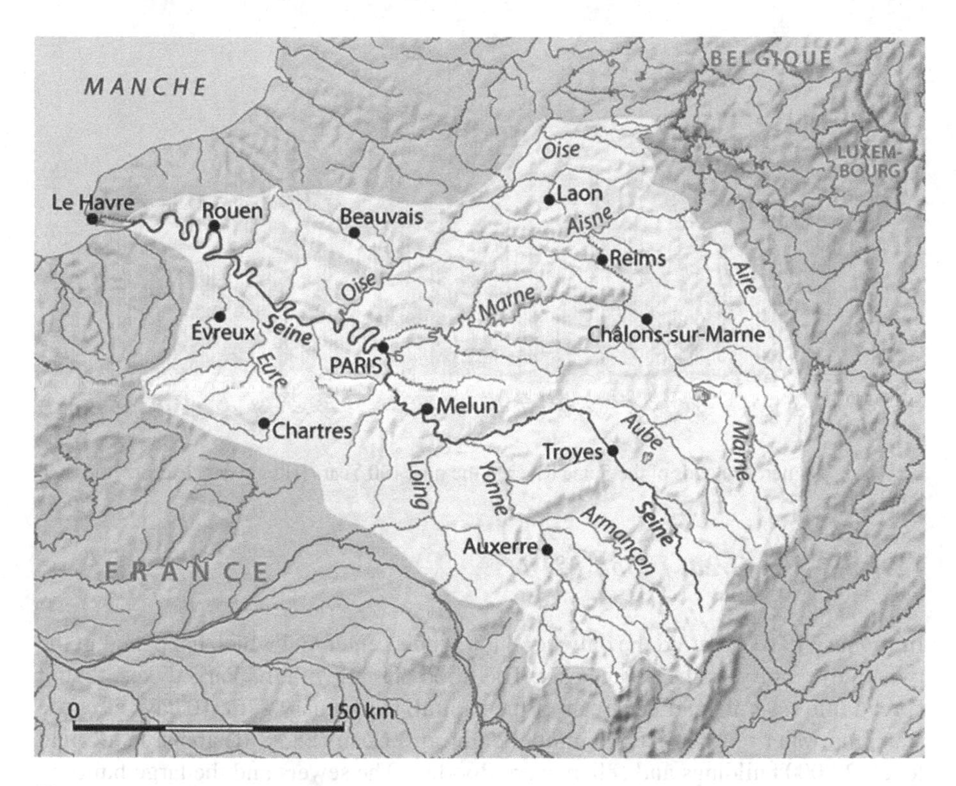

Fig. 12.1 Hydrological Seine basin

12.1.2 Floods of All Kinds

The floods which happened during the past 150 years were extremely diverse by their height, their frequency and their gravity. Their distribution in the time and the frequency of the major floods resulting from exceptional weather conditions varied also significantly year to year. During the twentieth century the major flood happened in 1910 and two significant ones but of less importance occurred in 1924 and1955. Taking into account its exceptional gravity, the flood of 1910 constitutes the event of reference, used in the inventory of known high waters. It is regarded as "centennial" i.e. that it is likely to be reached or exceeded on average once per century. However the variation of pluviometry can bring to more or less frequent floods standard "1910" or to even more important flood still. The most recent one happened end of May/beginning of June 2016 and reached 6.10 m above the normal compared to the 8.62 m of the 1910 reference flood (Fig. 12.2).

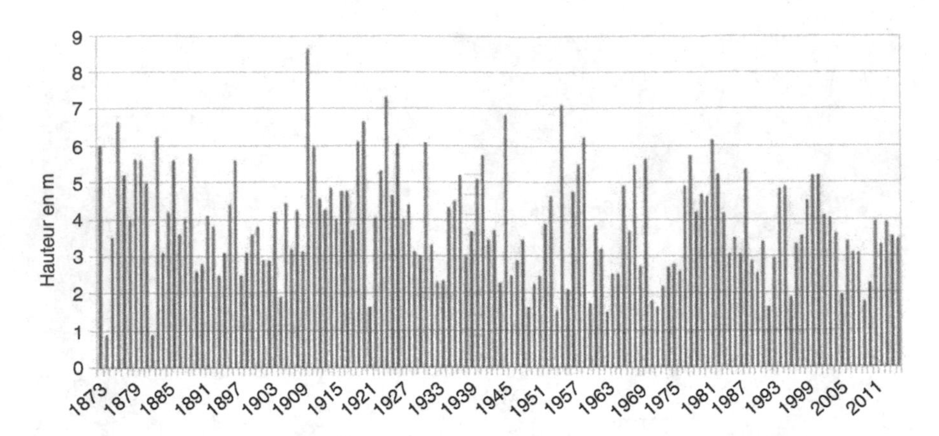

Fig. 12.2 Numerous floods of the Seine River in the past 150 Years (Height at Austerlitz bridge in Paris)

12.1.3 The Great Flood 1910

In January 1910 occurred the only flood known as centennial which happened in the twentieth century on the Seine basin. The flood started with the Paris dykes, then it spread into several hundred streets of the capital. On January 28th, 10 days after the start, the water reached its maximum height at 8.62 m (28.28 ft) above its normal level; 22,000 buildings and cellars were flooded. The sewers and the large building sites (among them the construction of the subway) made possible the flood to reach city districts far away from the normal river bed. The flood disturbed severely the transport, the electricity and gas distribution networks. Trams lines, 30 subways stations and 60 km of metro lines and 60 km of railroads were flooded. The household refuse, not being able to be evacuated, were thrown directly in the water. The Boucicaut hospital had to be evacuated. Paris was paralyzed. The damages were enormous, estimated at approximately 1.6 billion euros of 2009. After 35 days the water was completely gone but a nauseous mud covered the streets and the 30,000 dwellings which had to be evacuated.

The restart of the activities took more than 2 months and the normal functioning was reached only several months after.

12.1.3.1 The Ecosystem Basin

The Seine and its affluents constitute a living ecological domain largely overflowing the perimeter of the Ile-de-France region: it extends from the limits of the Parisian basin to the estuary of the Seine River.

On the east at the top of the basin, the natural environment principally made of wetlands, ensures the purification of surface waters before they flow in the river.

It contributes to the development of the flora and the aquatic life, as well as the regulation of the flows. The small floods are beneficial for the good performance of the waterway, but the biological state of the water is degraded due to its usage in economic activities and growing urbanization.

12.1.4 Which Would Be the Consequences of a Major Flood Comparable to the 1910 One?

Following the present development of Seine river basin more than 850,000 inhabitants are now directly exposed to the risk of flood, but the dysfunctions would not be limited to the directly flooded zone on the surface. They would impact indeed four to five million people to differing degree. Close to two millions people would be affected by cuts of electricity and 2.7 millions by cuts of drinking water. The flood would result in the deterioration of the services to the population, the damage to the equipment of the flooded companies, the disturbance of the supply and distributions chains, the mobility networks.

Nearly 170,000 companies would be impacted by a major flood (including 86,000 directly flooded) generating a regional paralysis which would affect the whole of the country activities.

In spite of the actions already taken on the protection infrastructures (lake-tanks, dams and walls), the direct damage resulting from a 1910s type of flood is estimated to a minimum of 17 billions Euros (value 2008) not taking into account the damage to the transport, electricity, gas, telecommunication networks, the local heating systems, etc. and the long term impact of the economic paralysis.

12.2 The EU Floods Directive, Water Directive and Communication on Resilience

12.2.1 The Challenge

Several EU Member States share a significant flood risk similar to the Seine River basin. Since the beginning of this century, in December 2003, France experienced a 100-year flooding of the Rhône tributaries downstream of Lyon, which caused seven deaths and one billion Euros in damage. In 2007, Great Britain experienced significant rainfall affecting Yorkshire which caused damage assessed at $7.2 billion dollars. In 2002 and 2013 in Central Europe, a region accustomed to 100-year floods, the Elbe and the Danube Rivers flooded. The 2013 floods seriously affected Germany, the Czech Republic, Austria, Hungary and Slovakia.

12.2.2 Alignment of the EU Members on Flood Risk Management

The recognition of the significant flood risk across many EU Member countries led to the establishment of the October 2007, European Directive which provided a general framework for public management of a flood. Member States were required to incorporate this Directive into national law. A key component was producing risk maps identifying the significant flood risk areas.

The Treaty of Lisbon, which came into effect in 2009, introduced several areas in which the European Union has power to act, including civil protection. This is a support power, meaning that the EU only intervenes to support, coordinate and complement the actions taken by the Member States.

12.2.3 Requirements of the Floods Directive

The Directive which applies to all kinds of floods (river, lakes, flash floods, urban floods, coastal floods, including storm surges and tsunamis), on all of the EU territory requires Member States to approach flood risk management in a three stage process whereby:

1. Member States have been ask to undertake in 2011 a preliminary flood risk assessment of their river basins and associated coastal zones, to identify areas where potential significant flood risk exists.
2. Where real risks of flood damage exist, they had by 2013 to develop flood hazard maps and flood risk maps. These maps identify areas with a medium likely hood of flooding (at least 1 in 100-year event) and extreme events or low likelihood events, in which expected water depths are indicated. In the areas identified as being at risk the number of inhabitants potentially at risk, the economic activity and the environmental damage potential are also indicated.
3. Finally, by 2015 **flood risk management plans** were drawn up for these zones. These plans include measures to reduce the probability of flooding and its potential consequences. They address all phases of the flood risk management cycle but focus particularly on prevention (i.e. preventing damage caused by floods by avoiding construction of houses and industries in present and future flood-prone areas or by adapting future developments to the risk of flooding), protection (by taking measures to reduce the likelihood of floods and/or the impact of floods in a specific location such as restoring flood plains and wetlands) and preparedness (e.g. providing instructions to the public on what to do in the event of flooding). Due to the nature of flooding, much flexibility on objectives and measures are left to the Member States in view of subsidiarity.

These steps are scheduled to be reviewed every 6 years in a cycle **coordinated and synchronised with the Water Framework Directive (WFD)** implementation cycle.

12.2.4 A Strategy to Support Implementation

To support the implementation of the Directive, a Working Group on Floods has been established under the Common Implementation strategy which focused on three pillars:

- Floods Directive Implementation: Development of reporting formats
- Water Framework Directive: towards joint implementation with the Floods Directive
- Flood risk management information exchange

The development of **reporting** formats responds to the requirement of the Directive, and is carried out via WISE (Water Information System for Europe).

12.2.5 EU Communication and Actions on Resilience

Natural disasters can strike anywhere at any time bringing devastation in their wake and presenting threats to long-term development, growth and poverty reduction, particularly in the poorest and developing countries. Good planning and preparation can limit the scale of impacts. Risk management policies save lives and enable growth and sustainable development. Building resilience is about helping communities withstand and recover from disasters, with the focus on tackling the root causes rather than dealing with the consequences.

In recent years, the EU have made huge progress towards strengthening disaster prevention and increasing its efficiency in dealing with disasters such as the devastating Typhoon Haiyan and the major earthquake in Haiti. The adoption of new Civil Protection legislation – with a strong DRR (Disaster Risk Reduction) focus – and the reinforced Emergency Response Coordination Centre were major milestones in this regard.

In October 2012, the European Commission presented a Communication – The EU Approach to Resilience: Learning from food crises, which provides the policy principles for action on helping vulnerable communities in crisis-prone areas. Increasing their resilience to future shocks will be a central aim of EU external assistance. An Action Plan, which followed the Communication, laid the foundations for more effective EU collaborative action on building resilience, bringing together humanitarian action, long-term development cooperation and on-going political engagement. The Action Plan adds value to previous commitments by maximizing the synergy between interventions across thematic areas. It also gives new, and necessary, impetus for the implementation of the strong commitments made in the Disaster Risk Reduction (DRR) Implementation Plan and the Nutrition and Food Security Action Plan, as well as in the 2012 Communication on Social Protection in EU Development Cooperation. It takes into account the principles of adaptation to climate change applied through the Global Climate Change Alliance

(GCCA), in particular with regard to policy dialogue and exchange of experiences, aid effectiveness and mainstreaming **across all relevant sectors, including agriculture, water and health**.

12.3 OECD Review of the Flood Management and Resilience Systems of the Seine River and the Ile de France Basin

12.3.1 Context

A major flood of the Seine River, similar to the historic event of 1910, is an important challenge for public policy. The economic, social and human issues at stake are considerable, given the large scale and high vulnerability of the floodplain and because the Paris – Île-de-France region is the principal economic hub in France. Over the last decade, many countries have experienced floods beyond historic records (Prague in 2002, the United Kingdom in 2007, Australia in 2011, Bangkok in 2011, New York in 2012, Germany in 2013). At the same time, vulnerabilities of modem societies and megacities in particular have steadily increased, calling for considering this major risk and its implications in all its multiple dimensions. As the economic impact of a major Seine flood could be significant on national output, it appears important to question the level of protection and the resilience of the Paris metropolitan area with an international perspective. This calls for a collective and voluntary reflexion on flood risk management policies in the Ile-de-France region.

This analysis is supported by an innovative flood risk assessment approach that considers the macroeconomic impacts that could be caused by a Seine flood in Ile-de-France. This assessment integrates the different cascading impact of such a major shock through the networks of critical infrastructures that are sustaining the Paris metropolitan functions. The macroeconomic effects on the national economy of different flood scenarios provide an indication of the issues at stake. In terms of public policies, the OECD review concentrated on the reduction of the risk over the medium to long term through prevention efforts and resilience and vulnerability reduction measures. It addresses the key governance issues related to flood risk prevention in Île-de-France. It analyses the different structural and non-structural prevention measures and their effective implementation to strengthen resilience in the region, and looks in depth at existing and potential financing mechanisms for the prevention of this major risk. The review shows overall that an effort to recalibrate, better co-ordinate and refocus public policies would decrease the consequences of the risk. It proposes new avenues for public policies to the French authorities that could support an ambitious integrated strategy for the management of the Seine flood risk in the Île-de-France region.

12.3.2 *Assessment of the Situation*

12.3.2.1 Despite Investments in Protection, Increasing Urban Development and the Interdependence of Critical Infrastructures Have Accentuated Vulnerability

The probability of a major Seine flood similar to that of 1910 cannot be neglected given the population and assets nowadays located in the floodplain and their vulnerabilities. If protection levels have increased since 1910, investments in infrastructures have been limited over the past decades. It appears that protection levels are not up to the standards of many other comparable OECD countries, particularly in Europe. In the meantime, exposure and vulnerability have significantly increased along with the massive urban development in Île-de-France and the increased interdependencies of critical infrastructures. A major flood could affect directly or indirectly up to 5 million people and thousands of businesses. It could severely disrupt the continuity of the state and many institutions as well as most of the networks of critical infrastructures (e.g. electricity, transport, communication, water) that are sustaining the largest metropolitan area of continental Europe. Given the hydrological conditions of the Seine basin, the effects of the flood could last over 3 months. There are in consequence major issues for public policies at stake, not only at the regional but also at the national level (Figs. 12.3 and 12.4).

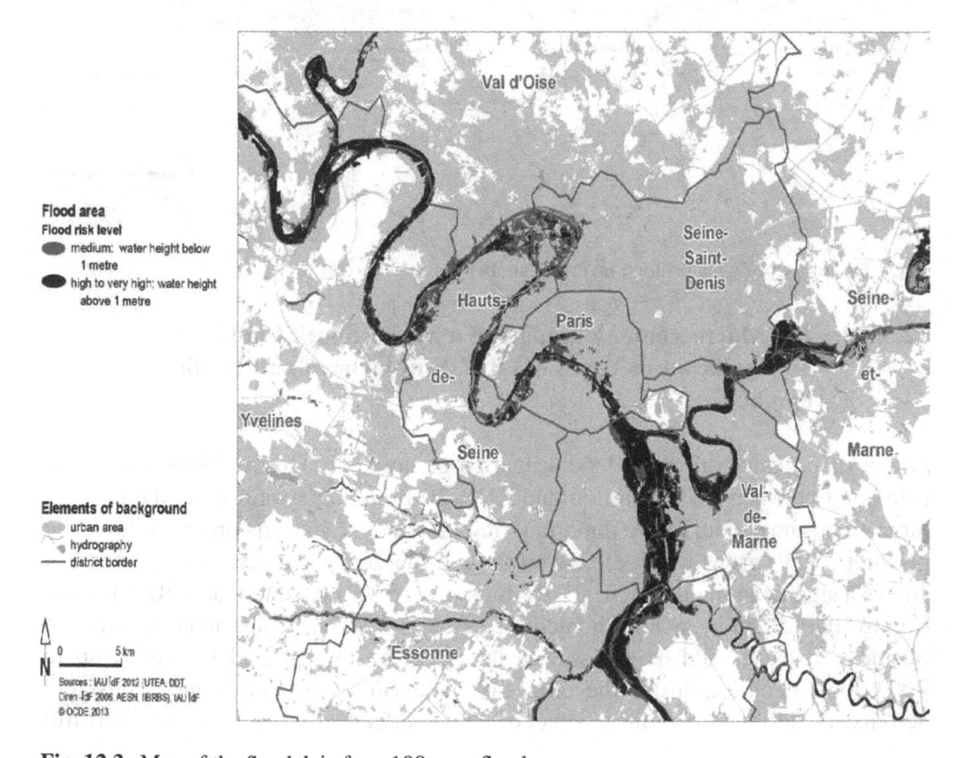

Fig. 12.3 Map of the floodplain for a 100-year flood

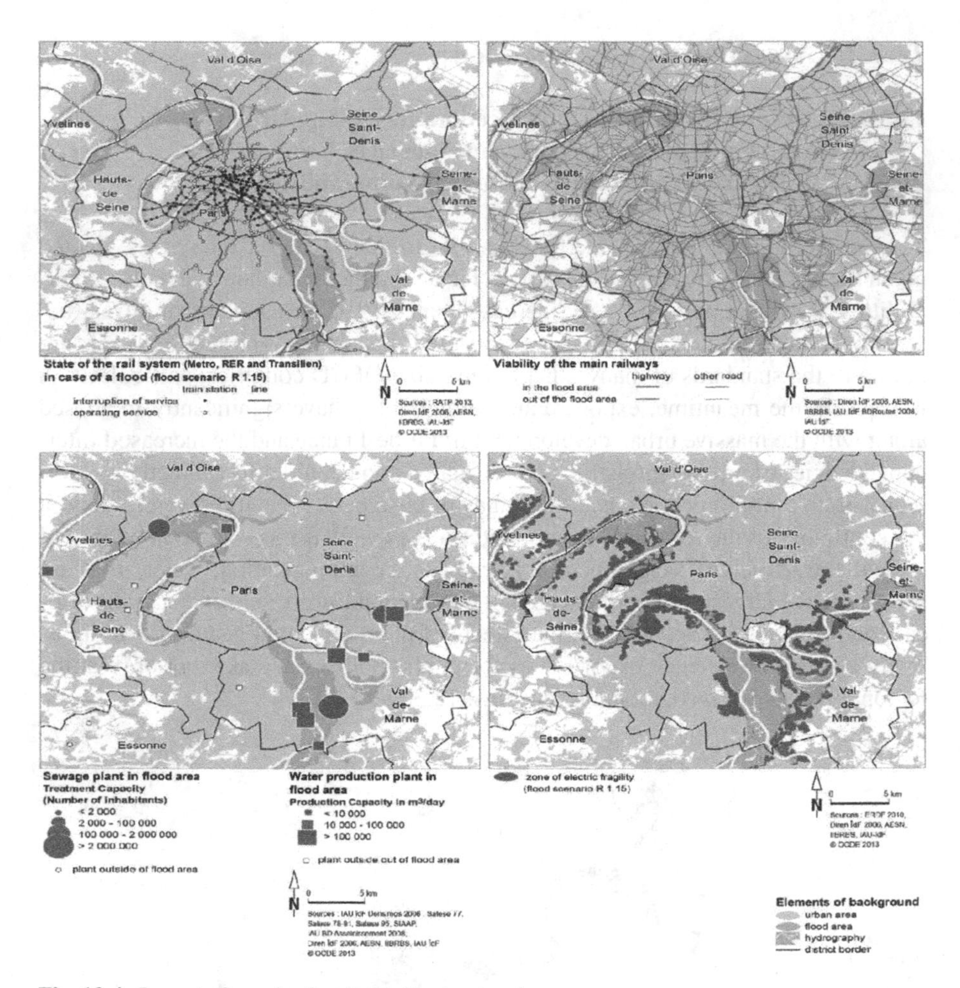

Fig. 12.4 Impact of a major flood on critical networks

12.3.2.2 The Macroeconomic Impact of a Major Shock Could Be Significant in Terms of GDP, Employment and Public Finances

Direct damages of different flood scenarios centered around the 1910 flood were estimated between EUR 3 billion and EUR 30 billion. Beyond direct damages, a large-scale shock could have important macroeconomic impacts on gross domestic product (GDP) growth with effects on the job market. The macroeconomic model shows a significant reduction in GDP which, over 5 years, could reach EUR 1.5 billion to EUR 58.5 , i.e. a consolidated total of 0.1–3%. The resulting contraction in business activity could have a significant effect on the demand for labor; up to 400,000 jobs could be lost in the worst case scenario. Even if a rebound in business activity could rapidly reduce some of these effects after a year, the harmful consequences of a major Seine flood could be felt over the medium to long term and weigh on public finances. In the case where the impact exceeds the reserves available through the national catastrophe compensation regime CATNAT and the Central Reinsurance Fund, the state could be called on to fully assume its role of ultimate guarantor.

12.3.2.3 Opportunities Are Emerging Today to Address Gaps in Governance That Should Be Seized to Better Prevent Flood Risk

Well-identified governance deficits have affected the design and implementation of flood prevention policies commensurate to the risk level in Île-de-France. These gaps could be filled if opportunities are properly seized. The institutional context has not favoured the emergence of an ambitious and coherent strategy for preventing this risk. The fragmented institutional framework in Île-de-France has been, in the past, a restraint on action. Resulting from successive waves of decentralization, this institutional and territorial fragmentation has not enabled the proper articulation of the different sectors of public policies for effective flood prevention (e.g. water policy, territorial and urban planning, crisis management).

The response to the major risk of a Seine flood in the Île-de-France region should be based on a revised governance framework. A more transversal and multi-level approach should aim to better align public policies to improve resilience. The on-going implementation of the European flood directive and the development of the Greater Paris project are offering a unique opportunity to revive the policy debate and promote innovative approaches.

12.3.2.4 A Coherent Resilience Strategy at the Metropolitan Scale Could Take Advantage of Synergies Between the Different Prevention Measures

The only way of reducing the Seine flood risk in the Île-de-France region is by means of practical measures aimed at increasing the territory's resilience. A broad range of measures contributes to preventing this flood risk, even if a certain diversity predominates. Whether these measures are regulatory or voluntary or are ensured by the state, local authorities, the public or businesses, this overview highlights the many opportunities for improvement. These opportunities relate to risk awareness and culture, resilience of communities, public services or businesses and hazard mitigation measures like protective or storage infrastructures.

Positive synergies leading to greater resilience have been identified and could be further exploited. This includes the incorporation of resilience into the development policies of the Greater Paris project, better linking the on-going development of a water culture and the culture of risk, strengthening the alignment between prevention and crisis management policies or the increasing awareness of businesses and network operators. This review also shows that the existing protective and water storage infrastructures are reaching their limitations. At a time when a new hazard mitigation project is proposed, key questions related to the financing of prevention, the prioritizing of actions, ensuring equity or the governance of these complex choices should be responded. The local flood management strategy is an opportunity to organize prevention measures as a whole and prioritize them in a coherent approach to improve resilience also based on innovation.

12.3.2.5 How to Finance Resilience Within a Constrained Budget Context?

Funding the preventive actions required to increase the level of resilience remains a major issue. Within a context marked by under-investment in the past, and current difficult economic conditions, investment in prevention is made under pressure, in view of the tight fiscal environment and the necessity to decide priorities in public spending both at the level of the state and the local governments.

If specific tools to finance prevention exist in France, particularly through the disaster compensation regime CATNAT, the system is faced with growing demands and has had a limited contribution to reducing the flood risk from the Seine in Île-de-France. Other strategic priorities have mobilized authorities and the available prevention funds in other regions. This led to a certain delay in funding the prevention of this major flood risk which represents a significant share of the total losses caused by flood risk in France. Up to now, the national allocation of resources has not been based on criteria giving priority to resources according to the level of risk. This is changing with the implementation of the EU Floods Directive which identified the most risk-prone territories, of which the Paris metropolitan area is one. There is room for progress in defining a financing approach for prevention which is adapted to the issues at hand. In the context of strained public finances, the question of additional resources and the sharing of efforts (state, local government, businesses, citizens, European funds) may be addressed by setting out a number of principles for an overall financing strategy. The general principle is that the beneficiaries of prevention measures should be the first to finance prevention.

Identification of the beneficiaries helps to determine the primary sources of funding to be raised for such a strategy. Prevention funding must aim at being most effective through an economic approach based on coherency, cost-effectiveness, long term and equity.

12.3.3 OECD Recommendations for Better Seine Flood Risk Prevention Policies in Ile-de-France

Increasing the resilience of Île-de-France region to Seine floods requires additional efforts to anticipate and invest in order to better limit this major risk. The objective to strengthen the capacity of the Île-de-France ecosystem to rapidly restore its basic capacities and functionalities from a social and economic perspective. The OECD review highlights several policy options and lines of action that the French authorities may wish to integrate in an ambitious comprehensive strategy for Seine flood risk management in Île-de-France:

12.3.3.1 Recommendations on Governance

1. Ensure the appropriate linkages between the various levels of flood prevention – from the exposed Ile-de-France metropolis to the river basin. This will mean engaging a differentiated approach with the stakeholders at local level in the Ile-de-France risk basin, and the upstream territories by means of a partnership from which they will also benefit, and which can also draw on the implementation of the EU Floods Directive. The governance structure envisaged between the State and the local contracting authorities at sub-basin level should be thoroughly explained to the local authorities and benefit from current developments in decentralization reforms to become well-established locally.
2. Define an ambitious and mobilizing global vision over the long term together with actionable principles. This global vision should be consistent with the ambitions of the Grand Paris project and will enable public decision-makers and citizens to mobilize beyond the regulatory obligations of the Directive and risk management policy. The principles for action in the national strategy for the management of flood risks may be adapted and formulated at the risk basin level (pooling risks, minimizing the moral hazard, proportionality of the charges and benefits, subsidiarity and role of the State, adaptability).
3. Break-down the global vision into precise objectives and make the stakeholders aware of their responsibilities. The local strategy's operational objectives and those of the PAPI should be aligned with each other and with this long-term vision. Economies of scale and greater effectiveness may be achieved by redefining the stakeholders' roles and responsibilities, as their numbers and diversity make co-ordination and efficiency more complex. The definition of performance criteria should make it possible to analyze the respective contributions made by the various stakeholders towards flood risk prevention; to monitor the performance of the various initiatives set up; and to establish more rational distributions of responsibilities and resources.
4. . Create effective gateways between the flood risk management strategy and related public policies. This involves incorporating the risk of floods in a multi-hazard approach with other aspects of resilience for the development of the

Grand Paris project (environment, green economy, well-being). This also means ensuring that the various initiatives and sectoral policies (water management, regional planning) actually incorporate the issue of flood risk management with a view to creating synergy and sharing benefits.

12.3.3.2 Recommendations on the Resilience Measures

5. Continue to improve and harmonize risk knowledge and ensure that risk information is made available. The collaboration between the prevention and crisis management stakeholders could be extended to other actors such as the insurance sector, in a coherent global risk assessment approach, particularly from an economic point of view. All information concerning the risks could be centralized whilst abiding to demands of confidentiality, security and competition. This could go hand in hand with the provision of modelling tools and related data according to needs, taking inspiration from the risk observatory established at the national level.
6. Reinforce the risk culture of citizens, decision-makers and companies. New communication approaches stressing the positive benefits of greater resilience, must aim at increasing risk awareness at all levels. Regular information, based on the best available knowledge and to the benefit of a common strategy could accompany the local flood risk management strategy. This communication strategy should use new technologies (3D imaging, virtual animation, social networks) for specific targets (companies, citizens, decision-makers, developers and architects) and its results be regularly assessed through regular surveys on risk perception.
7. Improve territorial resilience, using the opportunities offered by the Grand Paris project. The definition of a level of resilience for the Grand Paris project, particularly through the local Territorial Development Contracts could allow model resilient districts to emerge such as Les Ardoines. The harmonization and reinforcement of the Risk Prevention Plans at regional level will enable resilience to be improved towards this predefined level in the long term: these plans should use the latest risk assessments as a basis and their control should be improved. Incentives aiming to reduce the vulnerability of existing constructions could also be envisaged, by using opportunities such as the replacement of electricity meters.
8. Gradually improve the resilience level of critical networks and take steps towards preserving the continuity of business and public services. A predefined level of resilience should also be gradually applied to the networks operators to reinforce requirements. New infrastructures, particularly transport, should aim at the greatest resilience to floods. Establishing requirement levels and controlling them may become the responsibility of the sectoral regulator. A mechanism supporting companies in their business continuity approach, and particularly the SMEs, could also be developed, for instance the establishment

of a risk-diagnosis service, of a dedicated label or the development of risk awareness guides.

9. Place the flood protection infrastructures under the responsibility of a single contracting authority in charge of applying a pre-defined safety standard, based on a common cost/benefit approach, under an appropriate institutional structure. The management and organization of the maintenance, replacement and work requirements could also be assessed in accordance with common criteria and in comparison with potential new infrastructures. The feasibility of harmonizing the protection levels for the whole urban area should be assessed by planning the work over time giving priority to the most beneficial measures.
10. Encourage experimentation with regard to the La Bassée storage project. Rolling out the La Bassée project stage by stage should make it possible to adapt the approach through a process of learning by practice and to demonstrate its operational utility, beyond the theoretical cost-benefit studies. The question of the governance of such a structure should also be raised beforehand, particularly regarding decision-making in a time of crisis to guarantee its effectiveness.

12.3.3.3 Recommendations for Financing Prevention

11. Support the local of Seine flood risks management strategy in the Ile-de-France by a clear financial strategy, taking into consideration national specificities. This could focus on the following elements: sustainability and long-term vision; principles of responsibility and proportionality among the beneficiaries of the measures taken and the financiers; exploring the best effectiveness and considerations of equity in resource allocation; synergies with the other sectoral strategies (drought, water, development, crisis management).
12. Mobilize all the beneficiaries of preventive measures in a multi-level approach which would combine local government authorities and State funding, as well as the various network operators, the private sector and citizens by targeted incentives. Additional funding could come from positive incentive mechanisms in existing taxation raising systems, particularly by bringing together the insurance, real estate and water management sectors.
13. Strengthen efforts to clarify the priority criteria for prevention funding from State resources. This can also consider the possibility of European funding which can be mobilized for implementing the European Floods Directive in high flood risk areas such as the Ile-de-France region.
14. Re-examine the impact of the CatNat insurance regime on flood risk prevention. The bill aiming at reducing the system's dis-incentivizing effect could be revived, which would be an opportunity for a wider reflection on funding prevention.

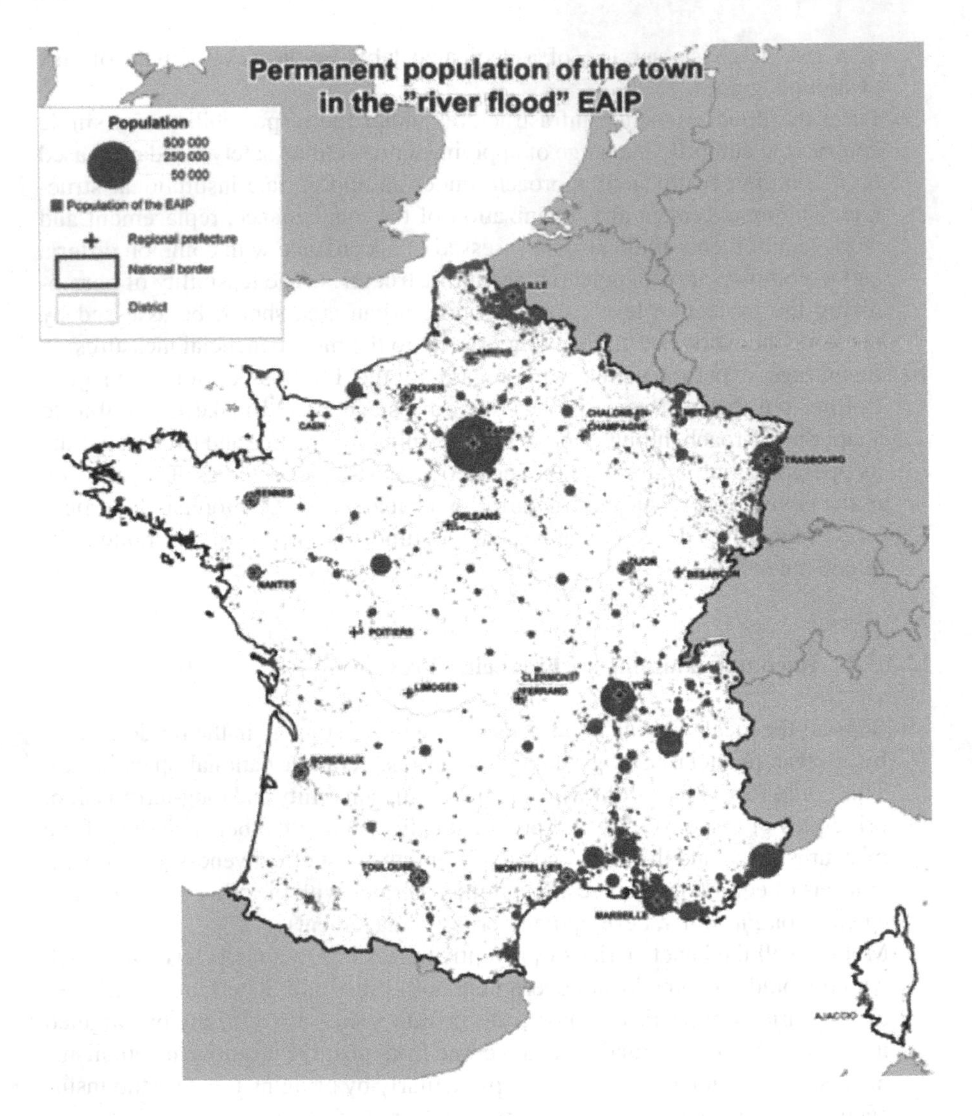

12.4 French National Policy for Flood Risk Management

12.4.1 Context

The annual average cost of economic losses over the last 30 years in France reaches about 650–800 million Euros, of which about half is covered by the Natural Disasters insurance system known as "Cat-Nat" implemented by the law of 13 July 1982. This average yearly cost could be much higher in the event of hazards of exceptional intensity.

Although France has been spared from any major disasters seriously impacting the national economy for many decades now, the preliminary flood risk assessment made in 2012 shows that almost one inhabitant in four and one job out of every three is today potentially exposed to these risks (see Fig. 12.5).

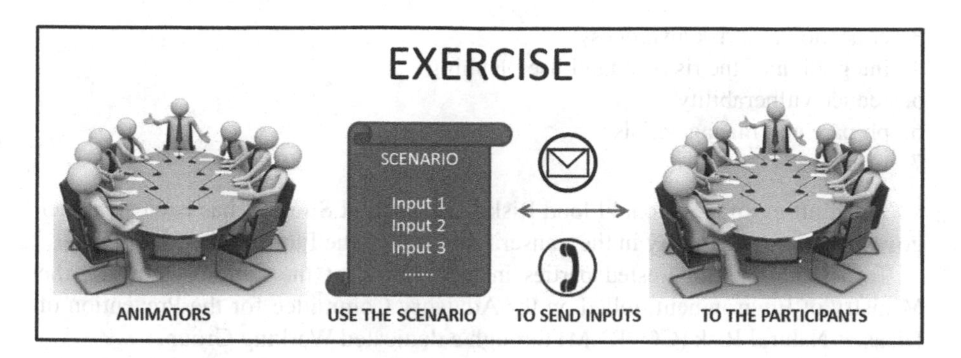

Fig. 12.5 The exercise concept

Additionally, the effects of climate change with a rise in the average sea level and a possible multiplication in the number of intense storms are all factors aggravating the risk for the coastline and areas surrounding estuaries.

Furthermore, the degree of vulnerability of the exposed populations depends on many factors: the degree of exposure of buildings and strategic locations used for crisis management, the concentration of assets, the complexity and the interdependency of the networks, production modes using just in time practices and therefore the absence of any stocks, the multiplicity of the public and private sector players in charge of services.

To respond to these issues and implement a renewed policy for flood management, Central government has chosen to set out a framework for the 13 French flood risk management plans (FRMP) promoted by the Directive 2007/60/EC of the European Parliament and the Council of 23 October 2007 on each district. This national strategy for flood risk management brings together measures applicable to fulfill the national policy and to highlight priorities.

The national strategy therefore responds to the significant expectations expressed by all of the partners, especially local authorities, seeking a shared framework to direct the national policy for managing flood risks.

12.4.2 Building a National Strategy

12.4.2.1 The Development of Flood Governance

Principle of a National Strategy

French flood risk management was previously part of the national disaster risk reduction (DRR) policy based on the following seven pillars:

1. understand and assess risk
2. forecast and early warning systems

3. education and risk awareness
4. integration of the risks in land-use planning
5. reduce vulnerability
6. prepare and manage crisis
7. feedback

The drafting of a National Flood Risk Management Strategy has been set under government responsibility in the transcription law of the Floods Directive.

To associate all interested parties in the writing of the national strategy, the Ministry of Environment, relied on the Advisory Committee for the Prevention of Disaster Natural Risk (COPRNM) through a dedicated Working Group.

This national Committee being the national platform for disaster risk reduction included parties interested in all natural hazards: flooding, earthquakes, landslides, etc.

A Flood Governance at All Levels

For the implementation of the Floods Directive, the Ministry of Environment engages in a national structuring of parties around three levels of flood governance:

- nationally, a Joint Flood Commission (CMI) has been created based for all parties (State, elected representatives, trade unions, syndicate and NGOs) from the COPRNM but also the National Water Committee (CNE);
- at the river basin district level, a Flood District Commission (CIB);
- locally with the Local Public River Basin Establishment (EPTB) and Local Public Water Management Establishment (EPAGE).

Thus the CMI has taken over the writing of the national strategy with the DRR, urban planning stakeholders and water management stakeholders.

12.4.2.2 Common Objectives

Major Objectives

The French National Flood Risk Management Strategy sets three major objectives:

- To improve the safety of exposed populations

- To stabilize in the short term and reduce in the medium term the cost of flood damage

- To significantly shorten the time required by affected areas to return to normal

After years of trade off, all stakeholders agreed that floods may cause populations to face serious and even mortal risk. Therefore, the first national priority aims to reduce any loss of human life.

Action Framework

The French National Flood Risk Management Strategy (SNGRI) also defines the guiding principles serving the three major objectives:

- Solidarity principle: Achieving these national objectives requires first and foremost, applying the solidarity principle when faced with risks. Governance applied to the management of flood risks, regardless of its geographical scale, reinforces this solidarity. Flood national policy is based on river basin level cooperation between stakeholders at the river basin level: up-stream and downstream stakeholders, urban and rural areas, the living basin and the risk basin, the run-off basin, the hydro sedimentary cell along the coast. The insurance national solidarity principle is also the basis of French disaster risk reduction solidarity.
- Subsidiarity principle: The subsidiarity principle aims to ensure that decisions are taken as closely as possible to the citizen and that each action is implemented in light of the possibilities available at national, regional or local level:

  - Central government guarantees public safety alongside Mayors and conducts the flood risk prevention policy through state actions.
  - Municipalities or the public establishments for inter-city cooperation with their own taxation powers (EPCI) are a part of the stakeholders, with responsibility in terms of urban development and territorial development.
  - Mayors are tasked with providing emergency services to the population. They develop municipal safeguard plans by informing and alerting community on any risks.

- Public policy synergy principle: Synergies between the various public policies relating to flood management lead to greater efficiency and cost rationalisation. It allows taking the impact of flooding into account in development projects or in the work done to preserve the natural habitat. Through this principle, the SNGRI established the flood risk management policy as an integrated policy addressing territorial policies (water, urban planning, infrastructure, etc.)
- Prioritisation and continuous improvement principle: To take into account the human, technical and financial resources that can be mobilised by each of the players involved, the national policy utilizes the prioritisation principle. The expected prioritisation-based scheduling is based on:

  - selecting territories facing a severe risk, the areas of potential significant flood risks (APSFR), within each major basin,
  - supporting relevant territorial projects addressing exposed territories with a cost-benefit and multiple criteria analyses to identify the measures that are economically sustainable.

Strategic Orientations: Four Challenges to Meet

Stakeholders also identify four challenges for the implementation of the major objectives:

- Developing governance and project management: The sharing of roles and responsibilities in risk management is finalised in an organisational and financial framework. If the creation of the CMI, CIB, EPTB and EPAGE has enabled to frame flood governance, locally municipalities can mobilize and cooperate in stronger structure.
- Sustainable territorial development: Taking into account the flood risk as part of a sustainable development approach for territories aiming to improve their competitiveness and getting more attractive.
- Increase knowledge to improve risk management: Looking beyond the expertise already gained, tools and methods have been developed to ensure an operational knowledge of territorial vulnerability when faced with flooding.
- Learning to live with floods: A mobilisation of all citizens, for which Mayors are the first representatives, requires being placed in a “crisis situation” illustrated by real life information and aimed at finding operational solutions.

An Inter-ministerial Strategy

The French National Flood Risk Management Strategy has been adopted on October, the 7th 2014 by an inter-ministerial decree from the Ministers of Environment, Interior, Agriculture and Housing. This adoption gives a strong strength to this framework in order to overtake barrier of policies, ministries and parties.

12.4.3 An Inclusive National Outline

12.4.3.1 A Tool for Each Scale

The strength of the French transposition of the Flood Directive was to rebuild the entire flood management national policy. Thus from a national framework, the SNGRI, a chain of tools has been created to implement the SNGRI and the Flood Directive at each level according to the Subsidiarity principle:

- At River basin District level, the national strategy is implemented by a Flood Risk Management Plan (FRMP) that defines the objectives and measures in terms of flood risk management for the district.
- At River basin scale, for each APSFR, a local strategy clarifies the common objectives and measures of stakeholders for flood risk management.
- Theses local strategies will be implemented through flood prevention actions programs (PAPI) providing state financing through a national call for proposals.
- At a municipal level with the urban planning, Flood Risk Prevention Plan (PPRI), administrative decisions in the field of water policy have to be compatible with the FRMP assuring the implementation of national objectives.

12.4.3.2 A Balance Between State Control and Local Governance

The Responsibility of Local Authorities

For decades, local authorities had the responsibility for flood management like taking into account flood risk in their urban plans, preparing and managing crises in case of a flooding. But the new national flood risk management policy consecrates the role of local authorities in flood management at the same time as confirming the subsidiarity principle. The creation of new administrative structures like EPTB and EPAGE, as well as the creation of a Water Management and Flood Prevention competency for intercommunal structures (GEMAPI) is the result of a long evolution. The turning point of this movement could be situated in 2002 with the first flood prevention actions programs (PAPI) call for proposals addressed to local governance to promote innovation actions for flood risk management at a river basin scale. The GEMAPI competency and the management by local authorities of the local strategies for flood risk management is the completion of this revolution.

State as Public Security Guarantor

But at the same time the State keeps the main responsibilities for flood risk management in association with stakeholders:

- FRMP is under the responsibility of the District Coordinator Prefect
- Flood and coastal risk prevention plans (PPRI&PPRL), the strongest tool at municipal scale to manage flood risk are under the responsibility of the Departmental Prefect, these tools can ban new constructions in floodplains but also control land use, activities, and order measures to reduce the vulnerability of existing assets (houses, building, factories…)

Thus the new French Flood Risk Management Policy succeeds in summarizing an integrated policy based on the subsidiarity principal with governance and tools at all levels by combating land pressure and encroachment of human activity (housing, production, services…) into floodplains.

12.5 How Has the EU SEQUANA 2016 Exercise Contributed to Improving the Resilience of the Ile de France Region to Its Major Risk?"

12.5.1 EU SEQUANA 2016, an Example of Crisis Simulation Exercise

Human resilience can be increased by the practice of exercises. Crisis-simulation exercises are trainings to manage complex situations. To simulate a crisis, the exercise is based on a crisis scenario, written in advance by a team of experts. During the exercise, the participants are coached by a team of experts linked to the simulated scenario. The experts, called animators, send an important amount of information, called inputs, to the participants throughout different telecommunication tools such as telephones, web applications, web mails, etc. Participants must then work together, take decisions and deal with the complex crisis situation.

The Paris Defense and Safety Zone covers the whole Ile-de-France area and performs missions relating to civil and economic safety as well as the security of sectors of vital importance through its Secretary General of the Defense and Safety Zone of Paris – the SGZDS. The SGZDS is then responsible for organizing the zonal exercises, such as EU SEQUANA 2016. It is an exercise simulating major Seine flooding in Ile-de-France. It lasted two whole weeks in 2016, March 7th–March 18th. More than 90 partners were involved in this project, most of them from vital importance sectors such as energy, transport, telecommunications, etc. The main objectives of this exercise were to:

- Gather the partners and the actors of the Ile-de-France around a crisis management exercise including the European scale to strengthen the coordination of their actions.
- Test the operational effectiveness of the European mechanism.
- Focus the population's attention on the flood phenomenon.[1]

EU SEQUANA 2016 was organized around the two phases of a flood. The first part simulated the raising of the floodwaters from March 7th to March 13th; and the second part, from March 15th to March 18th, featuring the deflooding. To answer the three fixed objectives, different types of exercises were implemented during those 2 weeks, table-top and field exercises.

During both phases, table-top exercises were organized with all the partners. The simulation of the flooding lasted 10 days. It was based on a complex scenario, detailing the consequences of the rise of water every 50 cm, written in advance with experts sent by the 90 exercise's partners. All levels and different sectors of crisis management were involved: from small cities in the suburbs of Paris to the French inter-ministerial crisis unit, the CIC; from France to the European Emergency Response Coordination Center – ERCC-; from small public agencies to international private companies, and from basic city halls to critical infrastructures (see Fig. 12.6). As for the deflooding, the partners worked on the subject throughout a case study focusing on two different periods: 5 days after the peak of the flood and 30 days after. The goal was to bring experts to work on structural rehabilitation and economical re-launch. The partners were divided into small study groups so they co better focus on their interdependencies.

On March 12th and 13th a field exercise was put in place in seven different sites (see Fig. 12.7) in order to test the operational effectiveness of the European modules sent by the four participating countries, Belgium, Italy, Spain and the Czech republic, in the context of the activation of the European Mechanism and its coordination team, the EU CP TEAM, sent by the ERCC. The field exercise was supervised by the Paris firemen brigade – BSPP – and mobilized more than a 1000 persons (Fig. 12.7).

Throughout the different phases of SEQUANA exercise, from the writings the scenarios to the implementation of the table-top and field exercises, the risk and crises department of the INHESJ National Institute of Advanced Studies in Security and Justice – supported and accompanied the SGZDS.

[1] Source: http://www.prefecturedepolice.interieur.gouv.fr/Sequana/EU-Sequana-2016

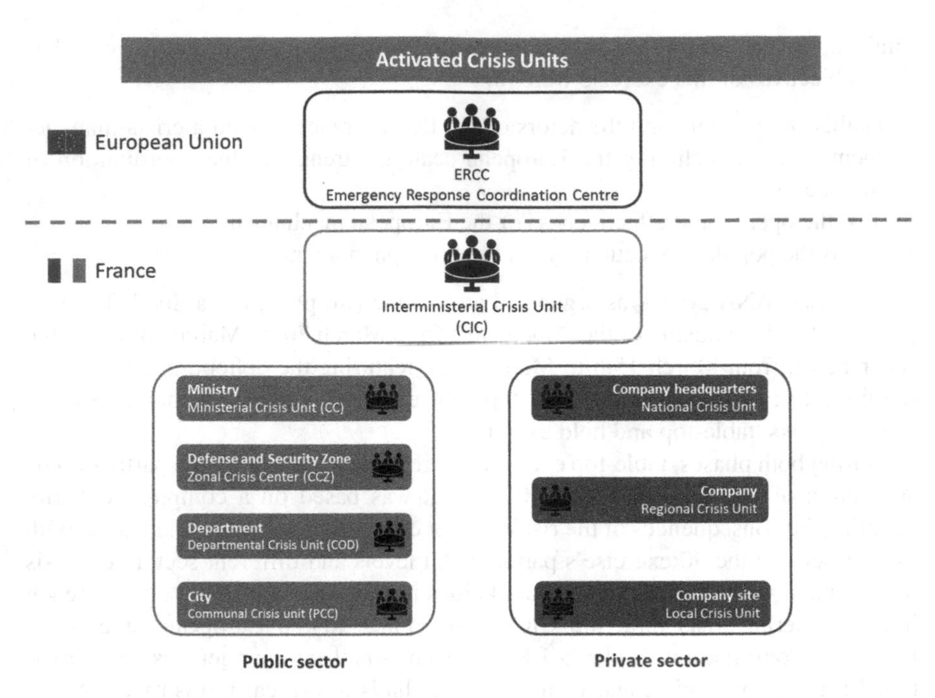

Fig. 12.6 Activated crisis units for SEQUANA

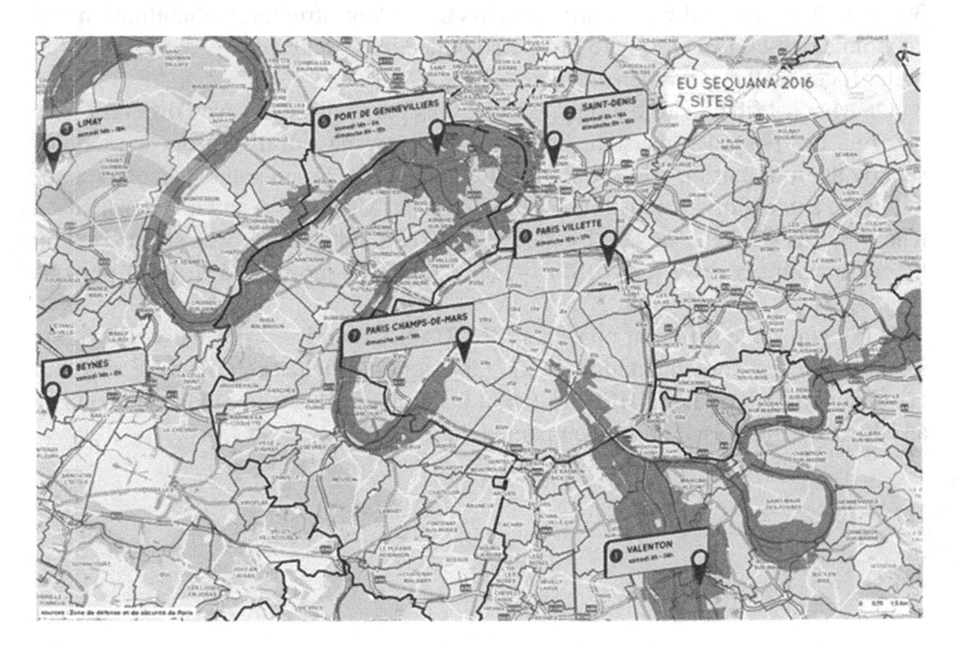

Fig. 12.7 Seven sites of the field exercise (Source: http://www.prefecturedepolice.interieur.gouv.fr/Sequana/EU-Sequana-2016)

12.5.2 Each Step of EU Sequana 2016 Exercise Has Helped to Increase Resilience

12.5.2.1 Preparation and Writing the Exercise's Scenario

Like most crisis simulation exercise, EU SEQUANA 2016 required a considerable amount of preparation time. Due to the scale of the exercise, this time period has been the most important. Indeed.Driven by the General Secretariat of the Paris Defense and Security Zone the organization began more than 2 years before the start of the exercise. As for the scenario, it took almost a year to finalize it. This length is explained by the fact that the creation of the scenario required a very important collaborative work. However, it is important to mention that thanks to this work the preparation of the exercise has greatly contributed to increase the resilience of Ile-de-France region. Indeed, a major flooding of the Seine (such as the one which occurred in 1910) constitutes one of the worst case scenarios that the Paris region could face. By impacting all the infrastructure networks (water, electricity, telecommunication, transport, roads, etc.), and considering current dependency on these networks, the consequences of this flood would impact all economic actors in Ile-de-France and also the majority of citizens.

In preparing for such an exercise and writing such a complex scenario, it was therefore essential that the actors involved worked together. This collaborative work initially allowed the partners to exchange on common issues and to increase their knowledge about the phenomenon and its consequences. Furthermore, throughout this work of several months, they were divided into working groups and were able to identify precisely:

- The interdependencies they have with other actors;
- The possible consequences that the absence of infrastructure networks could have on their own structure and therefore the constraints to be taken into account;
- Issues they had not thought of previously;
- And resource contacts on this subject in other structures.

These factors have enabled the partners to improve their preparation, to get to know each other better and to create common working practices and habits. This helped to develop, or at least facilitate, for the most experienced, coordination between actors. It also improved individual and collective planning of the various structures to respond to an event of this type. Indeed, by knowing its weaknesses, its correspondents, its interdependencies, a structure can improve its planning and its individual preparation. It acquires knowledge and automatisms that will enable it to respond more effectively and more quickly to a major event. The structure will thus recover more rapidly its normal operations. This is the most important in the case of structuring networks. By improving their preparation and thus their response, networks will be able to resume their service sooner. Since these services are essential for the functioning of other economic players, by improving their resilience, structuring networks improve collective resilience.

Collective resilience has also been increased directly by the effect of collective work. By identifying the exposure and working together, the organizations have learned to know each other and understood how they can and should work together. They thus fostered their cooperation and improved their coordination. Tomorrow, if a major flood occurs, the structures will have increased capacities to communicate together and therefore to coordinate better. This is how the structures can respond collectively, as efficiently and as quickly as possible. Whether they are the most experienced and most advanced actors on their planning or novice actors on this subject, the collective creation of the scenario has contributed to the improvement of resilience, both individually and collectively, in the various structures in Ile-de-France.

12.5.2.2 The Implementation of the Exercise

The exercise represents indeed a significant contribution to the improvement of the resilience. Like all crisis simulation exercise, regardless of its imperfections, it is an excellent training for the participants.

This training allows its participants to have a clear view of the efficiency of their internal procedures and to adjust them accordingly. They thus improve their ability to respond as efficiently as possible the day the actual event occurs. By improving their response, they will likely return sooner to routine operation, increasing their individual resilience.

The scope of the exercise also made it possible to test certain parameters specific to this exercise's dimension. One of the most important is probably the overall coordination of actors. Indeed, the participation of all levels of decision (national, zonal, departmental, communal) allowed to test the circulation of information, both transversal and vertical (descent and feedback), between the crisis cells. While in crisis management, difficulties of information circulation and coordination are regularly observed in the crisis cells, this phenomenon is even worse between several crisis cells. The different decision-making levels and the different actors do not approach the crisis from the same angle. They do not deal with issues in the same way. This makes coordination and sharing of information particularly complex, because at each decision level it is difficult to understand the needs of other cells. What information should be shared with the top or bottom level? With what level of details? By involving all levels of decision-making and so many actors at these different levels, EU SEQUANA 2016 was the first experience of a multi sector (private and public structures) global coordination in Ile-de-France. This experience highlighted the strengths and readjustments needed to improve the coordination and flow of information. But above all, it enabled actors to capitalize on how they should work with other structures and especially with the Paris defense and security zone. Collective resilience thus increased thanks to the capitalization in the experience of the actors and by the readjustments that will be implementedd.

Other dimensions of the exercise contributed to improve Ile-de-France resilience. Running the exercise, setting up a field exercise with the participation of

Fig. 12.8 The main animation unit of the SEQUANA exercise in the SGZDS (Paris)

the European modules and the dimension related to the deflooding are the three main ones.

The animation of the exercise allowed the actors in the central animation room (zonal crisis center [CCZ] of the zone of defense and security of Paris, Fig. 12.8) to have a global vision of exchanges and issues to be addressed by all structures. It also allowed them to see how the CCZ was organized and how it managed the many interactions with the actors of Ile-de-France. Having a shared vision of the situation is a major asset to improve coordination and information flow between actors. This, as pointed out earlier, contributes to collective resilience.

The field exercise, meanwhile, enabled testing the organization and the reception of European means as well as the coordination of French modules of civil security with European means (Spanish, Belgian, Italian and Czech). By carrying out joint operations, the teams on site were able to acquire the first common reflexes on different types of intervention: rescue clearing, rescue in flooded areas, water and soil pollution, NRBC accidents and installation of high capacity pumping. Coaching teams to work together allows for a faster, more efficient and less risky implementation of the means. EU SEQUANA 2016 thus helped to improve the preparation of French intervention teams to work with European teams and resources. It also encouraged the preparation of Ile-de-France to accommodate these European resources. On the day of a major flood in the Seine, the organization and implementation of resources will be facilitated, making collective resilience more effective.

Including the deflooding phase in the exercise EU SEQUANA 2016 enabled the actors to work on complex and little-known issues. Indeed, the exit from crisis, that is to say post-event management, is rarely included in crisis exercises. These phases have many uncertainties because the problems to be managed will depend on the consequences of the event and the decisions taken upstream by the organizations. By working on a new phase of the crisis, the structures were able to identify new issues to be dealt with, to see the emergence of solutions that had not yet been envisioned. They can thus improve their planning and preparation for the event, therefore helping to improve their own resilience to the event.

Finally, one of the most important contribution to improving the resilience of Ile-de-France in the face of a major flooding of the Seine, is the communication that has taken place during the full 2 weeks of the exercise. This communication was the vehicle for a significant improvement of the awareness among citizens of Ile-de-France region. It has helped to improve the flood risk culture of citizens by informing them about the hydrological phenomenon, about its consequences but also about behaviors to adopt in the case of a real event. By raising awareness among the citizens of the Ile-de-France, they become actors of their own security, which greatly contributes to improve collective resilience. Indeed, in the event of a major flood in the Seine, they will have greater ability to contribute to the return to normality.

12.5.3 Lessons Learnt from EU SEQUANA 2016 to Improve the Resilience of Ile-de-France in the Face of a Major Flood of the Seine River

It is a fact that the exercise itself and its preparation have contributed to the improvement of the individual resilience of the actors as well as the collective resilience of Paris area. In addition, this exercise has brought to light some improvements that need to be made in order to increase significantly global preparedness to major flooding in the Paris region.

Thus, the four main lessons learnt of the EU SEQUANA 2016 exercise are the following:

1. The importance of preparation and coordination of all actors. The work of preparation and coordination of the actors, partly initiated by the exercise, must be continued. Indeed, although the exercise has enabled the improvement of this dimension, existing links must persist and be developed. This is crucial to refine mutual knowledge and existing interdependencies. It is therefore essential to maintain collective training of concerned actors in order to continually improve their preparation and response planning.
2. The interest of tools shared by all actors managing the flood. Thanks to numerous exchanges with the structures following the exercise, it was often emphasized that shared tools would improve overall coordination in the event of a

major flooding of the Seine River. The tool which was most frequently mentioned was a shared mapping of data from networks of vital importance. It would allow actors to have a global and common view of the crisis, thus facilitating the coordination of actors, better coordination contributing to improve resilience.

3. The need to deepen the work carried out on the deflooding phase and on the organization of the return to usual functioning. While the deflooding case studies have made it possible to bring forward the planning of some structures, it has above all highlighted issues that were not very well known or unknown by the actors. In order to increase the resilience of the structures, it is therefore essential to continue working on the management of flood recession and the organization of the return to current functioning.
4. The need to involve all actors, public and private structures and, more broadly, all impacted citizens in the process. The sensitization carried out among the Ile-de-France citizens during the exercise must be continued. This will encourage the spreading of knowledge and will develop positive behaviors, or even reflexes, in the event of a major flooding of the Seine River.

12.6 May/June 2016 Seine River Flood – Analysis and Preliminary Conclusions

The hydrographic crisis of June 2016 only occurred 3 months after the development of the Sequana exercise. This simulation had followed a specific scenario assuming a much slower rise of the Seine river (50 cm per day), an "episode of cold" "directly inspired by the 1910 catastrophe (cold grounds and brutal rise in temperature) and considering only the overflows of the Marne and Yonne affluents.

The flood which occurred in fact in May /June 2016 has been very different. According to Météo France, Paris has known in the first quarter of 2016 its most rainy spring for 150 years (date of the beginning of measurements), breaking the record of 1928. In Paris. The rains over 3 months, cumulated to 320 mm, similar to the first quarter of 1873. After 8 days of fast and continuous rise of water in Paris, the Seine started to decrease on Saturday, June 4th, falling under the bar of the 6 m. The peak level reached 6.10 m in the night and was recorded as the highest since 1982 (6.18 m), very far however from the level reached in January 1910 (8.62 m).

Due to dysfunctionning of measurement stations there has been some inaccuracy of announced water levels and the releases by the responsible organizations of the security plans for the critical infrastructures were delayed. More over the four lake-tanks close to the Seine upstream from Paris were practically full as they were supposed to ensure their mission which is to support the summer low water level of the river. This situation could have been very problematic if the rains had not stopped.

According to a first estimate of the French Association of insurance (AFA), the flood has caused however between 900 million € and 1.4 billion € of ensured losses, recognizing that it is still very difficult to assess potential losses for the production sites.

About 18,000 households suffered electricity cuts in Loiret, the Loir-et-Cher and the Ile-de-France, and 20,000 people, most of them from Île-de-France, were evacuated.

The economic activities were severely affected due to strong disturbances of the transport networks with half of the Transilian lines seriously affected by landslides and mud flows.Flooding of the A10 motorway created critical traffic situations by trapping nearly 200 cars and 100 heavy weight trucks.

The tourist impact was also significant for the main Parisian cultural sites, since it was decided the preventive closing of the museums of Louvre, Orsay, Grand Palais and National Library.

Thanks to the recent SEQUANA exercise all the public and private actors involved participated very actively using the information and communication networks established, the tools developed for the exercise, the lessons learned identified during the 2 weeks joint work. A detailed analysis of the facts and returns of experience on resilience is being performed and will be published to establish future action plans.

12.7 Conclusion

Since few years Resilience is a term largely used and present in the political debates on policies dealing with the management of the risks and catastrophes. Resilience becomes the new frame of reference and tends to supplant the concept of vulnerability.

The State and the territorial communities see definively now the need to prepare the citizens to the catastrophes by implementing non structural measures. In order to reduce the vulnerability of the populations and their environment these measures need to priviledge the development of risk culture and the preparation to the risk of floods.

The existing reglementory tools participate already partly to the resilience of the populations involved, but the development of these official documents are still too often considered as a bureaucratic exercise rather than a means to fight against the flood and to improve resilience. Moreover, when they exist, these documents which aim at developing a risk and resilience culture remain largely ignored by the public who has consequently difficulty to become a real actor of its security.

The recommendations to improve resilience made above in the OECD study and the experience feedbacks from the Sequana exercise and the last 2016 flood are very important for the State administrations and the people and should be seriously considered in the development of future policies integrating better risk and resilience.

Further Suggested Readings

Aguire BE (2006) On the concept of resilience, Preliminary paper, Disaster Research Center, 10 p

Boin A, Kofman-Bos C, Overdijk W (2004) Crisis simulations: exploring tomorrow's vulnerabilities and threats. Simul Gaming 35(3):378–393

Bruisma G, Dehoog R (2006) Exploring protocols for multidisciplinary disaster response using adaptive work flow simulation. In: Proceedings of the 3rd international ISCRAM conference, Newark, NJ (USA), May 2006

Comfort LK, Sungu Y, Johnson D, Dunn M (2001) Complex system in crisis: anticipation and resilience in dynamic environments. J Conting Crisis Manag 9(3):144–158

Crichton M, Flin R (2011) Training for emergency management: tactical decision games. J Hazard Mater 88:255–266

Directive 2007/60/CE of the European Parliament and of the Council of October 23, 2007 related to the evaluation and management of flood risks

French national flood risk management policy. www.developpement-durable.gouv.fr/La-strategie-nationale-de-gestion,40051.html

French national flood risk management strategy, 7th October 2014, adopted by the Ministers of Environment, Interior, Agriculture and Housing, www.developpement-durable.gouv.fr/document152633

Kincaid JP, Donovan J, Pettitt B (2003) Simulation techniques for training emergency response. Int J Emerg Manag 1(3):238–246

November V, Créton-Cazanave L (2017) La gestion de crise à l'épreuve de l'exercice EU SEQUANA, la documentation française, 234 p (to be published Febrauaru 2017)

OECD Reviews of Risk Management Policies-Seine Basin, Ile de France :Resilience to major floods ISBN 9-78-92-64-20871-1

Serre D, Barroca B, Laganier R (2013) Resilience and urban risk management. Taylor and Francis Group, London

Wybo JL (2004) Mastering risk of damage and risk of crisis: the role of organization learning. Int J Emerg Manag 2(1–2):22–34

Chapter 13
Realising Critical Infrastructure Resilience

Jon Coaffee and Jonathan Clarke

> *Resilience is the ability of a system to survive and thrive in the face of a complex, uncertain and ever-changing future. It is a way of thinking about both short term cycles and long term trends: minimizing disruptions in the face of shocks and stresses, recovering rapidly when they do occur, and adapting steadily to become better able to thrive as conditions continue to change. A resilience approach offers a proactive and holistic response to risk management.*
>
> *Siemens, Toolkit for Resilient Cities (2013, p.3)*

Abstract The discourse of resilience has increasingly been utilised to advance the political prioritisation of enhanced security and to extend the performance of risk management in the Anthropocene. This has been notably advanced through integrated approaches that engage with uncertainty, complexity and volatility in order to survive and thrive in the future. Within this context, and drawing on findings from a number of EU-wide research projects tasked with operationalising critical infrastructure resilience, this paper provides a much-needed assessment of how resilience ideas are shaping how critical infrastructure providers and operators deal with complex risks to 'lifeline' systems and networks, whilst also illuminating the tensions elicited in the paradigm shift from protective-based risk management towards adaptive-based resilience. In doing so, we also draw attention to the implications of this transition for organisational governance and for the political ecologies of the Anthropocene that calls for more holistic, adaptable and equitable ways of assessing and working with risk across multiple systems, networks and scales.

Keywords Resilience • Anthropocene • Critical infrastructure • Adaptation pathways • Politics

---

J. Coaffee (✉) • J. Clarke
Resilient Cities Lab, Department of Politics and International Studies,
University of Warwick, UK
e-mail: J.coaffee@warwick.ac.uk; J.r.l.clarke@warwick.ac.uk

© Springer Science+Business Media B.V. 2017

I. Linkov, J.M. Palma-Oliveira (eds.), *Resilience and Risk*, NATO Science for Peace and Security Series C: Environmental Security,
DOI 10.1007/978-94-024-1123-2_13

13.1 Introduction

The Anthropocene can be viewed as an epoch which began in the mid-twentieth century through a rapid increase in technological change, population growth and consumption, and which is increasingly characterised by complex and dynamic system interaction, future volatility and ultimately an imperative to rethink the relationship of humans with nature, environment and technology. Concomitantly, in the early twenty-first century – catalysed by the devastating events of 9/11 and the release of the fourth Intergovernmental Panel on Climate Change (IPCC) report in 2007 highlighting unequivocal evidence of a warming climate – ideas and practices of *resilience* have become a central organising metaphor within policy-making processes and the expanding institutional framework of national security and emergency preparedness. For many, resilience offers an integrated approach for coping with all manner of disruptive events, as well as a new way to engage with future uncertainty (Chandler 2014; Coaffee et al. 2008; Walker and Cooper 2011; Zolli and Healy 2013). As we will argue in this paper, resilience thinking has subsequently been utilised to 'extend' established risk management approaches and to advance ways of surviving and thriving in the future through adaptation and long-term transformative action.

In many ways it is the spectre of unanticipated catastrophe that has driven the interest in resilience as a universal remedy for a range of 'natural' and human-induced risks (Aradau and van Munster 2011). Recent decades have been remarkable for the volume of high-impact anthropocentric disasters, such as the impact of Hurricane Sandy or the cascading effects following the Tohuku earthquake, and which have highlighted the vulnerability, complexity and interdependency of contemporary life. Pointing towards the new climatic norm of the Anthropocene, Fisher (2012, p. 3) has also highlighted the dramatic increase in 'weather-related catastrophes', such as floods, storms and drought, which have increased exponentially between 1900 and 2005. These events have foregrounded the political prioritisation of enhanced security – often badged as resilience – as a political imaginary of being 'insecurity by design' (Evans and Reid 2014). Such attention to governing insecurity has been highly related to historical and geographic contingency which sees governmental and corporate approaches to contingency planning, protection and resilience differentially applied in accordance to context (Lentzos and Rose 2009). The frequency and severity of recent crises have further channelled attention to vulnerable physical assets, with a particular focus on *critical infrastructure* whose disruption have the potential to significantly affect public safety, security, economic activity, social functioning or environmental quality.

Specifically post-9/11, there has been focus within security policy upon *critical infrastructure protection* using conventional risk management principles and on the interdependency and interoperability of these systems and, by extension, the cascading effects of a breakdown in one system on other interconnected systems.[1] The

[1] A further result of this trend has been the expansion of infrastructures considered to be critical and which has seen a shift from the line-based systems of public utilities, to more complex social

increased acknowledgement of such complex risk has, over time, led to a prioritisation of *critical infrastructure resilience.* However, despite the clear parallels between the emergence of critical infrastructure and resilience as mainstream anthropocentric policy concerns, there has been relatively little interconnection between theory and practice. Emerging approaches to improving critical infrastructure resilience are still in their infancy, with e orts focused predominantly upon single infrastructure sectors, across a number of easily compared critical infrastructure sectors and at limited spatial scales.

Drawing on the results from a number of EU-wide research projects tasked with operationalising critical infrastructure resilience,[2] this paper illuminates how resilience ideas are shaping the ways in which critical infrastructure providers deal with complex risk and the tensions elicited in the transition from protective based risk management towards adaptive-based resilience. In doing so, we will highlight the implications for this new way of working for organisational governance and for the importance of the political ecologies of the Anthropocene that call for more holistic and integrated ways of assessing risk and new modes of equitable governance across multiple systems, networks and scales (Biermann 2014; Crutzen and Stoermer 2000).

The remainder of the paper proceeds in three main sections. *First*, we frame our discussion through ideas of how to 'survive' anthropocentric challenges that require a different social–spatial framing, politics and ways of adapting to uncertainty. Here we view resilience as a supposed antidote – a new *biopolitical nomos* – to such anthropocentric destabilisation and insecurity, in contrast to a conventional probabilistic 'risk-based' world. *Second*, we operationalise ideas of change brought about by the Anthropocene and resilience discourse through the lens of critical infrastructure assessment and illuminate a normative paradigm shift from protection towards resilience. *Third,* we draw the key themes of the paper together in articulating how future critical infrastructure operations will need to adapt to the challenges of uncertainty and system interdependences in the Anthropocene. Drawing on detailed empirical survey work across Europe with a range of critical infrastructure providers, we also illuminate a series of interrelated barriers that has made the operationalisation of resilience approaches di cult to achieve in practice.

13.2 Survival in the Anthropocene

The Anthropocene presents a new role for humanity as the driving force behind planetary systems whilst at the same time operating within a world of 'persistent uncertainty' (Biermann 2014) – a condition where the broader security concerns of

infrastructures which safeguard the wellbeing of citizens and private enterprises performing societally significant roles.

[2] See acknowledgements for details.

nations have been increasingly rewritten to secure the conditions necessary for human life, for our very survival as a species.

In 2000, Crutzen and Stoermer first suggested that human-generated changes to the biosphere, including climate change, urbanisation, the deployment of nuclear weapons, large-scale biodiversity loss and accelerating landscape transformation, were creating a new geological epoch which they termed, the Anthropocene. Most recently, in a seminal paper in *Science,* Waters et al. (2016) concluded that 'the Anthropocene is functionally and strati- graphically distinct from the Holocene', and began with the 'Great Acceleration' in the mid-twentieth century. However, despite its geological basis, the Anthropocene has redefined critical human–environment relationships, the key role of risk in mediating this under- standing and illuminated how, in particular, the global climatic system is becoming more volatile, bringing new challenges for humanity (Oldfield et al. 2014). Despite dire warnings of increased storms, droughts and floods (IPCC 2014), some suggest that the primary challenge will be political and in how humanity collectively builds adaptable governance systems that tackle the challenges of climate change and enhance resilience (Biermann et al. 2015). Whilst concerns over climate change are most commonly used to articulate the nature and the impact of the Anthropocene, it can also be considered a much wider conceptual frame for understanding human–environmental relationship and their political significance. However, there is a paradox at the heart of these understandings of the Anthropocene: whilst humans are increasingly shaping the environmental conditions, the ability to do this in a conscious and deliberate way is hampered by our inability to tackle the complex interactions and interdependencies involved, and thus the true nature of anthropocentric risk to global society.

13.2.1 Managing Anthropocentric Risk

Social scientists for many years have studied the risks from natural hazards and the need to make contingency against their impact (Kates 1962; White 1942). However, accounts regarding the impact of technological and anthropocentric risk only became prevalent in the late 1980s and 1990s, which suggested that concerns about such risks had become de ning societal characteristics (Adams 1995; Beck 1992; Douglas 1994). This new range of 'risk theory' emerged primarily around concerns about global environmental hazards, the trans- national nature of such risk and the effect of such risk in challenging existing political governance configurations. Most notable amongst this canon of work was Beck's (1992) *Risk Society – Towards a New Modernity*. Published in the wake of the Chernobyl nuclear catastrophe in Ukraine, *Risk Society* considered what society might look like when disputes and conflicts about new types of risk produced by industrial society are fully realised.[3] *Risk Society* starkly illuminated the magnitude and boundless nature of the global

[3] Risk Society was first published in German as Risikogesellschaft in 1986.

risks, and how this is transforming the way in which risk is imagined, assessed, managed and governed, but not eradicated. Beck's work provided the impetus for further academic thought related to the impact of the emergence of a set of newly de ned and ubiquitous 'mega-scale' risks on the workings of global society that 'cannot be delimited spatially, temporally, or socially' (Beck 1995, p. 1). As Giddens (2002, p. 34) reiterated:

> ... whichever way you look at it, we are caught up in risk management. With the spread of manufactured risk, governments can't pretend such management isn't their business. And they need to collaborate, since very few new-style risks have anything to do with the borders of nations.

Risk Society is a story of survivability. As Blowers (1999, p. 256) commented, 'Risk Society is a pessimistic and conflictual diagnosis of modern societies ... that is exposed to risks from high technology ... that imperil our very survival'. new risk theory also further exposed the disenchanted world of formalised instrumental rationality abundant in the 'iron cage' of bureaucracy (Weber 1958) and the absence of social and cultural factors involved in discussions about risk that had been hidden beneath a preference for objective approaches to risk assessment – the 'possibility of calculation' (Giddens 2002, p. 28).

These new understandings of risk are echoed in more recent discourses on the critical thresholds of the Anthropocene. For example, current attempts to tackle climate change have exposed the failures of contemporary decision-making, highlighting that 'neither traditional risk management strategies nor conventional economic decision-making can be relied on to govern in the face of increasingly likely extreme events' (Dalby 2013a, p. 189). The Anthropocene ushers in an unknown future that requires policy-makers to shift their focus both from an appreciation of risk to one of criticality and in identifying and understanding those aspects essential for human well-being. This is particularly the case with regard to the long-term significance of system interdependencies and issues of social and spatial justice (Biermann 2014; Dalby 2013b; Mabey et al. 2013).

13.2.2 From Risk to Resilience

The anthropocentric view of risk has significantly contributed to the rise of *resilience* as the policy metaphor of choice for coping with and managing future uncertainty and the incorporation of 'the dynamic interplay between persistence, adaptability and transformability across multiple scales and time frames' (Davoudi 2012, p. 310). Whilst the concept of resilience is closely associated with an engagement with risk, a critical schism emerges between resilience and more established risk management practices (Baum 2015; Suter 2011); should resilience be considered as the end goal of traditional risk management approaches? Is it a new consideration for risk management? Does it extend current risk management practices? or does resilience require an entirely different paradigm for considering future uncertainties?

The shift towards resilience approaches is also not without critique, posing some fundamental questions of resilience for whom, by whom? (Coaffee and lee 2016). Much of this critical assessment concerns the alleged tarnishing of resilience ideas through 'neoliberal decentralisation' (Amin 2013) and a post-political landscape understood as the foreclosing of political choice, the delegation of decision-making to technocratic experts, growing public disengagement from politics and ultimately the closing down of political debate and agency (Flinders and Wood 2014). The emerging canon of work in 'critical resilience studies' has high- lighted the ways in which resilience policy and practice indicate a shift in the state's policies, reflecting a desire to step back from its responsibilities to ensure the protection of the population during crisis and to delegate to certain professions, private companies, communities and individuals.[4] Through the lens of resilience policy, we can therefore chart new forms of precautionary governance, attempts to create resilient citizens, the drawing in of a range of stakeholders to the resilience agenda and the corresponding adoption of new roles and responsibilities in enacting policy priorities. Whilst we are sympathetic to critical accounts and especially their powerful expose of who wins and who doesn't in neoliberal governance, we, like others, prefer to focus upon our analysis on a more inductive and performative approach which views resilience as a multiplicity of related, and often experimental practices. Like Brassett and Vaughan-Williams (2015, p. 34) in this paper we

> seek to reflect and develop upon a notion of resilience as an ongoing interaction between various (and often conflicting) actors and logics, one which can be viewed as far more contingent, incomplete and contestable in both its characteristics and effects than is usually acknowledged in the existing literature.

In resilience practice, as a consequence of anthropocentric uncertainty and the associated need to protect lifeline systems and infrastructures, there has been a growing interest in utilising the concept of resilience for critical infrastructure assurance. As Evans and Reid (2014, p. 18) note, 'critical infrastructure is now central to understanding living systems' and politically, the combined lifelines deemed necessary for security, survival and growth. But as Dalby (2013a) further argues, conventional approaches to designing critical infrastructure that leave too many key decisions to the market to decide are fundamentally awed, and policy-makers need to make large, far-reaching decisions if they are to avoid major disasters in the future. Moreover, the changing material politics, geographies and governance arrangements associated with critical infrastructure – the 'collective equipment' of state power (Foucault et al. 1996) by which control might be exerted, socio-economic restructuring advanced and inequity concretised – is also of critical concern. It is to such recent attempts to enhance the resilience of critical infrastructures that this

[4] Such a Foucauldian-inspired interpretation argues that resilience encourages individuals to autonomously act in the face of a crisis and which precipitates citizens behaving and adapting according to prescribed moral standards (Joseph 2013). As Welsh (2014, p. 16) highlighted, resilience policy is 'a post-political ideology of constant adaptation attuned to the uncertainties of neoliberal economy where the resilient subject is conceived as resilient to the extent it adapts to, rather than resists, the conditions of its suffering'.

paper now turns in order to articulate recent attempts to refocus the need to secure infrastructure through the lens of resilience rather than probabilistic risk management.

13.3 Enhancing Critical Infrastructure Resilience

The last 20 years have been remarkable for the volume of high-impact crises, disasters and global incidents with the ability of providers to assure the security and continuity of infra- structure becoming of high importance. Critical infrastructure assurance is therefore progressively moving away from a focus upon protection towards emphasising resilience.[5] It is perhaps the cascading effects of a breakdown in one system on other interconnected systems, which have provoked most significant concern – often articulated through the spectre of low probability–high consequence 'Black Swan' events. The failures of infrastructure illustrated during 9/11, the 2011 Tohuku earthquake in Japan or Hurricane Sandy in 2012 upon New York, highlight the vulnerability and potential weaknesses of our critical systems and man-made infrastructure and how such failures often have common roots, particularly around path dependencies and institutional failings (Dueñas-osorio and Vemuru 2009). A vivid example of cascade failure in Critical infrastructure, is provided by the events following the attack on the World Trade Centre in New York in 2001. As O'Rouke (2007, p.25) highlights, 'the WTC disaster provides a graphic illustration of the interdependencies of critical infrastructure systems. The building collapses triggered water-main breaks that flood rail tunnels, a commuter station, and the vault containing all of the cables for one of the largest telecommunication nodes in the world'.

Increasingly, infrastructural assemblages are being viewed as 'complex adaptive systems' with an emphasis on the ability to adapt to such conditions of uncertainty and volatility (Comfort 1994; Longstaff 2005). In turn, this has catalysed the emergence of *resilience* as a way to assess the complex challenges that critical infrastructure faces as well as providing a potential framework by which to respond. In the case of these non-linear, dynamic complex systems, a system is resilient when it can adapt or self-organise in the event of perturbation.

As the critical infrastructure sector has become a larger, more complex and an increasingly interconnected amalgamation of social, technical and economic networks, so, the risk of breakdown has risen. The growing interest in applying resilience methods in securing critical infrastructure has grown as traditional risk management methodologies have proved ineffective in the face of growing complexity and the unpredictability of threats, and growing knowledge about interdependency and cascade effects amongst critical infrastructure sec- tors. In the Anthropocene, where such volatility is a *leitmotiv* and where security is being con-

[5] Not all infrastructure is deemed to be 'critical', and thus infrastructure can be categorised using some form of 'criticality' scale to assess its value and the impact of its loss/disruption.

stantly recast as resilience, assuring the functioning of critical infrastructure against a range of known and unknown unknowns (notably the impacts of climate change being seen as an imminent security threat or threat multiplier) has become a core challenge of government. As Perelman (2007, p. 23) highlighted, in the post-9/11 age 'the allure of resilience is stoked by the contradictions and thorny trade-o s inherent in traditional concepts of 'national security' in an age of increasing social-technical complexity, transnational 'globalization,' and 'asymmetric' conflict'. Moreover, as national/homeland security has been reconfigured, so previously irreconcilable socio-political objectives (e.g. security against attacks vs. security against natural disasters, disease, accidents, etc., and centralised command and control versus communal collaboration) increasingly come into focus (*ibid.*). As security 'comes home' and becomes more localised (Coaffee and Wood 2006) so the impulse to completely eliminate risk and uncertainty and prevent harm is destabilised and security is recast. In many cases, the assumptions of positivist and instrumentally rational risk management have been turned on their head forcing the abandonment of the Modernist dream of total control, alongside a shift from traditional Euclidian, Cartesian and Westphalian notions of scale and territory.

By contrast, the current push for resilience increasingly highlights the importance of sub-national responses to new security challenges, 'placing the needs of the individual, not states, at the centre of security discourses' (Chandler 2012, p. 214). Resilience-thinking is thus increasingly forcing operators of infrastructure to work with the irreducibility of risk and uncertainty, to devise a range of alternative visions of the future, and to advance more deliberative and scalable methods that seek *adaptation* through flexibility and agility.

At the crux of the move *from* critical infrastructure protection *to* critical infrastructure resilience has been a struggle between what Perelman (2007, p. 24) referred to as *hard* and *soft* paradigms of security. Here, the hard paradigm represents the path of conventional security policies and practices associated with prevention and resistance, whilst by contrast, the soft paradigm is associated with adaptation and resilience – a move away from technocratic and techno-rational approaches and towards more socially grounded transformative approaches (Coaffee and lee 2016). Perelman (2007) further cites the work of influential American physicist Amory Lovins on future energy demand (Lovins 1976) who highlighted the advantages of the soft resilient path over the hard brittle path:

> [T]he soft path appears generally more flexible – and thus robust. Its technical diversity, adapt- ability, and geographic dispersion make it resilient and offer a good prospect of stability under a wide range of conditions, foreseen or not. The hard path, however is brittle; it must fail, with widespread and serious disruption, if any of its exacting technical and social conditions is not satisfied continuously and indefinitely. (ibid, p. 88, emphasis added)

13.3.1 Redefining the Protectionist Reflex Through Resilience

In times of vulnerability there is a natural impulse to evoke a 'protectionist re ex' in order to ensure safety (Beck 1999, p. 153). Such a reflex has been very evident in critical infrastructure protection programmes that have adopted approaches involving the 'hardening' of critical assets to increase 'resistance' and 'robustness'. Ironically, the net effect of such actions often leads to what are known as 'robust-yet-fragile' systems that are increasingly susceptible to unexpected threats and cascade failures (Carlson and Doyle 2000). An opposing approach is taken by those emphasising the 'soft' – more resilient – path: 'first, it takes a holistic view of 'infrastructure' as complex, dynamic, adaptive, even living *systems*, rather than discrete, concrete, fixed *assets*. And second, it aims at *softening* the brittleness of systems ...' (Perelman 2007, p. 28).

A protection-based approach to critical infrastructure is, in large part, a legacy of ingrained engineering-focused approaches to *risk management*, where an epistemic focus upon ordering and probability, a requirement for optimisation and control, and a near exclusion of social and human factors has created a very different reality from what is increasingly becoming known as *resilience management*. This emerging approach goes beyond risk management to address the complexities of large integrated systems and the uncertainty of future threats. As Linkov et al. (2014, p. 407) note, '... risk management helps the system prepare and plan for adverse events, whereas resilience management goes further by integrating the temporal capacity of a system to absorb and recover from adverse events, and then adapt'.

In terms of governance, the application of risk management for critical infrastructure is traditionally premised on a command and control approach from central government, and actualised through meta-strategies linked to national security or emergency management.[6] Such a static and often short-term approach to complex governance is what classic ecological resilience theory identifies as a 'rigidity trap' where such management can lead to institutions lacking diversity and becoming highly connected, self-reinforcing and in flexible to change (Gunderson and Holling 2002). In counter-response, and again drawing on established resilience ideas of Panarchy, 'adaptive management' is seen as necessary to enhance responsiveness, agility and resilience in interconnected systems. As such, we increasingly see critical infrastructure providers moving towards advancing horizontally integrated approaches where adaptability – 'the dynamic capacity to effect and unfold multiple evolutionary trajectories ... which enhance the overall responsiveness of the system to unforeseen changes' (Pike et al. 2010, p. 4) – is central to effective future action.

Concomitant to the shifting nature of governmental control is the central nature of technology in decision-making and the continual quest for technological 'silver bullets' to help cope with new security challenges. Here, Perelman (2007, p. 39)

[6] Many commentators argued that after 9/11 many states responded by returning to or reinforcing authoritarian command and control types approaches to managing aspects of emergency management or what was increasingly termed resilience (see for example, Alexander 2002).

argues that a more process-based viewpoint should dominate and that 'in place of the hard path's technocratic tunnel vision, the soft paradigm aims at investing in *social*-technical innovation processes ... [that] points toward managing technology and tangible infrastructure not as autonomous 'assets' but as dependent elements of complex, socioeconomic systems'.

13.3.2 The Resilience Turn

The so-called 'resilience turn' (Coaffee 2013) in the early 2000s saw resilience approaches and initiatives embedded within an array of global initiatives, national policies and more local practices, notably critical infrastructure. In critical infrastructure, early attempts to mitigate vulnerabilities tended to utilise conventional risk management approaches that struggled with accounting for complexity and interdependencies, and socio-economic and organisational issues. The US was amongst the first nations to develop a national strategy for the identification, management and protection of critical infrastructure through the 1997 President's Commission on Critical Infrastructure Protection (CCIP) and has been at the forefront of the shift from protection to resilience. The failures that followed the 9/11 attacks prompted a further addressing of vulnerabilities in the nation's critical infrastructure preparedness. In 2002, the US Congress funded the creation of the Critical Infrastructure Protection (CIP) Project to undertake applied research on critical infrastructure and to anticipate and reflect changes in the national risk environment (Mayberry 2013). Subsequently, a new way of perceiving and prioritising threats, vulnerabilities and consequences to critical infrastructures – based on ideas of resilience – was put in train in early 2006 when, in a presentation to the Homeland Security Advisory Committee (HSAC), the Critical Infrastructure Task Force (CITF) recommended 'Critical Infrastructure Resilience' as the top-level strategic objective – the desired outcome – to drive national policy and planning (Pommerening 2007, p. 10). Most recently in the 2013 Presidential directives on 'national Preparedness' (PPd-8) and on 'Critical Infrastructure Security and Resilience' (PPd-21) promote an all-hazards approach which stresses the importance of anticipating cascading impacts and high- lights the shared responsibility of critical infrastructure protection and resilience to all levels of government, the private sector and individual citizens (Obama 2013).

The US policy chronology noted above is by no means unique amongst advanced nations illuminating how critical infrastructure policy in many countries is incrementally shifting from being protection-focused towards the more integrated resilience paradigm. However, in spite of this expanding interest in the vulnerability of critical infrastructure, there are only a very small number of formal definitions for critical infrastructure resilience, currently in use. In the US, critical infrastructure resilience is now framed by the resilience cycle and defined as the 'ability to anticipate, absorb, adapt to, and/or rapidly recover from a potentially disruptive event' (DHS 2009). By contrast, Australia has an 'all hazards strategy' that provides a foundation for collaboration and organisational resilience building rather than a probabilistic risk management framework. It is contended that this better enables owners and operators to prepare for and respond to a range of unpredictable or unforeseen disruptive events. This is underpinned by two core objectives that treat foreseeable and unforeseen risks differently: adopting either a mature risk assessment approach to foreseeable risks to the continuity of their operations that underpinned prior critical infrastructure *protection* programmes, or extending this into an approach focused upon *resilience* so that 'critical infrastructure owners and operators are effective in managing unforeseen risks to the continuity of their operations through an organisational resilience approach' (Australian Government 2010, ibid., p. 14). This latter approach places an emphasis upon dealing with complexity and advancing adaptive capacities within organisations to respond to, recover from and prepare for a range of disruptive challenges.

13.3.3 The Transition in Critical Infrastructure Resilience Assessment

Whilst there is no agreed international measurement approach for critical infrastructure resilience, there is broad agreement on why we need to measure it. Such agreement focuses upon being better able to characterise resilience in context and to articulate its key constituents so as to be better able to raise awareness of where interventions might be placed in order to build resilience within organisations and networks. This allows additional focus upon allocating resources for resilience in a transparent manner and more broadly to monitor policy performance, as well as to assess the effectiveness of resilience-building policy through comparison of policy goals and targets against outcomes (Prior and Hagmann 2013, pp. 4–5).

This transition in critical infrastructure assurance from protection towards resilience can be represented as a continuous process of change, exemplified by the models of assessment adopted by critical infrastructure providers that are progressively shifting from a highly quantified metrics approaches towards emphasis on a cyclical and adaptive learning process. Schematically, we can conceptualised this transition as a series of overlapping phases that seek to assure the continuation of critical infrastructure lifelines (Fig. 13.1).

Within this conceptualisation, phase 1 is characterised by approaches that focus upon highly technical considerations (e.g. physical or informational) within a single critical infra- structure sector (e.g. energy, water or transport), at limited spatial scales (e.g. solely critical infrastructure facilities) and has typically led to enhanced physical characteristics notably:

- *Robustness*: the inherent strength or resistance in a system to withstand external demands without degradation or loss of functionality;
- *Redundancy*: system properties that allow for alternate options, choices and substitutions under stress;

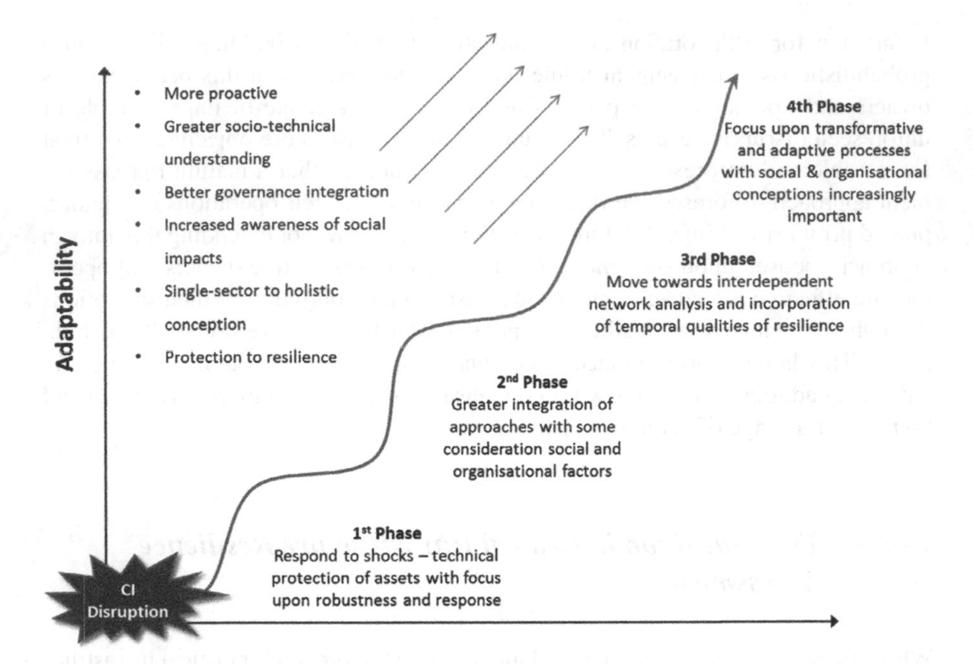

Fig. 13.1 Transitions in critical infrastructure assurance

- *Resourcefulness*: the capacity to mobilise needed resources and services in emergencies;
- *Rapidity*: the speed with which disruption can be overcome and safety, services and financial stability restored (Bruneau et al. 2003).

The technical emphasis of these critical infrastructure qualities understands resilience primarily as resisting and recovering from 'known threats'. By focusing upon the protection, preservation and recovery of single assets, these efforts to achieve resilience have, all too often, failed to account for cascading effects, unexpected events or the more integrated underpinnings necessary for critical infrastructure resilience.

In an attempt to evolve these characteristics into a more integrated understanding, sub- sequent approaches (phase 2) have sought to consider a range of social, economic and organisational factors alongside easily quantified protective criteria. For example, O'Rourke (2007, p. 27) proposed a 4 × 4 matrix that maps the four 'qualities' of robustness, redundancy, resourcefulness and rapidity, against technical, organisational, social and economic factors (Fig. 13.2).

In a practical example, the US national Infrastructure Advisory Council (NIAC) report that outlines resilience practices for the electricity sector utilised a similar approach around the headings of robustness, resourcefulness, rapid recovery and adaptability, subdivided into people and processes, infrastructure and assets and whether unintentional, intentional or cyber acts (NIAC 2010). despite this advancement, there has been widespread comment about the limitations of such approaches, including the lack of meaningful social and organisational considerations, that these

Dimension/Quality	Technical	Organisational	Social	Economic
Robustness	Building codes and construction practices for new and retrofitted structures	Emergency operations planning	Social vulnerability and degree of community preparedness	Extent of regional economic diversification
Redundancy	Capacity for technical substitutions and "work-arounds"	Alternate sites for managing disaster operations	Availability of housing options for disaster victims	Ability to substitute and conserve needed inputs
Resourcefulness	Availability of equipment and materials for restoration and repair	Capacity to improvise, innovate, and expand operations	Capacity to address human needs	Business and industry capacity to improvise
Rapidity	System downtime, restoration time	Time between impact and early recovery	Time to restore lifeline services	Time to regain capacity, lost revenue

Fig. 13.2 Matrix of critical infrastructure resilience qualities (Adapted from O'Rourke 2007)

technical-focused initiatives discourage necessary adaption (Fisher and Norman 2010) and that they are based upon a false idea of equilibrium and stability (Sikula et al. 2015). Notably, the measures utilised within typical phase 2 approaches are only measurable *after* an event has occurred.

In response, a third phase of critical infrastructure assurance has evolved around the idea of interdependent, network systems study; in effect adding an 'anticipatory' temporal dimension, including qualities that can be measured before an event or failure occurs (Linkov et al. 2014). For example, the US Army Corps of Engineers have produced a resilience measure (RM) consisting of a 4 × 4 matrix where one axis contains the major subcomponents of any system and the other axis lists the stages of a disruptive event (Fig. 13.3).

Collectively, these 16 cells provide a general description of the functionality of the system through an adverse event and assess resilience by assigning a score to each cell that reports the capacity of the system to perform in that domain and over time (Linkov et al. 2015).

Whilst phase 2 and 3 assessment methodologies have been advanced as workable mechanisms for resilience assessment and as a basis for making decisions about protective measures, they remain technologically orientated and facility-focused, with an assessment approach relying on workable yet cost and time intensive procedures performed via accompanying software. Despite this increasing sophistication of approaches to assessing critical infrastructure resilience, they have struggled to

Fig. 13.3 A resilience measure for critical infrastructure (Adapted from Linkov et al. 2015)

	Prepare	Absorb	Recover	Adapt
Physical				
Information				
Cognitive				
Social				

include 'organisational beliefs and rationalisations' (Boin and McConnell 2007, p. 56) and the path dependencies that have been increasingly identified as key barriers to enhanced resilience. It is these facets that have become central to an emerging fourth phase transition where organisational resilience is a key consideration and is understood as a property of an organisation that allows it to adapt proactively, following appropriate risk and resilience assessments. In some contexts, critical infrastructure operators are beginning to future-proof their decision-making by advancing a range of dynamic adaptive policy pathways in response to deep uncertainties about the future that can no longer be predicted by using traditional foresight and risk assessment methods. As Haasnoot et al. (2013, p. 485) highlight:

They develop a static 'optimal' plan using a single 'most likely' future (often based on the extrapolation of trends) or a static 'robust' plan that will produce acceptable outcomes in most plausible future worlds ... However, if the future turns out to be different from the hypothesized future(s), the plan is likely to fail.

Organisations need more than a Plan A. By contrast, an adaptation pathways approach provides an analytical approach – a form of 'iterative risk management' – to explore a set of possible actions based on alternative developments over time. Such an approach highlights potential lock-ins, path dependencies and tipping points which specify the conditions under which a pre-specified action to change the plan is to be taken (Coaffee and lee 2016; Haasnoot et al. 2012; Kwadijk et al. 2010). Whist such an approach is not novel in resilience studies with work focusing on experiential learning and adaptation central to ideas of adaptive management which formed the cornerstone of classic ecological resilience theory current adaptation pathway approaches take this one step further in grounding their work in the interdependencies and complexity of multiple interlocking infrastructures whilst presenting alternative ways of getting to a desired end point in the future. A focus upon such adaptation pathway processes essentially mainstreams resilience-thinking, adaptation and sometimes transformation into infrastructure planning rather than relying on short-term, incremental changes that will, in most cases, fail to shift organisation custom and practice from a protective risk-based mind-set.

Internationally, such an approach has been advanced predominantly in response to climate change, notably by the IPCC, who in a 2014 report, advanced the idea of climate resilient pathways (Fig. 13.4): 'sustainable-development trajectories that

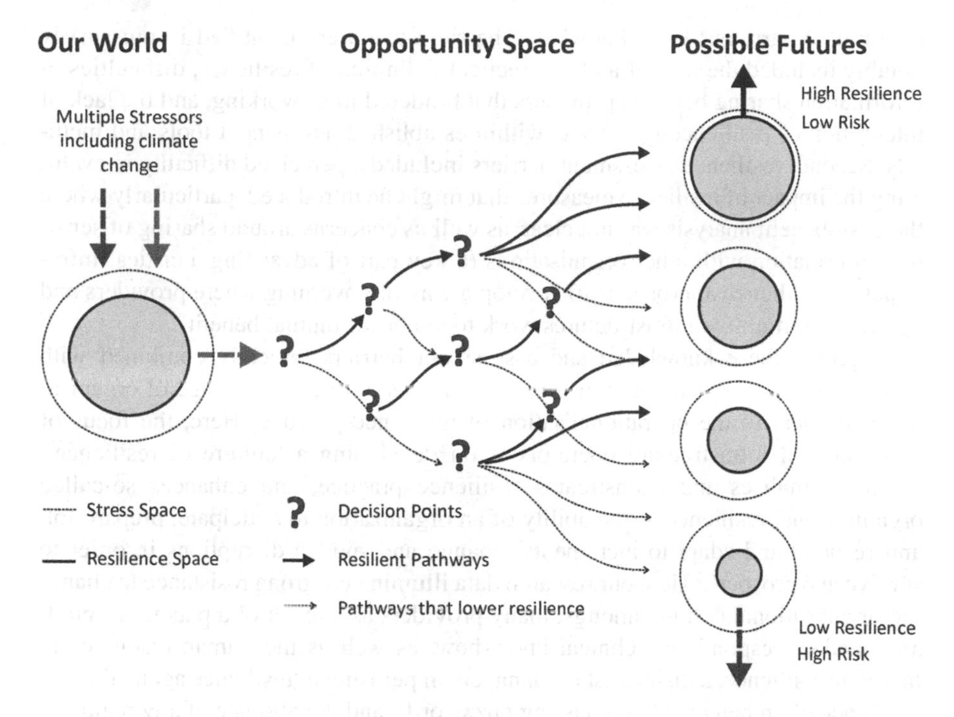

Fig. 13.4 Opportunity spaces and climate resilient pathways (Adapted from IPCC 2014)

combine adaptation and mitigation to reduce climate change and its impacts ... [including] iterative processes to ensure that effective risk management can be implemented and sustained' (IPCC 2014, p. 87). Such pathways can be either progressive or regressive; either leading to a more resilient world through adaptation and learning, or lower resilience as a result of insufficient mitigation and failure to learn; and which can be irreversible in terms of possible futures (ibid, p. 88).

In another example, adaptation pathway ideas have been fully integrated into the Dutch Adaptive delta Management (ADM) approach to enhancing climate resilience which seeks to be 'anticipative' of future conditions and stop 'tipping points' being reached. In the ADM has sought to understand how resilience ideas are shaping the ways in which critical infra- structure providers deal with complex risk, and the tensions elicited in the transition from protective approaches towards approaches founded on the basis of greater resiliency and how such assessment can be both accurate and fit of purpose.[7] An analysis of the survey data illuminates a range of barriers to implementation that we can categorise as knowledge, assessment and/or

[7]The formation of adaptation pathways are linked to the acknowledgement of uncertainty in climate change and thus bases much of its thinking on a scenario matrix which looks at the linkages between climate change and socio-economic development.

operational barriers.[8] First, knowledge barriers that were identified in our survey notably included the lack of a clear practical definition of resilience, difficulties of information sharing between providers that hindered joint-working, and the lack of integration of resilience-type ideas within established assessment tools and methods. Second, resilience assessment barriers included a perceived difficulty in evaluating the impact of resilience measures that might be introduced, particularly where the cost–benefit analysis was not clear, as well as concerns around sharing of sensitive information with other organisations (a key part of advancing a critical infrastructure resilience approach is to develop a way of governing where providers and operators of multiple infrastructures work together for mutual benefit).

Together, these knowledge and assessment barriers typically combined with shortcomings in institutional infrastructure and serve to create a range of organisational barriers to the operationalisation of resilience practice. Here, the focus of many critical infrastructure operators is on developing a 'culture of resilience', which normalises and mainstreams resilience practice, and enhances so-called organisational resilience – 'the ability of an organization to anticipate, prepare for, and respond and adapt to incremental change and sudden disruptions in order to survive and prosper'.[9] Here our research data illuminated strong resistance to changing organisational culture amongst many providers as a result of a paucity of guidance and corresponding technical know-how, as well as the human resources to facilitate resilience, a disinterest of managers in perusing a 'resilience agenda' (with resilience often perceived as a passing buzzword), and the absence of any regularly framework that prescribed and could be used to enforce change. From a more operational focus, many providers also reported difficulty in making a (financial) case for enhancing system redundancy – a key component of enhancing system resilience – and perpetual problems of balancing the requirement for short term efficiency and optimisation with the need to pro- vide resilience through flexibility and adaptability to cope with unexpected disruption.

13.3.4 Towards an Extended and Equitable Ecology of Anthropocentric Resilience

In her critique of how states respond to low probability–high consequence events – an increasingly common feature of the Anthropocene – Amoore (2013, p. 60) highlighted the ways in which new forms of risk calculus has extended traditional risk

[8]This research was carried out by way of two large scale survey of 40+ critical infrastructure operator's across Europe in 2015 and 2016. These surveys were conducted as part of the work underpinning the EU funded RESILENS project in which the authors we seeking to undertake a gap analysis of the current approaches to risk and resilience management within the infrastructure sector.

[9]This definition comes from the recently published British Standard on Organisational Resilience (2014).

management approaches based on probabilities by advancing a 'politics of possibly'. As she noted,

> the emerging security risk calculus is not a more advanced form of abstraction than one might find in statistical or prudential modes, but rather it is a specific form of abstraction that distinctively coalesces more conventional forms of probabilistic risk assessment with inferred and unfolding futures.

This does not infer a seismic break with tradition, but rather a transitionary process by which more intuitive and adaptively focused processes underpinning resilience are laminated over the calculative rationality of conventional risk management and assessment methodologies. In this paper, we have outlined such an ongoing transition towards resilience in critical infrastructure operations, highlighting the difficulties inherent in this process and the dearth of agreed ways forward within the sector. As Suter (2011, p. 5) noted, 'it remains often unclear if and to what extent the introduction of resilience changes the existing practices of critical infrastructure protection'. Moving from rhetoric to implementation in critical infrastructure, resilience is therefore not without its challenges and thus acknowledging and actualising how the role of resilience could drastically improve the security landscape for many critical infrastructure owners and operators, within their own context, is therefore crucial. Current practice in resilience assessment and promotion for critical infrastructure is too often limited to specific, endogenous technical factors. Unlike protection, resilience is not easily definable across all infrastructures, nor is it accurately measurable. As Pommerening (2007, p. 18) highlighted, 'there is a curious disconnect between recommending coping and adaptation strategies for new stages of stability, and the fact that we have just as little knowledge about how those stages will look'.

It can be seen from the foregoing discussion and emerging findings presented in this paper that there is a pressing need to address the shortcomings of traditional 'siloed' thinking and more 'traditional' views of 'hard' critical infrastructure protection that seeks bounce back to a pre-shock state rather than advancing more evolutionary 'bounce forward' pathways that seek to construct an approach more applicable to coping with increasingly complex and non-linear systems. This reflects a wider journey from the traditional, techno-rational approach with prescriptive, rigid methodologies, to a wider socially and organisationally informed extension of risk management that seeks a more transformative understanding of critical infrastructure resilience. Figure 13.5, provides a summary of this transition through number of criteria by which the appropriateness of protection measures versus resilience measures could be assessed.

Whilst enhancing critical infrastructure resilience is of vital concern given the nature of anthropocentric risk, we need to remain cognisant of its potentially inequities. Critical infra- structure resilience is not purely technocratic or value neutral and can have significant impacts upon social and spatial justice across a range of interlocking scales as new approaches, processes, actors and technologies are pulled together and deployed in the name of resilience. As Fainstein (2015, p. 157) notes, resilience in certain situations 'obfuscates underlying conflict and the distribution of benefits resulting from policy choices'. Critical infrastructure is far from inanimate

	Critical Infrastructure Protection	Critical Infrastructure Resilience
Aim	Equilibrium Existing normality Preserve Stability	Adaptive New normality Transformative Flexible
Focus	Endogenous Short term Reactive Hardened structures	Exogenous Long-term Proactive Redesigned processes
Critical Infrastructure Approaches	Techno-rational Technical Homogeneity Robustness Recovery Fail-safe Protection Optimisation Single-sector focus	Complex adaptive Socio-technical Heterogeneity Malleable Realign Safe-to-fail Predictive Greater redundancy/diversity Dependencies

Fig. 13.5 Differences between critical infrastructure protection and critical infrastructure resilience

and is increasingly imbued with agency (Evans and Reid 2014, p. 19). While often critical infrastructure resilience is masked in highly technical models showing complexity and indeterminacy there is a need to more fully engage with 'softer' approaches emphasising such agency, that, to date have been missing from critical infra- structure assessment and future planning. As such we should seek 'greater resilience *of the whole*, not just of what may be bureaucratically or politically deemed 'critical' to certain limited interests' (Perelman 2007, p. 40). Moreover, the advance of softer approaches, more grounded in social science methodologies, can assist critical infrastructure providers under- stand the complex multi-scalar and multi-institutional context in which they operate. As recent work on what has been termed 'resilience multiple' reminds us, context is vital and understanding how different perspectives and expertise in relation to resilience can be hybridised, and can help reconcile 'the tension between a desire for open, non-linearity on the one hand and a mission to control and manage on the other' as well as how different adoptions of resilience invoke 'differing spatialities, temporalities and political implications' is vital (Simon and Randalls 2016, p. 3). Such a combinational approach also talks to ongoing discussions about the changing nature of expertise linked the new *zeitgeist* of resilience. In current mobilisations within the critical infrastructures sector providers are confronting complex risk, necessitating that required expertise has to become more di use, pluralistic and integrative. This is leading to many viewpoints, methodologies and 'ways of doing' resilience being combined in operationalising it in practice. Given the slow nature of organisational change, this will however take time, patience and a willingness to embrace change and difference across the critical infrastructure sector. In particular it will require an inculcation and adoption of certain values, practices and research methodologies that focus upon more than the instrumentally rational and embraces adaptation, flexibility and grounded approaches that are more sociologically and politically informed.

Orchestrating such a coherent, sociotechnical and integrated approach to meeting the generational challenge of building resilient infrastructure is a significant challenge confronting the Anthropocene – and its academic theoreticians – over the coming decades. This is starkly represented in the Un Sustainable development Goals (SDGs) released in September 2015 where the discourse of resilience is utilised to highlight how global society should respond proactively to a range of shocks and stress and how we might collectively operationalise a joined-up response through developing 'quality, reliable, sustainable and resilient infrastructure' (Target 9.1) in order to advance global sustainable development.

Acknowledgments This chapter is an amended version of a prior paper published by the authors: Coaffee and Clarke (2015) "Critical infrastructure lifelines and the politics of anthropocentric resilience." Resilience: International Policies, Practices and Discourses, http://dx.doi.org/10.1080/21693293.2016.1241475. This work draws from empirical studies conducted during a range of research projects and notably the RESILENS project (Realising European ReSILiencE for Critical INfraStructure), which has funding from the European Union's Horizon 2020 Programme for research under grant agreement no 653260. We acknowledge the collaborative work of all consortium partners in this paper, details of which can be found at www.resilens.eu.

Further Suggested Readings

Adams J (1995) Risk. UCl Press, London
Alexander D (2002) From civil defence to civil protection – and back again. Disaster Prev Manag Int J 11:209–213
Amin A (2013) Surviving the turbulent future. Environ Plann D Soc Space 31:140–156
Amoore I (2013) The politics of possibility: risk and security beyond probability. Duke University Press, North Carolina
Aradau C, van Munster R (2011) Politics of catastrophe: genealogies of the unknown. Routledge, Abingdon
Australian Government (2010) Critical infrastructure resilience strategy. Retrieved Jan 27, 2016, from http://www.emergency.qld.gov.au/publications/pdf/Critical_Infrastructure_Resilience_Strategy.pdf
Baum SD (2015) Risk and resilience for unknown, unquantifiable, systemic, and unlikely/catastrophic threats. Environ Syst Decis 35:229–236
Beck U (1992) Risk society – towards a new modernity. Sage, London
Beck U (1995) Ecological politics in an age of risk. Polity Press, Cambridge
Beck U (1999) World risk society. Polity Press, Cambridge
Biermann F (2014) The Anthropocene: a governance perspective. Anthropocene Rev 1:57–61

Biermann F, Bai X, Bondre n, Broadgate W, Chen CTA, Dube o P, Erisman JW, Glaser M, van der Hel S, Lemos MC, Seitzinger S, Seto KC (2015) Down to earth: contextualizing the Anthropocene. Glob Environ Chang 39:341–350. doi:10.1016/j.gloenvcha.2015.11.004
Blowers A (1999) Nuclear waste and landscapes of risk. Landsc Res 24:241–264
Boin A, McConnell A (2007) Preparing for critical infrastructure breakdowns: the limits of crisis management and the need for resilience. J Conting Crisis Manag 15:50–59
Brassett J, Vaughan-Williams N (2015) Security and the performative politics of resilience: critical infrastructure protection and humanitarian emergency preparedness. Security Dialogue 46:32–50
British Standards Institute (2014) BS 65000: guidance for organizational resilience. BSI, London
Bruneau M, Chang SE, Eguchi RT, lee GC, O'Rourke T d, Reinhorn AM et al (2003) A framework to quantitatively assess and enhance the seismic resilience of communities. Earthquake Spectra 19:733–752
Carlson J, Doyle J (2000) Highly optimized tolerance: robustness and design in complex systems. Phys Rev Lett 84:2529–2532
Chandler D (2012) Resilience and human security: the post-interventionist paradigm. Security Dialogue 43:213–229
Chandler D (2014) Beyond neoliberalism: resilience, the new art of governing complexity. Resilience 2:47–63
Coaffee J (2013) Rescaling and responsibilising the politics of urban resilience: from national security to local place-making. Politics 33:240–252
Coaffee J, Clarke J (2015) On securing the generational challenge of urban resilience. Town Plann Rev 86:249–255
Coaffee J, Lee P (2016) Urban resilience: planning for risk crisis and uncertainty. Palgrave, London
Coaffee J, Wood D (2006) Security is coming home – rethinking scale and constructing resilience in the global urban response to terrorist risk. Int Relat 20:503–517
Coaffee J, Murakami Wood d, Rogers P (2008) The everyday resilience of the city: how cities respond to terrorism and disaster. Palgrave, London
Comfort LK (1994) Risk and resilience: inter-organizational learning following the Northridge earthquake of 17 January 1994. J Conting Crisis Manag 2:157–170
Crutzen PJ, Stoermer EF (2000) The 'Anthropocene'. IGBP Newslett 41:17–18
Dalby S (2013a) The geopolitics of climate change. Polit Geogr 37:38–47
Dalby S (2013b) Biopolitics and climate security in the Anthropocene. Geoforum 49:184–192
Davoudi S (2012) Resilience: a bridging concept or a dead end? Plann Theory Pract 13:299–333
Department of Homeland Security (2009) Critical infrastructure resilience. DHS, Washington, DC
Douglas M (1994) Risk and blame – essays in cultural theory. Routledge, London
Dueñas-osorio l, Vemuru SM (2009) Cascading failures in complex infrastructure systems. Struct Saf 31:157–167
Evans B, Reid J (2014) Resilient life: the art of living dangerously. Polity, Cambridge
Fainstein S (2015) Resilience and justice. Int J Urban Reg Res 39:157–167
Fisher T (2012) Designing to avoid disaster: the nature of fracture-critical design. Routledge, London
Fisher R, Norman M (2010) Developing measurement indices to enhance protection and resilience of critical infrastructure and key resources. J Bus Continuity Emerg Plann 4:191–206
Flinders M, Wood M (2014) Depoliticisation, governance and the state. Policy Polit 42:135–149
Foucault M, Guattari F, deleuze G, Fourquet F (1996) Equipments of power: towns, territories and collective equipments. In: Lotringer S (Ed), and (Hochroth I, Johnston J, Trans.), Foucault live: michel Foucault collected interviews, 1961–1984. Semiotext(e), New York, pp. 105–112
Giddens A (2002) Runaway world: how globalisation is reshaping our lives. Pro le, London
Gunderson I, Holling CS (eds) (2002) Panarchy: understanding transformations in human and natural systems. Island Press, Washington, DC
Haasnoot M, Middelkoop H, Olermans A, van Beek E, van Deursen WPA (2012) Exploring pathways for sustainable water management in river deltas in a changing environment. Clim Chang 115:795–819

Haasnoot M, Kwakkel JH, Walker WE, ter Maat J (2013) dDnamic adaptive policy pathways: a method for crafting robust decisions for a deeply uncertain world. Glob Environ Chang 23:485–498

IPCC (2014) Summary for policymakers. In: Field CB, Barros VR, dokken d J, Mach k J, Mastrandrea M d, Bilir TE, Chatterjee M, Ebi k l, Estrada Y o, Genova RC, Girma B, kissel ES, levy A n, MacCracken S, Mastrandrea PR, White l l (eds) Climate change 2014: impacts, adaptation, and vulnerability. Part A: Global and sectoral aspects, Contribution of Working Group II to the Fifth Assessment Report of the Intergovernmental Panel on Climate Change. Cambridge University Press, Cambridge, pp 1–32

Joseph J (2013) Resilience as embedded neoliberalism: a governmentality approach. Resilience 1:38–52

Kates RW (1962) Hazard and choice perception in ood plain management, Paper 78, department of Geography, University of Chicago

Kwadijk JCJ, Haasnoot M, Mulder JPM, Hoogvliet MMC, Jeuken ABM, van der Krogt RAA et al (2010) Using adaptation tipping points to prepare for climate change and sea level rise: a case study in the netherlands. Wiley Interdiscip Rev Clim Chang 1:729–740

Lentzos F, Rose n (2009) Governing insecurity: contingency planning, protection, resilience. Econ Soc 38:230–254

Linkov I, Bridges T, Creutzig F, decker J, Fox-lent C, Kröger W et al (2014) Changing the resilience paradigm. Nat Clim Chang 4:407–409

Linkov I, Larkin S, lambert JH (2015) Concepts and approaches to resilience in a variety of governance and regulatory domains. Environ Syst Decis 35:183–184

Longsta PH (2005) Security, resilience, and communication in unpredictable environments such as terrorism, natural disasters, and complex technology. Harvard University, Cambridge, MA

Lovins A (1976) Energy strategy: the road not taken? Foreign A airs 55:65–96

Mabey n, Schultz S, Dimsdale T, Bergamaschi I, Amal-lee A (2013) Underpinning the MENA democratic transition: delivering climate, energy and resource security. E3G, London

Mayberry J (2013) The evolution of critical infrastructure protection. Continuity Insights [online]. Retrieved Nov 10, 2015, from http://www.continuityinsights.com/articles/2013/11/evolution-critical-infrastructureprotection

National Infrastructure Advisory Council (2010) A framework for establishing critical infrastructure resilience goals final report and recommendations by the council. Retrieved Dec 29, 2015, from https://www.dhs.gov/xlibrary/assets/niac/niac-a-framework-for-establishing-critical-infrastructure-resilience-goals-2010-10-19.pdf

O'Rourke TD (2007) Critical infrastructure, interdependencies, and resilience. Bridge Linking Eng Soc 37:22–30

Obama B (2013) Presidential policy directive 21: critical infrastructure security and resilience. Washington, DC

Oldfield F, Barnosky A d, Dearing J, Fischer-Kowalski M, Mcneill J, Ste en W, Zalasiewicz J (2014) The Anthropocene review: its significance, implications and the rationale for a new transdisciplinary journal. Anthropocene Rev 1:3–7

Perelman IJ (2007) Shifting security paradigms: toward resilience. Critical Thinking, Moving from Infrastructure Protection to Infrastructure Resilience, George Mason University, 23–48

Pike A, Dawley S, Tomaney J (2010) Resilience, adaptation and adaptability. Camb J Reg Econ Soc 3:59–70

Pommerening C (2007) Resilience in organizations and systems: background and trajectories of an emerging paradigm. Critical Thinking, Moving from Infrastructure Protection to Infrastructure Resilience, George Mason University, 8–22

Prior T, Hagmann J (2013) Measuring resilience: methodological and political challenges of a trend security concept. J Risk Res 17:281–298

Reuters (2015) Britain needs complete rethink on flood defences after swathes of England hit. Retrieved Dec 29, 2015, from http://uk.reuters.com/article/us-britain- oods-idUkkBn0UB16I20151229

Scranton R (2015) Learning to die in the Anthropocene. City light, San Francisco

Siemens (2013) Toolkit for resilient cities. Retrieved Sept 22, 2013, from http:// les.informatandm.com/uploads/2015/4/Toolkit_for_Resilient_Cities_Full_Report.pdf

Sikula N, Mancillas JW, Linkov I, McDonagh JA (2015) Risk management is not enough: a conceptual model for resilience and adaptation-based vulnerability assessments. Environ Syst Decis 35:219–228

Simon S, Randalls S (2016) Geography, ontological politics and the resilient future. Dialogues Hum Geogr 6:3–18

Suter M (2011) Resilience and risk management in critical infrastructure protection. Center for Security Studies (CSS), ETH Zürich, Zürich

Walker J, Cooper M (2011) Genealogies of resilience: from systems ecology to the political economy of crisis adaptation. Secur Dialogue 42:143–160

Waters CN, Zalasiewicz J, Summerhayes C, Barnosky A d, Poirier C, Gałuszka A et al (2016) The Anthropocene is functionally and stratigraphically distinct from the Holocene. Science 351:aad2622

Weber M (1904–05/1958) The protestant ethic and the spirit of capitalism. Pearson Education, New Jersey

Welsh M (2014) Resilience and responsibility: governing uncertainty in a complex world. Geogr J 180:15–26

White GF (1942) Human adjustment to floods: a geographical approach to flood problems in the United States, Research Paper 29, department of Geography, University of Chicago

Zolli A, Healy A (2013) Resilience: why things bounce back. Headline, London

Part V
Cyber

Chapter 14
Bridging the Gap from Cyber Security to Resilience

Paul E. Roege, Zachary A. Collier, Vladyslav Chevardin, Paul Chouinard, Marie-Valentine Florin, James H. Lambert, Kirstjen Nielsen, Maria Nogal, and Branislav Todorovic

Abstract This chapter describes an evolution of practices in community and business assurance from protective programs based upon *risk* management to the emerging strategy of *resilience*. The chapter compares and contrasts these two basic approaches, identifying notable gaps where cyber security lags in the larger transformation. Recommendations address concepts, techniques, and strategies for integration of the cyber world with the physical and human worlds, and opportunities for future research.

Keywords Cyber security • Resilience • Risk management • Protection • Enterprise • Critical infrastructure • Digital word

---

P.E. Roege (✉)
Creative Erg, LLC, Corvallis, MT, USA
e-mail: paul.roege@alum.mit.edu

Z.A. Collier • J.H. Lambert
University of Virginia, Charlottesville, VA, USA

V. Chevardin
Ministry of Defense of Ukraine, Kiev, Ukraine

P. Chouinard
Defence R&D Centre for Security Science, Ottawa, ON, Canada

M.-V. Florin
EPFL International Risk Governance Center, Lausanne, Switzerland

K. Nielsen
Sunesis Consulting, LLC, Alexandria, VA, USA

George Washington University Center for Cyber and Homeland Security, Washington, DC, USA

M. Nogal
Trinity College Dublin, Dublin, Ireland

B. Todorovic
University of Belgrade, Belgrade, Serbia

© Springer Science+Business Media B.V. 2017

I. Linkov, J.M. Palma-Oliveira (eds.), *Resilience and Risk*, NATO Science for Peace and Security Series C: Environmental Security,
DOI 10.1007/978-94-024-1123-2_14

14.1 Introduction

Persistent and adaptive cyber (digital) threats have become the new normal in the modern, highly-connected economy. While there is a diversity of methods regarding how cyber security attacks are characterized and counted, the Identity Theft Resource Center estimates that in the first 5 months of 2016, there already have been 430 data breaches exposing over 12 million personal records in the United States (Identity Theft Resource Center 2016). The true number may be higher since many breaches either are not detected or not reported. Vulnerable personal information may include credit card and bank account numbers, social security numbers, personal health information, and biometric data. In 2015, malware capable of stealing money was detected on 1.9 million computers in the banking industry (Kaspersky Lab 2015). The number of ransomware attacks, in which a user's computer or personal files are locked until a payment is rendered, is on the rise (Savage et al. 2015).

According to the Ponemon Institute (2016), the average total organizational cost of a data breach in the United States is over $7 million, and the trend is growing. Moreover, they cite increasing consequences averaging $158 per compromised record, with health care records more than twice as costly at $355 per record – an increase of $100 over their 2013 assessment. Meanwhile, exploitation tools are available online, and personal information is surprisingly inexpensive on the black market. Prices for credit card details can range from $2–$90 (Panda Security 2010), depending upon the type of card, expiration, and credit limits. Black market prices are subject to the dynamics of supply and demand – for instance, after the Target breach, the influx of personal data caused the cost per record to temporarily fall to approximately $0.75 (Ablong et al. 2014).

In addition to data breaches and directly associated financial impacts, cyber attacks can compromise critical infrastructure such as utilities, communications, financial systems, and even transportation systems. The high level of connectivity and dependence upon cyber systems for the operations and management of critical infrastructure services can have the unintended consequence of widespread failures involving dependencies within and between multiple economic sectors (Kelic et al. 2013; Rinaldi et al. 2001). The connection between cyber-enabled systems and physical infrastructure has prompted an interest in *cyber-physical security* (DiMase et al. 2015). For instance, the power grid, comprising a network of connected substations, is susceptible to cascading failures (Thorisson et al. 2016; Shakarian et al. 2014; Sridhar et al. 2012; Lambert et al. 2012, 2013a; Karvetski and Lambert 2012). Industrial control systems (ICS) and supervisory control and data acquisition (SCADA) systems represent another example of vulnerable cyber-physical systems (Collier et al. 2016). These systems may be older, designed for functionality rather than security, and span a diverse array of requirements for operation of critical services (US Department of Energy 2002; Pollet 2002). ICSs differ from traditional information technology systems in that failures in ICSs may result in a threat to lives, environmental safety, or production output where, in some cases, halting or downtime may be unacceptable (Stouffer et al. 2011; McIntyre et al. 2007).

President Obama signed Executive Order 13636 in 2013 calling for improved cyber security through the establishment of risk-based standards and frameworks to protect critical infrastructure. Here, we distinguish between *risk* as the generic possibility of loss or injury, and *risk management* as an established discipline that seeks to optimize decisions on protective measures, seeking to maximize expected value in the face of known risks, largely based upon historical data. Classical risk management begins with assessment in terms of an interrogatory triplet: "*What can go wrong?*", "*How likely is it to happen?*", and "*What are the consequences?*" (Kaplan and Garrick 1981). When historical data or forecasts are not available, risk analysts and risk managers often must supplement their studies with data in the form of expert judgment. Once risks have been assessed, the next challenge lies in how best to manage the risks. Similar to the questions posed by Kaplan and Garrick (1981) for risk assessment, Haimes (1991) proposed a series of questions for risk management: "*What can be done and what options are available?*"; "*What are the associated trade-offs in terms of all relevant costs, benefits, and risks?*"; and "*What are the impacts of current management decisions on future options?*". Teng et al. (2012, 2013) describe that risk/safety/security programs address: "*What sources of risk are in the scope of the program?*"; "*What are the allocations of resources to program units, geographies, time horizons, topics, interfaces, etc.?*"; and "*What are the monitoring and valuation of program impacts?*".

A complementary but distinct concept is that of *resilience*. Presidential Policy Directive 21, published in concert with EO 13636, sets forth U.S. policy to "*strengthen the security and resilience of its critical infrastructure against both physical and cyber threats.*" (US White House 2013). The document distinguishes security and resilience, describing the former focus on protection of systems from threats or events, and the latter as the ability to "*prepare for and adapt to changing conditions and withstand and recover rapidly from disruptions. Resilience includes the ability to withstand and recover from deliberate attacks, accidents, or naturally occurring threats or incidents.*" (US White House 2013). Together, the documents prescribe security to protect against cyber (and physical) threats but propose resilience as the broader goal for critical infrastructure.

The National Academy of Sciences (2012) define resilience as "*(t)he ability to prepare and plan for, absorb, recover from, or more successfully adapt to actual or potential adverse events*". Whereas risk management selects and prioritizes potential measures to prevent or mitigate impacts to the status quo based upon understanding of the threat, resilience seeks to enhance the system's inherent capacity to respond throughout the process of inevitable change – both long and short duration (Linkov et al. 2014), thus invoking a fundamentally temporal perspective (Fig. 14.1).

Tierney and Bruneau (2007) emphasize a shift from event orientation and specific protective measures toward a capacity-centric view, stipulating four key attributes of resilient systems: robustness (withstand disruptive forces), redundancy (satisfy functional requirements with substitutable system elements), resourcefulness (effectively leverage resources to diagnose and solve problems), and rapidity (recover quickly from a disruption). In contrast with a risk management-based view, which is focused on preventing or protecting against intrusions (avoiding the risk)

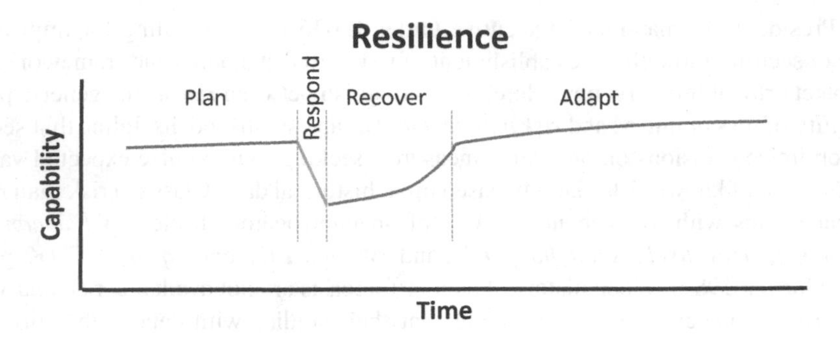

Fig. 14.1 Conceptual model of the stages of resilience as a function of time

or mitigating an event's negative consequences, a resilience-based approach is concerned with ensuring continuity in critical functions and services with minimal disruption (Young and Leveson 2014). Resilience may be applied to the cyber domain; a proposed definition is "*the ability to anticipate, withstand, recover from, and adapt to adverse conditions, stresses, attacks, or compromises on cyber resources*" (Dessavreand and Ramirez-Marquez 2015; Bodeau and Graubart 2016).

The remainder of this chapter further exposes the contrast between protection and resilience; explores how they respectively are applied in the cyber security context; proposes concepts and methods to integrate the cyber world into community and enterprise resilience; and identifies needs for future research.

14.2 Protection Versus Resilience – Different Postures for Facing Change and Uncertainty

Cyber security has gained attention in large part because we are constantly surprised by the pervasive nature, persistence, diversity, and consequences of attacks. Modern information systems exhibit characteristically dynamic design, and the broad availability of associated technologies creates fertile ground, not only for unanticipated behaviors (bugs), but also for innovative malicious actors to find and exploit new failure modes and attack vectors. Compounding these issues, personal, financial, and security information today have such intrinsic value that loss, diversion or alteration can rapidly yield unanticipated consequences. Hazard modes, probabilities and consequences each are poorly characterized and rapidly changing. The International Risk Governance Council (IRGC 2015a) identifies the major analytical challenges of cyber security risk management as (a) to assess the probability across a large set of possible attacks, (b) to determine how much to invest over time in protective designs and measures and/or in resilience and rapid response and recovery, and (c) how to allocate those investments.

Collier et al. (2014) point out several difficulties in applying the triplet approach outlined by Kaplan and Garrick (1981). For one, the dynamic and rapid development of threats renders it infeasible to maintain a library of possible threats in the cyber domain. Given this, it is difficult to answer even the first question "*vulnerable to what?*". Tversky and Kahneman (1973) identify the availability heuristic in which people's ability to envision future events depends heavily upon their experience of similar situations. This inhibits imagination of new and different vulnerabilities, or even those which simply have gone undetected or unpublicized. Further, it is difficult to estimate the consequences of successful cyber attacks due to the interdependent nature of cyber-physical systems. Additionally, the vast number of potential threats makes the prioritization of threats and countermeasure difficult to accomplish (Young and Leveson 2014). Branlet et al. (2011) note that difficulties in cyber defense arise because it is very collaborative (within the organization), adversarial (external to the organization), and the operating environment is highly uncertain, making individual and group decision making difficult. The adversarial nature of cyber threats poses a distinct challenge, in that attackers, in response to defensive measures, will either attack other parts of the system or utilize more dangerous attack methods (IRGC 2015a). Whereas traditional risk analysis estimates deterministic or stochastic risks such as failures or natural hazards, "adversarial risk analysis" models risk as adaptive and seeks to understand attacker and defender behavior utilizing tools from decision analysis and game theory (IRGC 2015a).

Recognizing the inevitable shortcomings in a "blockade" approach of firewalls and passwords, the IRGC notes three components to a cyber security strategy (IRGC 2015b):

- Defensive measures: Traditional, protection-based methods which focus on maintaining the cyber system *as is* by preventing unauthorized system access. Examples of defensive measures include access restrictions, basic hygiene, and patching.
- Proactive measures: Measures which seek to anticipate threats and either counter a threat or prepare the system for attack. Examples include creating false databases and network connections to distract malicious intruders, and even (potentially controversial and risky) preemptive cyber attacks on adversaries (Maitra 2015).
- Retroactive measures: Follow-up actions to deal with breaches and their consequences. For instance, forensic techniques may identify shortcomings, and monitoring of data exfiltration can help to assess damage, inform notification requirements, and possibly indicate perpetrators.

This taxonomy may invoke elements common to a resilient approach, such as diversity in solutions and consideration of pre- and post-event actions; however, like most cyber security programs, it retains an external threat (versus system capacity) orientation, and generally focuses on the information systems and cyber phenomena of concern.

Beyond the challenges of out-guessing cyber adversaries, estimates of risk probability often are skewed by overconfidence which may reflect an optimism that negative events can only happen to "the other guy". Research in the cybersecurity domain shows that in game settings, participants tend to under-invest in security because they believe that they can avoid a cyber attack, even when they learn that other players had fallen victim (Horowitz and Crawford 2007; Pfleeger and Cunningham 2010). Recognizing that under-investment is common, it is difficult to identify precisely what level of investment is warranted. Another source of optimism may come from the implementation of security controls. The fact that one has taken protective measures may actually create overconfidence that one will not be the target of an attack, assuming that the protective measures will successfully protect them, and actually increase risk-taking behavior (Pfleeger and Cunningham 2010). As it relates to this context, an organization's policies regarding information security, risk management, business continuity, and insurance may all contribute to a false sense of security derived from overconfidence that suffering a cyber attack is unlikely, thus underestimating the true probability.

Additionally, the actuarial-style risk management process generally involves selecting from among multiple alternatives with various costs and benefits in order to reduce the remaining risk to an acceptable level. Defining this risk threshold and selecting among alternatives is not a straightforward exercise, especially considering uncertainties about real effects, and where acceptability of impacts may be illiquid or nonlinear. Lowrance (1976) defined the concept of safety by stating that "*(a) thing is safe if its risks are judged to be acceptable*", therefore ultimately based upon personal and social values. Commercial organizations, especially, place different weights upon various types of consequences; some of which may include reputation, financial, legal liability, or intellectual property (Pfleeger and Cunningham 2010; Garcia and Horowitz 2007). Entities should first and foremost attempt to determine their risk appetite- what they will accept, what they will avoid, what they will transfer, and what they will manage. Success is defined not merely as the reduction of risk to an acceptable level; rather, an appropriate balancing of investment in preventative measures that reflect the organizational risk appetite based upon its values.

While cyber security guidance and programs continue to apply traditional risk management techniques, complications such as those described herein indicate the need to change strategies. Natural disasters, political conflict, and other disruptions produce unanticipated interactions and event frequencies, and produce extreme health and socio-economic consequences for which protective measures and insurance seem inadequate (Lambert et al. 2013b; Parlak et al. 2012). Research into resilience as a more robust alternative was underway and gradually increasing when Hurricane Sandy (October 2012) struck and may have triggered a tipping point, at least in the United States. Earlier that year, the National Infrastructure Advisory Council (NIAC) had formed a Regional Resilience Working Group, which received increasing support and visibility in the wake of the storm. The investigation informed recommendations (NIAC 2013) which strongly influenced introduction of resilience into PPD-21 and the subsequent revised National Planning Frameworks (US Department of Homeland Security 2016a) pursuant to Presidential Policy Directive 8.

Many public and private organizations, for example, Symantec and the World Economic Forum, have embraced resilience as a guiding principle for cyber security, recognizing the interrelationships among information systems, people and business processes. Many have adopted the five concurrent and continuous Functions from the NIST Cybersecurity Framework (2014) which, when considered together, provide "*a high-level, strategic view of the lifecycle of an organization's management of cybersecurity risk*":

- Prepare/identify
- Protect
- Detect
- Respond
- Recover

Research gaps exist regarding how to identify the correct harmony between risk management and resilience approaches, and how these value-based decisions can be aided (IRGC 2015a). Resilience is often mistaken as just another word for robustness or flexibility – two terms which seem logically contradictory. Husdsal (2010) relates flexibility (ability to divert) and robustness (stability) to resilience, as the overarching ability to survive disruptive changes despite severe impact. A recent National Research Council (NRC) report on "Nexus of Cybersecurity and Public Policy" recommends: "*acknowledging that defenses are likely to be breached, one can also seek to contain the damage that a breach might cause and/or to recover from the damage that was caused.*" It describes a resilient system as "*one whose performance degrades gradually rather than catastrophically when its other defensive mechanisms are insufficient to stem an attack. A resilient system will still continue to perform some of its intended functions, although perhaps more slowly or for fewer people or with fewer applications. Features of resilient systems include redundancies and the absence of single points of failure*" (Clark et al. 2015, page 69).

Woods (2012) identifies several desirable traits for resilient systems, citing their ability to:

- Recognize the signs that adaptive capacity is falling;
- Respond to the threat of exhausting buffers or reserves;
- Shift priorities across goal tradeoffs;
- Make perspective shifts and contrast diverse perspectives that go beyond their nominal position;
- Navigate changing interdependencies across roles, activities, levels, goals; and
- Learn new ways to adapt.

Notably, each of these statements addresses capacities of the system under consideration, not the nature of the hazards. In other words, resilience considers the system's response to change pre-, during and post-event, not simply an evaluation of protective, prevention and mitigation design alternatives. Assessment inevitably requires consideration of various potential events and trends to reveal patterns and key nodes. Roege et al. (2014) describe use of a generic enterprise analysis process that involves stakeholders with diverse expertise and scenario-based assessments, similar to those used to develop military operational capability requirements. This type of analysis may be applied to very specific systems, but it delivers the greatest benefit when applied at an enterprise level, where strategies, force structures, and capabilities such as survivability, flexibility and agility can be examined and designed into the overall entity.

The International Standard Organization (ISO) addresses security, risk and resilience in at least three series of standards:

- First a series of standards on cybersecurity, including the ISO/IEC 27000 family of standards on information security management. For example ISO/IEC 27032:2012, which in practice is about Internet security;
- Risk management is addressed in the ISO technical committee (TC) 262, which produced ISO 31000:2009, to provide principles, framework and a process for managing risk.
- TC 292 deals with standardization in the field of security to enhance the safety and resilience of society. TC 292 published in April 2016 a new set of guidelines for organizational resilience (ISO/DIS 22316). According to members of the development team, the standard will adopt the change and adaptation focus and other factors that distinguish resilience from traditional risk management.

The proliferation of standards and guidelines illustrates the diversity of views on how to ensure good cybersecurity risk management and resilience. Forward thinking among private and public sector institutions is beginning to transcend the established cyber security paradigm that emphasizes countermeasures against specific or known types of threats to maintain system integrity as designed. Nevertheless, the predominance of cyber security literature limits its focus to information technology (IT) roles and the IT system. The present opportunity includes not only expanding resilience principles in the context of IT systems, but sparking thought, conversation, and action outside the cyber "box" to include perspectives and requirements from the full enterprise.

14.3 The Cyber World and Its Interactions with Physical and Human Worlds

Risks are commonly seen as originating from one of two types of sources: those caused by natural hazards or risks of human origin. In addition, a third category has emerged as a result of interaction between the first two categories. This category is illustrated by technology being created by human action, but becoming increasingly difficult to be controlled by humans (IRGC 2010). We can now say that, in addition to the natural/physical world and the human world, there exists a cyber world. This world should perhaps be considered as having its own life, rules and autonomy.

This view has been cited in the literature. Smirnov et al. (2013) refer to cyber, physical and human systems (CPH), meanwhile Xiong et al. (2015) deal with cyber, physical and social systems. Below, we define these different worlds.

- Natural and Physical World: refers to those objects and living beings subjected to laws of physics (gravity, oxidation, etc.), without rational-thinking abilities.
- Human World: refers to the human being, individually and in groups, conforming societies. Its main characteristic is the ability of rational thinking, individually and collectively.
- Cyber World: refers to the environment in which communication over computer networks occurs. It is not subjected to natural laws and does not have thinking abilities, though it is able to make decisions based on pre-established rules. The cyber world develops in interaction with nature, from which it takes energy and materials such as rare earth elements, and to which it contributes, for example when it creates waste due to obsolescence; and with humans, from which it takes initial impetus and instructions, and which it nowadays supports in most of their activities.

Cyber systems that inhabit the cyber world have ascended to the level of necessity in all aspects of human life including indispensable roles in critical infrastructure systems. However, cyber systems are usually referred to as IT or Information Control Technologies (ICT), and their rapid technological development has not been followed by other organizational elements. In other words, the organizational structure of major owners and providers of critical infrastructure services (public or private) in most cases treat the cyber world as a separate and distinctive segment. The consequence is that cyber-related activities and procedures are coordinated and applied independently from other segments (e.g. physical), while execution can take place either within a dedicated cyber structure or remotely (e.g. Internet repeaters, hubs, data loggers, remote servers, cloud, etc.). The collection of all mentioned cyber-related elements includes enterprise hardware and software (the backbone of the cyber world), as well as various monitoring, control and programming units and devices, networks, elements and units dispersed within the critical infrastructure – often including devices that interact with the cyber structure, but not a part of it (e.g. personal smart phones in use by staff). This concept is shown in Fig. 14.2.

The cyber world is becoming more autonomous, manifested in, for example, the Internet of Things (IoT), robotics, smart interactive networks, data sciences, artificial intelligence, and self-healing systems that are trained to remedy their deficiencies. As this world becomes more independent of direct human intervention, the rules of behaviors must be created and understood. Not only should such rules deal with governance issues, such as what can be done and what must not be done, there also remain issues such as: allocation of authority to machines; how they interact with humans; and how issues of ethics or values would be identified and decided.

Issues of cyber risk follow. Susceptibility to malicious human intervention is a well-recognized cyber security risk. Yet, might we consider cyber risk as a deficiency in the current world order exhibiting a signal that something must be changed? Could such indicators be seen, as in the health field, as the precursor or the symptom of a disease? We live not in science-fiction, but in the real world, yet the latter sometimes looks very similar to the former (Clark et al. 2015).

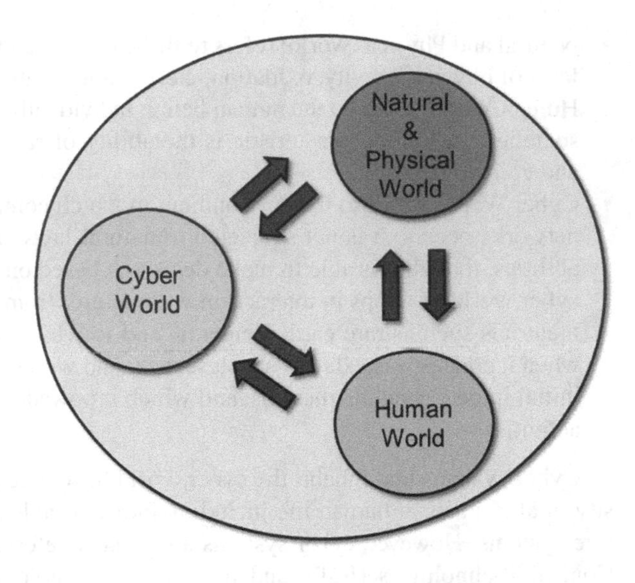

Fig. 14.2 Relationship between natural/physical, human, and cyber worlds

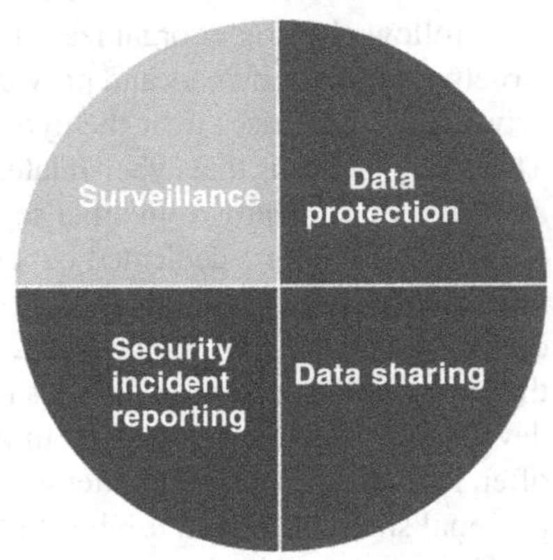

Fig. 14.3 Objectives and obligations for critical infrastructure providers, in the context of public policy objective and requirements

Providers of critical services and related stakeholders face multiple concurrent needs and obligations reflecting public policy missions and private strategies and goals. Figure14.3 illustrates four major categories of cyber-related obligations and objectives which require deconfliction and prioritization:

- **Obligation to report on cyber security incidents:** Many regulated industries and many countries require reporting on cyber security incidents today. This is already the case in the United States and Germany (BIS), and is expected to expand within Europe (NIS Directive). While these requirements may be intended to protect affected individuals and businesses, they may actually conflict with, for example, data protection, privacy, or intellectual property protection. The difficulty arises in defining the boundaries of incident reporting schemes: (a) which industries are concerned (countries have different approaches and different definitions of a critical infrastructure), (b) how "cyber incident" is defined, (c) where the mandatory reporting threshold is set, and (d) what type of information must be released (to what extent the information can be anonymized).

- **Data protection laws:** Such laws deal with access and use of personal data, to better protect citizens in all major democracies. However, even the most compliant organization can see the data it protects being released to external actors when (a) governments require the reporting of cyber security incidents (as seen above); (b) the organization shares the data with others for better analytics and business performance, or simply for research; or (c) government engages in surveillance activities. The second point is particularly sensitive, at a time when private data is aggregated in large databanks, such as biobanks that collect genotypic data, medical records and other health and lifestyle data for research, clinical trials and gene therapies.
- **Data sharing:** In sectors where the benefit of Big Data is established, companies may wish to engage in data sharing for improved analytical potential. In these cases, issues of ownership, consent, collection, access and use must be considered. If not done in accordance with data protection laws, data sharing may violate privacy and confidentiality. At the same time, some of the data protection laws may need to be revised, in particular to reflect changing societal values and preferences. Various national initiatives or policies for precision medicine (such as those in the US, UK or China) encourage the constitution of large data bases with genotypic data of hundreds of thousands of individuals.
- **Surveillance:** National governments engage in intelligence gathering for the purpose of restoring, maintaining and developing national security. Although surveillance is not an inherent obligation of critical infrastructure itself, it may be that data that is protected by a critical infrastructure asset ends up being collected by a government agency. One cannot neglect the possibility that government collection of private data may conflict with privacy or data protection rights.

This situation of potential conflicts between public policy objectives and requirements for critical infrastructure creates a context of uncertainty and ambiguity, subject to a variety of interpretations and behaviors. New risks may emerge from this situation and affect critical infrastructure and relevant stakeholders. As in physical and human domains, policy makers and managers must be cognizant and consider interactions to avoid unintended outcomes.

One option would be to design overarching principles for establishing the cyber world as a public good, with the stipulation that one's actions should not harm others. This could provide the basis for rules that apply to the Internet and cloud computing. Critical infrastructures increasingly rely on these broader information networks to deliver dramatic improvements in information communication, business efficiency and overall performance. However, the reliability, availability and security of data cannot be expected to be provided by security measures only (first component: encryption, shielding, patching, redundancies, backup, security by design…), or risk management measures (second component: education, cost-benefit analyses, transfer to insurance…). Resilience strategies (third component) are needed to emphasize

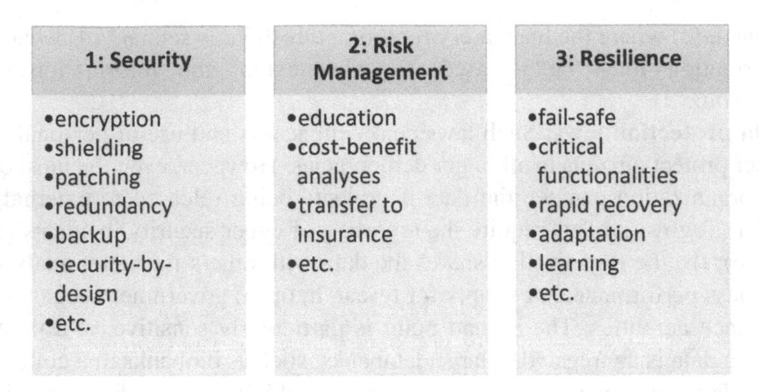

Fig. 14.4 Components of cyber security

performance in the face of unexpected disruptions and breaches (Fig. 14.4). Ultimately, strategies should focus on maintaining or recovering a sufficient level of functioning for the delivery of minimum critical services, even if ICT fails. The literature on fail-safe strategies might provide some help here. For instance, Abramovici and Bradley (2009) suggest the combination of different techniques to the recovery process, such as provision of fail-safe states, spare logic to replace misbehaving logic, returning to last safe checkpoint, and even wiping out critical data.

14.4 Progress in the Transition from Cyber Protection to Resilience

A number of risk-based methods and guidance exist within the cyber domain. Prompted by the enactment of Executive Order 13636, the US National Institute of Standards and Technology (NIST) created the "Cybersecurity Framework", a voluntary risk-based document containing standards and best practices for use by private and government organizations of all sizes and types (NIST 2014). The Cybersecurity Framework consists of a Framework Core, which is a set of five continuous and concurrent functions (identify, protect, detect, respond, recover). Implementation can range in levels of sophistication; a non-resilience approach focuses on protecting the existing system design, rather than going after fundamental capacity, whereas a resilience-oriented approach uses the concepts of the framework to design a holistic solution to resilience. A key difference in resilience thinking is that compliance does not equal security. The five functions are further decomposed into categories and sub-categories. The Framework Core represents a compilation of accepted standards and practices which serve as a minimum or baseline set of cyber security activities and processes by which an organization can understand their current security posture across all of the elements and sub-elements, and identify gaps to achieving a target security profile.

Within the US Federal Government, NIST Special Publication 800–39 (NIST 2011) provides guidance on risk assessment and management for information security systems. The guidance recommends that a threat and vulnerability assessment be conducted, followed by a risk determination, "*considering the likelihood that known threats exploit known vulnerabilities and the resulting consequences or adverse impacts (i.e., magnitude of harm) of such exploitations occur*" (NIST 2011). Note that the emphasis here is on *known* threats and vulnerabilities, which as we discuss later, presents a particular challenge in cyber risk assessment.

Another recommendation, in this case related specifically to energy systems, was published by Sandia National Laboratories. The report focused on microgrid security, proposing a defense-in-depth approach that leverages the natural geographical and logical segmentation of industrial control systems (Veitch et al. 2013). The microgrid control architecture is segmented into functionally distinct "enclaves" facilitating necessary data exchange and limiting access. Isolating functions into enclaves localizes the effect of malicious or accidental impacts thereby complementing a resilient strategy of limiting adverse impacts (regardless of cause) to the affected enclave. Access restrictions are also recommended to help ensure that only trusted actors with necessary credentials have access to protected system components.

In the European Union, five strategic priorities have been identified to achieve an open, safe, and secure cyberspace:

- Achieving cyber resilience
- Drastically reducing cybercrime
- Developing cyber defence policy and capabilities related to the Common Security and Defence Policy (CSDP)
- Develop the industrial and technological resources for cybersecurity
- Establish a coherent international cyberspace policy for the European Union and promote core EU values (European Commission 2013).

Notably, the first item on the list is resilience. To determine the progress of EU Member States in meeting these strategic priorities, the European Union Agency for Network and Information Security (2014) compared national strategies and found differences in many specifics; however, the report depicts substantial emphasis on outcomes, characterized into five categories:

- Critical information infrastructure and services;
- Business and innovation;
- Rights and society;
- Public-private relations; and
- General.

The breadth of these foci, and especially the emphasis on broad socio-economic outcomes, suggests a strongly positive trend toward adopting resilience as an emphasis in European cyber security efforts, at least at the strategic level. Additionally, the report describes a logic model for the evaluation of national cyber security strategies and enumerates several key performance indicators.

A popular assessment tool is the Common Vulnerability Scoring System (CVSS), which provides a scale based upon the "exploitability" of a vulnerable component and the impact of a successful exploitation (FIRST 2015). The metrics, when scored, provide a final rating on a 0–10 scale, which can be used to compare vulnerabilities across multiple systems, where a high score indicates a critical vulnerability (FIRST 2015). Optional metrics related to the temporal and environmental factors can be used as well.

Risk-based methods are essentially design for anticipated conditions, with decisions guided by actuarial-style, historically-based methods. Such practices guide prudent preparation for previously experienced conditions, which may contribute to resilience, but they also can divert focus from the more fundamental, holistic focus of resilience-building. Several resilience-based frameworks have been created. For example, the CERT Resilience Management Model (CERT-RMM) defines 26 process areas organized under four main resilience categories (Engineering, Enterprise Management, Operations, and Process Management) (Caralli et al. 2010). Each of the process areas are scored on a four point scale representing capability levels in that domain (Incomplete, Performed, Managed, Defined). Similar to the NIST Cybersecurity Framework, these 26 process areas can be used as a way to score the current organizational capabilities and compare it to target levels of resilience, identifying the gaps between the two.

The US Department of Energy (2014) published the Cybersecurity Capability Maturity Model for the electricity subsector, which defines four levels of maturity with the most mature level manifested when "*the cybersecurity program strategy is updated to reflect business changes, changes in the operating environment, and changes in the threat profile*". Ten domains relevant to cybersecurity resilience are identified:

- Risk Management
- Asset, Change, and Configuration Management
- Identity and Access Management
- Threat and Vulnerability Management
- Situational Awareness
- Information Sharing and Communications
- Event and Incident Response, Continuity of Operations
- Supply Chain and External Dependencies Management
- Workforce Management
- Cybersecurity Program Management

A similar method developed by the Department of Homeland Security (DHS) called the Cyber Resilience Review (CRR) consists of 297 assessment questions and is usually administered in a facilitated 6-h workshop (DHS 2016b). The CRR model defines 10 "domains" (Asset Management, Controls Management, Configuration and Change Management, Vulnerability Management, Incident Management, Service Continuity Management, Risk Management, External Dependencies Management, Training and Awareness, Situational Awareness), each of which are broken into several goals, and further into multiple practices. Each domain is scored on a six point scale (Incomplete, Performed, Planned, Managed, Measured, Defined). The result of the assessment is a red/yellow/green dashboard which identifies areas of strengths and weaknesses. The report suggests using the identified gaps to perform cost-benefit studies in order to further prioritize investments in process improvement (DHS 2016b).

Linkov et al. (2013) developed a matrix-based assessment procedure for cyber resilience based on the National Academy of Sciences definition of resilience, identifying four steps in the disaster management cycle (Plan/Prepare, Absorb, Recover, Adapt) and four operational domains based on Command and Control literature (Physical, Information, Cognitive, Social) (Alberts 2002). By placing the resilience steps on one axis, and the domains on the other axis, a 4 × 4 grid (the "resilience matrix") is created that helps decision makers answer the question "*How is the system's ability to [plan/prepare for, absorb, recover from, adapt to] a cyber disruption implemented in the [physical, information, cognitive, social] domain?*" (Linkov et al. 2013). While the metrics that may be identified within each cell of the resilience matrix will vary across organizations and contexts, the approach facilitates the identification and assessment of organizational capabilities and resources that contribute to cyber resilience.

Bodeau and Graubart (2016) note that no single resilience metric or set of metrics will be applicable across all systems, and that metrics need to be tailored to the specific system, mission, and stakeholders. They also note that defining the system and its boundaries can be a difficulty in implementing resilience-based approaches. Further, challenges exist in how to quantify and report cyber resilience. Ford et al. (2012) argue against representing resilience as a single figure of merit, which might obscure the contributing factors: "*resilience is not a 1-dimensional quantity*". Additionally, difficulties exist in determining how to integrate the resilience of systems across a multitude of different disruptions and scenarios, as resilience measurements are dependent on the type of perturbation (Ford et al. 2012). Some propose a multi-dimensional approach, such as cyber value-at-risk, which incorporates considerations of vulnerability, assets at risk, and profile of the attacker (World Economic Forum 2015).

14.5 Examples and Illustrations

14.5.1 Business Processes and Management Functions for a Resilient Enterprise

A key theme of this chapter is the often poor interaction between the cyber, operational and management components of an enterprise, and in particular, a critical infrastructure enterprise, which is a consequence of the human, physical and cyber domains often being seen as distinct and separated organizationally. This section looks at a generic enterprise in terms of generalized day-to-day functions to

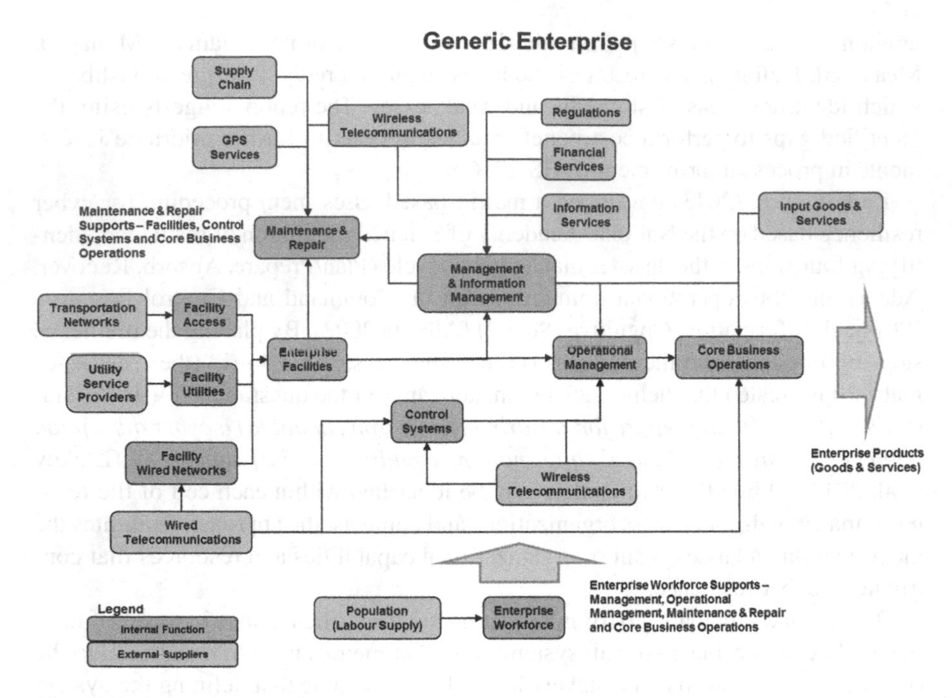

Fig. 14.5 A generic enterprise

illustrate the mutual dependence of the cyber, operational and management domains. However, day-to-day functions alone are insufficient to understand the resilience of an enterprise with respect to the cyber domain. In the event of an incident, whether a cyber-incident or otherwise, several response and recovery functions are instantiated. The last part of this section provides some insight into issues related to the cyber domain's role in a non-cyber-incident and to those when it is a cyber-incident. For the purposes of this section, 'cyber' refers to cyber information and to the functions that create, transform, store or transmit cyber information.

Figure 14.5 illustrates a generic enterprise with generalized functions necessary for the operations of that enterprise. Functions internal to the enterprise are shown in blue while those that are external are shown in a gold-yellow color. The arrows depict the dependencies amongst the functions with the direction of the arrow pointing towards the dependent function. These functions will be described in the context of a critical infrastructure enterprise, such as one responsible for the generation and distribution of electricity, with particular attention paid to the cyber-domain.

Core Business Operations The 'Core Business Operations' consists of the essential processes by which an enterprise transforms 'Input Goods & Services' into 'Enterprise Products' provided to consumers. In the case of electricity, these would include the processes for generating electricity from various fuels or other energy sources such as provided by water, wind and solar sources, and the storage, transmission and distribution of that electricity to consumers.

Management Functions The core business operations are supported directly by an 'Operational Management' function, which oversees day-by-day activities. For an electrical power enterprise, this would include the operational control of power generation facilities and balancing of the electrical grid to ensure effective and efficient transmission and distribution of electricity. Operational Management heavily depends on 'Control Systems' which include industrial control systems (ICS) and, for critical infrastructure industries, supervisory control and data acquisition (SCADA) systems (Stouffer et al. 2011).

Today ICS are supported by an operational information and communications architecture that has merged with the public information and communications infrastructure. The diagram shows this dependency on both the operational and public information and communications services as dependencies on 'Wired Telecommunications' and 'Wireless Telecommunications'. While the merging of operational and public systems has provided increased efficiencies for industries it has also meant there are new vulnerabilities as cyber-threats now have access to operational control systems that were previously isolated from public information and communications systems.

In addition to operational management, the core business operations are supported by a 'Management and Information Management' function, which includes general direction and oversight of strategic, non-operational business functions as well as stewardship of enterprise information or data. It also includes corporate support functions of human resource management, legal affairs, financial administration, etc. Effective management requires access to 'Financial Services', current government 'Regulations' and a variety of 'Information Services' related to the industry, including supplier and customer data – all of which are predominantly provided digitally. This implies a dependency on information and telecommunications systems, which might be provided by wireless or wired networks. Note that, in the latter case, the wired networks will likely be a standard service provided by the enterprise facilities and not dedicated networks.

Maintenance and Repair The 'Maintenance & Repair' function includes all enterprise facilities and systems including information technology (IT) systems. This function is an essential sustainment function to ensure that day-to-day operations run smoothly or, in the event of a more serious situation, that operations are restored quickly. While facility and equipment maintenance and repair are usually well integrated, the IT maintenance and repair is often situated in a separate division of a typical enterprise, which can lead to coordination issues should a problem transcend the physical-cyber divide. At a minimum all elements require an integrated, topographic appreciation of the enterprise. As with any other enterprise function, maintenance and repair are dependent upon access to information and telecommunications networks as well as spare parts, software patches, system expertise, etc. provided by various suppliers ('Supply Chain'). In some cases, such as utility companies, there may be a requirement for Global Positioning System ('GPS') data in order to locate and affect a repair of a damaged system component.

Facilities In the diagram 'Enterprise Facilities' refers to the provision of standard 'hotel' services for employees and housed equipment, which include accessibility (i.e., the facilities have 'Transportation Networks'), reasonably comfortable working environment (shelter) plus a number of typical utility services such as electricity, water and waste removal ('Utility Services') and information and telecommunications networks ('Wired Telecommunications'). While most employees expect access to wired networks, 'smart' facilities exploit those same networks for other aspects of facility operations, such as electricity distribution, climate control, etc. Use of external suppliers with access to facility IT services raises an issue of third-party vulnerability.

Workforce Of course, enterprises depend upon employees ('Enterprise Workforce'). At first glance one might assume that this is a straight forward non-cyber issue. However, increased automation may come at a cost and result in enterprise risks such as having insufficient depth in trained personnel in the event of a serious crisis such as a pandemic flu. Outsourcing of employee services and expertise is becoming common practice but it increases the dependency on information and communication infrastructure, and extends the list of potential threat vectors. The increased reliance on information and communications technology throughout enterprises, as shown above, implies a requirement for access to IT expertise which increasingly is not within the enterprise. The 'Enterprise Workforce' should, therefore not be seen as restricted to direct employees but also to include outsourcing and access to professional expert services. Finally, the 'workforce' can be a significant threat to an enterprise whether an employee has malicious intentions or does so inadvertently. For example, frequently, senior executives are given full access rights to an enterprise's IT systems even if they lack the need for such rights or adequate training. This can far too easily result in a compromised system. Unfortunately, human resources departments responsible for hiring and training of employees often do not understand an enterprise's IT security needs.

The above description of a generic enterprise is meant to illustrate ways in which the cyber domain is intertwined with physical and human domains. Unfortunately, in many cases the IT operations, management, maintenance and repair are organizationally separated by the other enterprise functions (Ernst and Young 2014). This presents enough of a challenge for day-to-day operations which can only be exacerbated in the event of a crisis.

Forbes magazine notes that adversarial cyber actors have three significant advantages (Forbes 2014):

> Cyber attackers have an edge on you. Just look at recent incidents of credit card information being stolen from Target and SnapChat users' names and cell phone numbers being published online.
>
> They've got the innovation. Normally innovation is a great thing. However, we would personally do without it being applied to introducing new cyber threats.
>
> They've got the timing. Surprise is on their side, and they take total advantage of it.

> They've got the target. They know where your crown jewels are kept, and they are going for the prize.

> Breaches and break-ins are, quite simply inevitable. And the best cyber security program in the world isn't going to change that hard fact.

Resilience includes preparedness, prevention, protection, mitigation, response and recovery. However, given the inevitability of security breaches, the discussion will first discuss response and recovery in the event an incident occurs. There are a number of functions which are instantiated in the event of a crisis, termed 'incident' below, to respond to the crisis and to recover normal operations. These functions are listed below and apply whether the incident is a cyber-incident or otherwise:

Response and Recovery Functions:

- Collective decision making and cross-discipline coordination that spans and includes all of the organizations involved in response and recovery;
- Incident communications that connects all of the response and recovery organizations;
- Incident situation awareness;
- Threat or hazard risk assessment;
- Provision of public information;
- Protection operations to contain, reduce and eliminate the threat or hazard;
- Responder safety;
- Incident logistics, including the provision of transportation, supplies, volunteers and surge emergency management and responder personnel;
- Fatality management;
- Community sustainment through the provision of emergency shelter, food, water, medical treatment and psycho-social support;
- Incident forensics; and
- Recovery operations.

In the case of a cyber-incident, critical response and recovery functions necessarily involve IT personnel. These include 'Collective decision making and coordination', 'Protection operations', 'Incident Logistics' (e.g., the provision of software patches, anti-virus software, etc.), 'Incident Forensics' and 'Recovery' (e.g., the recovery of critical data). However, the cyber domain is a key enabler during any contingency, providing for incident communications, timely situation awareness, data processing and risk assessment and forensic investigation, collaborative and co-creation of response plans and dissemination of alerts and warnings and public information. The exploitation of social media for providing public information is but one example. In addition to these functions, cyber underpins other functions as well. For example, fatality management depends on the cyber-infrastructure to quickly aid people in determining the fate and location of family members. Numerous other examples can be found in the emergency management literature.

A distinguishing feature of these functions in a cyber-incident versus other types of incidents is a general lack of experienced coordinating bodies such as emergency operations centers (EOC) with proven protocols, established practices, clearly defined accountability frameworks, well understood distribution of authority, and typically a collection of ad-hoc response capabilities. For something like a natural disaster, an EOC would be activated to an appropriate level to coordinate the response. An emergency may be declared with relevant political authorities designated and traditional response organizations activated. In the case of a cyber-incident, a government EOC would likely focus initially on the protection of government cyber-assets and the provision of information to enterprises, affected or potentially affected by the incident; telecommunications' EOCs will focus on maintaining the ability of communications systems to operate; utilities' EOCs will focus on alternative routes and sources; and law enforcement agencies will focus on identifying the threat actors. Attribution may be problematic restricting immediate response to reaction. The ability to designate a specific, responsible political authority will be challenging, since a cyber incident will likely not lie within a given jurisdiction.

Response and recovery will mostly fall to individual enterprises. Yet a recent survey shows that only 37% of organizations have a formal cyber incident response plan (PwC 2016). The above paints a bleak picture which might overstate the situation, but an effective collective response and recovery to a serious cyber-incident lags that of more traditional threats and hazards. Improving collective response to and recovery from large scale cyber incident requires collective preparedness. This is recognized in the US Presidential Policy Directive 41 (US White House 2016). While some progress has been made, it is uneven; and collective preparedness faces several obstacles such a reluctance to share information and competing with day-to-day enterprise demands. With respect to coordinating protection, prevention and mitigation, there are not only all of the above challenges but the additional challenge of sharing the financial burden when vulnerabilities are owned by specific critical infrastructure enterprises but potentially significant risks associated with those vulnerabilities go well beyond the enterprise.

As noted by PPD 41:

> While the vast majority of cyber incidents can be handled through existing policies, certain cyber incidents that have significant impacts on an entity, our national security, or the broader economy require a unique approach to response efforts. These significant cyber incidents demand unity of effort within the Federal Government and especially close coordination between the public and private sectors.

14.5.2 Cyber Security (Protection) Principles

As information technology becomes increasingly integrated with physical infrastructure operations, the potential grows for disruptions in the cyber world to impact critical infrastructure capabilities. Wide-scale or high-consequence events can cause harm or disrupt services upon the economy, energy sector, transport sector, defense, etc., impacting millions of citizens' daily lives. We must understand the nature and extent of threats in order to successfully build a threat model. Modern cyber threats can be modeled as random or deterministic processes, the definition and parameterization of which is difficult.

The nature of cyber threats has evolved greatly over the last century. Beginning from Claude Shannon's Mathematical Theory of Communication (Shannon 1948), this theory gave us the means to describe any telecommunication channel in terms of probability information measures. Later there was the Communication Theory of Secrecy Systems (Shannon 1949), which described the fundamental terms and models in cryptography. This work was the foundation of modern cryptography – namely the mathematical structure of systems which govern the theory and practice of transmission of secret messages. In this work, the theory is focused on a telecommunication channel with passive noise and a cryptanalyst. But this model did not consider the issue of authenticity of the sender and recipient. With advent of computer networks, the cryptographical channel model of Shannon became insufficient.

The demand for cryptosystems without channels for sending keys resulted in the appearance of asymmetric cryptography, involving paired public and private keys (Diffie and Hellman 1976). That made financial, telecommunication, and information services available in every home. The next step was authentication theory/coding theory (Simmons 1985). It complemented the Shannon mathematical model and provided an apparatus to describe the authentication model using redundancy. In parallel with this, the development of information security systems was facilitated with the appearance of the LaPadula model (Bell and LaPadula 1973) and other access management and control models.

Given the many parallel advancements in cryptography and security, a set of standards was developed. For example, ISO 7498 provides a general description of security services (authentication, access control, data confidentiality, data integrity, non-repudiation) and related mechanisms (encipherment, digital signature mechanisms, access control mechanisms and so on). ISO 15408, provides evaluation criteria for information security. Given the scale of today's cyber systems, and the wide variety of different software and hardware, different user errors can create a diverse array of system vulnerabilities. As a rule, contemporary cyber security system vulnerabilities are primarily related to modern Denial of Service (DoS) attacks in comparison with classical threats (confidentiality, integrity). Stakeholders should strive to continuously search for optimum security measures, understand their vulnerabilities, and minimize their own risks (Fig. 14.6). ISO 27032 recommendations offer possibilities for estimating risks, an overview of Cybersecurity, and an explanation of the relationship between Cybersecurity and other types of security. Also it provides a framework to enable stakeholders to collaborate on resolving Cybersecurity issues.

For common cyber systems we can define and categorize modern cyber threats according to the following attributes:

- cyber system type which is exposed to a threat (technical, social, ecologic);
- cyber system component which is impacted by a threat (software, hardware, process);

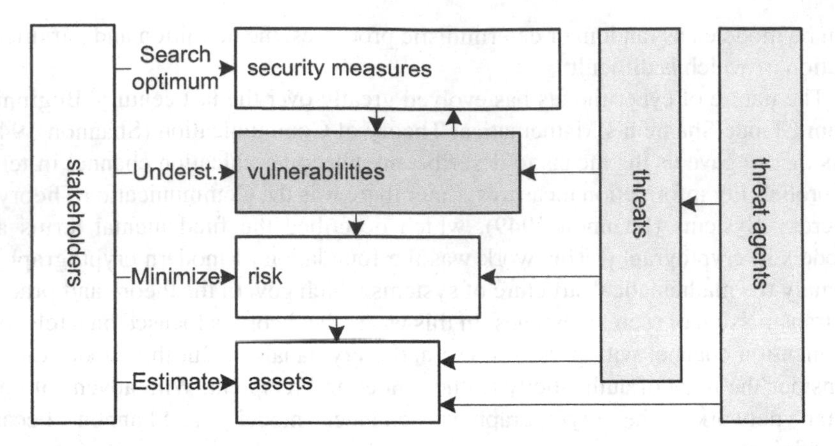

Fig. 14.6 Interests of stakeholders, system vulnerabilities, threats, and risks based on ISO 27032:2012, ISO 15408–1: 2009, ISO 27032

- vulnerability of cyber system or its elements impacted by a threat:
  - bed cryptographic key/bed DRBG/bed cipher;
  - backdoor in software;
  - vulnerability of routers and network devices to Denial of service attack (Dos), Distributed Denial of service attack (DDos) etc. There are more than 4000 different attacks today;
- placement of threat source (WAN, DMZ, LAN); in other words, clients, servers, networks and people;
- process for threat effects to manifest (malware; failure hardware; human factor; ecological process and so on);
- distribution environment
  - network (mailing, routing, social networks, games ...),
  - information (electronic documents, sites, games, ...),
  - technical (touch memory, drivers for devices, CD, DWD, ...),
  - social-technical (web, social networks,...);
- degree of targeting (random or deliberate);
- source of threat;
- threat frequency;
- threat stealthiness;
- magnitude of consequences after threat;
- threat hierarchy (low, middle, high);
- threat advisability;
- emerging time;
- realization conditions.

It is a common approach for organizations to define and classify threats for cyber systems. Some threats require more information than others for the identification of

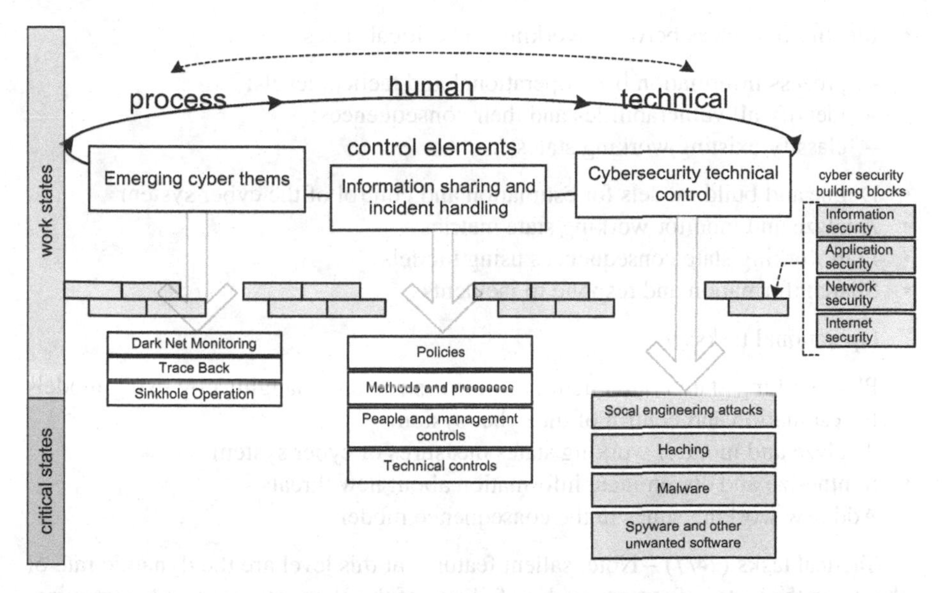

Fig. 14.7 Interconnections of cyber system elements, security system elements, and critical system states

relevant risks to the cyber system. So cyber security systems must continuously analyze vulnerabilities and threats using active and proactive methods on all critical system levels (Fig. 14.7).

Suppose we have a common critical system with three levels (Strategic, Operational, Tactical). We can prioritize some cyber security system tasks based on principles and procedures providing cyber resilience services and activities which provide the most effective protection against threats. For this we could use term working state. *Working state* is a set of telecommunication services, security services and cyber security services.

Strategic tasks:

- Identify threats
  - process information from operational and tactical level;
  - provide cyber intelligence services;
  - specify software security (e.g., Application Security Verification Standard 3.0.1, 2016);
  - specify hardware security (normative documents in technical security information, etc.);
- Identify vulnerabilities
  - process information from operational and tactical levels;
  - identify cyber system working states;
  - identify links between working states;
  - identify conditions which create critical situations;

- Identify interfaces between working and critical states
  - process information from operational and tactical levels;
  - identify all vulnerabilities and their consequences;
  - classify existing working states;
- Define and build models for estimation and control of the cyber system
- Analyze and monitor working state metrics
- Plan working state consequences using models
- Share information and respond to incidents

Operational tasks:

- Plan working state consequences (system structure changing) by using models for estimation and control of the cyber system
- Analyze and monitor working states measures of cyber system
- Synthesize and disseminate information about new threats
- Add new working states in the consequence model

Tactical tasks (24/7) – Note: salient features at this level are the dynamic rate of change in the states of system such as failures of the elements, gaps in communication with the control elements, and gaps in communication with subordinates. Consequences should be identified (e.g., routers destroyed, link destroyed, radio network lost) using models for estimation and control of the cyber system. Follow-up measures include:

- Analyze and monitor working states measures of cyber system.
- Identify new threats.
- Add new work state in the working state consequence model.

Further research in this area should include: assessment of the adequacy of certain threat categories, defining sets of services for cybernetic systems, definition of working states of cybernetic systems, development of new models of cyber system resiliency, and others.

14.5.3 Integrated Modeling of Cyber Systems with Operational Contingency Management

Given the convergence among cyber and other segments of critical infrastructure systems, there is a growing need for modelling methods which accommodate respective process characteristics yet facilitate collaborative development of integrated solutions. The following discussion illustrates such treatment in the context of contingency planning, response, and recovery seeking:

- Synchronization of cyber with other emergency response activities; and

- Optimization of cyber recovery procedures with the focus on re-establishing the critical infrastructure service/operation.

In order to support these tasks, critical infrastructure resilience may be quantified through dimensionless analytical functions related to the variation of functionality during a period of interest, including the losses in the disaster and the recovery path. The resilience function captures the effect of the disaster, but also the results of response and recovery, the effects of restoration and preparedness (Cimellaro et al. 2010). The resilience of the cyber structure, as an integral part of the current critical infrastructure system, then can be defined using the existing critical infrastructure resilience frameworks as a platform.

The first step would be to define the relationships and dependencies of cyber with other segments from the specific critical infrastructure system. Starting with the critical infrastructure operability (*Op*), as the main targeted output we want to reinforce by building resilience, we can define a subset of cyber security (*CS*) functions needed to maintain *Op* at 100%, namely:

$$CS\_F_j; j = 1 \ to \ n \tag{14.1}$$

Each *CS* function can be assigned a weight indicator *k*, representing its importance in the overall system:

$$Op = k_1^* CS\_F_1 + k_2^* \ CS\_F_2 + k_3^* \ CS\_F_3 + \ldots + k_n^* \ CS\_F_n \tag{14.2a}$$

or

$$Op = \sum k_j^* \ CS\_F_j; j = 1 \ to \ n \tag{14.2b}$$

Emergency response and critical infrastructure recovery steps for example, as the key elements of resilience, can be defined and elaborated based on the priority list. Since the higher goal is to enhance the resilience of critical infrastructure systems in general, therefore covering also the unforeseen threats, the cyber segment must embody flexibility to adapt to any conditions occurring during and after the specific incident. As noted in Roege et al. (2014), such dependencies and their importance may be exposed through real or simulated event analysis, but they ultimately should be distilled to reflect system characteristics, not simply as event-specific mitigation. Once all vital functions and outputs of the critical infrastructure are prioritised and their dependencies with the cyber structure defined, it is possible to create a finite list of worthwhile steps relevant to physical, human, and cyber domains before, during and after a change (whether emergency or long-term). The exact sequence and pattern of steps in a specific case will depend on type, volume and intensity of the event.

In order to define a list of indicators capable of assisting the cyber structure decision support system (DSS)/operator to prioritise tasks in case of a critical infrastructure threat, *CS* functions described with the Formula (1) have to be expanded taking into consideration, besides the cyber element, also the direct influence of that cyber component to the critical infrastructure operation in general (as discussed in Formula 2) and the interdependencies in the system as indicated in the scheme of cyber world. That can be achieved with the following parametric form:

$$CS\_F_j(c,o,i) \tag{14.3}$$

with parameters

- c – indicator of cyber importance
- o – indicator of critical infrastructure operational importance
- i – indicator of importance for interdependencies

Critical infrastructure resilience from the point of cyber structure can in such case be defined as:

$$\text{Re} = \sum CS\_F_j(c,o,i); j = 1 \text{ } to \text{ } n \tag{14.4}$$

Equation (14.4) lays the groundwork for the creation of emergency response of the cyber system and control over its influence on critical infrastructure recovery measures in case of any type of threat. Prioritisation based on threat parameters allows the cyber DSS/operator to obtain the initial list of steps necessary to restrain and eliminate the threat, as well as to adapt the sequence during the process. Once that phase is completed, Eq. (14.4) provides guidelines for optimal cyber and critical infrastructure mitigation and recovery activities.

Taking into consideration the specific roles of various cyber systems, a possible methodological approach to building resilience within the cyber domain could be to define and describe the related critical activities and procedures by using the language of the cyber world, i.e. through algorithms. Such an approach would be in line with standardized operations performed on a day-to-day basis by IT/ICT personnel and complement existing cyber activities taking place within critical infrastructure systems. Figure 14.8 depicts a generic algorithm for response & recovery procedures in cyber systems.

The model distinguishes three types of actions:

1. Standard maintenance and repair operations: dealing with routine issues and foreseen threats;
2. Supporting actions: taken by cyber personnel to enhance emergency response and critical infrastructure recovery in case of an impairment that is not of cyber origin;
3. Cyber emergency response and recovery operation: dealing with threats that include cyber attacks in whichever form.

For listed actions of type 2 and 3, the parametric Function (3) should be used to define and prioritise steps to be taken, namely:

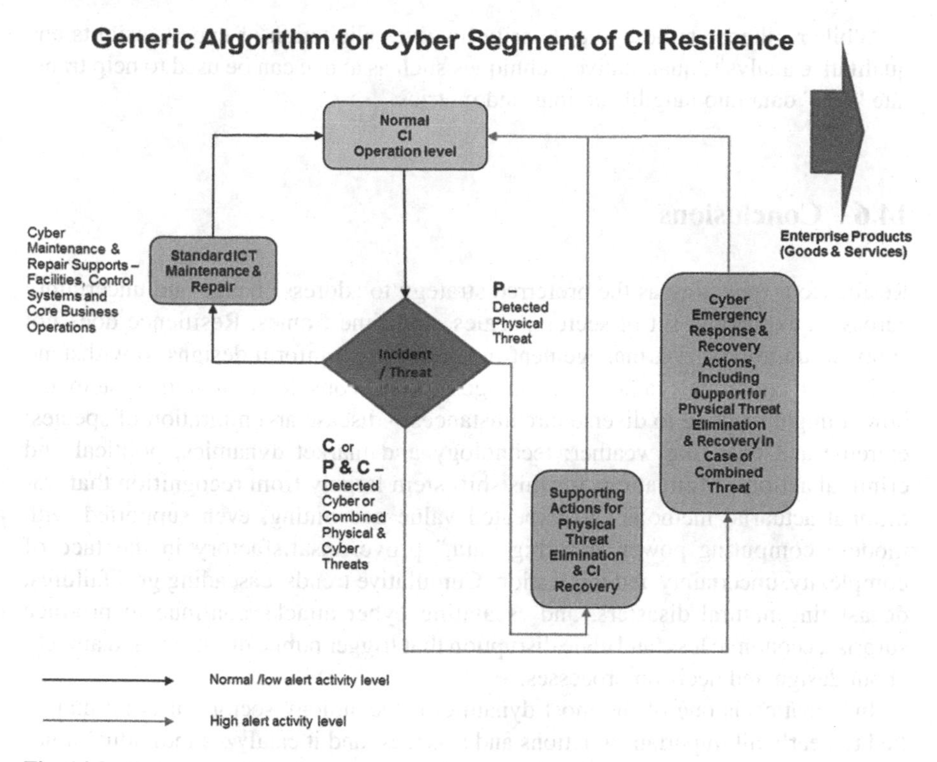

Fig. 14.8 Generic algorithm for response and recovery in cyber systems

- Type 2 (supporting actions in case of physical threat) would require that *CS* functions are prioritised and enhanced through the parameters *i* and *o* (interdependencies and critical infrastructure operational importance, respectively). Cyber related additional activities would therefore be focused on capabilities that most strongly assist recovery of the most important critical infrastructure services and at the fastest pace possible. Such activities might include the engagement of more ICT personnel, enhancement of wireless network(s), deployment of new routers, activation of auxiliary computer systems to provide more power, etc.
- Type 3 (emergency response and recovery in case of cyber threat) would require a balanced prioritisation using all three parameters (*c*, *o* and *i*). Depending on type of threat/incident, in particular in case of combined cyber and physical attack, required cyber related steps and tasks will have to be balanced to ensure the overall and the fastest critical infrastructure recovery, starting with the most important services. It is necessary to emphasize the importance of interdependency parameter (*i*) for optimisation of the sequence of actions, bringing to the forefront the cyber activities with the highest overall impact on the critical infrastructure operation, directly or indirectly.

While resilience building necessarily involves elicitation of expert insights and qualitative analysis, quantitative techniques such as above can be used to help translate “soft” data into tangible actions and metrics.

14.6 Conclusions

Resilience is emerging as the preferred strategy to address change and uncertainty across an expanding set of sectors, scales, and time frames. Resilience does not supplant traditional risk management methods which inform designs to withstand anticipated conditions; rather, it encourages holistic consideration of the system and how it might respond to diverse circumstances – disease and migration of species; extreme and changing weather; technology and market dynamics; political and criminal actions. Motivations for this shift stem largely from recognition that traditional actuarial methods and expected value accounting, even supported with modern computing power and “big data,” prove unsatisfactory in the face of complexity, uncertainty and innovation. Cumulative trends, cascading grid failures, devastating natural disasters, and escalating cyber attacks continue to produce surprise, economic loss, and civic disruption that trigger public questions and anxiety about design and decision processes.

Information is one of the most dynamic of technology sectors; it is intimately tied to nearly all important functions and services, and it catalyzes individual innovation and effectiveness. Moreover, given the sophistication of computing and communications capabilities in conjunction with growing levels of authority, it can be useful to conceptualize a digital world that operates distinctly, if in collaboration with the physical and human worlds. Given these factors, the cyber world would be a logical candidate for early adoption of resilience concepts. Instead, cyber security lags behind physical and human system management in this migration and remains largely segregated from these other domains in practice, even at the national policy level. Whether this distinction stems from stakeholder segregation due to the specialized nature of cyber practice or other reasons, there fortunately are recent indications of resilience inroads into cyber security literature. A small number of proponents, including the European Commission, Symantec, and MITRE Corporation, have begun to recognize and assert the need to employ resilience principles within the cyber domain, and to integrate treatment of the cyber world with physical and human reality, especially in relation to critical services and infrastructure, where the value proposition relates to urgent socio-economic outcomes.

Leaders, stakeholders, and subject matter experts should reflect on the prospect of stronger integration among cyber security and broader resilience development efforts. Those involved in the operational (human and physical) aspects of critical infrastructure should seek to enlist information professionals and cyber security groups as full participants in business and contingency planning. Those responsible for cyber security should examine how resilience thinking might alter approaches for managing risks explicitly in the cyber world; they also should assertively participate in enterprise-level analyses to better understand overall business processes and vulnerabilities, seeking to identify cyber-domain measures that support more resilient enterprise processes. This chapter builds upon recent work to offer conceptual aids, models and analytical techniques which may assist in integration of resilience across the physical, human, and cyber worlds. Practitioners and researchers alike face opportunities to expand ideas, tools, and experience further in order to advance our society's capacity for survival and sustainment through flexibility, adaptation and learning.

Acknowledgements The authors are grateful for discussion with members of the Cyber Working Group in the NATO Advanced Research Workshop "Resilience-Based Approaches to Critical Infrastructure Safeguarding", convened in Ponta Delgada, Azores, Portugal, 26–29 June, 2016. The organizers of the workshop were Igor Linkov, Bojan Srdjevic, and José Palma-Oliveira.

Further Suggested Readings

Ablong L, Libicki MC, Galay AA (2014) Markets for cybercrime tools and stolen data: hackers' bazaar. RAND Corporation Report RR-610-JNI. http://www.rand.org/content/dam/rand/pubs/research_reports/RR600/RR610/RAND_RR610.pdf

Abramovici M, Bradley P (2009) Integrated circuit security: new threats and solutions. In: Proceedings of the 5th annual workshop on cyber security and information intelligence research: cyber security and information intelligence challenges and strategies. ACM, p 55

Alberts D (2002) Information age transformation: getting to a 21st century military. DOD Command and Control Research Program, Washington, DC

Bell DE, LaPadula LJ (1973) Secure computer systems: Mathematical foundations (No MTR-2547-VOL-1). MITRE Corporation, Bedford

Bodeau D, Graubart R (2016) Cyber resilience metrics: key observations. Case No. 16–0779. The MITRE Corporation

Branlat M, Morison A, Woods DD (2011) Challenges in managing uncertainty during cyber events: lessons from the staged-world study of a large-scale adversarial cyber security exercise. Human Systems Integration Symposium, Vienna VA, 10–25 to 10–27, 2011

Caralli RA, Allen JH, Curtis PD, White DW, Young LR (2010) CERT resilience management model, Version 1.0: Improving Operational Resilience Processes. http://www.sei.cmu.edu/reports/10tr012.pdf

Cimellaro GP, Reinhorn AM, Bruneauc M (2010) Framework for analytical quantification of disaster resilience. J Eng Struct 32(2010):3639–3649

Clark D, Berson T, Lin H (2015) At the Nexus of cybersecurity and public policy, some basic concepts and issues. National Research Council, The National Academies Press, Washington, DC. http://www.nap.edu/catalog/18749/at-the-nexus-of-cybersecurity-and-public-policy-some-basic

Collier ZA, Linkov I, DiMase D, Walters S, Tehranipoor M, Lambert JH (2014) Cybersecurity standards: managing risk and creating resilience. Computer 47(9):70–76

Collier ZA, Panwar M, Ganin AA, Kott A, Linkov I (2016) Security metrics in industrial control systems. In: Colbert EJM, Kott A (eds) Cyber-security of SCADA and other industrial control systems. Springer, Cham, pp 167–185

Dessavreand DG, Ramirez-Marquez JE (2015) Computational techniques for the approximation of total system resilience. In: Podofillini L, Sudret B, Stojadinovic B, Zio E, Kröger W (eds) Safety and reliability of complex engineered systems. CRC Press, Boca Raton, pp 145–150

Diffie W, Hellman M (1976) New directions in cryptography. IEEE Trans Inf Theory 22(6):644–654

DiMase D, Collier ZA, Heffner K, Linkov I (2015) Systems engineering framework for cyber physical security and resilience. Environ Syst Decis 35(2):291–300

Ernst & Young (2014) The DNA of the CIO: opening the door to the C-suite. http://www.ey.com/Publication/vwLUAssets/ey-the-dna-of-the-cio/$FILE/ey-the-dna-of-the-cio.pdf

European Commission (2013) Cybersecurity strategy of the European Union: an open, safe and secure cyberspace. https://eeas.europa.eu/policies/eu-cyber-security/cybsec_comm_en.pdf

European Union Agency for Network and Information Security (2014) An evaluation framework for National Cyber Security Strategies. ISBN: 978-92-9204-109-0, DOI: 10.2824/3903

FIRST (2015) Common vulnerability scoring system v3.0: specification document. CVSS v3.0 specification (v1.7). https://www.first.org/cvss/cvss-v30-specification-v1.7.pdf

FORBES (2014) Why cyber security is not enough: you need cyber resilience. http://www.forbes.com/sites/sungardas/2014/01/15/why-cyber-security-is-not-enough-you-need-cyber-resilience/#461e9a695799. Retrieved 7 November, 2016

Ford R, Cavalho M, Mayron L, Bishop M (2012) Toward metrics for cyber resilience. In: 21st EICAR (European Institute for Computer Anti-Virus Research) annual conference proceedings

Garcia A, Horowitz B (2007) The potential for underinvestment in internet security: implications for regulatory policy. J Regul Econ 31(1):37–51

Haimes YY (1991) Total risk management. Risk Anal 11(2):169–171

Horowitz B, Crawford J (2007) Application of collaborative risk analysis to cyber security investment decisions. Fin Ser Technol Consorti Innov J 2(1):2–5

Husdal J (2010) A conceptual framework for risk and vulnerability in virtual enterprise networks. In: Ponis S (ed) Managing risk in virtual enterprise networks: implementing supply chain principle. IGI Global, Hershey, pp 1–27. doi:10.4018/978-1-61520-607-0.ch001

Identity Theft Resource Center (2016) Data breach reports. May 31, 2016. http://www.idtheftcenter.org/images/breach/DataBreachReports_2016.pdf

IRGC (2010) Emerging risks: sources, drivers, and governance issues. International Risk Governance Council, Geneva. https://www.irgc.org/risk-governance/emerging-risk/irgc-concept-of-contributing-factors-to-risk-emergence/sources-drivers-and-governance-issues/

IRGC (2015a) Comparing methods for terrorism risk assessment with methods in cyber security. Workshop report, International Risk Governance Council, Lausanne. https://www.irgc.org/wp-content/uploads/2016/01/Terrorism-Cyber-Security-28-29-May-2015-Workshop-Report.pdf

IRGC (2015b) Cyber-security risk governance, workshop report, International Risk Governance Council, Lausanne https://www.irgc.org/wp-content/uploads/2016/01/Cyber-Security-Risk-Governance-29-30-October-2015-Workshop-Report.pdf

Kaplan S, Garrick BJ (1981) On the quantitative definition of risk. Risk Anal 1(1):11–27

Karvetski CW, Lambert JH (2012) Evaluating deep uncertainties in strategic priority-setting with an application to facility energy investments. Syst Eng 15(4):483–493

Kaspersky Lab (2015) Kaspersky security bulletin 2015: overall statistics for 2015. https://securelist.com/analysis/kaspersky-security-bulletin/73038/kaspersky-security-bulletin-2015-overall-statistics-for-2015/

Kelic A, Collier ZA, Brown C, Beyeler WE, Outkin AV, Vargas VN, Ehlen MA, Judson C, Zaidi A, Leung B, Linkov I (2013) Decision framework for evaluating the macroeconomic risks and policy impacts of cyber attacks. Environ Syst Decis 33(4):544–560

Lambert JH, Keisler JM, Wheeler WE, Collier ZA, Linkov I (2013a) Multiscale approach to the security of hardware supply chains for energy systems. Environ Syst Decis 33(3):326–334

Lambert JH, Parlak AI, Zhou Q, Miller JS, Fontaine MD, Guterbock TM, Clements JL, Thekdi SA (2013b) Understanding and managing disaster evacuation on a transportation network. Accid Anal Prev 50(1):645–659

Lambert, J.H., C.W. Karvetski, D.K. Spencer, B.J Sotirin, D.M. Liberi, H.H. Zaghloul, J.B. Koogler, S.L. Hunter, W.D. Goran, R.D. Ditmer, and I. Linkov 2012. Prioritizing infrastructure investments in Afghanistan with multiagency stakeholders and deep uncertainty of emergent conditions. ASCE J Infrastruct Syst 18(2): 155–166.

Linkov I, Eisenberg DA, Plourde K, Seager TP, Allen J, Kott A (2013) Resilience metrics for cyber systems. Environ Syst Decis 33(4):471–476

Linkov I, Bridges T, Creutzig F, Decker J, Fox-Lent C, Kröger W et al (2014) Changing the resilience paradigm. Nat Clim Chang 4(6):407–409

Lowrance WW (1976) Of acceptable risk: science and the determination of safety. William Kaufman Inc.

Maitra AK (2015) Offensive cyber-weapons: technical, legal, and strategic aspects. Environ Syst Decis 35(1):169–182

McIntyre A, Becker B, Halbgewachs R (2007) Security metrics for process control systems. SAND2007-2070P. Sandia National Laboratories, U.S. Department of Energy, Albuquerque

National Infrastructure Advisory Council (2013) Strengthening regional resilience through national, regional, and sector partnerships: DRAFT report and recommendations. November 21, 2013. http://www.dhs.gov/sites/default/files/publications/niac-rrwg-report-final-review-draft-for-qbm.pdf

NIST (2011) Managing information security risk: organization, mission, and information system view. NIST Special Publication 800–39. National Institute of Standards and Technology, US Department of Commerce, Gaithersburg

NIST (2014) Framework for improving critical infrastructure cybersecurity, version 1.0. National Institute of Standards and Technology, US Department of Commerce, Gaithersburg

Panda Security (2010) The cyber-crime black market: uncovered. http://www.pandasecurity.com/mediacenter/src/uploads/2014/07/The-Cyber-Crime-Black-Market.pdf

Parlak A, Lambert JH, Guterbock T, Clements J (2012) Population behavioral scenarios influencing radiological disaster preparedness and planning. Accid Anal Prev 48:353–362

Pfleeger SL, Cunningham RK (2010) Why measuring security is hard. IEEE Secur Privacy 8(4):46–54

Pollet, J. (2002, November 19–21) Developing a solid SCADA strategy. Sicon/02 – sensors for industry conference. Houston, Texas, USA

Ponemon Institute (2016) 2016 cost of data breach study: global analysis. Ponemon Institute Research Report, Published June 2016

PwC (2016) Global economic crime survey 2016. http://www.pwc.com/gx/en/economic-crime-survey/pdf/GlobalEconomicCrimeSurvey2016.pdf. Accessed 29 June 2016

Rinaldi S, Peerenboom J, Kelly T (2001) Identifying, understanding, and analyzing critical infrastructure interdependencies. IEEE Control Syst Mag 21(6):11–25

Roege P, Hope T, Delaney P (2014) Resilience: modeling for conditions of uncertainty and change. MODSIM World 2014, Newport News, VA, 15–17 April 2014, paper no. MS1476. http://daviescon.com/wp-content/uploads/2012/08/Final-Energy-Resilience-MODSIM-2014--Paper_14-Mar-14.pdf

Savage K, Coogan P, Lau H (2015) The evolution of ransomware. Symantec. http://www.symantec.com/content/en/us/enterprise/media/security_response/whitepapers/the-evolution-of-ransomware.pdf

Shannon CE (1948). A mathematical theory of communication. Bell Syst Tech J 27(3):379–423

Shannon CE (1949) Communication theory of secrecy systems. Bell Syst Tech J 28(4):656–715

Shakarian P, Lei H, Lindelauf R (2014) Power grid defense against malicious cascading failure. Presented at 13th international conference of autonomous agnets and multiagent systems, Paris, France, 5–9 May 2014, arXiv:1401.1086

Simmons GJ (1985, April). The practice of authentication. In: Workshop on the theory and application of of cryptographic techniques (pp. 261–272). Springer, Berlin/Heidelberg

Smirnov A, Kashevnik A, Shilov N, Makklya A, Gusikhin O (2013, November) Context-aware service composition in cyber physical human system for transportation safety. In: ITS Telecommunications (ITST), 2013 13th international conference on (pp 139–144). IEEE

Sridhar S, Hahn A, Govindarasu M (2012) Cyber–physical system security for the electric power grid. Proc IEEE 100(1):210–224

Stouffer K, Falco J, Scarfone K (2011) Guide to Industrial Control Systems (ICS) security. Special Publication 800–82. National Institute of Standards, Gaithersburg. http://csrc.nist.gov/publications/nistpubs/800-82/SP800-82-final.pdf

Teng K, Thekdi SA, Lambert JH (2012) Identification and evaluation of priorities in the business process of a risk or safety organization. Reliab Eng Syst Saf 99:74–86

Teng K, Thekdi SA, Lambert JH (2013) Risk and safety program performance evaluation and business process modeling. IEEE Transac Syst Man Cybernetics Part A 42(6):1504–1513

Thorisson H, Lambert JH, Cardenas JJ, Linkov I (2016) Resilience analytics for power grid capacity expansion in a developing region. To appear in Risk Analysis

Tierney K, Bruneau M (2007) Conceptualizing and measuring resilience: A key to disaster loss reduction. TR News 250:14–17

Tversky A, Kahneman D (1973) Availability: a heuristic for judging frequency and probability. Cogn Psychol 5(2):207–232

US Department of Energy (2002) 21 steps to improve cyber security of SCADA networks. US Department of Energy, Washington, DC. http://energy.gov/sites/prod/files/oeprod/DocumentsandMedia/21_Steps_-_SCADA.pdf

US Department of Energy (2014) Electricity Subsector Cybersecurity Capability Maturity Model (ES-C2M2). Version 1.1. http://energy.gov/sites/prod/files/2014/02/f7/ES-C2M2-v1-1-Feb2014.pdf

US Department of Homeland Security (2016a) National Planning Frameworks web site: http://www.fema.gov/national-planning-frameworks

US Department of Homeland Security (2016b) Cyber Resilience Review (CRR): method description and self-assessment user guide. US Department of Homeland Security. https://www.us-cert.gov/sites/default/files/c3vp/csc-crr-method-description-and-user-guide.pdf

US White House (2013) Presidential Policy Directive 21 – critical infrastructure security and resilience. https://www.whitehouse.gov/the-press-office/2013/02/12/presidential-policy-directive-critical-infrastructure-security-and-resil

US White House (2016) Presidential Policy Direction 41 – United States cyber incident coordination. https://www.whitehouse.gov/the-press-office/2016/07/26/presidential-policy-directive-united-states-cyber-incident

Veitch CK, Henry JM, Richardson BT, Hart DH (2013) Microgrid cyber security reference architecture, Version 1.0. SAND2013-5472. Sandia National Laboratories, Albuquerque, New Mexico

Woods DD (2012) Chapter 9: Resilience and the ability to anticipate. In: Pariès MJ, Wreathall MJ, Woods DD, Hollnagel E (eds) Resilience engineering in practice: a guidebook. Ashgate Publishing Ltd, Farnham

World Economic Forum (2015) Partnering for cyber resilience: towards the quantification of cyber threats. http://www3.weforum.org/docs/WEFUSA_QuantificationofCyberThreats_Report2015.pdf

Xiong G, Zhu F, Liu X, Dong X, Huang W, Chen S, Zhao K (2015) Cyber-physical-social system in intelligent transportation. IEEE/CAA J Automat Sin 2(3):320–333

Young W, Leveson NG (2014) An integrated approach to safety and security based on systems theory. Commun ACM 57(2):31–35

Chapter 15
Cyber-Transportation Resilience. Context and Methodological Framework

Maria Nogal and Alan O'Connor

Abstract Cyber systems are gaining relevance in the Transportation field, to the point of being embedded in every level of traditional transportation systems. This new paradigm, defined as cyber-transportation, represents an opportunity for improving system performances in mobility, safety and environment; nevertheless, it also opens a door to new threats and risks. In this chapter, the interactions, the dependences and the synergies created by the cyber and the transportation systems are analysed under the perspective of resilience analysis.

15.1 Introduction

Traditionally, research on Transportation Systems (TS) used to involve infrastructure, vehicles, travellers and operators, and their interactions. Later on, advances in telecommunication allowed the implementation of the first Intelligent Transportation Systems (ITS) in the second quarter of the twentieth century. At present, with the introduction of more intelligent and interactive operations, the concept of ITS has shifted to the concept of Cyber-Transportation Systems (CTS). The most important aspect distinguishing ITS from the CTS is that the first focuses on the transportation aspects, using technology as a tool, meanwhile a CTS perspective involves consideration of the interactions, the dependences and the synergies in both directions.

The relationship between the Cyber Systems (CS) and the physical environment is already highlighted by Suo et al. (2012), who point out the recent evolution of machine-to-machine (M2M) interactions by the introduction of more intelligent and interactive operations, under the architecture of the internet of things (IoT) resulting in cyber-physical systems. The term M2M refers to the communications between computers, embedded processors, smart sensors, actuators, and mobile devices without or with limited human intervention (Chen et al. 2012), and IoT refers to uniquely identifiable objects, things, and their virtual representations in an internet-like structure (Weber 2010).

M. Nogal (✉) • A. O'Connor
Trinity College, Dublin, Ireland
e-mail: nogalm@tcd.ie

© Springer Science+Business Media B.V. 2017

I. Linkov, J.M. Palma-Oliveira (eds.), *Resilience and Risk*, NATO Science for Peace and Security Series C: Environmental Security,
DOI 10.1007/978-94-024-1123-2_15

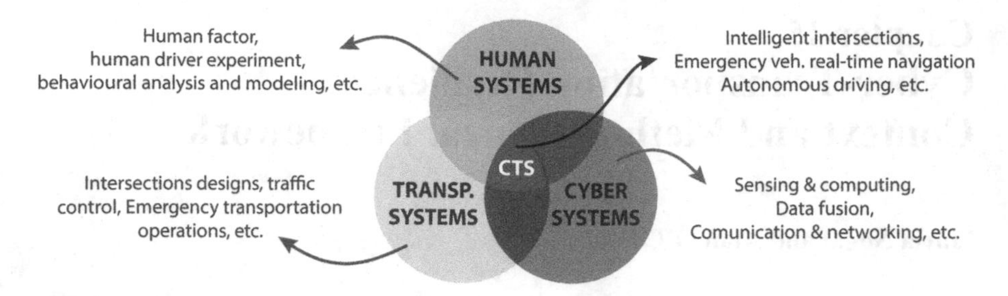

Fig. 15.1 The interactions among the cyber system, transportation system and human element (Qiao et al. 2010)

Accordingly, CTS can be considered as a specific case of the cyber-physical systems (Sadek et al. 2016), that is, systems with tight coupling between computational and physical resources. Nevertheless, CTS should be understood as the complex system resulting from the tight interactions between the cyber, the physical, and also the human world. It is noted that TS already includes the human factor, being travellers and operators as essential elements of the overall system. Figure 15.1 depicts these relations, presented in Qiao et al. (2010).

The potential implications of this evolution are complex and profound. It is undeniable that the CTS represent an opportunity for improving system performances in mobility, safety, environment, and resilience. However, they also imply new risks and threats. Traditionally the maximum concern of the TS has been the physical threats; however, with the introduction of the CS into the TS, the cyber-threats are gaining attention. For instance, the intended or unintended failure of the control system of the metro line can result in a disaster of considerable magnitude; trains might crash, causing an explosion, which in turn could result in the structural failure of tunnels and devastating impacts for potentially-crowded stations. The physical consequences of the cyber breaches are innumerable.

Studies on resilience considering the CS and the TS as a whole system are very uncommon, and to the best of the authors' knowledge, there is no scientific literature concerning cyber-transportation resilience. This might be due to the novelty of the topic, the multidisciplinary character required, and the difficulty of dealing with the interactions of such complex systems. In this sense, this chapter presents a step forward towards the study of the resilience of CTS based on the analysis of the individual transport and cyber systems, and the evaluation of their interactions to determine the vulnerabilities and redundancies of the CTS.

This chapter is organised as follows; Sect. 15.2 presents the main functions, vulnerabilities and security requirements of the TS and CS, independently. CTS is also introduced based on the interaction between TS and CS, discussing their dependences and their synergies. Section 15.3 discusses the resilience of interconnected systems, presenting an innovative approach to address systems with complex interconnections. In Sect. 15.4, the previous framework is applied to the case of CTS. Finally, conclusions are drawn in Sect. 15.5.

15.2 Cyber-Transportation Systems. Dependences and Synergies

15.2.1 Transportation Systems

The function of the TS is to enable the movement of people and goods, and to guarantee the supply chain, in terms of safety and security at the basic level, in terms of reliability and sustainability at a medium level, and in terms of efficiency at a highest level. The classification of the system functions at different levels is important from a resilient point of view. When the system is exposed to any disruption, a resilient system should minimize the consequences, maintaining the adequate level of performance according to the current situation.

Transportation Systems encompass three components, namely, hard infrastructure, vehicles for freight and passengers, and operations of travellers, traffic management and logistics.

The hard infrastructure includes network infrastructure (road, railways, etc.) and components such as signals and catenaries. The network infrastructure, which is the essential asset of the TS, is characterised by their physical continuity. On the other hand, traffic components, despite playing a secondary role in most of the cases, provide the system with a larger level of efficiency. For instance, self-explaining roads (Tingvall and Howarth 1999) look for designs consistent with the function of the road, through features such as width of carriageway, road markings, etc., in order for drivers to naturally adopt the adequate behaviour, which results in less traffic control devices such as additional traffic signs to regulate traffic behaviour. Traffic components are numerous and usually present with a high level of redundancy. The hard infrastructure is sensitive to physical damage and ageing. The loss of continuity of the network infrastructure represents the main risk given that it would affect the basic functions; meanwhile the consequences of damage or malfunction of one or several components might affect less important functions.

Vehicles for freights and passengers, in addition to the hard infrastructure, make up the physical domain of the TS. They are elements without a fixed position, numerous and discontinuous, which provide the TS with flexibility to adapt to and recover from disruptive situations. On the other hand, they can be the cause of disruptive events, such as traffic jams and accidents affecting other vehicles and even the infrastructure. They are also subjected to the obsolescence and very sensitive to the environmental conditions.

The last element, i.e. travellers and operators, traffic management and logistics, constitutes the domains of information, cognitive and social. Users and operators are characterized by their capacity of acting individually, that is, *stochastically*. Nevertheless, when looking at the average behaviour, group dynamics can be identified. A universal background on the traffic dynamics exists in each user based on principles such as caution, imitation and other world-wide basic rules (e.g. a red colour or a cross implies to stop). In general, traffic management and logistics aim

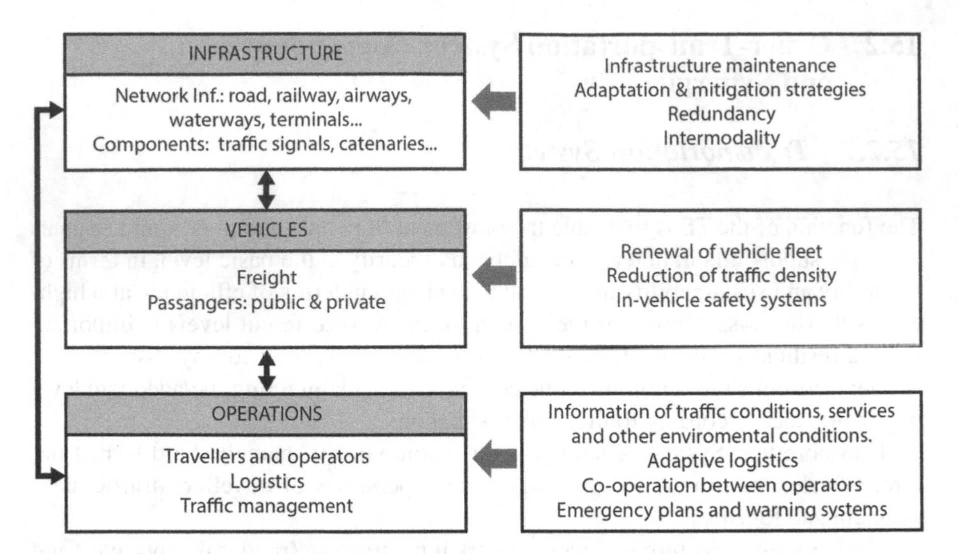

Fig. 15.2 TS structure and security requirements

at reducing the random behaviour of users and operators to improve the traffic performance, by means of the information management. They are a very sensitive component when dealing with the TS resilience, as they represents the main potential risk and, at the same time, the main capability of the system.

The security requirements to minimize the aforementioned risks are presented in Fig. 15.2. A further analysis of the resilience of the system is given in the next section.

15.2.2 Cyber Systems

The main function of the CS is to provide personalized services based on information flow, in terms of confidentiality, integrity and availability at the basic level, in terms of accountability, auditability, authenticity/trustworthiness, non-repudiation and privacy, at a medium level, and in terms of efficiency at a highest level. The degree to which each of these properties is considered in one or other level varies from one application to another. The reader is referred to Cherdantseva and Hilton (2013) for a detailed discussion of the previous terms.

Based on Yang et al. (2010), the IoT, as the structure that underpins the CS, can be divided into four layers, namely, perceptual, network, support, and application layers. The perceptual layer links the physical and the cyber world through the physical equipment. This layer is characterised by low computing power and storage capacity, therefore setting up security protection systems becomes very difficult. The network layer transmits information from the perceptual layer, conducting initial processing, classification and polymerization of information. Computer viruses and the collapse due to data overload represent the main risks of this layer, though

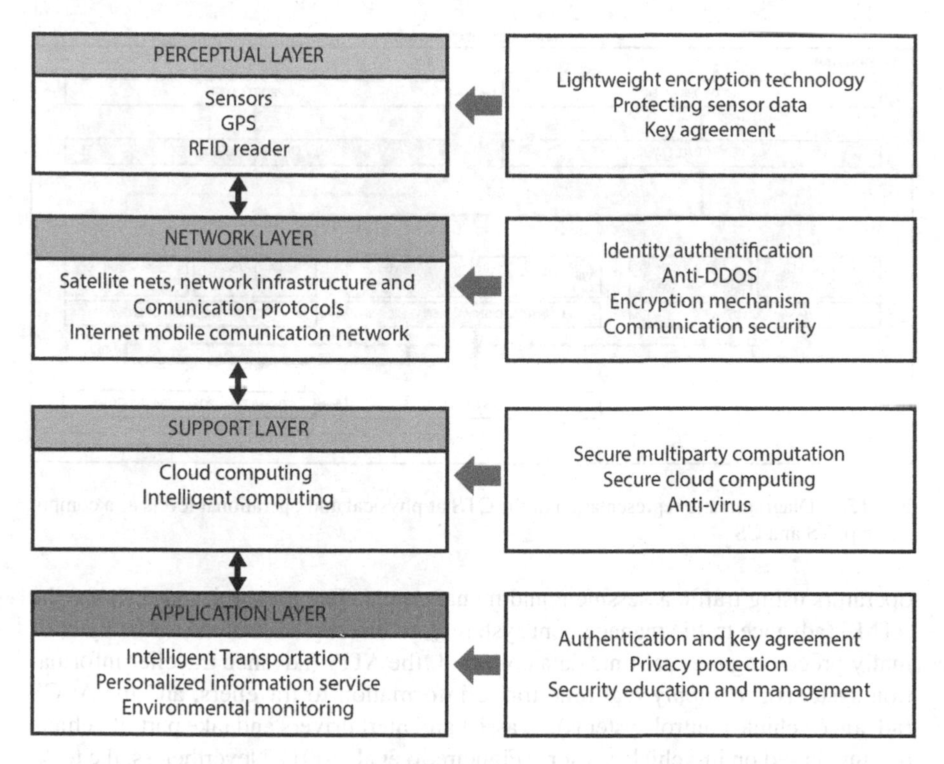

Fig. 15.3 IoT architecture (Yang et al. 2010) and security requirements (Suo et al. 2012)

Distributed Denial of Service (DDoS) attack is also very common. The support layer evolves the intelligent computing (network-grid based or cloud based), which is potentially exposed to malicious information. Finally, the application layer links with users and the physical world to provide specific services. In this case, problems of data privacy due to data sharing, access control and disclosure of information are the main concerns. In addition, the hardware (sensors, satellites, data center hardware, etc.) can suffer from physical damage, resulting in a complete interruption of the information flow.

Suo et al. (2012) presents some security requirements to minimize the aforementioned risks (see Fig. 15.3).

15.2.3 Cyber Transportation Systems

Cyber Transportation System implies a multi-disciplinary approach which includes cyber technologies, transportation engineering and human factors.

CS started penetrating into the Transportation field via Intelligent Transportation Systems (ITS), which used synergistic technologies and systems engineering concepts for a more efficient and safer mobility. The ITS aimed at assisting human

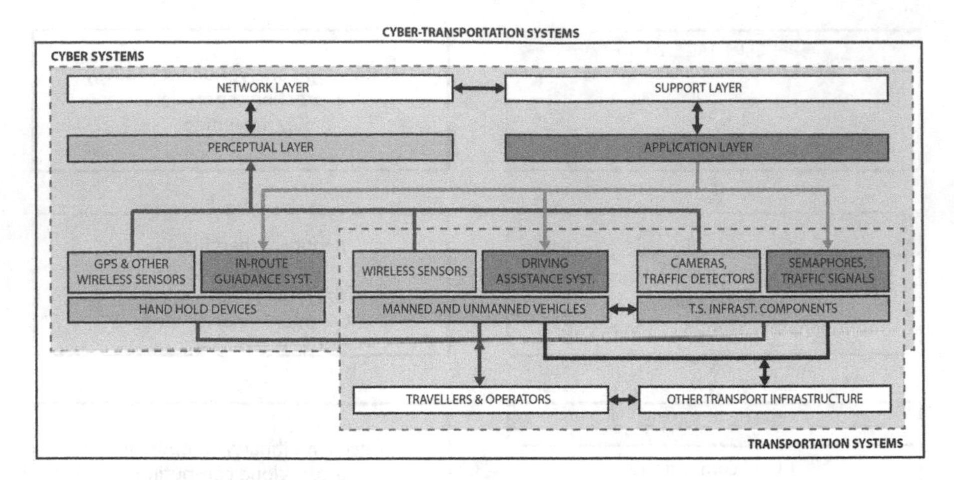

Fig. 15.4 Diagrammatic representation of the CTS at physical and operational levels as a composition of TS and CS

operators using traffic assessment and management. The most relevant ITS are the ATMS (advance traffic management systems), to directly control the traffic by optimally processing the real-time data collected, the ATIS (advance traveller information systems), to supply real time traffic information to travellers, and the AVCS (advance vehicle control systems), to assist and alert drivers and take part of vehicle driving, based on in-vehicle sensors (Figueiredo et al. 2001). Nevertheless, the technology has gone far beyond, and nowadays CS are present within the TS in (a) named and autonomous vehicles, in the driving assistance systems and other wireless sensors; in (b) transportation system infrastructure components, including automatic protection systems, computer controlled semaphores and traffic signals, cameras and traffic detectors, etc.; and in (c) hand-held smart devices providing in-route guidance systems, GPS, etc.

The interaction between both systems is represented in Fig. 15.4. The CS obtains information on the traffic conditions from wireless sensors, traffic detectors, cameras and other devices. Then, the data set is computed and returned to provide processed information, which is enriched with other external data sources, such as the current environmental conditions and the traffic forecasting. As a result of this synergistic interaction, the TS improves the performance of all their functions and the CS in turn services other systems (business, health services, etc.) with the acquired information. The proposed interaction scheme is valid for every transport system, namely, aviation, railway, roads and shipping.

One of the most relevant examples of the synergy created between CS and TS is seen in the intermodality. CS enable the geo-location of vehicles and goods, the updated characteristics of passengers and freight, and the available transport capacity of different modes, among others. The management of this information results in smoother switching between transport modes and allows adaptive logistics to complex, uncertain and ever-changing situations.

The other side of the coin, as indicated above, is that this interaction between systems implies new threats and risks. Thus the list of possible (physical) hazards affecting the TS has to be completed with cyber-threats, and how these risks potentiate each other.

15.3 Resilience of Interconected Systems

The advantages of a resilience-based approach instead of a risk-based perspective are widely discussed throughout this book. In fact, assessment and improvement of resilience are becoming increasingly common in many social and engineering fields.

According to Vugrin and Turgeon (2013), the approaches evaluating cyber resilience can be classified into two types, i.e., design methods, which aim at improving the system architecture and activities to cyber-attacks (e.g. Norman et al. 2016); and operational methods, which include business operations into the analysis, allowing, in addition, the consideration of physical threats (e.g. Linkov et al. 2013a). The qualitative and semi-qualitative outputs of these assessments provide an identification of the system weaknesses and the resilience enhancement opportunities, rather than a result that allows the objective prioritization of use of scarce resources and the validation of the effect of investments.

Similarly, the resilience of physical systems, and therefore of the TS, can be assessed by indices that rely on subjective assessments (e.g., diversity or adaptability, proposed by Murray-Tuite (2006)), or by indicators (e.g., reliability, proposed by Ip and Wang (2009)) that quantify system attributes, which are assumed to be related to the resilience of the system. Moreover, in the case of physical systems, quantitative methods have been also developed. The so-called performance-based methods quantify resilience by measuring the performance of a system in a particular disturbing scenario (e.g., Nogal et al. 2016, 2017). These last methods, though they present the advantage of proving a framework to objectively compare different cases, are usually less holistic approaches.

In research and engineering literature, studies on resilience considering the CS and the TS as a whole system are uncommon, and to the best of the authors' knowledge, there are none concerning CTS.

When analysing a system it is important to initially define the context in terms of time and scale. Under the perspective of a functional approach, the resilience evaluation implies to identify the system functions at different stages, that is, the required functions in a pre-disaster stage might be different from those in a post-disaster stage. Given the fractal nature of the systems, they can in turn be split into subsystems, on which the same functional analysis can be conducted. Therefore, every system k has a resilience associated with a specific time and scale, $R_k(t)$.

In order to isolate the system k from the rest, neglecting their interactions, an ideal performance of the other connected systems can be assumed, obtaining the

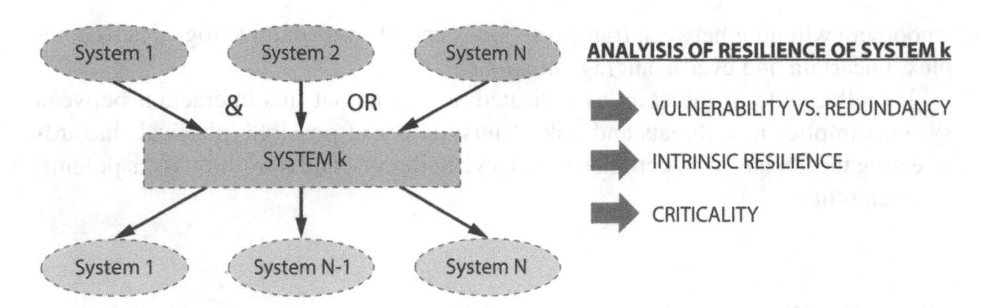

Fig. 15.5 Analysis of resilience of a system at a given scale considering its relation with other systems

intrinsic resilience the system k , IR_k. Nevertheless, given that (sub-) systems are functionally related between each other, their inter-relations must be considered.

Figure 15.5 shows a diagrammatic representation of the analysis of the resilience of the system k at a given scale and stage, considering its relation with other systems. System k functionally depends on a set of systems (upper row). These dependencies can be of two types, "AND" and "OR". The larger the number of dependences of type "AND", the more vulnerable, and less resilient the system is. On the contrary, a large number of dependences of type "OR" implies a high redundancy, resulting in a more resilient system. Finally, the criticality of system k can be determined by the number of systems depending of it.

Mathematically this can be expressed as follows;

$$R_k(t) = f\left(IR_k(t), V_k(t)\right), \tag{15.1}$$

where $V_k(t)$ denotes the vulnerability derived from the dependencies of the system k with other systems at a given time or stage t.

Considering Eq. (15.1), the following clarifications should be made; (a) the term vulnerability will have a negative impact upon $R_k(t)$; (b) the term redundancy is not included, as it is already considered in IR_k, with a positive impact, by assuming the ideal performance of the interconnected systems; and (c) criticality is not included in the evaluation of resilience, however its analysis is relevant in order to design strategies and prioritize resources.

It is important to highlight that the interactions with other systems do not necessarily make the system weaker by adding vulnerability. Figure 15.5 shows how a good design of the dependence structure may result in a more resilient system.

15.4 Cyber Transportation Resilience

For the sake of clarification, CTS resilience is analysed according to the framework presented in the previous section. In this case, the scale of the system selected is a traffic network of a small/medium city. This system includes the road-based transportation and the tram, autonomous and manned vehicles, electric and non-electric vehicles, users, managers, and operators. Table 15.1 shows the list of elements included in this example. All those sub-systems not considered within this scale should be introduced in the analysis of dependences. In this case, the CS, power systems (PS) and other transportation systems (OTS), such as train and marine transport, are considered as external related networks.

Table 15.1 shows the analysis of dependences of the defined system, for the physical domain. It is noted that similar analysis should be conducted for other domains, such as the cognitive, the social, etc. The interactions and dependences should be in accordance with the domain considered.

As indicated before, under a resilient perspective, the optimal system performance depends on the stage analysed. In other words, systems should guarantee different functions according to the exhausting reserves of the system. This implies that the structure of dependence will likely change through the different stages, which shows the importance of considering the context when analysing the resilience of the system.

The following comment regarding Table 15.1 can facilitate its understanding;

1. There exists iteration between the hard infrastructure and OTS, materialized by junctions and drawbridges. This interaction can be understood as a potential risk but not as vulnerability, given that any physical dependence does not exist, that is, the elimination of the elements conforming the physical domain of OTS would not affect the performance of any of the functions of the system analysed.
2. Semaphores, cameras, traffic sensors might be physically independent of the PS because it can be replaced by solar panels, autonomous battery, etc.
3. The tram is independent of the PS, given that this interaction has already been introduced in the overhead lines.

Table 15.1 shows how the CS represents a clear vulnerability in the physical domain during the disaster stage, with the autonomous vehicles causing this dependence. Given that autonomous vehicles are hardly introduced in the present traffic networks, there is still room to figure out solutions to reduce this physical dependence, such as mechanisms to man these vehicles, independent of the CS.

The analysis of the vulnerability associated with the dependence to other systems, $V_k(t)$, according to Eq. (15.1) is given by Table 15.1. The intrinsic resilience, $IR_k(t)$, should be studied by assuming the perfect performance of the PS, CS and OTS, using tools such as the Resilience Matrix proposed by Linkov et al. (2013b).

Table 15.1 Study of dependences of specific CTS

SCALE: Traffic network of a given small/medium city					
DOMAIN: Physical					
STAGE:			Pre-disaster	During disaster	Post-disaster
FUNCTION: Movement of people and animals, supply chain…			In terms of efficiency	In terms of safety & security	In terms of reliability & sustainability
Elements		**External interactions**	**External dependences**	**External dependences**	**External dependences**
Hard infr.; (a) Network infrastructure	Roads, intersections, bridges, tunnels				
	Tram railway				
	Overhead lines	PS	PS	PS	PS
	Junctions	OTS			
	Bicycle lane				
	Drawbridges	CS, PS, OTS	CS, PS		PS
Hard infr.; (b) Components	Traffic signs				
	Lighting systems	CS, PS	CS, PS		PS
	Semaphores	CS, PS	CS, (PS)		CS, (PS)
	Cameras	CS, PS	CS, (PS)		
	Traffic sensor	CS, PS	CS, (PS)		
	Electric veh. Charging depots	CS, PS	CS, PS	PS	PS
Vehicles	Non-electric manned (cars, trucks, bicycles)	CS	CS		
	Electric manned vehicles	CS	CS		
	Autonomous vehicles	CS	CS	CS	CS
	Bus	CS	CS		
	Tram	CS	CS		
Users and operators	Operators	CS	CS		
	Travellers	CS	CS		
	Pedestrians	CS	CS		
	Traffic controllers	CS	CS		

PS Power systems, *CS* Cyber systems, *OTS* Other transportation systems

15.5 Conclusions

Given that the inclusion of CS into TS has been steadily increasing over the years, it is necessary to be aware of the consequences of a poorly conceived design plan. The interactions among systems should be set up in a manner that they generate redundancies, increasing the resilience, instead of vulnerabilities. CTS need to be designed with resilience in mind right from the start.

The study of the structure of dependences allows a general picture of the resilience of complex systems to be drawn. These interdependences will depend on the stage, the scale and of course, the domain. With the aim of allowing the comprehensive perspective required, this chapter has presented an operational framework to analyse these interrelations in depth.

As an example, in the physical domain, autonomous vehicles represent a clear vulnerability in terms of safety and security for a CTS of a small/medium city. Solutions to reduce the physical dependence of the vehicles to the CS should be considered in the design of the autonomous vehicles.

Acknowledgements The research leading to these results has received funding from the European Union's Horizon 2020 Research and Innovation Programme, under Grant Agreement no 653260. The authors are also grateful for the valuable discussion and insight gained from the participants of the Cyber Working Group during the NATO Advanced Research Workshop "Resilience-Based Approaches to Critical Infrastructure Safeguarding", held in Ponta Delgada, Azores, Portugal, 26–29 June, 2016.

References

Chen M, Wan J, Li F (2012) Machine-to-machine communications: architectures, standards and applications. KSII Trans Internet Inform Syst, 6(2):480–497

Cherdantseva Y, Hilton J (2013) A reference model of information assurance & security. In *Availability, reliability and security (Ares), 2013 eighth international conference* on (pp 546–555). IEEE

Figueiredo L, Jesus I, Machado JT, Ferreira J, de Carvalho JM (2001, August) Towards the development of intelligent transportation systems. In *Intelligent transportation systems* (Vol. 88, pp 1206–1211)

Ip W, Wang D (2009) Resilience evaluation approach of transportation networks. *Computational sciences and optimization, 2009*. CSO 2009. International joint conference on, Vol. 2, pp 618–622

Linkov I, Eisenberg DA, Plourde K, Seager TP, Allen J, Kott A (2013a) Resilience metrics for cyber systems. Environ Syst Decisions 33(4):471–476

Linkov I, Eisenberg DA, Bates ME, Chang D, Convertino M, Allen JH, Flynn SE, Seager TP (2013b) Measurable resilience for actionable policy. Environ Sci Technol 47(18):10108–10110

Murray-Tuite P (2006) A comparison of transportation network resilience under simulated system optimum and user equilibrium conditions. *Proceedings of the Winter Simulation Conference, WSC 06*, 1398–1405

Nogal M, O'Connor A, Caulfield B, Martinez-Pastor B (2016) Resilience of traffic networks: From perturbation to recovery via a dynamic restricted equilibrium model. Reliab Eng Syst Saf 156:84–96

Nogal M, O'Connor A, Martinez-Pastor B, Caulfield B (2017) Novel probabilistic resilience assessment framework of transportation networks against extreme weather events. ASCE-ASME J Risk Uncertain Eng Syst Part A Civil Eng 3(3):04017004

Norman S, Chase J, Goodwin D, Freeman B, Boyle V, Eckman R (2016) A condensed approach to the cyber resilient design space. Insight 19(2):43–46

Qiao C, Sadek AW, Hulme K, Wu S (2010) Addressing design and human factors challenges in cyber-transportation systems with an integrated traffic-driving-networking simulator. In Workshop, October (Vol. 28, p 29)

Sadek AW, "Brian" Park B, Cetin M (2016) Special issue on cyber transportation systems and connected vehicle research. J Intell Transp Syst 20(1):1–3

Suo H, Wan J, Zou C, Liu J (2012) Security in the internet of things: a review. In *Computer science and electronics engineering (ICCSEE)*, 2012 international conference on (Vol. 3, pp 648–651). IEEE

Tingvall C, Howarth N (1999) Vision Zero: an ethical approach to safety and mobility. *The 6th institute of transport engineers international conference on road safety and traffic enforcement:* Beyond 2000. Melbourne

Vugrin ED, Turgeon J (2013) Advancing cyber resilience analysis with performance-based metrics from infrastructure assessments. Int J Secure Software Eng 4(1):75–96

Weber RH (2010) Internet of things – new security and privacy challenges. Comput Law Secur Rev 26:23–30

Yang G, Xu J, Chen W, Qi ZH, Wang HY (2010) Security characteristic and technology in the internet of things. Nanjing Youdian Daxue Xuebao (Ziran Kexue Ban)/*Journal of Nanjing University of Posts and Telecommunications* (Natural Nanjing University of Posts and Telecommunications), 30(4)

Chapter 16
Resilience and Fault Tolerance in Electrical Engineering

Niels P. Zussblatt, Alexander A. Ganin, Sabrina Larkin, Lance Fiondella, and Igor Linkov

Abstract As a result of the increased importance of engineered electrical systems to modern civilization, it is necessary to design systems that sustain ideal levels of performance despite the potential for internal faults and external attacks. Designing systems that exhibit resilience, also known as fault tolerance, is the primary method by which optimal performance is preserved despite adverse conditions. This paper is a review of a variety of computational and electromechanical fault tolerance techniques from the literature in order to evaluate the state of the art and identify potential areas for improvement. Our findings suggest that the existing literature has only focused on a limited number of resilience challenges, and that no single resilience-enhancing solution, either hardware- or software-based, is capable of addressing all of the major types of possible faults. Further, we classify the papers using the resilience matrix, which combines four resilience phases put forth by the National Academy of Sciences and four Network Centric Warfare domains.

N.P. Zussblatt
Environmental Laboratory, US Army Engineer Research & Development Center, Concord, MA, USA

Department of Chemical Engineering, University of California, Santa Barbara, CA, USA

A.A. Ganin
Department of Systems and Information Engineering, University of Virginia, Charlottesville, VA, USA

S. Larkin
Environmental Laboratory, US Army Engineer Research & Development Center, Concord, MA, USA

L. Fiondella
Department of Electrical & Computer Engineering, University of Massachusetts, Dartmouth, MA, USA

I. Linkov (✉)
US Army Corps of Engineers Research and Development Center, Concord, MA, USA
e-mail: Igor.Linkov@usace.army.mil

© Springer Science+Business Media B.V. 2017

I. Linkov, J.M. Palma-Oliveira (eds.), *Resilience and Risk*, NATO Science for Peace and Security Series C: Environmental Security,
DOI 10.1007/978-94-024-1123-2_16

We identify the matrix components insufficiently addressed: particularly, we have found no relevant literature on the cognitive and social domains. Even within the parts of the resilience matrix that have received attention in the literature to date, we observe that there is relatively less emphasis placed on the adaptation of the computational and electromechanical systems so that a repeated fault will not incur significant disruption in subsequent occurrences. Therefore, based on this review, we find that while significant and sustained attention has been dedicated to enhance the resilience of engineering electrical systems, substantial work remains to fully address resilience challenges that instill confidence in our ability to engineer resilient systems.

Keywords Resilience • Electrical engineering • Fault tolerance • Risk • Safety by design

16.1 Introduction

Computers and engineered electrical systems have become ubiquitous in the modern world. Moreover, society's increased dependence on computers in both critical and everyday applications necessitates that continuous operation of computational resources be preserved despite the presence of external threats. Of particular concern are the Critical Infrastructure Sectors identified by the United States Department of Homeland Security, which include chemical facilities, financial service institutions, and transportation systems, (Presidential Policy Directive n.d.; Department of Homeland Security n.d.) many of which are heavily reliant on computerized or electrical systems. Despite increased attention to threats directed against computerized systems, these critical systems are not fully protected from compromises that could result from accidents or deliberate malevolent acts. For example, in recent years, the aviation industry has suffered a series of debilitating incidents resulting from failure of computerized systems. In 2014, an air traffic control system in the southwestern United States suffered a failure when a single aircraft with a complex flight path overwhelmed the memory of the computers (Scott and Menn 2014). On another occasion, Delta Air Lines experienced a collapse of its flight management and passenger reservation systems when a piece of electrical equipment failed and the automatic backup systems failed to engage (Stelloh and Gutierrez 2016). As indicated by these examples, the design of computer systems that are capable of maintaining operation through faults is of the utmost importance.

Resilience is defined as the ability to anticipate and adapt to changing conditions as well as to withstand and recover rapidly from disruptions (Presidential Policy Directive n.d.). As a result of these varied requirements of resilience, four stages of resilience goals have been described: anticipate, withstand, recover, and evolve (Bodeau and Graubart 2011). Alternatively, these goals have been labeled by the National Academy of Science (NAS) as plan/prepare, absorb, recover, and adapt (Resilience 2012). A truly resilient system will have mechanisms in place to address each of these goals. Additionally, a complex system will have several distinct domains whose preparedness against adverse operation will need to be considered. The common method of dividing a system into domains, originally described by the U.S. Department of Defense Network Centric Warfare (NCW) is to consider its physical, information, cognitive, and social components (Alberts 2002). Here, the physical domain refers to the physical components and capabilities of the system, the information domain refers to the information and data within the physical domain, the cognitive domain refers to the use of other domains for decision-making processes, and the social domain refers to the robustness of the organizational structure and the ability of the system to communicate information (Alberts 2002).

As a result of these distinct perspectives on the problem of evaluating resilience, a unified method to characterize the preparedness of a system was desired. To meet this need, Linkov, et al. introduced the concept of the "resilience matrix" where the four stages of resilience defined by the NAS and the four NCW resilience domains are placed on separate axes to generate a 4x4 matrix (Linkov et al. 2013). A system with robust safeguards where all elements of the resultant matrix have been addressed can be considered to be highly resilient. In contrast, a lack of attention to one or more elements in the resilience matrix would indicate a point of vulnerability, which may be used to direct attention to improve the security of the system as a whole. Recent publications have evaluated the state of resilience in a variety of fields, including cyber security (DiMase et al. 2015) and the energy sector (Roege et al. 2014). The concept of resilience is not foreign to computer scientists, who often know it by the term "fault tolerance." Regardless of the term used, resilience is an important attribute that must be implemented in electrical engineering systems and circuits so that they can provide stable service even in the face of errors. The importance of designing computers to exhibit resilience was first described by Avižienis in 1967, (Avižienis 1967) but to the best of our knowledge there does not exist a review of the field according to modern multi-criteria resilience principles. In this work, we examine the literature on resilience in electromechanical and computational systems according to the concept of the resilience matrix. In addition to identifying how the field of computer science has addressed resiliency (and failed to do so with respect to parts of the resilience matrix), the scope of robustness challenges examined within the field and the general methods employed are also examined.

16.2 Materials and Methods

In this work, we reviewed 61 papers that discussed resilience or fault tolerance within a computational system or from an electromechanical engineering standpoint and evaluated the degree to which strategies to assess and enhance resilience were articulated. Papers were identified by Google Scholar search in October 2014 for "electrical engineering", "resilience", "robustness", and "fault tolerance," and then sorted manually for relevance to the field of resilience in electrical engineering. Relevance was determined by interpretation of the abstract and the main text, with effort made to ensure selection of papers for review was performed without bias. The review was not strictly exhaustive and it is possible that certain relevant papers were left out of its scope for reasons such as a different terminology, low ranking in the search results, and our misinterpretation of their abstracts. In addition, certain older works (such as Avižienis 1967) have been included as well if they were highly cited or identified as having made a lasting impact in later work.

The final count of 61 papers published between 1967 and 2014 was primarily from within computer science journals and conference proceedings, although some were found in other disciplines such as aerospace engineering (Chen and Trachtenberg 1991; Alena et al. 2008, 2011), where robustness of computer systems is also considered a priority. While most of the papers focused on the robustness of computer architecture (e.g., logic gates) and internal memory or data to faults or corruption, some of the works extended their scope to tolerating faults of internal mechanical components or mechanical systems controlled by the computer systems as well (Pradeep et al. 1988; Maciejewski 1990). Hence, we consider this to be a review of fault tolerance in both computational and electromechanical systems.

Once identified, papers were divided according to the type of problem they sought to guard against and the general methodology of their proposed solution(s). In each paper, the type of failure that was to be guarded against was noted. To make this work accessible to a general audience, the types of problems were grouped into a small number of distinct categories, including manufacturing variation and external malicious attacks. Following this, the method of the solution proposed in each work was identified, whether it required hardware alterations, changes to software or internal coding, or a combination of both. Finally, each paper was evaluated according to the extent to which the solutions they proposed addressed the National Academy of Science and Network Centric Warfare components of resilience. The complete list of papers examined and their assignments according to the resilience problem(s) addressed, method of solution (hardware, software, or combination), and the resilience phases and domains that were considered can be found in Tables 16.1, 16.2, and 16.3, respectively, which are placed after the conclusion of this chapter.

16.3 Results and Discussion

To appreciate the kinds of resilience of concern to the computational and electromechanical engineering research communities, the papers examined for this work were organized according to the general types of problems they considered. The percentage of papers addressing each general type of problem is given in Fig. 16.1, and the specific assignment of each paper examined to problem types is provided in Table 16.1. Of the types of failure examined, single event upsets, which are the flipping of a single bit of computer memory, primarily induced by cosmic rays or other radiation (Ziegler and Lanford 1979), was the most common problem that resilience considerations sought to guard against. Following single event upsets, failure of circuits or mechanical components such as logic gates stuck in the "on" or "off" state and voltage decreases that could result in failure to convert a bit state were the next most common impediment to achieving a resilient computer system. Another

Table 16.1 Resilience problems and solution types in reviewed publications

Author/s (Year)	Identified problem						
	Voltage errors or droops	Timing errors	Manufact-uring variation	Failure of circuits and mechanical components	Single event upsets	Human-machine inter-action faults	Malicious attacks
Agarwal et al. (2007)		x		x			
Alena et al. (2008)				x		x	
Alena et al. (2011)				x	x		
Avižienis (1967)	x			x	x		
Avižienis (1997)			x	x	x	x	x
Banerjee et al. (2007)	x		x				
Bartlett and Spainhower (2004)				x		x	
Bau et al. (2009)			x	x	x		
Bowman et al. (2009a)	x	x					
Bowman et al. (2009b)	x	x					
Bowman et al. (2011)	x	x					
Breuer (2005)		x	x				
Brunina et al. (2012)			x	x	x		
Chakrapani et al. (2006)			x				
Chen and Trachtenberg (1991)				x	x		
Chippa et al. (2010)	x			x			
Dolev and Haviv (2006)	x				x		
Fang et al. (2014)					x		
Galster et al. (1998)					x		
Gaubatz et al. (2008)							x
Hayes and Polian (2007)					x		
Hazucha et al. (2003)					x		
Hsieh et al. (2008)			x				

(continued)

Table 16.1 (continued)

Author/s (Year)	Identified problem						
	Voltage errors or droops	Timing errors	Manufact-uring variation	Failure of circuits and mechanical components	Single event upsets	Human-machine inter-action faults	Malicious attacks
Huang et al. (2000)				x			
Kang and Kim (2007)	x	x	x				
Leem et al. (2010)	x		x	x	x		
Li and Yeung (2006)						x	
Lima et al. (2001)					x		
Liu and Whitaker (1992)					x		
Maciejewski (1990)				x			
Merlin et al. (2014)	x						
Meshram and Belorkar (2011)			x	x	x		
Mitra et al. (2005)					x		
Mitra et al. (2007)					x		
Mukherjee et al. (2002)	x				x		
Nassif et al. (2010)			x	x	x		
Nickel (2001)	x	x			x		
Nicolaidis (1999)	x				x		
Normand (1996)					x		
Oh et al. (2002a)	x	x					
Oh et al. (2002b)					x		
Pradeep et al. (1988)				x			
Reddi et al. (2012)	x		x	x	x		
Rennels (1978)			x	x			
Richardeau et al. (2002)	x						
Roche and Gasiot (2005)					x		

Rockett (1992)					x		
Rotenberg (1999)	x	x					
Sanda et al. (2008)					x		
Saxena et al. (2000)	x		x	x	x		
Seshia et al. (2007)	x		x	x	x		
Touba and McCluskey (1997)				x			
Tschanz et al. (2009)	x	x		x			
Ullah and Sterpone (2014)				x	x		
Vishwanath and Nagappan (2010)				x			
Visinsky et al. (1994)				x		x	
Walters et al. (2011)					x		
Wong (2006)					x		
Yoshimoto et al. (2012)					x		
Yu et al. (2000)				x			
Zhang et al. (2006)					x		

Table 16.2 Reviewed publications sorted by use of "fault tolerance" or "resilience" and proposed solution type(s)

Author/s (Year)	Fault tolerance (FT) or Resilience (R)	Proposed solution	
		Hardware	Software/Calculations
Agarwal et al. (2007)	FT	x	
Alena et al. (2008)	FT	x	x
Alena et al. (2011)	FT	x	
Avižienis (1967)	FT	x	x
Avižienis (1997))	FT	x	x
Banerjee et al. (2007)	R	x	x
Bartlett and Spainhower (2004)	FT	x	x
Bau et al. (2009)	R	x	
Bowman et al. (2009a)	R	x	
Bowman et al. (2009b)	R	x	
Bowman et al. (2011)	R	x	
Breuer (2005)	R		x
Brunina et al. (2012)	R	x	x
Chakrapani et al. (2006)	R	x	
Chen and Trachtenberg (1991)	FT		x
Chippa et al. (2010)	R	x	
Dolev and Haviv (2006)	FT		x
Fang et al. (2014)	R		x
Galster et al. (1998)	R	x	
Gaubatz et al. (2008)	R		x
Hayes and Polian (2007)	R		x
Hazucha et al. (2003)	R	x	
Hsieh et al. (2008)	R		x
Huang et al. (2000)	FT	x	x
Kang and Kim (2007)	R	x	
Leem et al. (2010)	R	x	x
Li and Yeung (2006)	FT		x
Lima et al. (2001)	R		x
Liu and Whitaker (1992)	R	x	
Maciejewski (1990)	FT	x	
Merlin et al. (2014)	FT	x	
Meshram and Belorkar (2011)	FT	x	
Mitra et al. (2005)	R	x	x
Mitra et al. (2007)	R	x	
Mukherjee et al. (2002)	FT		x
Nassif et al. (2010)	R	x	x
Nickel (2001)	FT	x	x
Nicolaidis (1999)	FT	x	x
Normand (1996)	R	x	x
Oh et al. (2002a)	R		x

(continued)

Table 16.2 (continued)

Author/s (Year)	Fault tolerance (FT) or Resilience (R)	Proposed solution	
		Hardware	Software/Calculations
Oh et al. (2002b)	FT		x
Pradeep et al. (1988)	FT	x	
Reddi et al. (2012)	R	x	x
Rennels (1978)	FT	x	x
Richardeau et al. (2002)	FT	x	
Roche and Gasiot (2005)	R	x	
Rockett (1992)	R	x	
Rotenberg (1999)	FT		x
Sanda et al. (2008)	R	x	x
Saxena et al. (2000)	FT	x	x
Seshia et al. (2007)	R		x
Touba and McCluskey (1997)	FT		x
Tschanz et al. (2009)	R	x	
Ullah and Sterpone (2014)	FT	x	
Vishwanath and Nagappan (2010)	R	x	
Visinsky et al. (1994)	FT		x
Walters et al. (2011)	FT		x
Wong (2006)	R		x
Yoshimoto et al. (2012)	R	x	x
Yu et al. (2000)	FT	x	x
Zhang et al. (2006)	R	x	x

commonly identified problem is variation in manufacturing that causes components to behave differently, including failures at different rates which has become a more serious problem as the physical size of circuits have become smaller and manufacturing tolerances harder to meet (Borkar 2005). A variety of additional types of failures such as timing errors, human-machine interaction faults, and malicious attacks were discussed in a smaller number of papers. In many cases, the solutions proposed by a paper are applicable to a variety of problems, as is often indicated by the authors of each work. For instance, an error-correcting technique based on dynamic bit steering can be applied to address errors which are due to radiation-induced bit flips, and permanent hardware defects which may result from initial manufacturing variation or failure at a later time (Brunina et al. 2012). This broader applicability of many resilience techniques is indicated by the summation of the percentages in Fig. 16.1 being greater than 100%. Specifically, it was found that the 61 papers proposed solutions to 112 problems, indicating that the average proposed method to improve resilience was determined to be applicable to approximately two major types of fault-inducing errors.

Table 16.3 Types of resilience strategies in reviewed publications

	National Academy of Sciences (NAS) phases				Network Centric Warfare (NCW) domains			
Author/s (Year)	Plan	Adsorb	Recover	Adapt	Physical	Information	Cognitive	Social
Agarwal et al. (2007)	x				x			
Alena et al. (2008)	x			x	x	x		
Alena et al. (2011)	x		x			x		
Avižienis (1967)	x	x		x	x	x		
Avižienis (1997)	x	x	x		x			
Banerjee et al. (2007)	x	x			x	x		
Bartlett and Spainhower (2004)		x	x	x	x	x		
Bau et al. (2009)	x		x		x			
Bowman et al. (2009a)	x		x			x		
Bowman et al. (2009b)	x	x	x			x		
Bowman et al. 2011)	x		x		x	x		
Breuer (2005)	x	x			x	x		
Brunina et al. (2012)	x		x	x	x	x		
Chakrapani et al. (2006)		x			x			
Chen and Trachtenberg (1991)		x			x	x		
Chippa et al. (2010)	x				x	x		
Dolev and Haviv (2006)	x	x			x	x		
Fang et al. (2014)	x	x	x		x			
Galster et al. (1998)	x	x			x			
Gaubatz et al. (2008)	x		x	x		x		
Hayes and Polian (2007)	x	x	x		x	x		
Hazucha et al. (2003)	x	x			x			
Hsieh et al. (2008)		x		x		x		

Huang et al. (2000)	x	x	x	x	x			
Kang and Kim (2007)	x	x			x			
Leem et al. (2010)	x	x			x	x		
Li and Yeung (2006)	x	x		x	x	x		
Lima et al. (2001)		x		x	x			
(Liu and Whitaker 1992		x			x			
Maciejewski (1990)	x	x	x	x	x			
Merlin et al. (2014)	x		x		x			
Meshram and Belorkar (2011)	x	x			x			
Mitra et al. (2005)	x		x		x	x		
Mitra et al. (2007)	x		x		x			
Mukherjee et al. (2002)		x			x	x		
Nassif et al. (2010)	x	x			x	x		
Nickel (2001)		x	x		x			
Nicolaidis (1999)	x	x			x	x		
Normand (1996)	x				x			
Oh et al. (2002a)		x			x	x		
Oh et al. (2002b)	x	x				x		
Pradeep et al. (1988)	x		x	x	x			
Reddi et al. (2012)	x	x	x	x		x		
Rennels (1978)	x	x	x	x	x			
Richardeau et al. (2002)	x	x	x	x	x			
Roche and Gasiot (2005)	x	x			x			
Rockett (1992)	x	x	x		x	x		
Rotenberg (1999)	x	x			x	x		

(continued)

Table 16.3 (continued)

Author/s (Year)	National Academy of Sciences (NAS) phases				Network Centric Warfare (NCW) domains			
	Plan	Adsorb	Recover	Adapt	Physical	Information	Cognitive	Social
Sanda et al. (2008)		x	x		x	x		
Saxena et al. (2000)	x	x	x	x	x			
Seshia et al. (2007)	x		x			x		
Touba and McCluskey (1997)	x	x	x		x			
Tschanz et al. (2009)	x	x	x			x		
Ullah and Sterpone (2014)	x		x		x			
Vishwanath and Nagappan (2010)	x	x	x	x	x	x		
Visinsky et al. (1994)	x		x	x	x			
Walters et al. (2011)	x		x	x	x	x		
Wong (2006)	x		x		x	x		
Yoshimoto et al. (2012)	x	x				x		
Yu et al. (2000)	x	x			x	x		
Zhang et al. (2006)	x		x		x	x		

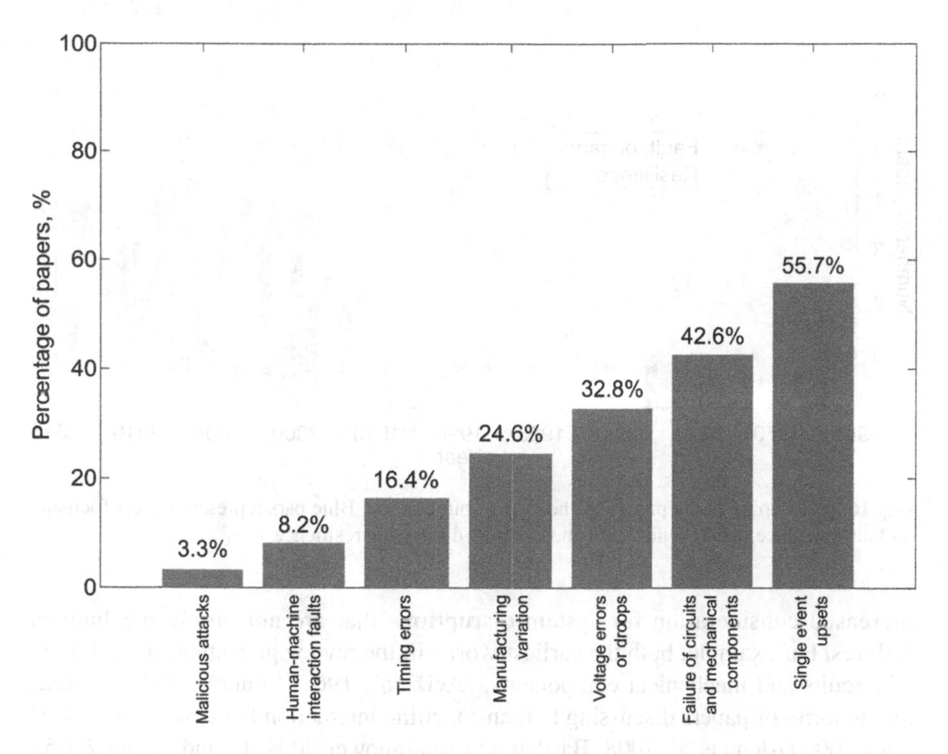

Fig. 16.1 Resilience problems identified in the literature, ranked according to the number of papers. The sum of percentages (184%) is greater than 100% because many papers address multiple resilience issues

There has been increasing interest in resilience of electromechanical systems in recent years. Figure 16.2 shows the number of papers in the review, sorted by year of publication. Further, the papers are also sorted by whether they use the terminology "fault tolerance" or "resilience," with the number of papers using each of these terms indicated by different colors. The complete listing of papers by year of publication and terminology used is available in Table 16.2. For many years, the number of publications included in the review was zero. There is however, an unmistakable trend toward increased numbers of papers with a maximum observed in 2007. Although a decline in interest may be suggested by the fewer papers in subsequent years, it is important to note that the number for 2014 only reflects those papers published during part of the year, and that the number of papers included from 2010–2014 (Alena et al. 2011) outpaced the number from 2000–2004 (Avižienis 1967), indicating a continued rise in attention to resilience concepts.

During the course of the review, it was noted that the publications referred to their attention to the resiliency of computer systems as either "fault tolerance" or "resilience." When papers were sorted by which term used, it was evident that in early years "fault tolerance" was the normal descriptor, while the term "resilience" became more common after 2000. This shift in terminology may be a reflection of

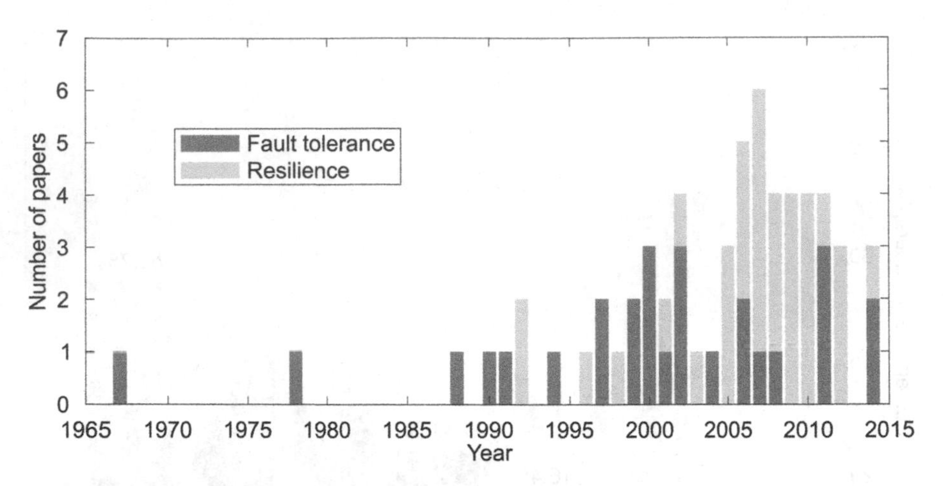

Fig. 16.2 Grouping of the papers by the year of publication. Blue bars represent papers focusing on fault tolerance, while yellow bars show papers discussing resilience

increased consideration for system disruptions that are not purely mechanical failures. For example, both the earliest works in the review give attention to failure of circuits and mechanical components, (Avižienis 1967; Rennels 1978) whereas the majority of papers discussing human-machine interaction faults were published after 2000 (Alena et al. 2008; Bartlett and Spainhower 2004; Li and Yeung 2006). From this simple analysis, it is clear that as computer and electromechanical systems have increased in importance, there has been an increase in attention given to their resilient- and fault-free-operation. The transition to the term "resilience" suggests that the field is examining the robustness of these systems to failure modes not initially considered when discussing "fault tolerance."

The papers examined were also indexed according to the primary methodology they proposed as a solution for the particular resilience problem identified. For the sake of simplicity, methodologies were divided according to their focus on hardware, software, or a combination of both. The results summarized in Fig. 16.3 suggest that there is no preferred methodology within the computer and electrical engineering communities, with roughly equal attention dedicated to hardware-based methods (Merlin et al. 2014), software-based methods such as error-correcting codes or multi-threading of computational operations (Rotenberg 1999; Mukherjee et al. 2002), and mixed methods combining elements of both hardware and software (Brunina et al. 2012). A full summary of the high-level assignments of the resilience solution proposed by each paper to hardware, software, or a combined methodology is provided in Table 16.2.

Finally, papers were sorted by how they addressed the NAS Phases (Resilience 2012) and NCW Domains (Alberts 2002) of resilience. The resilience procedures that make up the four phases of resilience are called planning, absorbing, recovering, and adapting. For example, a resilience plan may focus attention on the prevention of operational failures. Alternatively, mechanisms could be incorporated to preserve

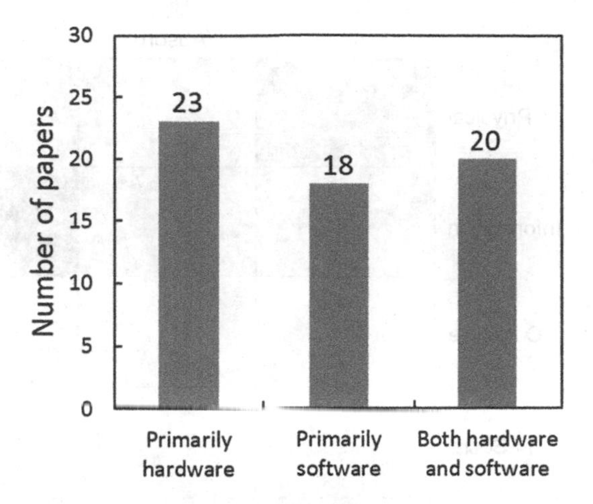

Fig. 16.3 Number of papers by methodology of proposed resilience solution(s)

the function of circuits or other components despite a particular impairment. Still other options include focusing on correcting the problem at hand in hopes of returning to ideal state of function as soon as possible, or providing a mechanism whereby it was possible to learn more about the failure that had occurred in order to prevent similar occurrences in the future. These are the four resilience procedures of planning, absorbing, recovering, and adapting, respectively.

In addition to evaluating papers according to the NAS Phases of resilience, the solutions they presented were also indexed according to the four domains of resilience: physical, informational, cognitive, and social. Actions to improve a system's resilience can be categorized in terms of these domains. A physical solution could be a change to the design of the circuit; an informational solution could involve the way the circuit communicates information to those programming it; a cognitive solution could involve the engineers processing the physical and informational outputs to better inform future design decisions, and a social solution would be a way to share the learned cognitive conclusions.

Complete classification for all papers in the review into NAS Phases and NCW Domains is included in Table 16.3, while a summary of the results, presented in the form of a resilience matrix (Linkov et al. 2013), is shown in Fig. 16.4. In this figure, the percentage of papers in the review addressing each element of the matrix generated by permutations of the NAS Phases (columns) and NCW Domains (rows) is indicated. In the review, most papers either evaluated resilience strictly in the physical or informational domains; only 26 of the 61 articles considered both types of solutions in their resilience efforts. Additionally, no paper took either the cognitive or social domains into consideration. Ultimately, more solutions for faults in electrical engineering systems and circuits could potentially be determined if more of these resilience domains were taken into account.

There is a clear bias toward the planning and absorbing phases within the two resilience domains addressed in the articles reviewed. Further, a clear reduction in the percentage of papers addressing later stages of resilience (recover and adapt)

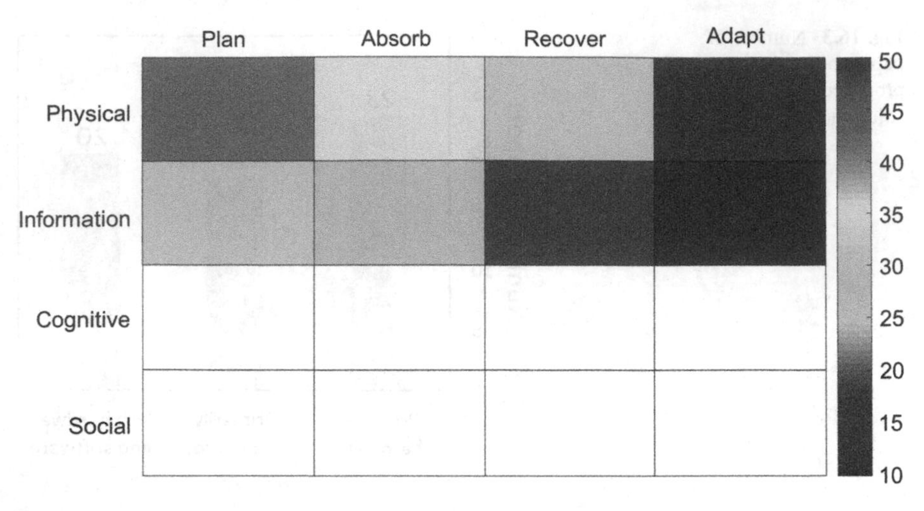

Fig. 16.4 Number of papers reviewed by classification of the phases and domains of resilience addressed

can be observed. Specifically, every reviewed paper included fault prevention (i.e. planning) or absorption as a stage in their resilience strategy, and fully 22 of the 61 reviewed papers included only these two phases. One particular exception, Bartlett and Spainhower, focused strictly on what should happen after an error is detected: damage should be minimized, and a quick recovery plan initiated (Bartlett and Spainhower 2004). While this paper touched upon both the recovery and adaptation phases, most of the remaining papers that addressed the latter two phases only considered one or the other, with a greater number including recovery as an aspect of their resilience solution. In total, less than one-third (specifically 18/61) papers attempt to consider adaptation – how to improve system function and performance to prevent similar errors from occurring in the future. One recent example of adaptation as a resiliency phase is included in Reddi et al. (2012) where software is used to control the frequency of operation of individual circuits. This control is linked to measures of the operation of the circuits, and so it is possible for the system to predict optimal settings for individual components and adjust these to maximize error-free performance despite manufacturing differences, voltage droops, and transient single-event upsets. Ultimately, resilience efforts will be most effective if all four tenets are considered. However, most electrical engineering and circuit-related resilience studies only considered on average two of the four categories.

Resilience in electromechanical systems focused primarily on the plan and absorb phases may be a consequence of how the field of how fault tolerance and resilience in this field developed. Many of the earliest works addressed resilience problems via hardware-only or hardware-and-software methods (Pradeep et al. 1988; Maciejewski 1990; Rennels 1978; Liu and Whitaker 1992; Rockett 1992). The types of resilience issues that primarily hardware-based solutions may be suited to address appear to be more suited to planning and absorbing problems, rather than adapting to them. Examples of such issues and their proposed hardware-based solutions include introducing additional joints that allow a computer-controlled robot to retain function during partial mechanical failure (Pradeep et al. 1988), and improved diodes that better absorb radiation-induced upsets (Galster et al. 1998). By contrast, software-based resilience solutions, such as error-correcting codes (Nickel 2001) or software that can produce qualitatively correct outputs despite errors in inputs or calculations (Li and Yeung 2006; Wong 2006), are somewhat more modern and have the ability to address neglected resilience phases of recovery and adaptation. However, it is important to note that hardware-based approaches can be used to address recovery or adaptation. For instance, Tschanz, *et al.* describe "tunable replica circuits" that can dynamically respond to timing and voltage errors (Tschanz et al. 2009). Addressing all phases of resilience is an important goal for future work within the computer and electrical engineering fields, and it will be exciting to see new software-based solutions presented that are suited to recovery and adaptation. However, it will be necessary for the field to begin to address the cognitive and social domains of resilience, which they appear to not have done to date. Once these portions of the resilience matrix are explicitly considered, computer systems will be much more robust against faults, errors, and unexpected disruptions.

16.4 Conclusions

Creation of resilient electrical systems (circuits) is challenged by the fact that throughout the years engineers and scientists have focused on the processing efficiency of computing systems and their optimization, while resilience calls for adaptive and flexible structure of the system. Current practices lead to unnecessarily rigid streamlined algorithms of how the circuits process information. The general approach to combat errors occurring in these systems have been focused on error detection and correction. Recovery is accomplished by recalculation of the originally miscalculated values, while adaptation is difficult in many cases because the system is not flexible. This tendency is highlighted in how the phases of resilience rank by the number of the reviewed papers addressing them: plan (51/61 papers), absorb (41/61), recover (33/61), and adapt (18/61). Moreover, the inflexibility of the systems defined the predictable and probabilistic nature of the fault events considered.

Although significant attention was given to the physical (51/61 papers) and information domains (36/61), we found no papers on the cognitive and social domains of the resilience matrix. Understandably, this may be caused by the fact that electrical engineering as a field of research lies in the physical and, to a lesser extent, in the information domains. On the other hand, our review also shows the need for an integrated design and deployment practices which encompass all four domains.

The main contribution of this work is a comprehensive review of the electrical engineering research papers on resilience and fault tolerance published to date and a new approach to classify resilience research papers with the resilience matrix. The next steps may include the development of guidelines and recommendations for resilient design of electrical circuitry and a methodology for resilience quantification in the field of electrical engineering.

Further Suggested Readings

Agarwal M, Paul BC, Zhang M, Mitra S (2007) Circuit failure prediction and its application to transistor aging. In 25th IEEE VLSI test symposium (VTS'07); pp 277–286

Alberts DS (2002) Information age transformation. getting to a 21st century military, Revised.; Washington, DC

Alena R, Ellis SR, Hieronymus J, Maclise D (2008) Wireless avionics and human interfaces for inflatable spacecraft. In IEEE aerospace conference proceedings

Alena R, Gilstrap R, Baldwin J, Stone T, Wilson P (2011) Fault tolerance in ZigBee wireless sensor networks. In IEEE aerospace conference proceedings; pp 1–15

Avižienis A (1967) Design of fault-tolerant computers. In Proceedings of the November 14–16, 1967, Fall joint computer conference; pp 733–743

Avižienis A (1997) Toward systematic design of fault-tolerant systems. *Computer (Long Beach Calif) 30*(4):51–58

Banerjee N, Karakonstantis G, Roy K (2007) Process variation tolerant low power DCT architecture. In Proceedings – Design, Automation and Test in Europe, DATE'07; Vol. 7, pp 630–635

Bartlett W, Spainhower L (2004) Commercial fault tolerance: a tale of two systems. IEEE Trans Dependable Secur Comput 1(1):87–96

Bau J, Hankins R, Jacobson Q, Mitra S, Saha B, Adl-Tabatabai A-R (2009) Error resilient system architecture (ERSA) for probabilistic applications. In Proceedings of the international symposium on low power electronics and design

Bodeau DJ, Graubart R (2011) MITRE cyber resiliency engineering framework, MTR110237; Bedford

Borkar S (2005) Designing reliable systems from unreliable components: the challenges of transistor variability and degradation. IEEE Micro 25(6):10–16

Bowman KA, Tschanz JW, Kim NS, Lee JC, Wilkerson CB, Lu S-LL, Karnik T, De VK (2009a) Energy-efficient and metastability-immune resilient circuits for dynamic variation tolerance. IEEE J Solid-State Circuits 44(1):49–63

Bowman K, Tschanz J, Wilkerson C, Lu S-L, Karnik T, De V, Borkar S (2009b) Circuit techniques for dynamic variation tolerance. In Design Automation Conference, 2009. DAC '09. 46th ACM/IEEE; pp 4–7

Bowman KA, Tschanz JW, Lu S-LL, Aseron PA, Khellah MM, Raychowdhury A, Geuskens BM, Tokunaga C, Wilkerson CB, Karnik T, De VK (2011) A 45 Nm resilient microprocessor core for dynamic variation tolerance. IEEE J Solid-State Circuits 46(1):194–208

Breuer MA (2005) Multi-media applications and imprecise computation. In Proceedings – DSD'2005: 8th euromicro conference on digital system design – architectures, methods and tools; pp 2–7

Brunina D, Lai CP, Liu D, Garg AS, Bergman K (2012) Resilient optically connected memory systems using dynamic bit-steering [Invited]. J Opt Commun Netw 4(11):B151

Chakrapani LN, Akgul BES, Cheemalavagu S, Korkmaz P, Palem KV, Seshasayee B (2006) Ultra-efficient (Embedded) SOC architectures based on probabilistic CMOS (PCMOS) technology. In Proceedings – design, automation and test in Europe, DATE'06; pp 1110–1115

Chen M, Trachtenberg EA (1991) Permutation codes for the state assignment of fault tolerant sequential machines. In Proceedings Of The 10th digital avionics systems conference; pp 85–90

Chippa VK, Mohapatra D, Raghunathan A, Roy K, Chakradhar ST (2010) Scalable effort hardware design: exploiting algorithmic resilience for energy efficiency. In Design Automation Conference (DAC), 2010 47th ACM/IEEE; pp 555–560

Department of Homeland Security. Critical Infrastructure Sectors (n.d.) https://www.dhs.gov/critical-infrastructure-sectors

DiMase D, Collier ZA, Heffner K, Linkov I (2015) Systems engineering framework for cyber physical security and resilience. Environ Syst Decis 35(2):291–300

Disaster Resilience: A National Imperative (2012) The National Academies Press: Washington, DC

Dolev S, Haviv YA (2006) Self-stabilizing microprocessor: analyzing and overcoming soft errors. IEEE Trans Comput 55(4):385–399

Fang L, Yamagata Y, Oiwa Y (2014) Evaluation of a resilience embedded system using probabilistic model-checking. In Electronic proceedings in theoretical computer science; Vol. 150, pp 35–49

Galster N, Frecker M, Carroll E, Vobecky J, Hazdra P (1998) Application-specific fast-recovery diodes: design and performance. In Power Conversion April 1998 Proceedings; pp 1–14

Gaubatz G, Savaş E, Sunar B (2008) Sequential circuit design for embedded cryptographic applications resilient to adversarial faults. IEEE Trans Comput 57(1):126–138

Hayes JP, Polian I, Becker B (2007) An analysis framework for transient-error tolerance. In Proceedings of the IEEE VLSI test symposium; pp 249–255

Hazucha, P.; Kamikl, T.; Walstra, S.; Bloechell, B.; Tschanzl, J.; Maiz, J.; Soumyanath, K.; Demer, G.; Narendra, S.; De, V.; Borkar, S. (2003) Measurements and analysis of SER tolerant latch in a 90 nm Dual-Vt CMOS Process. In IEEE 2003 custon integrated circuits conference; pp 617–620

Hsieh T-Y, Lee K-J, Breuer MA (2008) An error rate based test methodology to support error-tolerance. IEEE Trans Reliab 57(1):204–214

Huang W-J, Saxena N, McCluskey EJ (2000) Reliable LZ data compressor on reconfigurable coprocessors; pp 249–258

Kang K, Kim K, Roy K (2007) Variation resilient low-power circuit design methodology using on-chip phase locked loop. In ACM/IEEE Design Automation Conference; pp 934–939

Leem L, Cho H, Bau J, Jacobson QA, Mitra S (2010) ERSA: error resilient system architecture for probabilistic applications. In Design, Automation Test in Europe Conference Exhibition, 2010; pp 1560–1565

Li X, Yeung D (2006) Exploiting soft computing for increased fault tolerance. In Workshop on architectural support for Gigascale integration

Lima F, Rezgui S, Carro L, Velazco R, Reis R (2001) On the use of VHDL simulation and emulation to derive error rates. In 6th European Conference on Radiation and Its Effects on Components and Systems; pp 253–260

Linkov I, Eisenberg DA, Bates ME, Chang D, Convertino M, Allen JH, Flynn SE, Seager TP (2013) Measurable resilience for actionable policy. Environ Sci Technol 47:10108–10110

Liu N, Whitaker S (1992) Low power SEU immune CMOS memory circuits. IEEE Trans Nucl Sci 39(6):1679–1684

Maciejewski AA (1990). Fault tolerant properties of kinematically redundant manipulators. In IEEE Conference on Robotics and Automation; pp 638–642

Merlin MMC, Green TC, Mitcheson PD, Trainer DR, Critchley R, Crookes W, Hassan F (2014) The alternate arm converter: a new hybrid multilevel converter with DC-fault blocking capability. IEEE Trans Power Deliv 29(1):310–317

Meshram SS, Belorkar UA (2011) Design approach for fault tolerance in FPGA architecture. Int J VLSI Des Commun Syst 2(1):87–95

Mitra S, Seifert N, Zhang M, Shi Q, Kim KS (2005) Robust system design with built-in soft-error resilience. Computer (Long. Beach. Calif). No. February, 43–52

Mitra S, Zhang M, Seifert N, Mak TM, Kim KS (2007) Built-in soft error resilience for robust system design. In IEEE International Conference on Integrated Circuit Design and Technology; 2007; pp 1–6

Mukherjee SS, Kontz M, Reinhardt SK (2002) Detailed Design and Evaluation of Redundant Multithreading Alternatives. In 29th Annual International Symposium on Computer Architecture; pp 99–110
Nassif SR, Mehta N, Cao Y (2010) A resilience roadmap. In Design, Automation &Test in Europe Conference & Exhibition; pp 1011–1016
Nickel JB, Somani AK (2001) REESE: a method of soft error detection in microprocessors. In Proceedings of the International Conference on Dependable Systems and Networks; pp 401–410
Nicolaidis M (1999) Time redundancy based soft-error tolerance to rescue nanometer technologies. In Proceedings 17th IEEE VLSI Test Symposium
Normand E (1996) Single event upset at ground level. IEEE Trans Nucl Sci 43(6):2742–2750
Oh N, Mitra S, McCluskey EJ (2002a) ED4I: error detection by diverse data and duplicated instructions. IEEE Trans Comput 51(2):180–199
Oh N, Shirvani PP, McCluskey EJ (2002b) Error detection by duplicated instructions in super-scalar processors. IEEE Trans Reliab 51(1):63–75
Pradeep AK, Yoder PJ, Mukundan R, Schilling RJ (1988) Crippled motion in Robots. IEEE Trans Aerosp Electron Syst 24(1):2–13
Presidential Policy Directive – Critical Infrastructure Security and Resilience. https://www.whitehouse.gov/the-press-office/2013/02/12/presidential-policy-directive-critical-infrastructure--security-and-resil. n.d.
Reddi VJ, Pan DZ, Nassif SR, Bowman KA (2012) Robust and resilient designs from the bottom-up: technology, CAD, circuit, and system issues. In Proceedings of the Asia and South Pacific Design Automation Conference, ASP-DAC; pp 7–16
Rennels DA (1978) Architectures for fault-tolerant spacecraft computers. Proc IEEE 66(10):1255–1268
Richardeau F, Baudesson P, Meynard TA (2002) Failures-tolerance and remedial strategies of a PWM multicell inverter. IEEE Trans Power Electron 17(6):905–912
Roche P, Gasiot G (2005) Impacts of front-end and middle-end process modifications on terrestrial soft error rate. IEEE Trans Device Mater Reliab 5(3):382–395
Rockett LR (1992) Simulated SEU hardened scaled CMOS SRAM cell design using gated resistors. IEEE Trans Nucl Sci 39(5):1532–1541
Roege PE, Collier ZA, Mancillas J, McDonagh JA, Linkov I (2014) Metrics for energy resilience. Energy Policy 72:249–256
Rotenberg E (1999). AR-SMT: A microarchitectural approach to fault tolerance in microprocessors. In Twenty-Ninth Annual International Symposium on Fault-Tolerant Computing; pp 84–91
Sanda PN, Kellington JW, Kudva P, Kalla R, McBeth RB, Ackaret J, Lockwood R, Schumann J, Jones CR (2008) Soft-error resilience of the IBM POWER6 processor. IBM J Res Dev 52(3):275–284
Saxena N, Fernandez-Gomez S, Huang W, Mlra S, Ya S-Y, Mccluskey EJ (2000) Dependable computing and online testing in adaptive and configurable systems. IEEE Des Test Comput 17:29–41
Scott A, Menn J (2014) Exclusive: air traffic system failure caused by computer memory shortage. Reuters
Seshia SA, LiW, Mitra S (2007) Verification-guided soft error resilience. In Proceedings of the conference on design, automation and test in Europe; pp 1442–1447
Stelloh T, Gutierrez G (2016) Georgia power company disputes "Outage" behind delta's system failure. NBC News
Touba NA, McCluskey EJ (1997) Logic synthesis of multilevel circuits with concurrent error detection. IEEE Trans Comput Des Integr Circuits Syst 16(7):783–789
Tschanz J, Bowman K, Wilkerson C, Lu S-L, Karnik T (2009) Resilient circuits – enabling energy-efficient performance and reliability. In Proceedings of the 2009 International Conference on Computer-Aided Design - ICCAD '09; pp 71–73

Ullah A, Sterpone L (2014) Recovery time and fault tolerance improvement for circuits mapped on SRAM-based FPGAs. J Electron Test Theory Appl 30(4):425–442
Vishwanath KV, Nagappan N (2010) Characterizing cloud computing hardware reliability. In Proceedings of the 1st ACM Symposium on Cloud Computing - SoCC '10; pp 193–203
Visinsky ML, Cavallaro JR, Walker ID (1994) Expert system framework for fault detection and fault tolerance in robotics. Comput Electr Eng 20(5):421–435
Walters JP, Kost R, Singh K, Suh J, Crago SP (2011) Software-based fault tolerance for the maestro many-core processor. In IEEE aerospace conference proceedings; pp 1–12
Wong V, Horowitz M (2006) Soft error resilience of probabilistic inference applications. In IEEE workshop on silicon errors in logic; pp 1–4
Yoshimoto S, Amashita T, Okumura S, Nii K, Yoshimoto M, Kawaguchi H (2012) Bit-error and soft-error resilient 7T/14T SRAM with 150-Nm FD-SOI Process. IEICE Trans Fundam Electron Commun Comput Sci E95–A (8), 1359–1365
Yu S-Y, Saxena N, McCluskey EJ (2000) An ACS Robotic control algorithm with fault tolerant capabilities. In IEEE Symposium on FPGAs for custom computing machines, Proceedings pp 175–184
Zhang M, Mitra S, Member S, Mak TM, Seifert N, Wang NJ, Shi Q, Kim KS, Shanbhag NR, Patel SJ (2006) Sequential element design with built-in soft error resilience. IEEE Trans Very Large Scale Integr Syst 14(12):1368–1378
Ziegler JF, Lanford WA (1979) Effect of cosmic rays on computer memories. *Science (80-.). 206* (4420), 776–788

Part VI
Applications

Chapter 17
Building Resilience Through Risk Analysis

Philip F. O'Neill

Abstract Resilience is the ability of systems and organizations to maintain an acceptable level of service in spite of crises or adverse operating conditions and to recover quickly in the event that service falls below acceptable standards. By creating network models of infrastructure, resources and processes, it is possible to prioritize risks and identify critical vulnerabilities in processes, organizations and systems. Consequently, network models can be used to create effective plans for improving resilience. The main idea is to use a directed graph to construct a network risk model that can be used to identify the most effective options for improving resilience. Methods and tools for building and analyzing such models will be presented along with an actual case study of a supply chain in Afghanistan.

Keywords Risk • Risk analysis • Risk synthesis • Resilience • Critical infrastructure • Network models • Dependency relationships • Graph theory • Cumulative impact • Cumulative vulnerability

Improving the resilience of critical infrastructure has become a major thrust in the area of emergency planning. The National Infrastructure Advisory Council (NIAC) has defined resilience as "the ability to reduce the magnitude and/or duration of disruptive events", (National Infrastructure Advisory Council 2010). Although the definition was put forward in the context of infrastructure resilience, it can be applied equally well in the context of resilience of systems, processes, organizations and enterprises. Because of its abundant applicability, this chapter will make use of the NIAC definition to develop a modeling paradigm for building resilience.

The NIAC definition implies that resilience includes the ability to anticipate, absorb, adapt and recover from disruptive events. Indeed, in National Infrastructure

---

P.F. O'Neill (✉)
Chief Scientist, RiskLogik, Almonte, Ontario, Canada
e-mail: Phil.oneill@risklogik.com

© Springer Science+Business Media B.V. 2017

I. Linkov, J.M. Palma-Oliveira (eds.), *Resilience and Risk*, NATO Science for Peace and Security Series C: Environmental Security,
DOI 10.1007/978-94-024-1123-2_17

Advisory Council (2010), the NIAC has identified four outcome-focused characteristics of resilience:

1. Robustness – the ability to absorb shocks and keep operating
2. Resourcefulness – the ability to manage a crisis as it unfolds, as skillfully as possible
3. Rapid Recovery – the ability to bring service back to acceptable standards, as quickly as possible
4. Adaptability – the ability to incorporate lessons learned from past events to improve resilience

The modeling paradigm developed in this chapter will make use of a network model that describes the functional parts of the resilience environment under consideration and the dependency relationships that exist among them. In order to create a network model, we will make use of a mathematical object known as a ***directed graph***. This will be defined more formally in the next section. In fact the chapter will make use of several concepts taken from graph theory, especially the concept of a ***path*** in a graph. An excellent resource for more extensive information on the subject is found in Ganin et al. (2016).

Our modelling paradigm will make use of the 2 components that characterize directed graphs: (i) nodes and (ii) edges. Nodes will be used to represent entities in the resilience environment that we are modelling. Here, an ***entity*** is something that has an existence in its own right or a perceived existence in the resilience environment. For example, people, organizations, resources, processes, plans and activities can be represented in a model as entities. Edges will be used to represent direct dependency relationships that exist among the entities. A ***direct dependency*** exists when one entity provides something directly to another entity. The providing entity is called the ***source*** and the receiving entity is called the ***dependent*** in such a relationship. For example, some equipment involved in a manufacturing process might depend on diesel fuel or natural gas. In this case there would be source entities for natural gas and diesel fuel; as well, there would be dependent entities representing the equipment that requires each source of fuel.

Note that entities need not be physical objects. Abstractions, concepts and other intangibles are permitted as entities when they play a functional role in a resilience environment. For example, knowledge, information or data might constitute an entity or entities. Likewise psychological stress could be taken as an entity in a resilience environment involving the performance of human resources. Similarly, dependency relationships might involve non-physical exchanges between entities. Consider, for example, command and control relationships that exist in a military context.

In order to measure resilience, we will make use of ***risk analysis***. More specifically, we will make use of the risk analysis framework put forward by William D. Rowe in his landmark book of 1977 which was revised and augmented in 1988 Rowe (1988). Accordingly, we define *risk* as ***the possibility of loss.*** This definition of risk implies that risk has 2 dimensions: a *loss* dimension and a *possibility* dimension.

In order to measure resilience we will, therefore, measure the risk dimensions of the entities in the model. The loss dimension will be measured according to the impact that could ensue if any risk is realized as an actual event. The model will include defined metrics based on its context, specific activities and potential events that are of interest. The possibility dimension will be measured according to the likelihood of events actually occurring. This estimation might be subject to a high margin of error. In such cases, we can explore any potential events that are deemed to be of interest if their estimated likelihood is low.

For example, suppose that we want to measure the resilience of a particular city with regard to flooding during the spring. One metric for impact would be the expected cost of damage and cleanup if a flood occurs. Suppose that there is, furthermore, the possibility of human injury and loss of life. In this situation, the expected value of insurance claims for death and injury should be included in the cost estimate as well as the costs of property damage and cleanup. The likelihood calculations would involve some or all of the following:

- Expert knowledge and opinions, including weather forecasts and flood predictions
- Historical data
- Customized mathematical models
- Simulations

Under Rowe's framework, emphasis is placed on identifying pathways of exposure to risk and pathways that propagate consequences if a potential risk is realized as an actual event. The modeling paradigm presented in this chapter makes use of path analysis in the directed graph model to take all pathways of exposure and all pathways of consequence into consideration. The cumulative impact and cumulative likelihood of failure of every entity in a model is calculated algorithmically.

Other authors have proposed graph models of resilience. An excellent example is found in Ganin et al. (2016). The authors propose a framework for modelling resilience that uses directed graphs and includes a time dependent metric for measuring resilience. They use simulation to obtain results from the model. However, there methods require that the directed graph be acyclic. The methodology of this chapter does not have that restriction.

Typically, risk analysis of systems with complex dependency relationships is carried out by means of constructive simulation. However, such systems are often governed by mathematical equations that are difficult to calibrate against the real world. Extensive data gathering and complex computer programming might be required to create a useful model. Consequently, simulation models are often time-consuming and costly to develop.

We will describe how graph theory can be used to model dependency and interdependency among entities in systems. Such a model can be used for risk analysis, process analysis or capability analysis. Systems can be described in as much detail as suits the analysts' purposes. If values for level of impact and likelihood of failure are available for each entity together with a level of dependence for each direct dependency relationship then measures of cumulative dependence can be derived for every pair of entities.

The modeling approach described herein has been used to model infrastructure risk for the province of Ontario in Canada and continues to be developed. It has also been used to model supply chains in the province of New Brunswick. Under this approach, pathways of exposure to risk are characterized by dependency relationships that exist among entities. The entities and relationships are modeled using a directed graph. By using path analysis in the graph, a strategic ***risk synthesis*** can be modeled from expert knowledge about the individual entities and their direct dependency relationships. Decision makers can use the model to prioritize risk and to explore risk mitigation strategies.

17.1 Risk and Resilience

At this point, it is useful to elaborate on how risk analysis relates to the evaluation of resilience using a directed graph model.

Recall that "resilience is the ability to reduce the magnitude and/or duration of disruptive events", (National Infrastructure Advisory Council 2010). Therefore, as with risk, *resilience* has 2 dimensions: (1) robustness and (2) recovery time.

In the resilience environment being modeled, the model architect can create a node that represents the primary purpose of all the entities in the functioning in the environment and can formulate direct dependency relationships that model its role. Furthermore, the model can include auxiliary entities and dependency relationships that relate to acceptable recovery time of the primary entity in the event that any of the other entities fall below acceptable performance standards. http://permalink.lanl.gov/object/tr?what=info:lanl-repo/lareport/LA-UR-04-6927 provides background on how this can be accomplished.

17.2 The Modeling Paradigm

Systems can be viewed as collections of networks that overlap and interact with each other. There are transportation networks, information and communications networks, energy networks, supply chains, distribution networks, social networks and so on. Each network takes inputs from other networks and by means of its own enablers and activities, produces outputs that are in turn taken as inputs by other networks. As well, there are controls and regulations that govern the activities of any network along with monitoring and verification agents who oversee the activities and processes.

The risks in our society that result from ***interconnectedness*** can be characterized as stemming from ***dependency relationships*** which exist at the level of the ***connections***. In our modeling paradigm, the connections are relationships and hence dependency relationships will be represented as edges in a directed graph and the strength of each relationship will be represented by a weight called the ***level of dependence***.

A dependency relationship is a special kind of relationship in which something is passed from one entity to another. The *something* need not be material (such as gasoline or water) but can be immaterial (such as data or instructions). Entity ***B*** depends on entity ***A*** if and only if ***A*** provides ***B*** with something. So if edge ***(A, B)*** exists in the graph then ***B*** depends on ***A***.

Direct dependency relationships are usually well understood by domain experts; indirect dependencies are not. While experts have much insight and intuition about direct dependencies, analysis is needed to verify or correct intuition and to synthesize expert knowledge into a comprehensive view of indirect dependencies. There are two main challenges in this (1) capturing expert knowledge (2) constructing a complete strategic picture with prioritization of indirect dependences for purposes of contingency planning.

17.3 Directed Graphs

A *directed graph* $\boldsymbol{G}$ is a set of *nodes* $\boldsymbol{N} = \{x_1, x_2 \ldots x_n\}$ and a set of *directed edges* $\boldsymbol{E} = \{(x_i, x_j), \forall i \in N_O, j \in N_D\}$, where N_O is the subset of $\boldsymbol{N}$ that are origin nodes of edges and N_D is the subset of $\boldsymbol{N}$ that are destination nodes of edges.

We will use nodes to represent entities and directed edges to represent dependency relationships in our modeling paradigm. An edge that is directed from node $\boldsymbol{x}$ to node $\boldsymbol{y}$ is defined by the ordered pair $(\boldsymbol{x}, \boldsymbol{y})$. For our purposes loops, *i.e.* edges of the form $(\boldsymbol{x}, \boldsymbol{x})$, are not needed and will not be allowed. Each of the edges can be assigned a *weight*. For our purposes the weight will represent the *strength* of the associated relationship. In order to develop the analytical techniques that follow, it is sufficient to use *high*, *medium* and *low* as weights. We will use ***h***, ***m*** and ***l*** to abbreviate *high*, *medium* and *low* throughout the remainder of the chapter.

Figure 17.1 illustrates a directed graph with weighted edges.

Lower case italic letters such as $\boldsymbol{x}$, $\boldsymbol{y}$ and $\boldsymbol{z}$ will be used to represent *nodes*. Following the usual convention, $(\boldsymbol{x}, \boldsymbol{y})$ will represent the *directed edge* from node $\boldsymbol{x}$

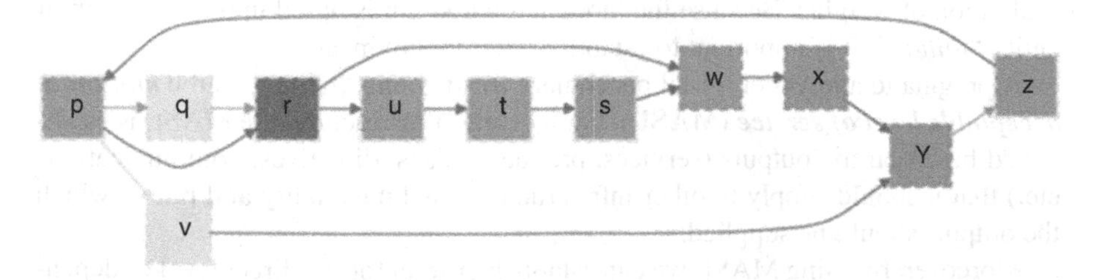

Fig. 17.1 A directed graph with weighted edges (*red* = high dependence, *orange* = medium dependence, *yellow* = low dependence)

to node y. In our modeling construct, directed edges of the form (x, x), (sometimes called loops), are not allowed; in other words, x and z must be *distinct* nodes.

A ***directed path*** in G is identified by a set of nodes $\{x_1, x_2 \ldots x_k\} \subseteq N \wedge (x_i, x_{i+1}) \in E, \forall i = 1, 2 \ldots k-1$. In plain language, a directed path is a sequence of nodes with the property that each node is connected to its respective successor by an edge in G. For example, in Fig. 17.1, $\{w, x, y, z\}$, is a directed path but $\{v, w, x, y, z\}$ is not because the edge (v, w) does not exist.

Note, that according to this definition, any edge $(x, y) \in G$ is also a path of length 1. Furthermore, if $(x, y) \in G$, we say that y is ***adjacent*** to x. And in the context of modeling dependency relationships, we also say that y is a ***direct dependent*** of x.

Because we will only consider directed paths, we will simply use ***path***, to mean *directed path*. The set of all paths from x to y in G will be represented by the notation $[x, y]$. Bearing in mind that $(x. x)$ is not allowed, all paths of the form $\{x = x_1, x_2 \ldots x_k = x\text{: } k \geq 2\}$ are called *cycles*; therefore, we use the notation $[x, x]$ to refer to all of the cycles passing through x. The notation $[x, G]$ represents the set of all paths from x to all nodes in G (including x itself). Similarly, $[G, x]$ represents the set of all paths from all nodes in G, (including x itself), to x.

In order to derive a method for estimating the impact of every node in G on all nodes in G, we will use path analysis. In particular, the analysis will be based on *paths of strongest impact*, from any node x to any node z (including $z = x$). Next, we will define what we mean by "paths of strongest impact", we will refer to them as ***strongest paths***, and we will use the notation $[[x, z]]$ to mean the set of strongest paths connecting x to z.

Each of the edges of G can be assigned a ***weight***. For our purposes the weight will represent the ***strength*** of the associated relationship. In order to develop the analytical techniques that follow, it is sufficient to use ***high***, ***medium*** and ***low*** as weights which will be associated with numerical values. However, the general method can support any finite collection of positive numerical weights.

For our purposes, a subgraph of G, call it G', is a graph determined by a subset E' of E such that only end-points of edges in E' are in N', the node set of G'. It will be useful to identify particular subgraphs of G that are determined by the weights of the edges in G. For example, we might want to consider the subgraph G' in Fig. 17.1 determined by the edges of level h. In this case, $E' = \{(v, y), (p, r), (t, p), (u, t), (t, s), (s, w)\}$ and $N' = \{r, p, v, y, t, u, s, w\}$

Depending upon the context of the model, it is necessary to define criteria for the evaluation of weights. Because the modeling paradigm is based upon the notion of entity "*failure*", it is important to establish what failure means.

A pragmatic and widely-used benchmark for defining "failure" is the ***minimum acceptable level of service*** (**MASL**). For a given infrastructure, the MASL is established based on the outputs (services, products, plans, directives, communications etc.) that it should supply to other infrastructures and the quality and rate at which the outputs should be supplied.

Moreover, by using MASL we can establish criteria for the direct level of dependence of one entity on another. For in as much as an entity "*fails*" if it falls below its defined MASL, we define the following criteria:

- if failure of entity x inevitably leads to failure of entity y, then y has a ***high direct dependency*** on x, and conversely x has a ***high direct impact*** on y.
- otherwise, if failure of entity x leads to degradation of entity y to the extent that y must enact a contingency plan or resort to alternate operating procedures in order to stay above MASL, then y has a ***medium direct dependency*** on x and conversely x has a ***medium direct impact*** on y;
- otherwise, if failure of entity x leads to significant degradation of entity y, but y can stay above MASL without significantly changing its operating procedures, then y has a ***low direct dependency*** on x and conversely x has a ***low direct impact*** on y.
- otherwise, if failure of entity x does not lead to significant degradation of entity y, then y has ***0 direct dependency*** on x, and conversely x has ***0 direct impact*** on y.

In the worst case, a medium direct dependency relationship could result in marginal acceptable level of service, under a contingency plan or alternate operating procedures. Similarly, in the worst case, a low direct dependency could result in marginal acceptable level of service under normal operating procedures.

If x is adjacent to y, then these criteria provide relatively straightforward rules for measuring the level of direct dependence of y on x and conversely the level of direct impact of x on y. The challenge now, is to enlarge them in order to generalize from *direct* to *indirect* dependencies. The definition of level of direct dependence is given with respect to a path of length 1, so by making a few more considerations we can use the same criteria for the level of indirect dependence by assessing impact propagated over paths of any length.

The criteria describe the effect of a *high* impact event (*i.e.* less than minimum acceptable level of service) on a direct dependent. However, we also need to estimate the effect of a *medium* impact event (*i.e.* possibly marginal acceptable level of service under alternate operating procedures) and a *low* impact event (*i.e.* possibly marginal acceptable level of service under normal operating procedures) on a direct dependent. There are two dimensions for this estimate: (1) the level of the triggering impact event; and, (2) the level of the direct dependency relationship. It is reasonable to expect that a strong triggering event will have little impact if the level of dependence is low; while even a relatively weak triggering event will be felt if the level of direct dependence is high.

Thus, we estimate that the propagated impact can be no higher than the lesser of the triggering level of impact and the level of dependence. For example, a medium level triggering impact acting over a low level of direct dependence will cause a low impact because the level of direct dependence is low; whereas, a medium impact trigger acting over a high level of direct dependence may cause a medium impact because of the high level of direct dependence.

In summary, we will use the following rule to estimate the propagation of impact along each edge in a path. For edge (x, y) with triggering impact at x of level = $I(x)$ and level of direct dependence of y on $x = D(x, y)$ then the level of the triggering impact at y is given by:

$$I(y) = \min\{I(x), D(x,y)\} \tag{17.1}$$

Just as the direct level of dependence of $\boldsymbol{y}$ on $\boldsymbol{x}$ is based on the presumption of failure of $\boldsymbol{x}$, the estimate of the level of indirect dependence of $\boldsymbol{y}$ on $\boldsymbol{x}$ resulting from a particular path $\{x = x_1, x_2 \ldots x_k = y\}$ will be based on the presumption of failure of $\boldsymbol{x}$.

Assume, therefore, that $\boldsymbol{x}$ has failed and that $I(x) = h$. Then by using the propagation rule, we find $I(x_2) = min\ \{h, D(x, x_2)\} = D(x, x_2)$ because $D(x, x_2) \leq h$. Subsequently, $I(x_3) = min\ \{D(x, x_2), D(x_2, x_3)\} = D^*$; $I(x_4) = min\ \{D^*, D(x_3, x_4)\}$, and so on As we proceed to $\boldsymbol{y}$ beyond the first edge in the path, each subsequent application of the propagation rule compares the lowest level edge yet encountered with the level of the next edge on the path and sets the triggering level of impact to the lower value. Therefore, the indirect dependence of $\boldsymbol{y}$ on $\boldsymbol{x}$ resulting from the selected path is the level of the lowest level edge along that path.

Of all paths from $\boldsymbol{x}$ to $\boldsymbol{y}$, we want to find the path or paths that propagate the greatest level of impact from $\boldsymbol{x}$ to $\boldsymbol{y}$. Equivalently, we want to determine the path or paths that determine $\boldsymbol{y}$'s greatest dependence on $\boldsymbol{x}$. By the conclusion of the previous paragraph, we know that the greatest level of impact is carried by the path or paths whose lowest level edge is the highest among all paths connecting $\boldsymbol{x}$ to $\boldsymbol{y}$. Any such path of shortest length will be referred to as a ***strongest path***.

For example, consider the graph in Fig. 17.1 and all paths connecting p to y. These five paths are as follows:

- Path 1 = $\{p \rightarrow q \rightarrow r \rightarrow u \rightarrow t \rightarrow s \rightarrow w \rightarrow x \rightarrow y\}$
- Path 2 = $\{p \rightarrow q \rightarrow r \rightarrow w \rightarrow x \rightarrow y\}$
- Path 3 = $\{p \rightarrow r \rightarrow u \rightarrow t \rightarrow s \rightarrow w \rightarrow x \rightarrow y\}$
- Path 4 = $\{p \rightarrow r \rightarrow w \rightarrow x \rightarrow y\}$
- Path 5 = $\{p \rightarrow v \rightarrow y\}$

Using the propagation rule, we find that the impact of p on y from Path 1 is medium by virtue of (p, q) and (q, r), the impact of p on y from path 2 is medium by virtue of (p, q) and (q, r), the impact of p on y from Path 3 is high by virtue of all of its edges being high level, the impact of p on y is high from Path 4 by virtue of all of its edges being high level, the impact of p on y from path 5 is low by virtue of (p, v). Therefore, the indirect dependence of t on s is high and the strongest paths are Path 3and Path 4.

In addition to the level of dependence that is assigned to the edges of the graph, the nodes of the graph are assigned two values: (1) ***direct level of impact***, and (2) ***direct likelihood of failure***. Respectively, these values are intended to measure the relative importance of a node in and of itself and the relative likelihood that it will fail independently.

At this point suffice it to say that the level of impact assigned to a node is a weight that reflects its importance relative to the others and that the likelihood of failure is a weight that reflects the relative likelihood that a node will fail because of random internal causes or non-random external threats that target it.

In order to accomplish the risk analysis according to Rowe's framework, we need to estimate the cumulative level of impact between every pair of nodes in the graph. While the strongest paths constitute the most significant influence of one node on another, we need to take into account all of the paths that connect a pair of nodes in order to assess the relative risk of every node.

To do this, we will define several utility functions. In what follows, we will use naming conventions, based on letters, for these functions:

- ***D***, level of ***Dependence***;
- ***I***, level of ***Impact***;
- ***F***, likelihood of ***Failure***; and,
- ***L***, path ***Length***.

There are four possible cases for the argument of each function:

1. the argument is a single node, x;
2. the argument is a single edge, (x, z);
3. the argument is a single path, $\{x_1, x_2 \ldots x_k\}$;
4. the argument is a set of paths, such as $[x, y]$, $[x, G]$, $[G, x]$, bearing in mind that any of these sets may be the empty set, $\varnothing$.

If the argument is a single node, x, then $I(x)$ is the direct impact of x on the general environment without regard for any of its explicit dependency relationships in the model; $F(x)$ is the believed likelihood of failure of x, without regard for any of its explicit dependency relationships in the model; $D(x)$ and $L(x)$ have no meaning (except trivially in terms of loops) and are consequently undefined (because loops are forbidden in our paradigm).

For example, $I(x)$ is the direct impact of node x on the general environment without regard to any of its explicit dependency relationships in the model; $I(x, z)$ is the direct level of impact of x on z from the direct relationship (x, z); $I([x, z])$ is the cumulative level of impact of x on z taken over all paths; $I([[x, z]])$ is the level of impact of x on z taken over a strongest path from x to z; $I(x, G)$ is the global impact of x on the entire graph G.

We have noted that all paths from x to y in G will be represented by $[x, y]$. Additionally, the notation $[[x, y]]$ will be used to represent any strongest paths from x to y in G, bearing in mind not only that such a path might not exist, but also that there might be more than one.

1. Strongest Path Impact of x on z:

$$I([[x,z]]) = D([[x,z]]) \times I(x)^{L([[x,z]])} \tag{17.2}$$

The strongest path impact of x on z is the strongest path level of dependence of z on x multiplied by $I(x)$ raised to the power of the length of a strongest path. Recalling that $I(x) \in [0, 1]$, the latter term acts as a damping factor. In other words, we allow that the ripple effect of propagated impact will decrease in severity as it moves away from x. Note also, that if $[[x, z]] = \varnothing$, then $D([[x, z]]) = 0$ and therefore $I([[x, z]]) = 0$.

2. The *direct* impact of x on z can be extended to define the *cumulative impact* of x on z as $I[x, z]$ as follows:

$$\text{if}\,[x,z] = \varnothing, \text{then}\, I([x,z]) = 0; \text{otherwise},$$

$$I([x,z]) = 1 - \Pi_{(y,z)\in E}\left(1 - \min\left\{D([[x,y]]), D(y,z)\right\}\right) \times I(x)^{L([[x,y]])+1} \quad (17.3)$$

The cumulative impact of x on z includes a term for every pathway that exists from x to z. Let $y \in N$ be the set of nodes on which z is directly dependent. For each of z's direct dependencies, the strongest path impact of z on x through y, is factored into the function. The contribution of each of these paths is damped by the factor $I(x)$ raised to the power of the path length. Similar to the binomial probability function, it compounds the effects of the terms. Note as well, that if $[x, z] = \varnothing$, then $I([x, z]) = 0$.

3. The term "interdependency" is used by many authors regarding the mutual level of dependency between pair of entities in an infrastructure model. In order to quantify this value, we can define the mutual interdependency index of x and z as $M(x, z)$:

$$M(x,z) = I(x,z) \times I(z,x) \quad (17.4)$$

The mutual interdependency index of x and z is the product of the cumulative impact of x on z times the cumulative impact of z on x. This gives a measure of the ***level of interdependency*** that exists between x and z. Clearly, $M(x, z) = M(z, x)$.

4. Global Impact of x on G:

$$I([x,G]) = \Sigma_{z\in N}\left(I([x,z]) \times I([z])\right) / \Sigma_{z\in N}\, I([z]) \quad (17.5)$$

By "global" we mean the impact of any node x on the entire model. Consider the impact of a node, x, on every individual node, z, in the graph (including x itself). The cumulative impact of x on z is given by $I([x, z])$ and the relative importance of z itself is given by $I([z])$. The terms $I([x, z]) \times I([z])$ give the global significance of the impact of x on z. The sum of these products gives the impact of x on the entire model. This can be regarded as a sum of cumulative impact values weighted by direct impact values. In order to make a better relative comparison of different versions of a model, we normalize the weighted sum by dividing by the sum of the weights.

5. Cumulative Vulnerability of z from $\boldsymbol{x}$:

$$\boldsymbol{F}([x,z]) = \boldsymbol{I}([x,z]) \times \boldsymbol{F}(x) \tag{17.6}$$

There are two factors about $\boldsymbol{x}$ that influence the possible failure of z: the cumulative impact of $\boldsymbol{x}$ on z and the likelihood of failure of $\boldsymbol{x}$. Therefore, the cumulative vulnerability of z from $\boldsymbol{x}$ is the product of the cumulative impact of $\boldsymbol{x}$ on z and the likelihood of failure of $\boldsymbol{x}$.

6. Global vulnerability of **z** from $\boldsymbol{G}$:

$$\boldsymbol{F}([\boldsymbol{G},z]) = \boldsymbol{1} - \Pi_{x \in N}\left(\boldsymbol{1} - \boldsymbol{F}([x,z])\right) \tag{17.7}$$

The cumulative vulnerability of z from all nodes in the model is the binomial probability that a failure event of any node will cause z to fail.

7. The global risk index of $\boldsymbol{x}$, called $\boldsymbol{R(x)}$, is defined as the product of the global impact of $\boldsymbol{x}$ on all entities in the model and the global vulnerability of $\boldsymbol{x}$ from all entities in the model:

$$\boldsymbol{R}(x) = \boldsymbol{I}([x,G]) \times \boldsymbol{F}([G,x]) \tag{17.8}$$

The global risk index of an entity is the product of the global impact of an entity times its global vulnerability. This provides a single score for comparing risk among all of the entities in the model.

The analytical tools needed for risk analysis are now complete, except for a description of how to find the strongest path level of dependence between all ordered pairs of nodes. A modification of any "*shortest path*" algorithm suffices; we will call the algorithm, ***Strongest Path.***

It is the case that the problem of finding a strongest path in our modeling paradigm is equivalent to a number of other problems in graph theory such as the Widest Path problem (Donavalli 2013). Hu (1961) outlines the others. It is known that all can be solved by modifying Dijkstra's algorithm for finding the shortest path between any pair of nodes in a directed graph.

In graph $\boldsymbol{G}$ let $\boldsymbol{G^*}$ be the subgraph of $\boldsymbol{G}$ generated by edges of level $\geq$ *. Assume that **G** has a scale for level of dependence = $\{1,2, \ldots, k\}$. If $[x, z] \neq \varnothing$, let $S^*(x, z)$ be the length of the shortest path from $\boldsymbol{x}$ to z in $\boldsymbol{G^*}$; if $[x, z] = \varnothing$ then $S^*(x, z) \equiv \boldsymbol{0}$.

The following algorithm is intended to show that the level of the strongest path between every pair of nodes can be found in polynomial time. This algorithm can be implemented using any shortest path algorithm. Efficiencies with the choice of shortest path algorithm and the organization of the search can be exploited to reduce the execution time.

```
Algorithm: Strongest Path

    Set D([[x, z]]) ← 0 for all ordered pairs of nodes {x, z},
x∈N, z∈N
```

```
    For * = k to 1, step = −1
        For all ordered pairs {x, z}, calculate S*([x, z])
        Set D([[x, z]]) ← max {D([[x, z]]), S*([x, z])}
    Next *

End: Strongest Path
```

The strongest path algorithm initializes all ***D*([[*x*, *z*]])** to ***0***. It then searches for a shortest path connecting **x** to **z** in a sequence of subgraphs starting with G^k, then G^{k-1} and so on to G^1. For each ordered pair **{x, z}** it saves the first non-zero shortest path length that it finds as ***D*([[*x*, *z*]])**. Subsequently it does not change that value of ***D*([[*x*, *z*]])** throughout the remainder of the search, because it can only encounter ever decreasing values of ***S***([*x*, *z*])**. If a shortest path is not found, then there is no path from x to z and consequently, ***D*([[*x*, *z*]]) = 0**, its initial value.

17.4 Example

We will continue the discussion of the risk synthesis method with reference to a particular case study. In 2007, the Strategic Advisor to the Ministry of Education in Afghanistan designed a system for the printing and distribution of textbooks to schools throughout the country (Hu 1961). The example presented here was developed in consultation with Mr. Rory Kilburn and is based on his knowledge of the eventual implementation of the system in Afghanistan.

There are 11 entities in the model based on the description of the supply chain given in Hu (1961):

- Annual Planning
- Estimation of Required New Textbooks
- Printing
- Registration of New Books in the Inventory Management System
- Delivery to Regions
- Delivery to Districts
- Distribution to Schools for School Year Beginning
- Collection at School Year End
- Annual Assessment of Inventory
- Storage
- Annual Report

The next step is to establish criteria for level of impact, likelihood of occurrence and level of direct dependence for these entities.

The primary purpose of this supply chain system is to deliver the required textbooks to schools throughout the country for the beginning of each school year. Therefore, the level of impact has been established based on consequence to the textbook supply if an entity is deemed to have failed. Five levels of impact have been defined as shown in Table 17.1.

Likelihood of occurrence has been established based on observed outcomes in Afghanistan. The results are shown in Table 17.2.

Finally, criteria for level of dependence are given in Table 17.3.

The entities and dependency relationships were scored according to the defined criteria. Entity scoring appears in Table 17.4. Note that in Table 17.4, the risk index of each entity is the product of its level of impact times its likelihood of occurrence. Scoring of dependency relationships appears in Table 17.5.

The directed graph for the model appears in Fig. 17.2. The edges of the graph are color-coded according to the level of dependence that they represent. Red denotes 8, and orange denotes 5. There are no relationships with a level of 2. The nodes are colored according to their level of impact: red for 8, orange for 5 and yellow for 2.

Table 17.1 Criteria for level of impact

Level of impact	Value	Description
Very low	0	Little or no impact (≤ 20% of requirement)
Low	2	Minor impact (>20%, ≤ 30% of requirement)
Medium	5	Moderate impact (>30%, ≤ 40% of requirement)
High	8	Major impact (>40%, ≤ 50% of requirement)
Very high	10	Disastrous impact (>50% of requirement)

Table 17.2 Criteria for likelihood of occurrence

Likelihood of occurrence	Value	Description
Very low	0	Unlikely to happen (< 1 year out of 7)
Low	2	May happen occasionally (1,2 years out of 7)
Medium	5	May or may not happen (3,4 years out of 7)
High	8	Is likely to happen (5,6 years out of 7)
Very high	10	Is almost certain to happen (> 6 years out of 7)

Table 17.3 Criteria for level of dependence

Level of dependence	Value	Description
Low	2	Without invoking a contingency plan, the dependent entity will sustain low impact if the providing entity sustains a high level of impact
Medium	5	By invoking a contingency plan, the dependent entity will sustain low impact if the providing entity sustains a high level of impact
High	8	The dependent entity will unavoidably sustain high impact if the providing entity sustains a high level of impact

Table 17.4 Scores for entities based on the defined criteria

Entity name	Level of impact	Likelihood of occurrence	Risk index
Annual Planning	8	2	16
Estimation of Required New Books	5	2	10
Printing	5	2	10
Registration into Inventory System	5	2	10
Delivery to Regions	5	5	25
Delivery to Districts	5	5	25
Distribution to Schools	5	5	25
Collcction at Year End	5	2	10
Assessment of Inventory	5	2	10
Storage	2	2	4
Annual Report	2	2	4

Table 17.5 Level of dependence for each dependency relationship

Providing entity	Dependent entity	Level
Annual Planning	Printing	8
Annual Planning	Estimation of Required New Textbooks	5
Estimation of Required New Textbooks	Printing	8
Printing	Registration of New Books in the Inventory Management System	8
Registration of New Books in the Inventory Management System	Delivery to Regions	8
Delivery to Regions	Delivery to Districts	5
Delivery to Districts	Distribution to Schools for School Year Beginning	5
Distribution to Schools for School Year Beginning	Collection at School Year End	5
Collection at School Year End	Assessment of Inventory	8
Assessment of Inventory	Storage	5
Storage	Distribution to Schools for School Year Beginning	8
Assessment of Inventory	Annual Planning	8
Annual Report	Annual Planning	5
Assessment of Inventory	Annual Report	5

There are 11 entities and 14 direct dependency relationships in the model.

With traditional risk analysis, the entities would be prioritized according to their risk index. For the example, the entities "Delivery to Regions", "Delivery to Districts" and "Distribution to Schools" would be flagged as the highest risk because they each have a risk index of 25.

However, the inclusion of path analysis in the estimation of risk provides a different assessment. Figure 17.3 shows a plot of the cumulative impact (x-axis) versus the cumulative likelihood of failure (y-axis) for each entity.

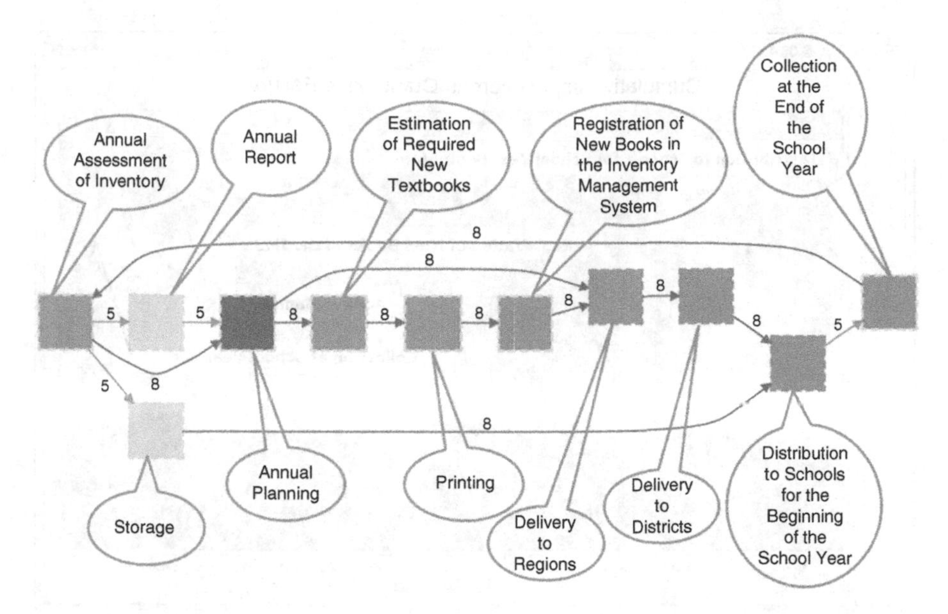

Fig. 17.2 The supply chain model

Recall that the cumulative risk index of an entity is its cumulative impact times its cumulative likelihood of failure in the model. By including path analysis in the risk assessment, the entities, "Collection at School Year End", "Annual Planning", "Printing" and "Registration of New Books in the IMS" have been identified as the highest risk.

The entity "Delivery to Schools for School Year Beginning" has been labeled with its cumulative likelihood of failure score, which is 73.5.

The main purpose of the supply chain is to deliver adequate supply of textbooks to schools for the beginning of the school year. Thus, from a resiliency point of view, our primary objective is to make the cumulative likelihood of failure score for the entity "Distribution to Schools for the Beginning of the School Year" (which is currently 73.5) as low as possible. How might this be accomplished?

Consider the four entities that have been rated as highest risk and consider if we might be able to take any actions to reduce either direct impact or likelihood of failure for each of them:

Annual Planning – this entity is necessarily high impact and currently has low likelihood of failure therefore no action needs to be taken

Registration of New Books in the IMS – this entity has high impact necessarily and currently has low likelihood of failure and therefore no action needs to be taken.

Collection at the End of the School Year – Action should be taken to lower its likelihood of failure which is currently 5.

Printing – By putting more books into storage at the end of the school year, we can lower the impact of printing new books.

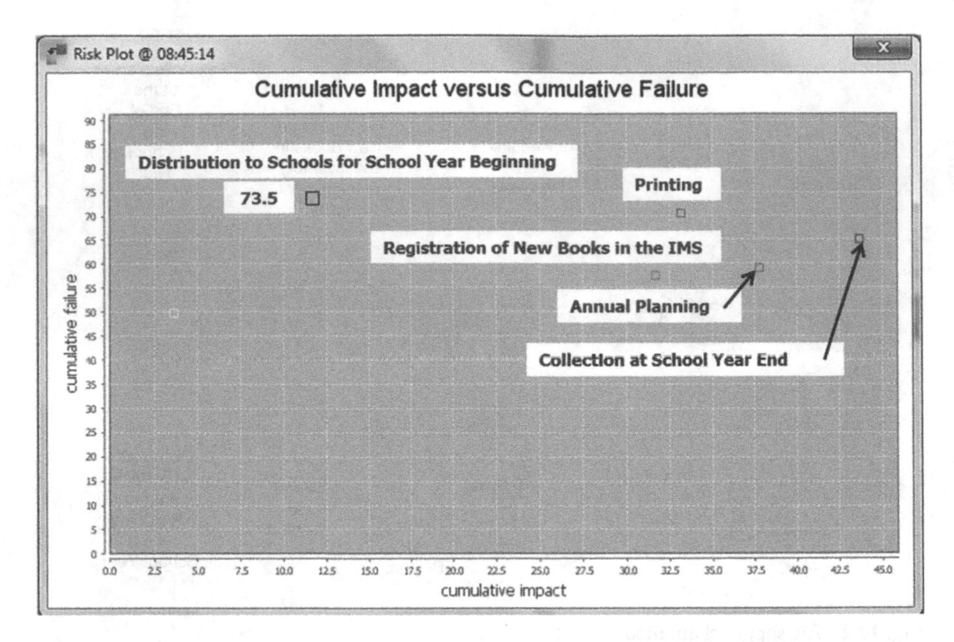

Fig. 17.3 Plot of cumulative impact versus cumulative likelihood of failure for the model shown in Fig. 17.2.

We will consider one other action for comparison. What would be the impact of lowering the likelihood of failure of the entity "Estimation of the Required Number of New Textbooks"?

The next step is to consider the effect of the three proposed actions on the cumulative likelihood of failure of the entity "Distribution to Schools for the Beginning of the School Year" by changing appropriate data in the model. The results are shown in Fig. 17.4.

The results of the proposed actions are revealing. If we improve the reliability of the "Estimation of Required New Textbooks" the effect is marginal. Lowering the impact of "Printing" has a small effect as does improving the reliability of "Collection at School Year End". However if both lowering the impact of "Printing" and improving the reliability of "Collection at School Year End" are carried out together then there is a dramatic improvement of resilience.

Therefore, decision makers should implement a more reliable process for collecting textbooks at the end of the school year and by putting more books into storage will lower the impact of printing new books. This will improve the resilience of the supply chain immensely.

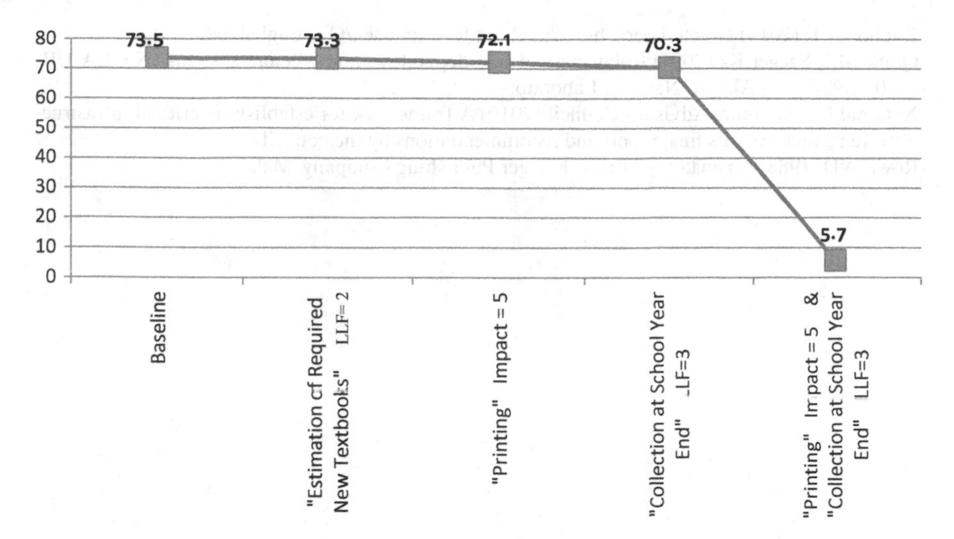

Fig. 17.4 Effect of the proposed actions on resilience

17.5 Conclusion

The risk synthesis method is a paradigm for modeling systems by using a directed graph. Models are constructed from entities that are assessed with a level of impact and a likelihood of failure together with dependency relationships that are scored for level of dependence according to well defined criteria.

The paradigm allows the knowledge of experts to be directly used for risk analysis. Results from other analytical models, such as simulations, can also be included in a model. As a result of performing the path analysis, such models reveal the potential consequences of the failure of any entity on all of the others. This enables contingency planners to anticipate all outcomes in any infrastructure damage scenario.

Resilience can be modeled with the risk synthesis paradigm.

Moreover, the four characteristics of resilience that the NIAC identified can be assessed with such a model.

The modeling paradigm has been implemented as commercial off-the-shelf software under the trade name RiskOutLook®. It is available for licensing from RiskLogik.

References

Donavalli VS (2013) Algorithms for the widest path problem

Ganin AA et al (2016) Operational resilience: concepts, design and analysis. Sci Rep 6:19540. doi:10.1038/srep19540

Hu TC (1961) Letter to the editor—the maximum capacity route problem. Operat Res 9(6):898–900

Ruohonen K (2013) Graph theory. http://math.tut.fi/~ruohonen/GT_English.pdf
Quirk MD, Saeger KJ (2004) Robustness of interdependent infrastructure systems (No. LA-UR-04-6927). Los Alamos National Laboratory
National Infrastructure Advisory Council (2010) A framework for establishing critical infrastructure resilience goals final report and recommendations by the council
Rowe WD (1988) An anatomy of risk. Krieger Publishing Company, Malabar

Chapter 18
Enhancing Organizational Resilience Through Risk Communication: Basic Guidelines for Managers

D. Antunes, J.M. Palma-Oliveira, and I. Linkov

Abstract RESILENS is a project funded by the European Union's Horizon 2020 Research and Innovation Programme for the development of tools that support managers with the improvement of resilience pertaining to Critical Infrastructure (CI). One of the tools developed by this project is the European Resilience Management Guidance (ERMG). This tool guides several topics (e.g. risk management, budget and financial issues, information management systems, business continuity, and risk communication) related to the practical application of resilience to all CI sectors. This paper, which was developed as an input for the ERMG, presents some guidelines on risk communication that can be useful to managers.

The research leading to these results has received funding from the European Union's Horizon 2020 Research and Innovation Programme, under Grant Agreement no. 653260 corresponding to the RESILENS project.

D. Antunes (✉)
Factor Social, Lda, Lisbon, Portugal
e-mail: Dalilaantunes@factorsocial.pt

J.M. Palma-Oliveira
University of Lisbon, Lisbon, Portugal

I. Linkov
US Army Corps of Engineers Research and Development Center,
Concord, MA, USA

© Springer Science+Business Media B.V. 2017

I. Linkov, J.M. Palma-Oliveira (eds.), *Resilience and Risk*, NATO Science for Peace and Security Series C: Environmental Security,
DOI 10.1007/978-94-024-1123-2_18

18.1 Introduction

RESILENS is a project funded by the European Union's Horizon 2020 Research and Innovation Programme for the development of tools that support managers with the improvement of resilience pertaining to Critical Infrastructure (CI). One of the tools developed by this project is the European Resilience Management Guidance (ERMG). This tool guides several topics (e.g. risk management, budget and financial issues, information management systems, business continuity, and risk communication) related to the practical application of resilience to all CI sectors. This paper discusses the benefits of ERMG for the field of risk communication.

The ERMG risk communication guidelines follow the same guidelines as those presented for ERMG resilience management: (i) Prepare, Prevent and Protect (prior to disruption); (ii) Mitigate, Absorb and Adapt (during disruption); and (iii) Respond, Recover and Learn (following disruption) (Fig. 18.1). We discuss this structure for risk communication.

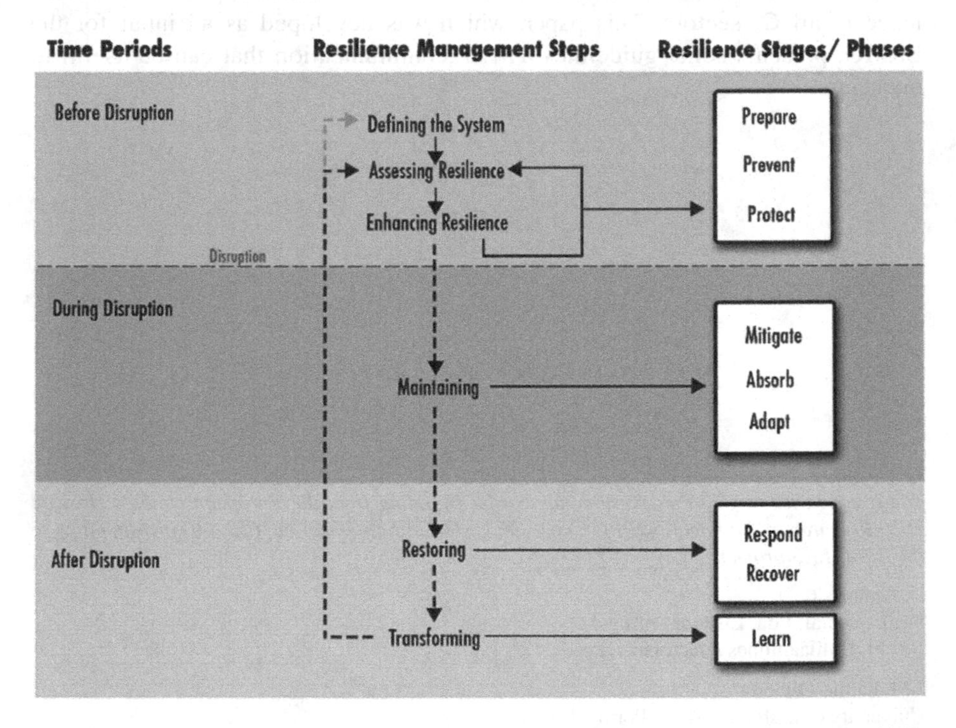

Fig. 18.1 Resilience management steps according to ERMG (Source: RESILENS 2017; Deliverable D3.2: Draft ERMG in www. Resilens.eu)

18.2 Prepare, Prevent and Protect

Organizations are sociotechnical systems with defined physical and social internal constructs that exist within more expansive social, physical and environmental external contexts. The establishment of risk communication standards at internal and external levels is useful to improve the detection of events that can lead to accidents and to provide guidance in the case of an incident or breakdown (Arvai and Rivers 2014). Internal and external communication activities are widely referenced by ISO 22301:2012 under the section regarding establishment and implementation of Business Continuity Procedures.; it is recognized in this section that poor internal and/or external communication can lead to crisis (e.g. BS 11200 – BSI 2014).

When planning for organizational resilience it is important to note that risk perceptions influence the perceived importance of risk and, therefore, the drive to identify and assess risks, as well as the seeking out of critical services and risk management (Palma-Oliveira et al. 2017). Risk acceptance levels also frame organizational risk management goals based on the willingness of the organization to undergo risk exposure. Consequently, risk management decision-making (defined as risk treatment, according to ISO 31000:2009) and budget allocation are also influenced by these perceptions (International Standards Organization 2009a, b).

It is important to understand whether risk perceptions are misleading decision-making processes regarding risk and business-continuity management. Furthermore, guidance on risk and business continuity management cannot be provided unless risk perceptions and technical analysis are understood and balanced. Communication of plans regarding risk and business continuity management must be made with stakeholders, as well as those inside and outside of the organization. As ISO 22301:2012 points out, the Business Continuity Procedures shall establish an appropriate internal and external communications protocol, among other issues.

Below, we posit several points to consider relative to internal and external communication:

18.2.1 Internal Communication

1. **Identify workers' risk perception and risk acceptance levels**

First, understanding risk perception and acceptance can help us predict how managers will behave regarding risk management decision-making. Secondly, understanding risk perception and acceptance within the context of the organizational system and leadership structure allow us to interpret how workers will react to communications on risk management actions including plans for achievement of Business Continuity Objectives. Individuals with low risk perception or high levels of risk acceptance may be inclined to neglect training and execution of preventative measures. It is important to understand if workers' low risk perceptions exist because they anticipate and are prepared to manage risks or because their perceptions are faulty. In some cases, previous misconceptions need to be corrected so that individuals recognize the usefulness of preventive and reactive measures. Hence, adequate levels of risk perception must be monitored and controlled throughout the organizational risk management process (i.e., low before planning, higher when destroying misconceptions, low after providing the right strategies to prevent and act in case of an accident).

Qualitative and quantitative methods can be used to monitor risk perception and acceptance within the organization. Risk perception assessment includes not only the identification of factors to be assessed, but also specificities related to different groups (e.g., gender, age, and culture) and sectors of activity (for instance risk perception on traveling may be assessed for transport sector; while on water or health sector risk perception towards biological contamination may be analyzed). Consultants or researchers familiar with the topic and experienced on assessing risk perception and acceptance levels can be specifically hired for such a task.

2. **Map internal stakeholders**

Workers are needed for an organization to function, but some workers are needed to make decisions regarding critical processes that need to be maintained in the case of an event or incident. These workers need to be identified through the evaluation of workers' contributions regarding the type and the priority of communications for different phases.

(a) **Critical groups**

Risk assessment of the organization will identify not only the most risky processes but also the groups of workers managing those processes. These are critical groups of workers to address regarding the fit of perceived risk and risk assessment and the recognition of safety and security procedures (preventive and reactive).

(b) **Vulnerable groups**

Organizations shall determine whether vulnerable groups and services exist in the organization that require specific attention in the case of an event. For instance, wheelchair-dependent individuals may need a specific plan for evacuation and deaf-toned individuals may require visual signs for the alert of an incident. Each group shall be considered when developing safety and emergency plans. Furthermore, special communication actions need to be considered for informing these groups and the people involved on measures specifically addressing them.

3. **Definitions of the procedures for the communication events**

Some minor incidents can be solved inside of teams while others require collaboration with other departments (i.e., interdependencies within organizations). Also, one incident may start within one team or department and escalate towards a bigger problem. Processes addressing how such analysis and communication occurs inside teams, and who has the authority and/or responsibility to recognize the need for taking the problem to a different department and/or level should be established in order to properly manage organizational resources and minimize disruption, as well as panic or distress.

For those organizations following ISO 2230:2012 these are some of the issues that should be covered by the internal communications protocol.

4. **Identification with the organization and work group**

Communication is not a simple and straightforward process. People have different values and mindsets that can mislead interpretation of messages during communication. Individuals will be more effective in communicating with others if they identify themselves with the organization and with the group.

When developing teams for risk management or business continuity management it is important to provide some time for the team to know each other work and their roles so the work between them can be more fluid and effective. This is also important when developing intervention teams. Team members should know each other, have a clear idea of their own role and other roles, and be able to act almost automatically in face of an event. In the case that emergency response includes outside services to guarantee the work of the organization it is important to establish previous contacts between the members of the organization and the outside services. If the organization has a Business Continuity Management System in place, people that should be considered for training communication for incident response should at least include those people identified on the incident response structure (ISO 22301:2012).

18.2.2 External Communication

1. **Map stakeholders**

All the stakeholders are important for the organization, but some of them are essential for the services of that organization to continue in the case of an incident. It is important to not only identify stakeholders but also to evaluate the contribution of the stakeholders and to group them regarding the type and the priority of communications for different phases. Examples of stakeholders are: workers, workers' families, customers, partners, local community, media, etc.

Several tools are available for mapping stakeholders. For this purpose the person responsible for mapping stakeholders needs to make sure each stakeholder listed is evaluated regarding two-ways interdependencies (stakeholder is critical for the organization, the organization is critical for the stakeholder) and that at least it considers all Critical Infra-structures on the external stakeholders list.

(a) **Stakeholders who can support management of external risks**

Special attention should be paid to external events that can pose risks to the organization. Not only the failure of subcontractors should be considered, but other environmental and man-made disasters which can affect the work of the organization should be foreseen. Communicating with managers of such menaces (nearby industries, road managers, government environmental services) to better understand the extension of risks and the organization role in case of an event is desired.

Articulated emergency plans and recovery plans should be developed together with relevant stakeholders. Regulators should be heard when developing organization risk management plans (several risk management plans require approval depending on national laws), and relevant subcontractors should be engaged when developing recovery plans. Civil Protection services shall be included on developing organization emergency and recovery plans. Nearby industries posing risks that can affect the organization and generate domino effects should be integrated into response plans. Integrated action between parts shall be envisaged on those plans.

(b) **Identify population affected by the incident**

Everyone who may be affected by an event is considered an interested part and communicate with stakeholders that can be affected in case the organization faces an accident so they can also be prepared may be part of the organization social responsibility. Still this is audacious and may not be adjusted to all organizations. Risks and benefits of such approach should be carefully considered beforehand.

In this case communication needs to be addressed properly in a way that does not raise panic or inappropriate risk perceptions towards the organization which will have a strong negative impact on its image. The communication must be different through different populations affected and different population characteristics including risk perception and risk acceptance. Such communication strategy also needs hard work on developing and pre-testing proper messages that support not only preparedness of surrounding communities but also a positive image towards the company. Analyzing case-studies must be valued but replication of messages used by other organizations or even by the same organization in other contexts without previous testing is strongly discouraged.

(c) **Media**

Media are important stakeholders that can be helpful on communicating when an event occurs, hence they should be considered as part of the strategy to prepare for the case of an event. Prepare some first communication messages to be delivered in case of an event is important as if an event happens priority focus is emergency and no one will really have much time for developing a proper communication message.

ISO 22301:2012 lists some issues of the organization's media response following an incident that should be contained in the Business Continuity Plans, but it does not provide a guidance on how to perform communication. Specialized consultants on risk perception and risk communication and/or in crisis management can be asked to support the organization on developing such plans, as communication departments are usually more specialized on brand communication (it is important to notice brand communication and risk communication are quite different regarding the way people interpret the messages) (British Standard Institute 2014). Legal departments may be engaged on this work regarding preventing messages that may commit the liability of the company.

Previous prepared messages should be general and applicable to all types of events as they aim to provide time to the company to focus on emergency procedures before communicating to general public.

2. **Take stakeholders as part of risk management system**

The organization exists in a context with includes local communities, regulators, and other external sources of risk. Hence, the organization's risk management plan should also consider the risks posed by local stakeholders, their acceptance levels of risk, and moreover their ability to respond to risks posed by the organization.

(a) **Definitions of the procedures and responsible people for performing communicating during crisis**

The procedure to communicate incidents between the organization and the external stakeholders should to be developed before the incident happens (also in accordance with international norms ISO 22301:2012 and BSI 11200:2014). This will help to avoid panic and minimize distress and to provide better phased and prioritized connection with stakeholders.

When defining the procedures, it is also essential to define who are the people responsible for establishing the contact between the organization and each group of external stakeholders. For instance, one Human Resources representative may be appointed to contact workers' families, an Health and Safety officer may be responsible for making contact with Firefighters or Medical Response teams, one representative of the Communication department may be appointed for communicating with media, etc.

Nowadays social media can support this relationship if properly managed, or completely destroy company plans. Procedures to be followed by employers and subcontractors regarding the use of social networks for communicating company related events should be developed.

(b) **Defining regular communications**

If someone only contacts a counterpart during emergency communications then the contact may not be successful. The person may be on holidays, has changed functions or moved places or changed phone numbers. Having a regular contact with stakeholders which are relevant for response in case of an accident may prevent not to find him/her when needed. Establishing the goal of regular messages; frequency of contacts; how to perform the contact; and a getting a link for an alternative contact in case of emergency is desirable.

18.2.3 Training

Training is part of all management systems, as people need competence to perform the roles they are assigned. Testing (also required by managing systems) is also part of that training.

Training on communication should be provided to all people engaged on managing emergency situations or any kind of risk communication. Competences on how to communicate risks effectively, how to communicate during a crisis situation, how to communicate with media, are important issues to be included on training the teams.

Effectiveness of training actions should be evaluated through tests or demonstrations of the process in case of an incident. The communication in this training exercises or tests shall be similar to a real incident. The use of a code word for tests and pilots is desired to prevent that people not engaged on an exercise who receive an information from the training can recognize it as not real, but if receiving a real one can understand that it is not generated by the test.

18.3 Mitigate, Absorb and Adapt

Risk communication during the incident requires triggering response mechanisms as well as coordinating actions (Lundgreen and McMakin 2013).

18.3.1 Internal Communication

1. **Codification of messages in case of warning**
   When an incident is happening it is necessary to keep simple codes to warn the start of the correct and sequential procedures in the organization. The message code may be different if referring to different procedures and meaning shall be previously known. It is always important to make sure that the code messages for training situations are never the same as the ones one would use on a real situation in order to reject any doubt involved.
2. **Implementing criteria for warning escalation**
   The definition of criteria for escalating an alert should be considered according with the different organization sectors and hierarchical levels and put in practice accordingly with the organization plan.
3. **Execution of the communication processes**
   On this phase the implementation of standardized procedures is required. Granting the availability of the information (names and contacts of people to contact) and communication channels for internal and external communication is required.
4. **Coordination team**
   In order to effectively respond to an accident, and whenever it involves the engagement of several departments there is often the need to establish a coordination team that can discuss the better response options of the situation and how to proceed within the existing plans available. This team needs to have connection to local information so they can perform decisions and inform about actions to display.

The coordination team should be adjusted to the situation. If the situation is significant and a crisis is declared a Crisis Management Team may be called to action. BS 11200:2014 may be used for more detailed information on how to differentiate an event from a crisis and for providing more information on crisis management team.

5. **Response recording**

   During response the teams shall be encouraged to develop annotations. A logging system for notes may be previously developed and trained so each person can make notes about their actions and instructions provided along time. These are important materials to be reviewed and analyzed in order to generate learning for the organization to improve its plans.

18.3.2 External Communication

1. **Crisis communication team**

   In face of an event that requires communication with external stakeholders, there should be a communication team prepared to respond to leaks of information about the event. The company may prefer to always use a crisis communication team or part of it as it can act fast if needed, on one hand; and fast action can prevent events to turn into crisis, on another hand. This team should manage at least the following task:

- monitoring media and social media information to check about communications related with the event,
- inform decision makers about public information displayed
- advise workers on how to inform family and friends and how to proceed if directly contacted by the media
- deployment of information accordingly with the instructions of the Crisis Management Team or Coordination Team

2. **Stakeholders communication – execution of the activation processes**

   The organization needs to assure the communication with the external stakeholders. The organization shall have guidelines and procedures to be activated in case of an incident. Who (inside the organization) talks to whom (outside the organization), when, using which channel (phone, radio, fax, mail, letter) and providing what kind of message (content – examples of sentences to complete with real time information can be made available) should be clearly defined for different type of scenarios.

   The manual containing such instructions should be easily available for people having responsibilities to communicate with external stakeholders.

(a) **Special cases**

   Usually, civil protection services take care of accidents affecting local communities. Nevertheless some organizations deal with specific processes or products which require special countermeasures and specific resources and knowledge. There should be specific communication programs for this cases in order to make sure usual response measures are not activated by civil protection teams and proper response is provided on time.

3. **Public communication**

   Sometimes there is the need to make an event public or to respond publicly as an event was made public by someone (most often). A few times there is the need to alert local surrounding communities.

   These kind of communications are highly sensitive so it is expected they are addressed either at top level of the company (CEO), by the Communication Director, or by a Public Relations representative (internal or subcontracted for the purpose). Because this communication is so sensitive there is the need to plan it previously and to properly train communicators to preserve organization image at critical times.

   Planning it in advance allows not only to train people for it but it may also allow testing different communicators to understand who is more trusted by the public and who provides clearer messages that enhance action.

   Legal representatives of the company may be engaged on defining risk communication messages or defining the scope of communications to prevent legal actions against the company based on company public communications. They may be engaged not only on defining pre-set messages and on training but may also be part of the crisis communication team.

4. **Communication with families and victims**

   When an incident occurs and if it becomes aware to the public, the families of the workers will be distressed, and will try to contact them or search for information. Some or most of workers will have a role on managing the event and will not be able to answer their phones. Nevertheless, it is important they can abstract from it by knowing someone will take care of it (talk to their family and eventually manage to take care of their family duties they will not be able to fulfil – for example get kids from school).

   Special services should be organized for taking care of these communications with workers' families and making sure the fact a worker is engaged on organization business continuity plan does not induce crisis on their private life.

   The information about the victims or given for the victims' family has to be treated carefully. National legal requirements about this issue should be considered when planning for it.

18.4 Respond, Recover and Learn

After an incident several scenarios may pose. Continuity may proceed on same place or not depending on the extension of the damages and on business continuity plans.

Communication has the goal of controlling workers expectations regarding what they are going to find – physical and social context, resources available, information about previous and new team – and motivate them for implementing recovery tasks or to continue working. Monitoring psychosocial status of workers is important as the crisis event may induce changes on risk perceptions, attitudes towards safety and the organization, and attitudes towards colleagues which can affect the way they will react to risk communication in the future.

18.4.1 Internal Communication

1. **Receiving recovery teams**

   After an incident recovery plans will be executed. This means that different teams will start working at different times. It is important to receive teams in a way to provide them confidence and focus them directly on goals to be attained for the recovery periods and critical tasks to be shortly executed. It is important to understand some of these teams will be working under stress conditions that enhances the risk for wrong assessments and bad decision-making. Special control, monitoring and support of these teams is recommended.

   Availability of resources that go beyond resources to perform tasks should be provided (e.g. water, food). Someone must have the responsibility to check people are having food as well as taking breaks and respecting rest periods.

   It is important for the organization to previously plan and control the information about the accident regarding what is happening or what happened on the event as teams will be working at different pace and depending differently from information about the event. Different messages shall be prepared for different sectors.

   If the organization decides no information should be provided to a recovery team measures will be needed to make sure team members have no access to information about the accident provided by third parties.

2. **Workers that are not part of the recovery teams**

   After an event, workers may have uncertainty regarding what will happen to the organization, if they still have a job, and certainly not all of them will have the same answer. Communication plans should be developed to communicate with workers along time. Different communication phases can be established

(a) **Vulnerable groups of workers**

   After an accident, there may be new vulnerable groups of workers requiring specific attention: injured workers, workers that will be dismissed (because of organization restructuring requirements due to the event or because of related reasons). It is important to understand the risks these groups face and communicate them in a way to minimize distress.

3. **Workers that were part of the emergency and/or crisis communication teams**

   Workers that were part of emergency teams and/or crisis communication teams should be paid special attention. On one hand, they may have useful information regarding the response procedures and lessons learned so debrief sessions may be developed with them. Information need to be registered and make sure it influences improvement of risk communication procedures.

On another hand, if subjected to significant distress situations, they may need psychosocial support and/or stress management.

18.4.2 External Communication

1. **Communicating with stakeholders engaged on crisis management**

   Communicating with other stakeholders engaged on crisis management is important to become aware of their notes and registers on the response which can support learning. Meeting with them for debriefing and analyzing how to improve plans together may bring organization's resilience to a new level.
2. **Communication with subcontractors**

   Both new and previous subcontractors required for working with the organization in implementing recovery plans may see the situation as an opportunity to inflate negotiations for their side. After all the organization is working to reestablish its operation functions and is deeply in need of support. Time is critical and there is not much time for searching or analyzing different bids. That is why subcontractors that are foreseen to be required during recovery phase should be contacted previously to accident (on prevent phase) and pre-contract agreements are performed that limit values charged for their work.

   If no previous work is done, special measures shall be taken to make sure the organization does not become dependent on the subcontracted organization after the recovery plan is implemented, and/or negotiation shall address the possibility to extend the contract after recovery period. Organization's legal advisors should be engaged on negotiations.

   The communications with general subcontractors not involved in recovery tasks implies to let them know how their organization will be affected by the event: cancel the contract, change the contract, reinforce the contract... Learning from crisis and improvement of plans for enhancing both organizations resilience should be discussed.
3. **Communication with insurers and regulators**

   After an even there may be the need to negotiate with insurance companies for getting refund from losses and/or new insurance contracts. The company shall consider tactics and negotiation option for accelerating refunding in case it is critical for recovery process. Again, legal advisors should be engaged on this.

   Usually regulators also need to be informed on accidents. Different kind os information is required accordingly with different types of accidents and their consequences. It is important to understand that the way organization addresses the regulator will frame their risk perceptions and attitudes towards the organization which can undermine organization future plans. Risk communication guidelines should be considered when planning for this communication.

4. **Communication of the recovery plans with individuals and services directly affected and stakeholders**

   When an incident has an impact over a local community, a recovery plan for the local site may be required. Because of its liability communication between the organization and the affected individuals and/or regulators may be required. Again, legal services are an asset to be part of the communication team to plan how to address such communications and/or negotiations.

   Moreover, organization learning can lead to a reinforced management plan that has an impact over the local area. Such a plan shall include hearing all parts involved. The communication process should be transparent to improve resilience for everyone and reduce the risk.

5. **Communication with the media and social networks**

   After an accident the organization image may be weakened. It is important to assess the impact of the event over the organization image in order to understand the implications of the event for the organization strategy. Risk perceptions and attitudes towards the organization shall be reassessed. Specific risk communication plans to improve trust and organization image should be developed and executed. Monitoring of results is advised.

18.5 Further Resources

Research on risk perception, risk communication and crisis management has flourished in the most recent years (for a revision see Àrvai and Rivers 2014 and Lundgreen and McMakin 2013) There are also some international groups organizing, producing and sharing knowledge on business continuity management and risk communication and crisis management. British Standards Institution (BSI), Business Continuity Institute (BCI), International Organization for Standardization (ISO), International Consortium for Organizational Resilience (ICOR), and CIP Institute, among others, are some of these organizations.

References

Àrvai J, Rivers L III (eds) (2014) Effective risk communication. Routledge, London

British Standard Institute (2014) BS 11200, Crisis management – guidance and good practice

International Standards Organization (2009a) ISO 31000:2009, Risk management – principles and guidelines, Geneva

International Standards Organization (2009b) ISO 22301:2012, Societal security – business continuity management systems – requirements, Geneva

Lundgreen R, McMakin A (2013) Risk communication a handbook for communication environmental, Safety and Health Risks. Willey, Hoboken

Palma-Oliveira J, Trump B, Wood M, Linkov I (2017) The tragedy of the anticommons: a solutions for a "NIMBY" post-industrial world. Risk analysis

Chapter 19
Integrative Education Model for Resources and Critical Infrastructure Protection Based on Risk Assessment, Resources Valorization and Threat Ranking

Dejan Vasovic, Goran L. Janackovic, and Stevan Musicki

Abstract Efficient and effective resource protection from the perspective of society usually involves minimizing costs and capital commitment in any way, while maximizing the percentage of resource utilization. Effective resource management from the perspective of the environment means leaving them in their natural, intact, state. On the other hand, effective and efficient resource protection management activities from the perspective of the army involve efficient resource protection measures during peacetime, emergencies, and even wartime. The aim of this paper is to help develop an adaptive, integrative education model for resource protection, addressing the contemporary needs both within the Ministry of Defence (MoD) and Ministry of environmental protection (MEP) of the Republic of Serbia. The paper offers an in-depth analysis of related core terms: improvement of the current approach to the subject area, nature of the resources that are subject to protection, and different modalities of protection. The applied methodology consists of comparative analysis, statistical methods, and multi-criteria evaluation and assessment. Core determinants (ranking) within the defined model are specified by means of expert judgments method. Obtained results are intended to be used in further implementation processes regarding the developed model.

Keywords Education model • Resources • Environmental protection • Critical infrastructure

---

D. Vasovic (✉) • G.L. Janackovic
Faculty of Occupational Safety in Nis, University of Nis, Carnojevica 10A, 18000 Nis, Serbia
e-mail: djnvasovic@gmail.com; janackovic.goran@gmail.com

S. Musicki
University of Defence, Military Academy, Pavla Jurisica Sturma 33, 11000 Belgrade, Serbia
e-mail: mustmilenko@yahoo.com

© Springer Science+Business Media B.V. 2017

I. Linkov, J.M. Palma-Oliveira (eds.), *Resilience and Risk*, NATO Science for Peace and Security Series C: Environmental Security,
DOI 10.1007/978-94-024-1123-2_19

19.1 Introduction

Contemporary life style and production methods lead to the significant usage of all kinds of natural resources in the industrial, agricultural, and other production activities depending on energy and material demanding technological systems and consumer society (Haake 2000). Risk and potential harmful effects derived from emergencies only arises when hazards interact with people, material assets or elements of living environment (Nikolic and Zivkovic 2010). An emergency (flood, storm, drought, landslide, terrorist attack) striking an uninhabited (unsettled) area without any material structures or human individuals cannot be considered as causing risk (from human perspective). Basic prerequisites for such interaction are different kind of vulnerabilities:

- physical,
- social,
- economic and
- environmental vulnerabilities.

There are three core reasons that strengthen the abovementioned interaction:

- change of climatic conditions on Earth that inevitably lead to more frequent and intensive natural disasters on the one hand,
- increased human settling of the areas that have previously not contained any human settlements, so there are fewer and fewer unsettled areas, which means larger areas susceptible to disasters,
- complex political interaction between the states and social turmoil within the some states deriving the potential terrorist threat (Vasovic et al. 2016).

When defining critical infrastructure, the European Union distinguishes between national critical infrastructure and European critical infrastructure. Both terms refer to a property or a system in a Member State that is necessary to maintain key social functions, healthcare, safety, security, and economic and social well-being, the only difference being the ultimate effect. As regards national critical infrastructure, any destruction of or damage to critical infrastructure would significantly impact the Member State in which it is located, whereas in the case of European critical infrastructure, the impact refers either to two or more Member States or to one state which does not contain the critical infrastructure. European Programme for Critical Infrastructure Protection (EPCIP) refers to the approach or specific programs created as a result of the abovementioned European Commission's directive which designates European critical infrastructure facilities that could impact both the country where it is happened and at least one other European Member State. EU Member states are obligated to adopt this directive into their national statutes (EC 2005, 2006).

With no less importance are the contemporary approaches and analytical tools dedicated to the critical infrastructure protection within the United States of America. At the USA level, Critical Infrastructure Protection (CIP) stands for a concept that relates to the preparedness and response to serious disruptions that could involve the critical infrastructure of a region or nation. The United States of America CIP is a national program to ensure the security of vulnerable and interconnected infrastructures of the United States of America. Quite different from the EPCIP, the USA CIP defines critical infrastructure sectors as listed below:

- Banking and finance: The Department of the Treasury is responsible for coordinating the protection of not just systems but also maintaining public confidence, through industry initiatives such as the Financial Services Information Sharing and Analysis Center,
- Transportation: The Department of Transportation is responsible for protecting the road, rail, air, and water transportation infrastructure, including computer-controlled just-in-time delivery systems, optimization of distribution through hubs, and traffic and operations centers that are consolidated into key locations, and regulation of the transport of hazardous materials,
- Power: The Department of Energy oversees energy supplies including electricity, oil, and gas, and works with the Nuclear Regulatory Commission for the protection of nuclear materials and power. Note that CIP in this sector is different from energy security, which is the politics and economics of supply. Additionally, operating under the auspices of the Federal Energy Regulatory Commission is the North American Electric Reliability Corporation (NERC), a non-profit organization that defines and enforces reliability standards for the bulk power system,
- Information and communications: Overseen by the Department of Commerce, most areas of life rely on telecommunications and information technology,
- Federal and municipal services: Overseen jointly by Federal and State agencies. They guarantee continuity of government at the federal, state, and local levels to meet for provision of essential services,
- Emergency services: Overseen by the Health and Human Services, this includes emergency health services and public health,
- Fire departments: Overseen by the Federal Emergency Management Agency FEMA,
- Law enforcement agencies: – Overseen jointly by the Department of Jus-tice and the Federal Bureau of Investigation to ensure the orderly running of activities during times of threat or crises,
- Public works: Overseen by the United States Environmental Protection Agency. This includes safe water systems and drainage,
- Agriculture and food, with the Department of Agriculture overseeing the safe supply of meat, poultry, and egg products,
- National monuments and icons, under the Department of the Interior (PDD 1998).

While the EU and USA has made considerable efforts in analyzing critical infrastructures and has adopted a series of documents pertaining to the critical infrastructure protection (Yusta et al. 2011) no clear criteria for identifying critical infrastructure and supporting regulatory framework have been established in

Republic of Serbia, so far. The Law on Defence (Law on Defence 2007; 2009) defines facilities of special relevance for defense, which include certain critical infrastructure facilities, whereas the Law on Emergency Situations makes no mention of critical infrastructure even though it covers the establishment of protection and rescue systems for people and material and cultural wealth (Law on Emergency Situations, Official Gazette of the Republic of Serbia, 2009, 2011). However, in broader sense this previous regulation introduced the term critical infrastructure in Republic of Serbia, but still without a clear definition of which elements or areas of infrastructure the term refers to (Nikolic and Vasovic 2015).

19.2 Risk vs. Safety

Following contemporary needs and tendencies pertaining to more sustainable emergency risk reduction strategies, frameworks and practices, researchers who are engaged in this area in recent years are orientated towards viewing and reflecting on the issue of emergency risk reduction within the broader context of sustain-able development concept. An adverse impact of different kind of emergencies differs by:

- nations,
- regions,
- communities and (even),
- individuals,

because of differences in their exposure to disasters (susceptibility) and intrinsic vulnerability. States that develop policy, legislative and institutional frameworks for emergency risk reduction are able to manage emergency risks and to achieve broad consensus for risk reduction measures across all sectors of society. At the other hand, there is a clear-cut consensus that the states without educated professionals and citizens (safety culture), has insufficient capacity to respond to the threats posed by emergencies.

Hazard identification during task study is very significant for risk management (US Army 2007). If the hazard is not identified, it will not be taken into consideration, so the assessment of its consequences and probability of occurrence will not be conducted (Musicki et al. 2015). One of the most comprehensive and integrative is holistic approach to risk management, which is shown on Fig. 19.1. This approach integrates the deliberations regarding hazards, risks, control measures and (most important) exposed elements. Exposed elements are seen in form of complex dynamic system, as the environment and society certainly are.

Composite risk management is a process conducted through various stages, which are not discrete, but complementary:

- hazard identification,
- hazard assessment in order to determine risk level,

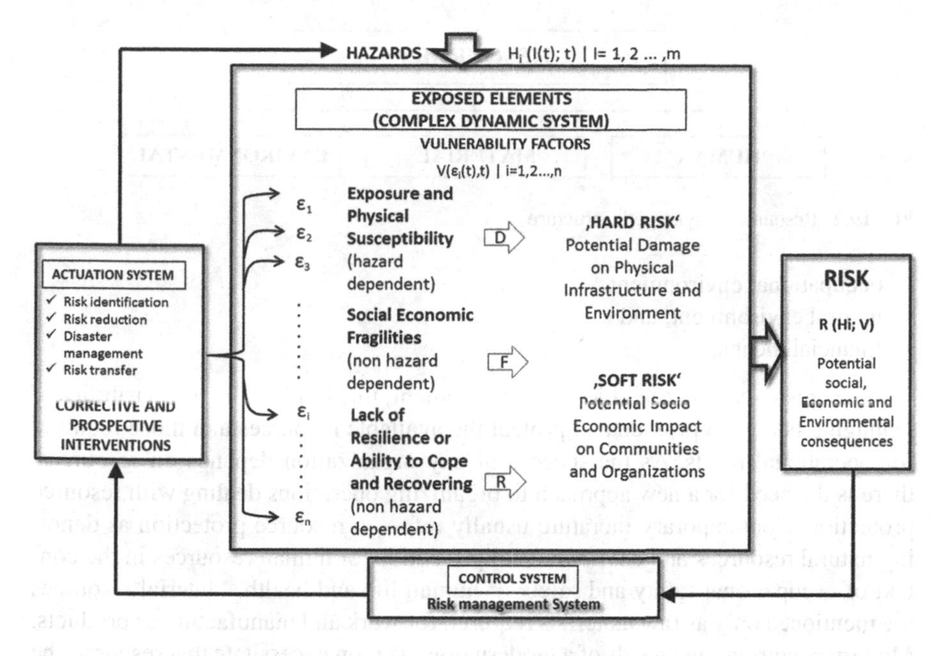

Fig. 19.1 Holistic approach in disaster management (Ciurean et al. 2013)

- preparation of control and decision-making measures,
- implementation of hazard control measures, and
- control and improvement.

At the other hand, the concept of safety is a highly complex social phenomenon and a scientific discipline within the social sciences. Safety is also a polysemic term. In the most general sense, it refers to absence of fear, threats, and physical violence. Nevertheless, safety also includes ethical, ideological, and normative elements, which impedes a precise definition. It is a socially constructed concept, which acquires a specific meaning within a given social context. After the analysis of numerous definitions, the concept of safety can be defined in the simplest terms as a state of protected value in which there is no potential or actual threat to the value, and also as a goal that cannot be fully realized but that should be strived for.

19.3 Education Model for Resources and Critical Infrastructure Protection

Resources are essential for the implementation and realization of an organizations goal. As such, they are classified into:

- human resources,
- infrastructure,

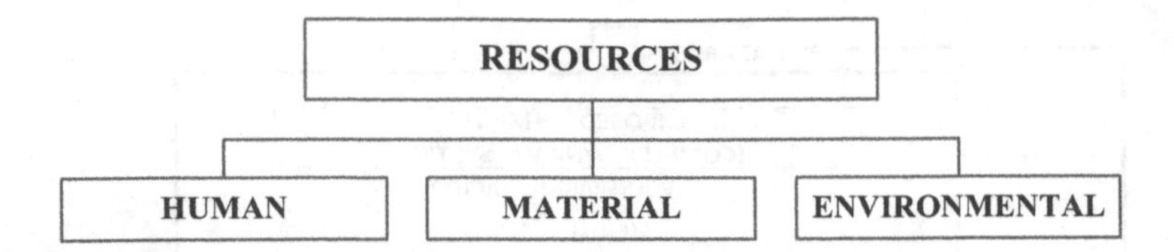

Fig. 19.2 Resources – synergetic structure

- occupational environment,
- natural environment, and
- financial means.

Since resources are the object of management, this implies the responsibility of managing bodies to provide and protect the available resources and those required for operational needs. As the success of any organization depends on resources, there is the need for a new approach to organizing operations dealing with resource protection. Contemporary literature usually refers to resource protection as denoting natural resources and environmental protection, or human resources in the context of occupational safety and safety of human life and health. Material resources are mentioned only as raw materials required for work and manufacture of products. Modern conditions and needs of a modern organization necessitate that resources be viewed as being in a synergy and, as such, forming a structural whole which is shown on Fig. 19.2 (Musicki et al. 2016).

The following terms are important for the topic of this paper:

- integrative model – involves a holistic approach, in this case a view of the model as a whole, which enables combining the best aspects, thus leading to the optimal solution for a given model,
- resource – fr. la ressource, means, source; lat. Resurgere, rise again, re-appear, be restored, is a means necessary for the undertaking or completion of an action. A resource may be material or non-material. The basic division of resources is into human, natural (renewable and non-renewable), and material resources,
- resource protection – utilization of resources on a scientific basis, identification of the ways to use resources rationally and complexly, and development and improvement of all forms of cooperation within scientific research,
- integrative model of resource protection improvement – a learning process used to define the model and the important features of a complex resource protection model required by the state authority (in this case experience of the MoD and MEP is used), and to provide scientific knowledge about the organizational structure of the bodies in charge of implementation and realization of measures and procedures of resource protection for the needs of MoD and MEP (Musicki et al. 2016).

Using SWOT analysis and Balanced Scorecard method (Malenovic Nikolic et al. 2015), we listed strengths, weaknesses, opportunities, and threats to resources protection both within the MoD and MEP, which allows the identification of positive

Table 19.1 SWOT analysis of sustainability for resource protection in the MoD and MEP

Strengths	**Weaknesses**
Clear vision, mission, and goals;	Insufficient number of professional personnel from the given field;
Operational efficiency;	Insufficient number of suitable teaching personnel;
Favorable educational structure of employees;	Inadequate training in the given field;
Existence of legal and normative acts for resource management;	Insufficient knowledge and skills in the given field;
Planning and organization of occupational safety and health;	Insufficient employee interest;
Implementation of occupational safety and health;	
Control of occupational safety and health;	
Opportunities	**Threats**
Improvement of the current state of the given field in the EU accession process;	A drop in the economic standard;
Promotion of the needs of protection implementation;	Lack of adequate material capacities;
Employee motivation for implementing resource protection measures;	Resistance to changes;
Control of training implementation and subsequent employee skills;	Insufficiently developed culture concerning the given field;
Introduction of mandatory classes at all education levels in the MoD and MEP;	Opposing views on the need for and scope of measures to be implemented;
Adequate training/education of current personnel;	Failure to understand the necessity of professional personnel at all levels;
Cooperation with university faculties from the same field;	Employee fluctuation;

and negative factors that affect the choice and balance between internal capabilities and external possibilities (Table 19.1).

Performed SWOT analysis indicates the question of how to define the goals of optimizing the structure, tasks, and functioning of resource protection in the MoD and MEP. Depending on the branch, service, or type of work performed in any of the MoD's and MEP's organizational units, the answers to the previous question would differ and would involve different approaches to defining the optimization of structure and tasks. For instance, the task of MoD's managing and command personnel within resource protection improvement is to provide uniform under-standing and agreement about what is achieved by the optimization of structure, tasks, and functioning of the resource protection system, where it would represent the general starting point.

19.4 Method

Multi-Criteria Decision Analysis, or MCDA, is a valuable tool that we can apply to many complex decisions. It is most applicable to solving problems that are characterized as a choice among alternatives. Decision-making process regarding evolving of adaptive, integrative education and management model for resource protection in the Republic of Serbia represents the process of identification and selection of alternatives based on the values and preferences of the decision maker that are important both in management process and education.

The main goal of such structured decision-making process is to select best possible alternative from analyzed alternatives, according to previously defined objectives, recommendations, preferences and desired results (Linkov and Moberg 2014). The best alternative is identified according to one or more criteria defined by decision-maker. Although the decision making can be done by analyzing only one criterion (e.g. time, costs, flexibility, compliance with the MoD's and MEP's needs, etc.), more adequate decision can be made by including more criteria. Thus, many methods for multi-criteria analysis and decision-making have emerged (Srdjevic 2012; Srdjevic and Cvjeticanin 2012). The general character of these methods allows to be applied in different fields, from education to safety and security, in order to define key factors influencing the efficiency and effectiveness of analyzed system.

The decision making is conducted with some risk and/or uncertainty included, based on the knowledge on the object of the decision-making and the available information from the environment. The decision is based on the accepted level of risk, available resources and social responsibility. The goal or objectives can be modified if there was some feedback information from the field that can improve the decision-making process in terms of efficiency or precision. According to Baker et al. (2002), the decision-making process consists of the following steps:

- problem definition,
- requirements specification,
- goal definition,
- alternatives identification,
- criteria definition,
- selection of the decision-method, and, finally,
- alternatives assessment.

The selection and ranking of the alternatives is either qualitative (based on the previous experience of the assessors) or quantitative (based on the values of key performance indicators). One or more persons can participate in the process of decision making (Janackovic et al. 2014). During the process of group decision making, the individual decision-making judgments or priorities can be aggregated in order to obtain the final decision. The participation of more people in decision making reduces subjectivity. However, it is important to identify the level of expertise of included experts and to assess the consistency in every step of the decision-making process.

As starting point within the MCDA process, we used existing literature algorithm that is modified for the purposes of this research (Fig. 19.3).

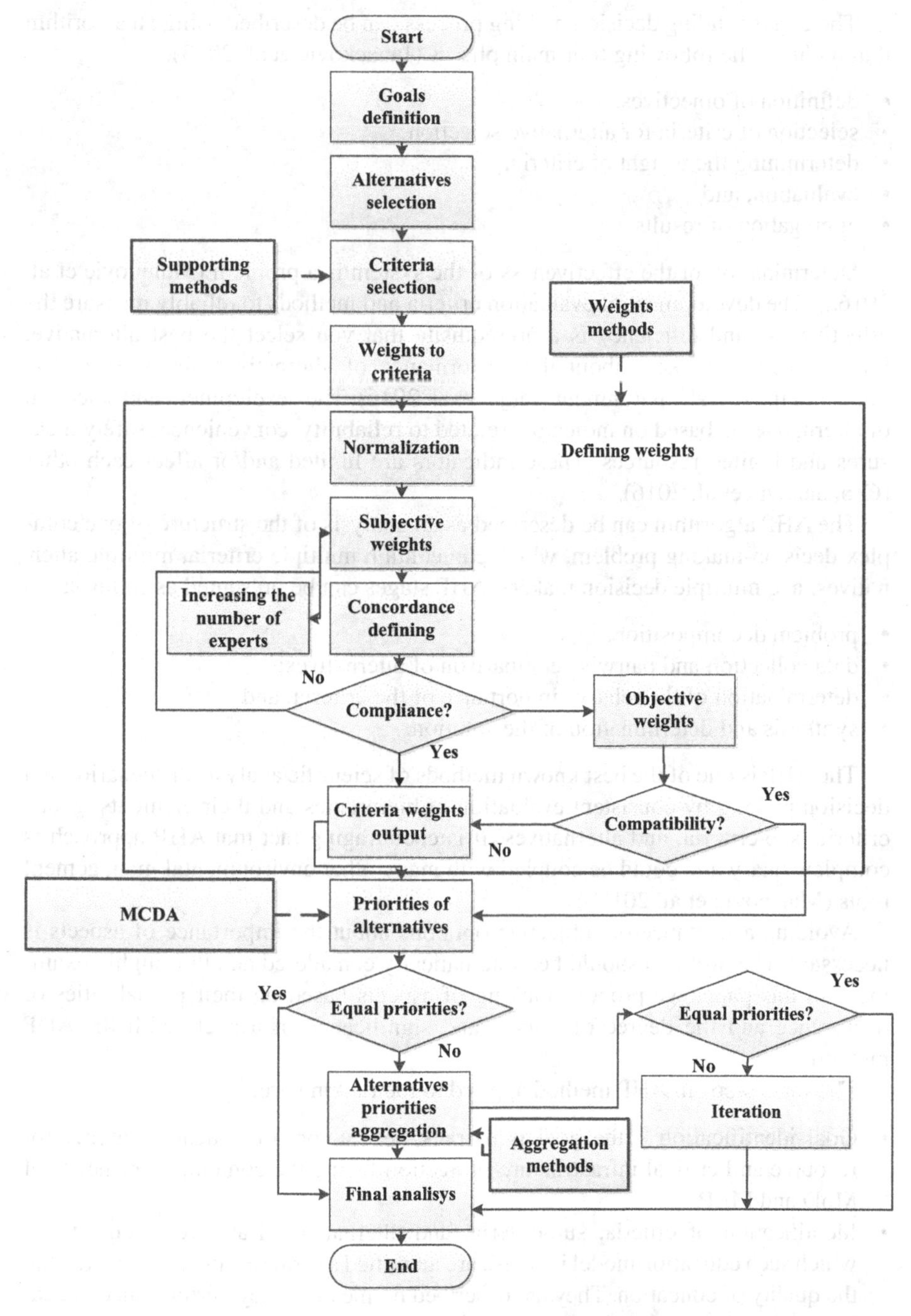

Fig. 19.3 Proposed decision making algorithm and MCDA (Adapted from Ustinovichius et al. 2007)

The corresponding decision making process can be described using an algorithm that includes the following four main phases (Janackovic et al. 2013):

- definition of objectives,
- selection of criteria for alternative selection,
- determining the weight of criteria,
- evaluation, and
- aggregation of results.

Determination of the effectiveness of the system is a problem (Malenovic et al. 2016a). The development of evaluation criteria and methods to reliably measure the effectiveness and efficiency is a prerequisite that you select the best alternative, inform decision-makers about the performance of alternatives and monitor the impact of the social environment (Janjic et al. 2016). The development and selection of alternatives is based on indicators related to reliability, convenience, safety measures and limited resources. These indicators are limited and/or affect each other (Grozdanovic et al. 2016).

The AHP algorithm can be described as an analysis of the structure of one complex decision-making problem, which can contain multiple criteria, multiple alternatives, and multiple decision makers. AHP stages can be presented as follows:

- problem decomposition;
- data collection and pairwise comparison of alternatives;
- determination of the relative importance of the criteria; and
- synthesis and determination of the solution.

The AHP is one of the best known methods of scientific analysis of scenarios and decision making by consistent evaluation of hierarchies and their elements: goals, criteria, sub-criteria, and alternatives. It is encouraging fact that AHP approach is complementary and could be coupled with many other environmental management tools (Malenovic et al. 2016b).

Avoiding acceptance of subjective opinions about the importance of aspects is necessary. The problem should be systematically considered and thoroughly examined. In this paper we propose ranking of aspects based on their probabilities of occurrence and the degree of impact and significance of aspects with the AHP method.

The main steps in AHP method applied to the ranking are:

- Goal identification – the goal is to rank key factors of education model for resource and critical infrastructure protection facing the contemporary needs of MoD and MEP,
- Identification of criteria, sub-criteria, and alternatives. Basic requirements on which such education model is based, are identified as criteria, and factors affecting the quality of education. They are described by means of key performance indicators as determined by experts from the list of the proposed indicators,
- Creation of hierarchical structure – the AHP method presents a problem in the form of hierarchy, where the top level presents the goal or objective (ranking of

the key performance indicators of education model for resources and critical infrastructure protection), the next level presents the criteria and the lowest level presents the indicators,

- Pair-wise comparison of criteria and indicators – pairs of elements at each level in a hierarchy are compared according to their relative contribution to the elements that are at the first hierarchical level above them; the relative contribution of each pair of criteria or alternatives to the main goal objective is described by 1–9 comparison scale, where 1 represent equal importance, and 9 strong dominance of one alternative over the other,
- Pair-wise comparisons at each level, starting from the top of the hierarchy, are presented in the square matrix form, in pair-wise comparison matrix with $n \times n$ dimension, where n is the number of compared criteria or indicators,
- Determination of relative weights based on the eigenvector approach, as presented in (Saaty 1978, 1980),
- Checking results consistency – the decision making procedure must be based on coherent judgments during the specifying the pair-wise comparison of criteria or alternatives, and the deviation of maximum eigenvalue λ_{max} from n describes the level of comparison consistency; two values, consistency index and consistency ratio are used for analysis of consistency of pair-wise comparison,
- Global priority determination – involves finding a vector of global priority, which shows the contribution of certain indicators to the achievement of the goal, i.e. to efficient or flexible education model; a vector of global priorities consists of relative priorities of all indicators at the lowest hierarchical level.

19.5 Case Study

Serbia is a Western Balkans country, which is the region where a lot of natural disasters happened, especially floods and earthquakes. Large natural and techno-logical disasters require responses to reduce negative effects on the functioning of a society. In these situations, it is necessary that all parts of the society react appropriately, including the army. In Serbia, there is no defined program of protection of natural resources and critical infrastructure (water supply, electric installations, roads), other than those defined in the context of corresponding public organizations and local communities.

Certain aspects of the participation of the Army of Serbia are defined by the Law on Emergency Situations, the Law on Serbian Armed Forces, and the Law on Defence. They allow the Army to be involved in the process of response to emergency situations. However, the response in these situations is based on military combat roles and strategies, which may not always be efficient. The army is neither educated nor trained to organize the system in case of natural disasters. In or-der to define the role of the Army of Serbia and the manner of reaction in detail, the development of the study program at the University of Defence, which would include the details on protection of natural resources and the critical infrastructure from natural disasters and terrorist attacks, is initiated.

The general methodology applied in the analysis is based on group decision-making, based on the following steps:

1. Selection of experts, assessment of their competences, and selection of key performance indicators of the quality of the study program based on the Delphi process;
2. Ranking key performance indicators based on the comparisons in pairs, i.e. analytical hierarchy process applied to determine the relative weights;
3. Consistency checking and final consultation with experts on obtained results.

In the process of ranking performance indicators, the experts from the University of Defence, MoD, MEP, Faculty of Technical Sciences in Novi Sad, and Faculty of Occupational Safety in Nis were included. The total competence coefficient for the group of experts is 0.80, based on the method described in (Vasovic et al. 2016; Musicki et al. 2016).

This study pertains to the design and selection of a education model for re-source protection and critical infrastructure safeguarding as required by the MoD and MEP, and its aim is to unify the requirements and needs and to append the existing model with necessary explanations and thus facilitate the revision of the old and design of a new, modern, model for resource protection. The study directly contributes to quality improvement. General propose is that to learning outcomes should be given special attention. For our research and selection of a suitable model for resource and critical infrastructure protection for the MoD and MEP we used the AHP approach throughout the Expert Choice software. Expert Choice is based on the AHP and it is used to solve semi-structured and non-structured decision-making problems. In order to facilitate the assessment, we used Expert Choice for pairwise comparison. After the assessment of all pairs of criteria, as well as all pairs of alternatives in relation to each criterion, the software aggregated the importance of alternatives in relation to the global goal, thus yielding the ranking of alternatives, which enables direct decision making.

The following hierarchy, presented in Fig. 19.4, is used during the decision-making. We compared the following criteria, which any modern and well-designed education model should comprise: compliance with the MoD and MEP needs (C_1); curriculum flexibility (C_2); curriculum aims (C_3); curriculum content (C_4); learning outcome (C_5); teaching methods (C_6); and practical training (C_7).

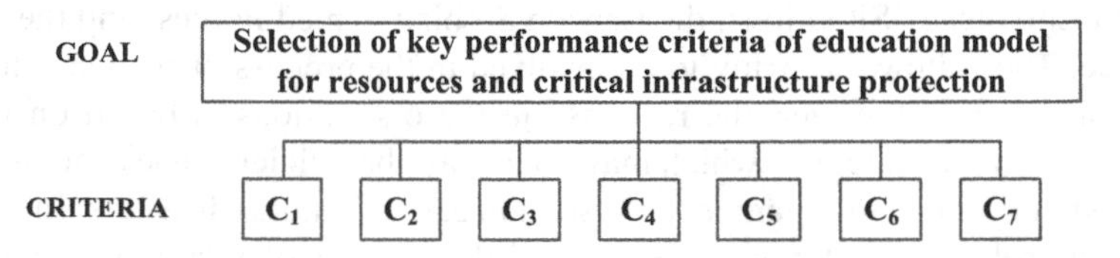

Fig. 19.4 Decision-making hierarchy

The concept of flexibility is more related to access to lifelong-learning. However, elements of flexibility can be offered in the curriculum. The flexibility of the curriculum should be seen in the context of the new courses, content, delivery and assessment system. The necessary changes relate to the development of clear and concrete learning outcomes in terms of knowledge, skills and competences that the cadets of the University of Defence hold at the end of studies.

The objectives of the curriculum presents a brief statement about the place and the role of objects, links to other articles and academic standards. They are written mainly from the perspective of teaching to show the general direction and content.

The learning outcomes contribute to the learning outcomes of the study program. There are views that the individual item should not contain more than six to eight learning outcomes. Too broad learning outcomes are difficult for the evaluation, but too narrow to jeopardize the ability to process comprehensive questions. For each subject, learning outcome must be given in order to establish a clear link with the content, teaching methods and evaluation. The learning outcomes of each course must appear somewhere (final competencies of the study program or as a prerequisite for another course).

To clearly describe learning outcomes, it is necessary to use adequate teaching methods that will ensure that the set of outcomes is achieved. Methods of teaching and learning that should be taken into account are: lectures, exercises, field work, practice in the combat units, seminars, group presentations, and others.

Practical training is an opportunity for learning in working conditions and strengthening professional competencies (professional development). It is especially important for army, and actual results undertaken in the workplace must be complementary to what is the output of the study program.

The purpose of practical training is reflected in the following:

- Learning about specific technologies, models and techniques;
- Improving the skills needed in the workplace;
- Development of new skills that facilitate teamwork, safe use of new technologies and troubleshooting;
- An opportunity to acquire skills through a gradual: from simple to complex;
- Evaluation in real conditions; and
- Forming an opinion on the relevance of the content of the program for the workplace.

Using the AHP (via the Expert Choice software) and based on expert opinions, we compared the given criteria by means of the matrix of relative importance in relation to the global goal and selected a suitable model curriculum for resource protection for the MoD (Fig. 19.5).

Having compared the criteria, we reached the final priority in terms of the aim of this study – an education model. According to the given expert judgments, the most important elements are ranked as follows (Fig. 19.6):

1. compliance with the MoD and MEP needs ($w_{c,1} = 0, 3381$);
2. learning outcome ($w_{c,2} = 0, 1857$);

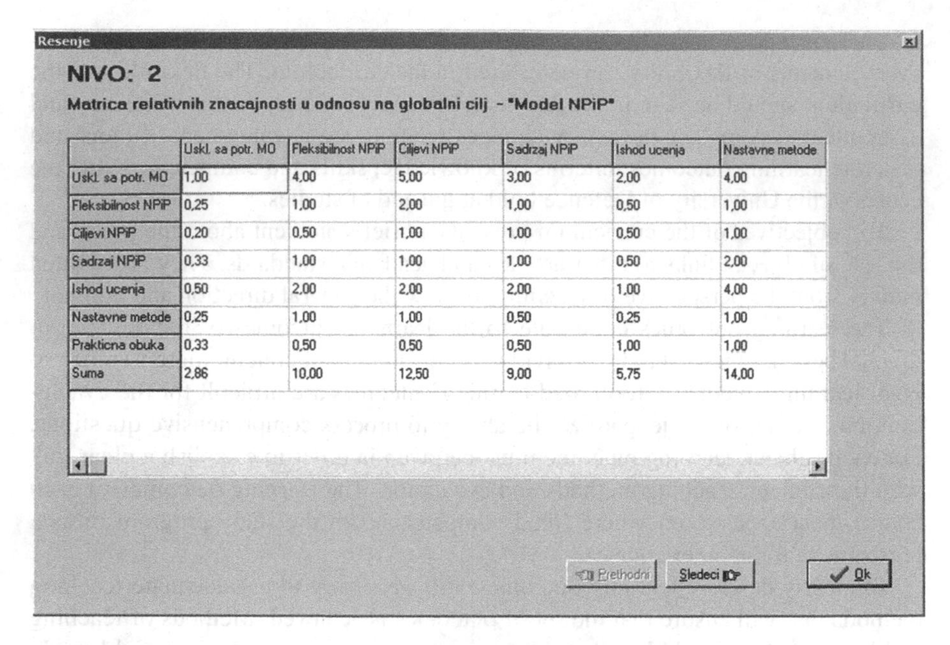

Resenje

NIVO: 2

Matrica relativnih znacajnosti u odnosu na globalni cilj - "Model NPiP"

	Uskl. sa potr. MO	Fleksibilnost NPiP	Ciljevi NPiP	Sadrzaj NPiP	Ishod ucenja	Nastavne metode
Uskl. sa potr. MO	1,00	4,00	5,00	3,00	2,00	4,00
Fleksibilnost NPiP	0,25	1,00	2,00	1,00	0,50	1,00
Ciljevi NPiP	0,20	0,50	1,00	1,00	0,50	1,00
Sadrzaj NPiP	0,33	1,00	1,00	1,00	0,50	2,00
Ishod ucenja	0,50	2,00	2,00	2,00	1,00	4,00
Nastavne metode	0,25	1,00	1,00	0,50	0,25	1,00
Prakticna obuka	0,33	0,50	0,50	0,50	1,00	1,00
Suma	2,86	10,00	12,50	9,00	5,75	14,00

Prethodni Sledeci Ok

Fig. 19.5 Matrix of relative importance of criteria in relation to the global goal

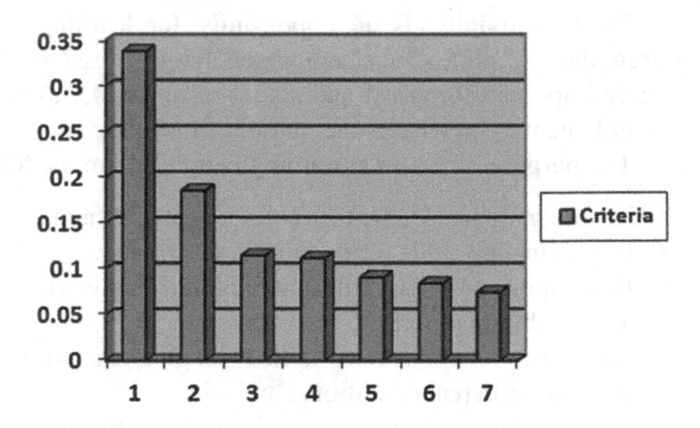

Fig. 19.6 Final ranking of criteria

3. curriculum content ($w_{c,3} = 0,1147$);
4. curriculum flexibility ($w_{c,4} = 0,1119$),
5. curriculum aims ($w_{c,5} = 0,0909$);
6. practical training ($w_{c,6} = 0,0842$); and
7. teaching methods ($w_{c,7} = 0,0745$).

The highest rank of the compliance with the MoD needs – criterion is explained by the fact that the education model must be based on the identified needs of resource protection in the MoD and MEP. In order to verify this, a scientific discussion is needed on the curriculum for resource protection for the MoD and MEP. As regards whether changing the current curriculum for resource protection is justified, the aims of the newly-proposed curriculum cannot be addressed separately, because its compliance with the MoD and MEP needs can only be considered when the curriculum is described in detail, in terms of the expected learning out-come, abilities, and the knowledge it offers.

19.6 Conclusion

Compatible, internationally recognized, certified education model for critical infrastructure protection at the different levels represents the basis for efficient critical infrastructure protection in a broader context: from the amenable organization and toward the amenable organization. Activities which are geared towards the rational recognition of the critical infrastructure protection aspects, ranking their impact and the definition of planned activities for their impact reduction or elimination can be defined as derivatives from an internal system defined according to the guidelines of international documents such CIP and EPCIP. In this sense, the proper application of education methods facilitates their effect.

At the other hand, simultaneous development of environmental, army and society friendly management practices is a necessary requirement for the development and improvement of an effective, inter-entity resource protection system at the national level. All of these represent dynamic social phenomena, which are the subject of research of many scientific disciplines. The organizational structure of the armed forces of any country represents a complex system, which within the wider community operates and exists under specific conditions and circumstances. Starting from its basic purposes, the regular armed forces of a state do not have unknowns regarding the rules and their core roles. However, the complex structure and the interdependence of different organizational structures within the country (like institutions in charge for environmental protection) and of the armed forces have an impact on the implementation of measures in the field of resource protection and management of risk interrelated with them. Therefore, there is a need for radical changes in the organization and professional management, starting from the administrative bodies that are subjected to a set of challenges in the field of transformation. Only professional management staff that is ready for constant changes and that has an effective quality management system can implement radical changes in order to improve the performance of organizations through: restructuring, reengineering, programs dedicated to quality, integration, strategic redirection, and application of resource management measures towards the creation of sustainable resource management system. Detailed analysis of all important indicators describing the efficiency of resources and critical infrastructure protection included geopolitical situation and climate changes tendencies, can affect better understanding of on-going or potential critical infrastructure problems, and faster reaction on diverse situation that can emerge.

Acknowledgments The presented research is a part of the projects Development of new information and communication technologies, based on advances mathematical methods, with applications in medicine, telecommunications, power systems, protection of natural heritage and education (Project code III 44006) and Research and development of energy efficient and environment friendly polygeneration systems based on renewable energy sources utilization (Project n III 42006), under the auspices of the Ministry of Education, Science and Technological Development, Republic of Serbia.

References

Baker D, Bridges D, Hunter R, Johnson G, Krupa J, Murphy J, Sorenson K (2002) Guidebook to decision-making methods. WSRC-IM-2002-00002, Department of Energy, USA

Ciurean RL, Schroter D, Glade T (2013) Conceptual frameworks of vulnerability assessments for natural disasters reduction, approaches to disaster management – examining the implications of hazards, emergencies and disasters. In: Prof. Tiefenbacher John (ed), InTech. doi:10.5772/55538

EC (2005) Green paper: on a European programme for critical infrastructure protection. Comission of the European Communities. Brussels (Belgium).

EC (2006) Commission of the European Communities from the commission on the European programme for critical infrastructure protection, COM (2006) 786, Final, Brussels (Belgium).

Grozdanovic M, Janackovic G, Stojiljkovic E (2016) The selection of the key ergonomic indicators influencing work efficiency in the railway control rooms. Trans Inst Meas Control 38(10). doi:10.1177/0142331215579948

Haake J (2000) Firm strategies for an environmentally friendly use of materials: an application of the dematerialization concept to the environmental management of industrial firms. Dissertation. Universite de Versailles, Saint-Quentin-en-Yvelines, Centre d Economie et d Ethique pour Environnement et le Development (C3ED)

Janackovic G, Savic S, Stankovic M (2013) Selection and ranking of occupational safety indicators based on fuzzy AHP: Case study in road construction companies. S Afr J Ind Eng 24(3). doi:10.7166/24-3-463

Janackovic G, Malenovic NJ, Vasovic D (2014) Effects of mining and thermal power plants and the key aspects of environmental quality ranking by AHP. Commun Depend Qual Manag 17(1)

Janjic A, Savic S, Janackovic G, Stankovic M, Velimirovic L (2016) Multi-criteria assessment of the smart grid efficiency using the fuzzy analytical hierarchy process, Facta Universitatis. Series: Electron Energ 29(4). doi:10.2298/FUEE1604631J

Law on Defence, Official Gazette of the Republic of Serbia, No. 116/2007 and 88/2009

Law on Emergency Situations, Official Gazette of the Republic of Serbia, No. 111/2009 and 92/2011

Linkov I, Moberg E (2014) Multi-criteria decision analysis: environmental applications and case studies. CRC Press/Tailor & Francis Group, Boca Raton

Malenovic Nikolic J, Ristovic I, Vasovic D (2015) System modelling for environmental management of mining and energy complex based on the strategy principles of sustainable balanced scorecard method (SBSC). J Environ Prot Ecol 16(3):1082–1090

Malenovic Nikolic J, Vasovic D, Filipovic I, Musicki S, Ristovic I (2016a) Application of Project Management process on environmental management system improvement in mining-energy complexes. Energies 9(12):1071

Malenovic Nikolic J, Vasovic D, Janackovic G, Ilic Petkovic A, Ilic Krstic I (2016b) Improvement of mining and energy complexes based on risk assessment, environmental law and sustainable development principles, J Environ Prot Ecol 17(3)

Musicki S, Kljajevic L, Milanovic M, Cebela M, Milanovic S, Nenadovic S, Nenadovic M (2015) Predicting of lead distribution and immobilization in soil of the region of lignite mining (Rudovci, Serbia). Acta Montanistica Slovaka 20(3):192–199

Musicki M, Nikolic V, Vasovic D (2016) Resource protection – the Serbian Army experience. VI International symposium engineering management and competitiveness (EMC 2016), Kotor, Montenegro.

Nikolic V, Vasovic D (2015) Tailor made education: environmental vs. energy security and sustainable development paradigm. In: Denis Caleta, Vesela Radovic (eds) Comprehensive approach as "Sine qua non for critical infrastructure protection & managing terrorism threats to critical infrastructure challenges for South Eastern Europe". IOS Press, Amsterdam/Berlin/Tokyo/ Washington, NATO SPS Series D: Information and Communication Security

Nikolic V, Zivkovic N (2010) Occupational and environmental safety, emergencies and education Faculty of Occupational Safety, University of Nis, Nis.

Presidential directive PDD-63 (1998) Washington

Saaty TL (1978) A scaling method for priorities in hierarchical structures. J Math Psychol 1:1–104

Saaty TL (1980) The analytic hierarchy process. McGraw-Hill, New York

Srdjevic Z (2012) Comparison of different multicriteria methods in selecting optimal biomechanical model parameters. ZAMM J Appl Math Mech 92(2):105–112

Srdjevic Z, Cvjeticanin L (2012) Identifying nonlinear biomechanical models by multicriteria analysis. J Sound Vib 331(5):1207–1216

U.S. Army Environmental Command (2007) U.S. Army Commander's Guide for Mission-Focused Environmental Management Systems

Ustinovichius L, Zavadskas EK, Podvezko V (2007) Application of a quantitative multiple criteria decision making (MCDM-1) approach to the analysis of investments in construction. Control Cybern 36(1):251

Vasovic D, Malenovic Nikolic J, Janackovic G (2016) Evaluation and assessment model for environmental management under the Seveso III, IPPC/IED and water framework directive. J Environ Prot Ecol 17(1):356–365

Yusta JM, Correa GJ, Lacal-Arantegui R (2011) Methodologies and applications for critical infrastructure protection: state-of-the-art. Energ Policy 39:6100–6119

Chapter 20
"Valuation of Imminience Analysis in Civil Aircraft Operations"

Eugeniusz Piechoczek and Katarzyna Chruzik

Abstract Presented article attempts to execute the first stage of risk management for civil air operations. Therefore it is necessarily to identify and classify external sources of hazards. Risk management is one of the most important tools of modern systemic approach to process modeling. Generally, it may be described as a system of interrelated actions and decisions aimed at integration and coordination of processes in the organization, in cooperation with the external parties. The aim of management is therefore to improve the proficiency, effectiveness and efficiency of the operation and – in the case of risk management arising from the transport processes – to improve flight safety. The publication describes a proposal of risk management method in air transport including analysis of threats to civil air operations.

Keywords Risk management • Source of hazard • Aviation • Safety management system • Helicopter operations

20.1 Introduction

The contemporary approach to transport assessment, based on the process approach, suggests that safety is the determining factor of properly functioning safety management systems. Authors introduce the entireness of safety management in aviation civil transport. The purpose of safety management is to eliminate, and if possible, reduce the size of damage that can be caused by transport incidents. Along with development of this area, one must bear in mind all possible actions employed upon occurrence of an incident (reactive actions) as well as those implemented preventively on the basis of case analysis (proactive actions) – Fig. 20.1. This analysis is also more and more and more frequently based on projective actions, allowing to identify little probable hazards (most commonly being a compilation of improbable hazard sources) which give rise to catastrophic results.

E. Piechoczek (✉) • K. Chruzik
Silesian University of Technology, Gliwice, Poland
e-mail: eugenius@it.pl

© Springer Science+Business Media B.V. 2017

I. Linkov, J.M. Palma-Oliveira (eds.), *Resilience and Risk*, NATO Science for Peace and Security Series C: Environmental Security,
DOI 10.1007/978-94-024-1123-2_20

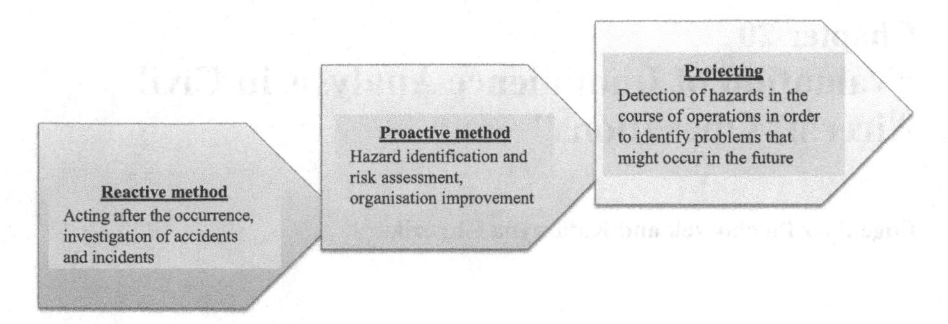

Fig. 20.1 Transport safety management model

The international aviation law imposes the obligation to manage the operating risk and, thus, to draft hazard registers and estimate the organisational risk upon the transport operators. The previously applicable Commission Regulation (EU) no. 185/2010 laid out the measures for implementation of the common basic standards on aviation security against illegal interventions hazardous for the safety of civil aviation, which had been amended 20 times before it entered into force. To ensure clarity and certainty of the law, it was revoked and superseded by a new act consolidating the original act and all its amendments – Commission Implementing Regulation (EU) 2015.1998 of 5 November 2015. The legal changes of the basic standards on aviation security against illegal interventions hazardous for the safety of civil aviation are introduced systematically and are not surprising for the aviation operators. The implementation of the basic standards requires short-term actions within the scope of system documentation as well as long-term actions within the scope of applied technical measures. The act comprises the following basic changes:

- detailing within the scope of documents invoked in the Regulation,
- detailing within the scope of interpretation of the provisions,
- changes within the scope of terms applied in the Regulation,
- references to third countries in the said Regulation and, in relevant cases, in the Commission Implementing Decision C(2015) 8005 (Projects in Controlled Environments PRINCE2 2009) final, covering other countries and territories in relation to which, pursuant to Art. 355 of the Treaty on the Functioning of the European Union, title VI in the third part of the said treaty is not applicable,
- extension of the scope of employed security control methods and standards in this area, including:

  (a) manual control,
  (b) metal detection gates (WTMD),
  (c) dogs detecting explosives,
  (d) explosives trace detectors (ETD),
  (e) x-ray devices not involving use of ionising radiation,
  (f) explosives trace detectors (ETD) in combination with hand-held metal detectors (HHMD)

- detailing within the scope of approval and monitoring of known suppliers,
- additional regulations within the scope of protection of on-board supply with fluids, aerosols and gels as well as bags secured in a manner allowing for easy statement of their manipulation,
- extension of requirements within the scope of training of persons performing searches and providing aircraft security services.

The history of implementation of the safety management system (SMS) in civil aviation dates back to 2006 when the International Civil Aviation Organisation commenced successive introduction of subsequent annexes (Chicago 1944) (within the areas of aviation training (Annex 1), aviation operations (Annex 6) and airworthiness (Annex 8), and then, successively, in the air navigation services (Annex 11), plane crash investigation (Annex 13) as well as airport operation (Appendix 14)), standards and recommended practices (SARPs), making the system (SMS) globally applicable. These annexes were announced in 2003 by way of a governmental declaration (Declaration of Polish Government 2003), and their translations as well as subsequent amendments (pursuant to Art. 3 section 2 and Art. 23 section 2 point 2 of the Aviation Law (Journal of Laws of 2013, item 1393)) are published by the CAO president in the CAO Official Journals. In 2013, the ICAO secretary general issued a new Annex 10 that was fully devoted to safety management in civil aviation. This resulted in introduction of relevant consequential amendments to annexes 1, 6, 8, 11, 13 and 14. The requirements of Annex 10 of ICAO SARPs charge the Member States with obligation to draft, publish and keep the State Safety Programme (SSP) in which the state is obliged to determine, inter alia, the acceptable levels of safety performance (ALoSP) as well as the related Safety Performance Targets (SPTs) and Safety Performance Indicators (SPIs) expressed in measurable values obtainable in a specific time horizon (Annex 19, para 3.2.1 The acceptable level of safety performance to be achieved shall be established by the State). The authorities have been charged with the obligation to conclude an agreement with the aviation organisation regarding safety indicators adopted by it, which must include the value of indicators determined in the governmental documents, i.e. in the state safety programme (SSProgramme) and annually revised state safety plan (SSPlan). The ICAO Secretary General supported implementation of the safety management system by issue of a comprehensive Manual (SMM) (Doc) in which he explained in detail the principles as well and processes of implementation of the safety management system on a state scale (NAA) and in service providers (SPs). This Manual was revised twice in 2009 (Edition II) and 2013 (edition III). This Manual (ICAO Doc 9859, edition III of 2013) provides detailed guidelines defining the principles and methods of preparation of the state safety programmes, as well as safety targets and indicators with their alarm thresholds as well as determination of safety indicators by service providers as well as their alarm thresholds. Furthermore, it contains detailed checklists for assessment of quality of implementation of the management system (SMS). The SMM Manual Doc. 9859 (edition II of 2013) has not been published in Polish yet in the CAO Official Journal. Implementation of safety management in the legal regime of the Union is supported with the so-called acceptable means of compliance (AMC) (Acceptable Means of Compliance (AMC) and Guidance Material (GM) of Part–ARA) and guidance materials (GM) (Acceptable Means of Compiliance (AMC) and Guidance Material (GM) of Part–ORA) issued by the EASA director, belonging to the "soft law" category. According to the provisions of Part ARA.GEN.120, Part ARO.GEN.120 and Part ADR.AR.A.01532, achievement of compliance with EASA AMC means meeting of all requirements of the implementing laws (Analisys of proces entry of samety management in civil aviation of Poland 2015).

Aviation transport is a complex system, combining advanced technical systems, operators and procedures. The sources of hazards can be found in all these closely related and mutually effective areas, operating in great spatial dispersion, within a short time horizon. A very important element of risk management is, thus, identification of hazard sources, not only within the scope of own risk (hazard sources and hazard activation regard the same transport organisation), but also within the scope of common risk (hazard sources outside the transport system the hazard activation regards) and external risk (hazard sources outside the transport system). Total risk of the transport organisation must include all three hazard areas. The article presents a proposal of a unified register of hazard sources in civil aviation operations and the resulting main areas of hazards as well as the detection methods applied contemporarily.

20.2 Risk Management

In practice, two basic approaches can be differentiated in transport risk management: operating (process and professional) and strategic. Operating management means that management occurs through goals and determination of what is expected from each entity, organisational unit and every employee, assessment of achievement in past results, and formulation and implementation of improvement plans. Strategic management indicates projective thinking about the phenomena and processes that will occur in the future and will create new operating conditions for the organisation and country. Thus, it requires continuous tracking of capture and on-going assessment of changes, assessment of the effect the changes have on the organisation/area, and determination of key problems requiring a solution (Krystek 2009).

The presented risk management method is based on the two most popular risk management methods based on a systematised process allowing to identify, plan and manage the hazard risk step by step in the course of planning and implementation of the project/process (Guide to the Project Management Body of Knowledge 2013; Projects in Controlled Environments PRINCE2 2009; Pritchard 2001):

- risk in "Guide of Project Management Body of Knowledge (PMBoK)" fifth edition by Project Management Institute Inc.;
- risk in "Projects in Controlled Environments (PRINCE2)" – process-based method for effective project management. Used extensively by the UK

Government, PRINCE2 is also widely recognised and used in the private sector, both in the UK and internationally.

The following is differentiated in the model based on the popular PMBoK Guide methodology (Guide to the Project Management Body of Knowledge 2013):

- risk management planning;
- hazard recognition;
- performance of qualitative analysis of hazards;
- performance of quantitative analysis of hazards;
- planning of reactions to hazards;
- hazard monitoring and control.

Risk management planning is the basic method of imminence protection. It means that the most important factor is to bring into practice in planning process all steps of the process with their specific requirements. The management plan must be implemented: in regular time intervals, prior to essential changes, prior to further stages as well as in relation to assessment of the project/process progress. According to the plan proposed by C. Pritchard (Pritchard 2001), the first elements of the analysis is the description and summary of the project/process, where goals, requirements and operating properties are established. The next step is determination of the risk management conditions – the current preventative measures as well as stakeholders' risk tolerance are discussed. The tools to be employed in risk measurement are then specified. The plan is concluded with the following points: other essential plans (where alternative solutions are proposed), methodology summary, literature and approval – i.e. a list of persons responsible for preparation and implementation of the plan (Pritchard 2001). The next step is identification of hazards as well as qualitative and quantitative analysis, and the planning of reaction to hazards as well as hazard monitoring and control.

Another essential risk management model is the one proposed by the PRINCE2 project management methodology (Projects in Controlled Environments PRINCE2 2009). It applies the M_o_R (Management of Risk) procedure which consists of the following five steps: identification, assessment, reaction planning, reaction plan implementation, as well as communication which is a repeatable task (Pritchard 2001). Presented above trends of risk management gives possibility to describe method of risk management presented later. These tasks have a similar course as in case of the PMBoK methodology. These methods differ in terms of examples of tools (Delphi method or SWOT analysis), with identification of common tools as well.

Due to the scope of impact and legal regulations determining it, the hazard assessment methods are described for the operating risk: process (short time horizons essential for the organisation in the main process aspect) and professional (combination of probability of occurrence of adverse hazards related to the performed work and causing losses as well as their effect on health or life of the employees – in the form of occupational diseases and accidents at work (Kadziński 2014)).

20.3 Methods of Risk Assessment in Transport

Risk assessment regards all processing having a direct or indirect impact on the main process of the transport organisation (carriage, infrastructure management, maintenance, production). The transport industry employs hazard risk estimation and valuation. The risk management method described in the publication (Fig. 20.2) is based on the hazard of aviation operator registers. Imminence register identify to air-operator hazardous area as result of source of imminence. Those registers verify and modify by air-operator periodically with estimation and valuation of hazard risk. These actions allow for direct hazard management and precise dedication of measures for areas valued as unacceptable, with concurrent hazard monitoring and communication.

Risk management in transport can be analysed in multiple regards, depending on the following considerations:

- sources of analysed hazards (technical, organisational);
- objectivity/subjectivity of assessment;
- risk assessment strategy (individual, social);
- assessment method (qualitative, quantitative);
- nature of losses;
- time horizon (operating risk, tactical risk, strategic risk);
- action admissibility criterion (acceptable risk, tolerable risk, unacceptable risk).

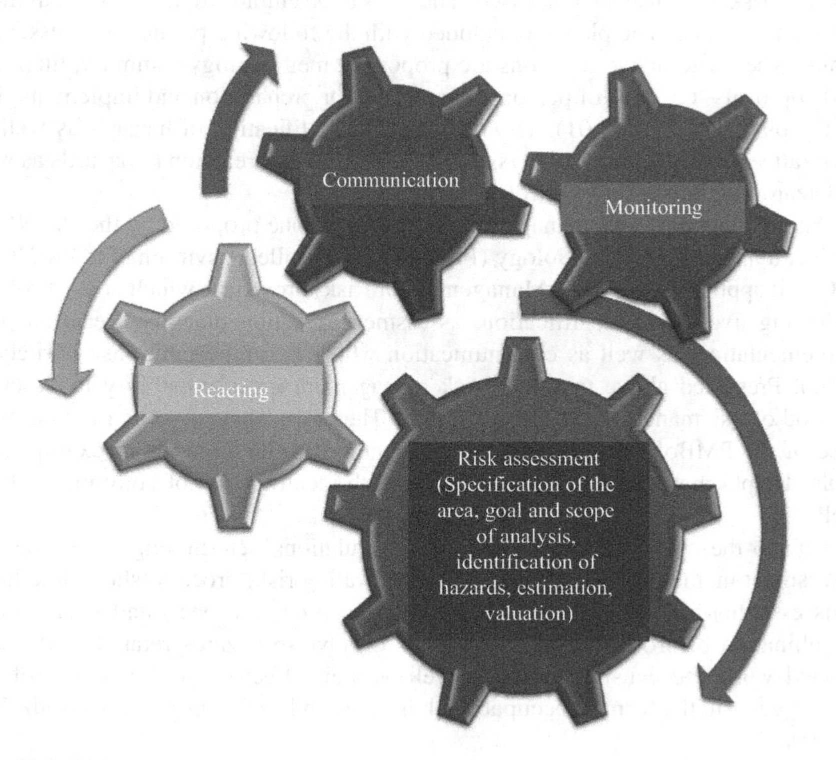

Fig. 20.2 Risk management model in transport.

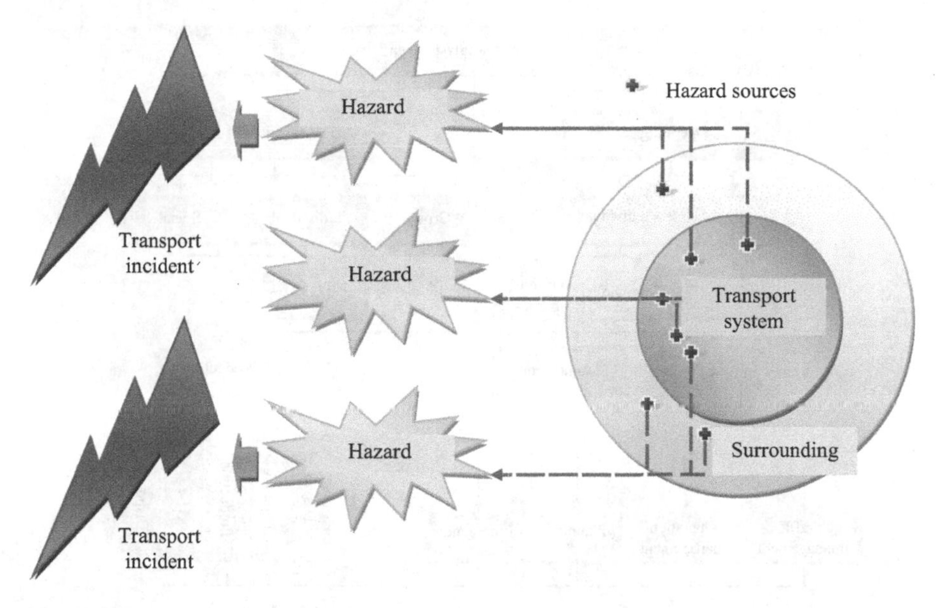

Fig. 20.3 Incident initiation diagram

The drawn conclusions of transport risk management gives possibility to take corrections or preventive action in the process.

In practice, we deal only with certain types of risk, e.g. we deal with technical means risk, we assess individual risk more frequently and the social one – less frequently which means that we analyse only short time horizons (operating risk) and we do not deal with longer time horizons (strategic risk) in the enterprises.

Risk management in transport is, thus, concentrated in two categories of goals:

- strategic, related to goals at the highest level of process management, supporting the mission (European Transport Agencies, Transport Offices),
- operating, addressing effective and efficient resource use (carriers, infrastructure administrators, entities responsible for maintenance).

Among studies focused on the application of risk assessment in transport (Krystek 2009; Kadziński 2014), safety is a condition of lack of hazards characterised with unacceptable risk (a hazard is a source of transport incident). The sources of hazards are constructs (e.g. physical, chemical, biological, psychophysical, organisational, personal) the presence of which in the indicated area of analyses, condition, properties can be the cause (source) of hazard formulation. Risk is a combination of probability of hazard activation into a transport incident and the resulting damage.

Any operation in transport creates imminence. Fusion of hazard can generate incident (Fig. 20.3). The awareness of hazard sources is the basis of risk management in transport. This consciousness we can create on the basis of historical data or experience. Action allows to estimate the identified hazards and to refer these values to the initially adopted ranges, and if the risk of hazards is exceeded (accept-

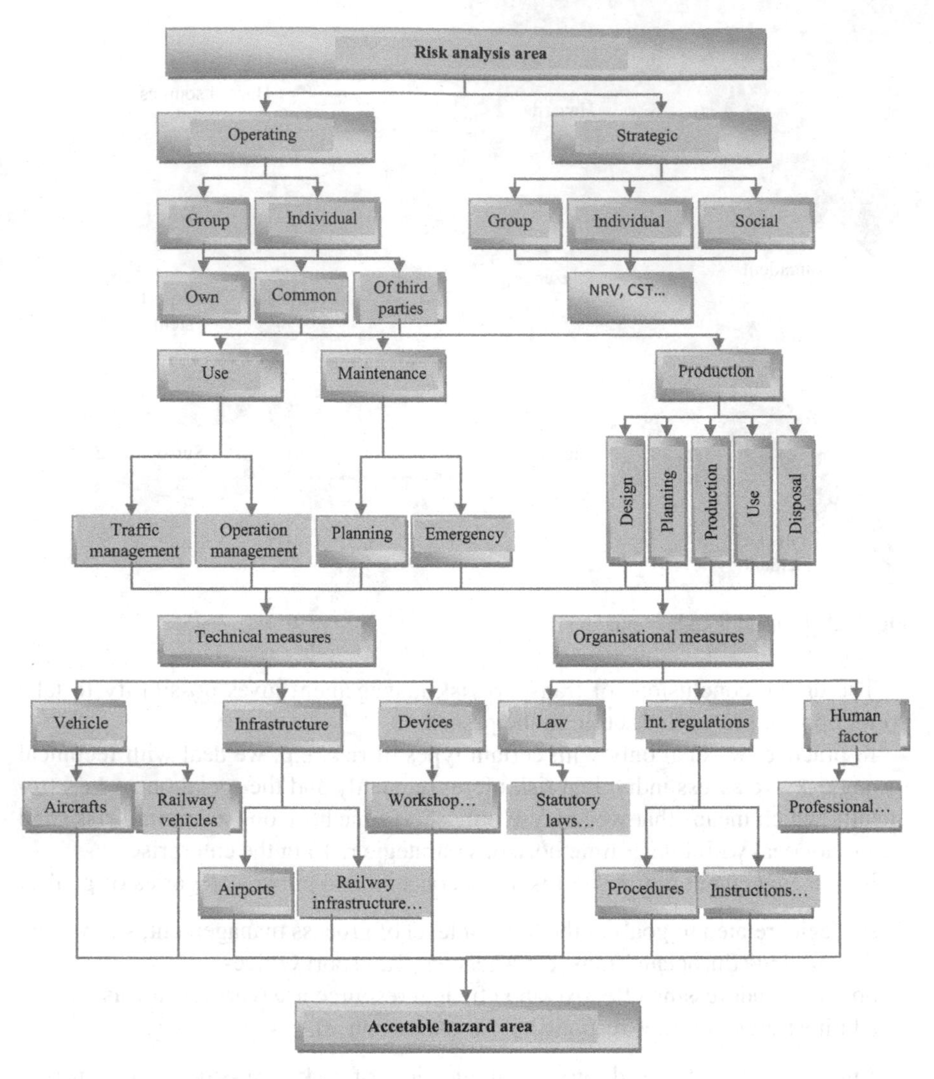

Fig. 20.4 Transport risk management scope and area

able, tolerable, unacceptable area) – to implement corrective and preventive actions in the processes. Monitoring of the entire process continuously, communicate the hazards to person engaged directly in the process (employees) as well as its recipients (passengers, third parties) present Fig. 20.2.

Figure 20.4 presents the risk analysis area. The schematic diagram specifies operational and strategic risk, participants, organizational and technical means, infrastructure and presents the relationships between each of the components. Finally, the air-operator creates an acceptable hazard area. Risk analysis is – in the specified transport system analysis area – systematic use of all and any available information

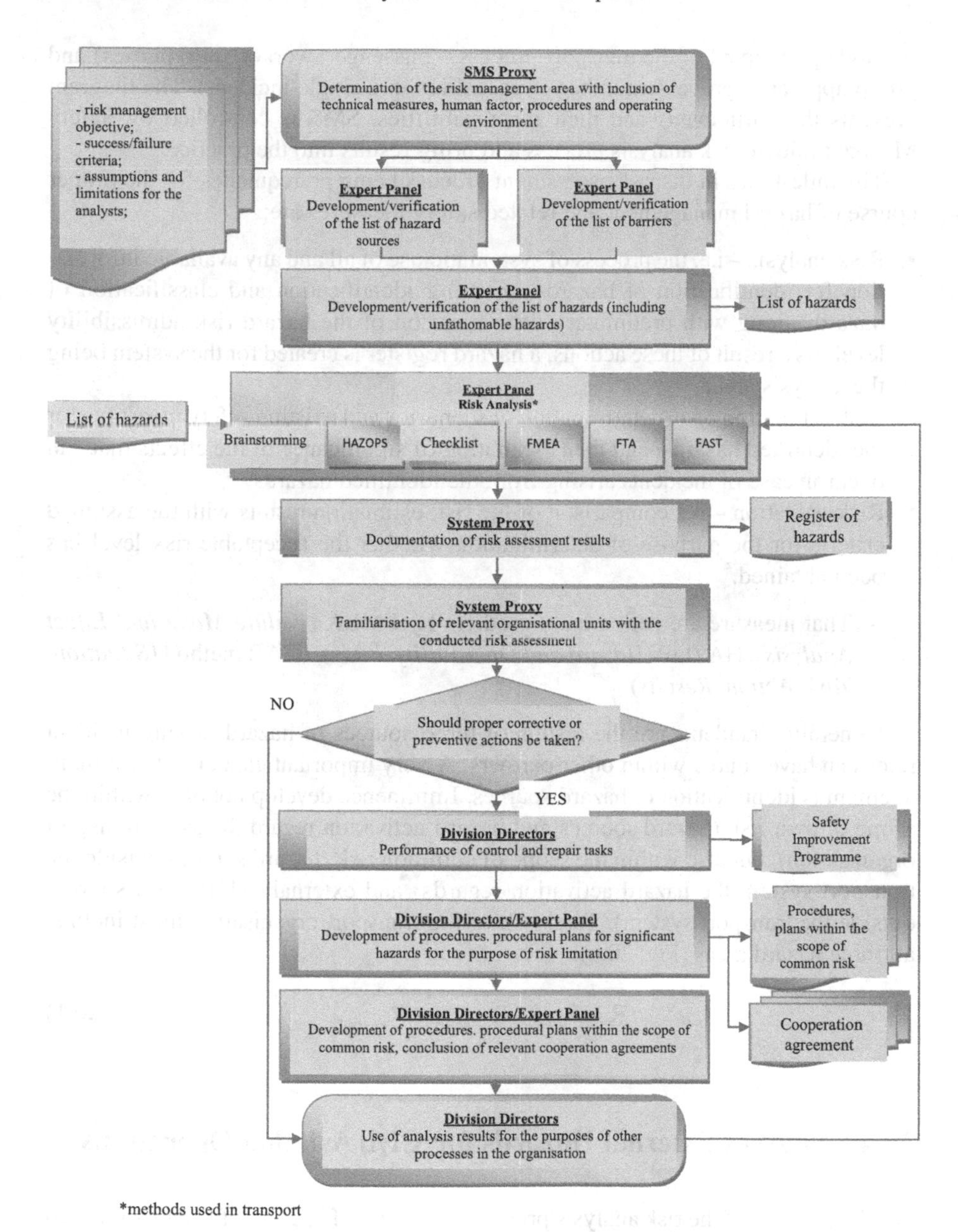

Fig. 20.5 Risk management procedure for the air-operator

for the purpose of hazard source identification, hazard formulation, risk estimation and hierarchismisation (Fig. 20.2). In practice, transport operators carry out short-term analyses arising from everyday activity of the organisation (operating risk).

Figure 20.5 present Safety Management System (SMS). In transport, this risk is related to technical and organisational hazards . It is analysed in individual approach (a single participant of the transport process – passenger, worker, third parties) and group approach (probable number of casualties in a single incident). The diagram presents the participants and their responsibilities. SMS is controlled by Safety Manager initiate risk analysis process and bring results into the practice.

The milestones in the risk assessment process being prerequisites for the proper course of hazard management and related safety measures are:

- Risk analysis – i.e. the process of systematic use of all and any available information for identification of hazards, covering identification and classification of hazards along with preliminary determination of the hazard risk admissibility level. As a result of these actions, a hazard register is created for the system being the analysis area;
- Risk estimation – i.e. determination of scenarios and existing safety measures for the identified hazards, and then estimation of significance of the effects that can occur in case of incidents arising from the identified hazards;
- Risk valuation – i.e. comparison of the risk estimation results with the assumed criteria for the purpose of determination whether the acceptable risk level has been obtained.

  – That measure are define among others by FMEA (*Failure Mode and Effect Analysis*), HAZOP (*Hazard and Operability Study*), STAR method (*Situation, Task, Action, Results*)

Generally, incidents are the results of three sources of hazard. Rising incident rates can have source within other partners. A very important element of risk management is identification of hazard sources. Imminence develop not only within the scope of own risk (hazard sources and hazard activation regard the same transport organisation), but also within the scope of common risk (hazard sources outside the transport system the hazard activation regards) and external risk (hazard sources outside the transport system). Total risk of the transport organisation must include all three hazard areas (Fig. 20.6):

$$R = R_{own} + R_{common} + R_{external} \quad (20.1)$$

20.4 Sources External Hazards for Civil Aviation Operations

For the purpose of the risk analysis process, a register of sources of hazards for civil aviation operations has been developed (Table 20.1) as well as possible scenarios of their combinations – hazard areas (Table 20.2).

Table 20.2 Identified hazard groups [own study on the basis of the European Aviation Safety Plan (EASp) and CAO Communication]

Hazard group contain: runway incursions, runway excursions, mid-air collision, controlled flight into terrain, loss of control in flight, controlled space trespass, fire,

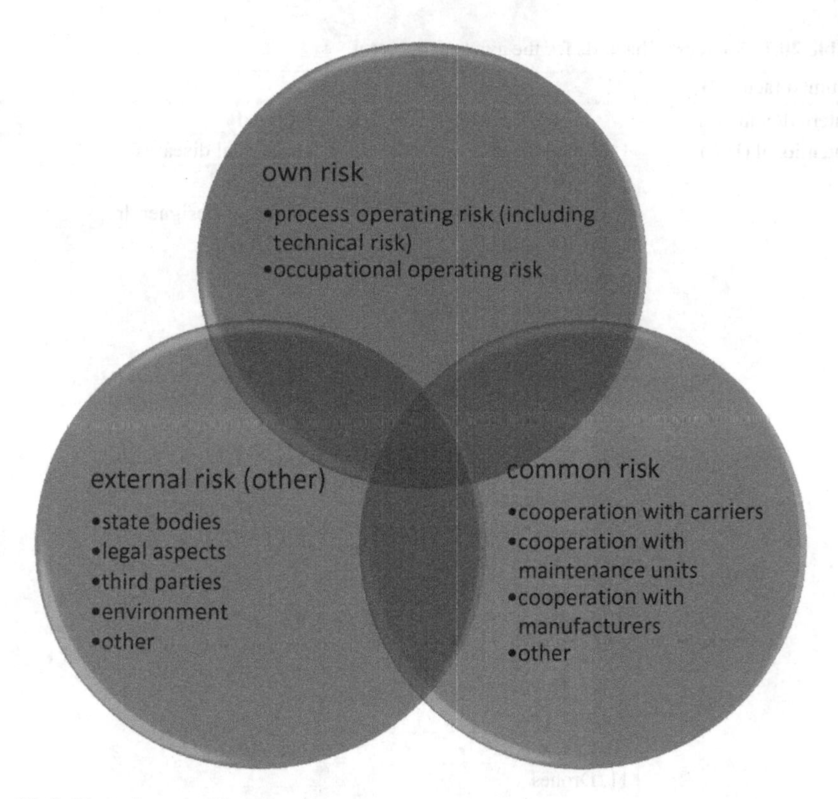

Fig. 20.6 Hazard source identification area

smoke, fumes, illegal restricted airport zone trespass, illegal restricted airport zone trespass with prohibited objects, planting explosives in airport facilities and devices, planting explosives in airport facilities and devices, hostages in the airport area, sabotage/diversion acts, order disturbance (public order disturbance, vandalism), plane crashes unrelated to human activity (weather anomalies), damage of technical means with consequences.

The hazard areas presented in Table 20.2 are a combination of probable hazard sources presented in Table 20.1. The value of the hazard is determined by the number and type of hazard sources (probability of their compilation) as well as the results they can cause, resulting in an incident. Detailed analysis of the particular hazard areas for the proposed hazard sources allows for full assessment of the aviation operator's risk.

Table 20.1 Sources of hazards for the aviation operator

Human factor (H)		
Internal/common		External
Intentional (H.I.)	1. Mental diseases	16. Mental diseases
	2. Political terrorism	17. Alcohol
	3. Financial terrorism	18. Drugs/designer drugs
	4. Fundamentalism	19. Bravado
	5. Alcohol	
	6. Drugs/designer drugs	
	7. Prohibited objects	
	7.1 Explosives	
	7.2 Short gun	
	7.3 Long gun	
	7.4 Sharp tools	
	7.5 Bacteria/viruses	
	7.6 Radioactive materials	
	8. "Overhead" utilities	
	8.1. Surface-to-air missile launchers	
	8.2. Grenade launchers	
	8.3. Hunter rifles	
	9. False alerts	
	10. Lasers	
	11. Drones	
	12. Theft of technical means	
	13. Hackers	
	14. Fires near the airport	
	15. Vandalism	
Unintentional (H.N.)	1. Nervousness during security control	10. Panic
	2. Willingness to "help" fellow passengers	11. Exhaustion
	3. Panic	12. Insufficient training
	4. Bacteria and viruses	13. Insufficient experience
	5. Radiation	14. Lack of knowledge of procedures
	6. Lack of consequence awareness (O.1)	15. Professional "burnout"
	7. Unauthorised flying objects	16. Errors in communication
	7.1 Drones	17. Low awareness of hazards
	7.2 Flying models	18. Failure to communicate hazards
	7.3 Ultralight trikes, paragliders, powered paragliders	
	7.4 Balloons/sky lanterns	
	8. Fireworks	
	9. Fires near the airport	

(continued)

Table 20.1 (continued)

Human factor (H)		
Technical factor (T)	1. Interferences resulting from external technical and protective measures	6. Unauthorised access to ground devices
	2. Unauthorised access to ground devices	7. Unauthorised access to the aircraft
	3. Unauthorised access to the aircraft	8. Structural errors in technical measures
	4. Structural errors in technical measures	9. Accumulation of flammable substances
	5. Accumulation of flammable substances	10. Incorrectly selected security control measures
		11. Computer system failures
		12. Station ergonomics
		13. Improper repairs
		14. Improper inspections
		15. Structural errors in technical measurements
Environmental factor (E)	1. Weather anomalies	9. Stress
	1.1. Violent wind blows	10. Amenity rooms
	1.2. Whirlwinds	11. Errors in works of handling companies
	1.3. Persistent fog	12. Failure to maintain the runway condition, signage or information
	1.4. Heavy rainfalls	
	1.5. Heavy snow fall/blizzards	
	1.6. Turbulences	
	2. Fires near the airport	
	3. Volcanic eruptions	
	4. Mammals	
	5. Birds	
	6. Flood hazards	
	7. Low grass at the airport	
	8. High grass at the airport	

(continued)

Table 20.1 (continued)

Human factor (H)		
Organisational factor (O)	1. Lack of "safety culture" awareness (H.N.6)	6. Lack of operating procedures
	2. Lack of knowledge of the airport requirements	7. Errors in operating procedures
	3. High density of people in a small area	8. Errors in training programmes
	4. "Important" events	9. Errors in loading
	5. Deficiencies in legal regulations	10. Errors in refuelling
		11. Errors in the safety control process
		12. Insufficient process monitoring
		13. Errors in crisis situation management
		14. Lack of hazard control and assessment
		15. Improper planning
		16. Insufficient funding
		17. Political situation
		18. Deficiencies in legal regulations

20.5 Hazard Source Detection by Air Operator (Selected Special Procedure of Operations/High Risk) Follow Commission Regulation (EU) 965/2012

Selected air operation as area of risk analysis are: survey operations (power and gas pipe line monitoring), helicopter contruction work, aerial photography flights, advertising flights, and training flights. All of them are an example of imminience analysis and valuation. Whole air-operation comprises tasks. Risk analysis has been performed for the following phase of action – tasks. Each of them has some specific characteristic and requirements. Results presents Table 20.3. Safety risk assessment involves an analysis of identified hazards that includes two components – the severity of a safety outcome as well as the probability that it will occur. Once risk have been assessed, the service provider will engage in a decision making process to determine the need to implement risk mitigation measures. This decision-making process involves the use of a risk categorization tool that may be in the form of an assessment matrix (ICAO Doc).

Table 20.2 Example of imminence identification, analysis, valuation

Faze of action						Rezults				Results after bringing int practice new procedures				
Lp. (Faze of action)		Risk description	Reason	Results	Existing checking procedures	Level of severity	Probability of appearance	Risk tolerance level	Future procedures reducig risk	Level of severity	Probability of appearance	Risk tolerance level	To whom or what risk appertain	Monitoring or verifcation of procedures redusig risk
1.	Preparation to flight	Roll over after landing	Soft groundwork – winter Snow cover irregularity of surface	Main rotor damage, helicopter fire, death, people injure, helicoper structural damage, property blastment	Verification and acceptance of landing place according to information from customer or employer – durring mission planning	4	4	Non-acceptable	Inspecion of landing place from air, landing place check by crew member or veryfication landing place by employe inspection	4	2	Analysis	Air operations	Abiding of standard oprational procedures, transfer of new information by radio, SMS, abiding ruleas concern to landing in unknown terrain
		Wrong management of system and mission	Rush, stres, not sufisiant or inaccesible informations,	Wrong navigation preparation, wrong fuel cosnumption computing, abnormal selection of landing fileld and route, incorrect decission making, risk of mission execution,	Briefing and mission management (mark out route and landing fileds, emergency landing field, fuel calculation, weather analysis), information for ATC about mission (flight plan)	2	3	Analysis	Early planning, map and satelite pictures analysis, weather condition and forecast – calculation of fuel endurance	2	1	Acceptable	Air operations	Abiding of standard oprational procedures Checking state preparation to mission

(continued)

Table 20.2 (continued)

Faze of action						Rezults				Results after bringing int practice new procedures				
Lp. (Faze of action)		Risk description	Reason	Results	Existing checking procedures	Level of severity	Probability of appeareance	Risk tolerance level	Future procedures reducig risk	Level of severity	Probability of appeareance	Risk tolerance level	To whom or what risk appertain	Monitoring or verifcation of procedures redusig risk
1.	Preparation to flight	Flying with not updated GPS	Not updated GPS data base	Impossibilty of communication, inaccesible worning informations durring flight	GPS update in accordance to update schedule.	2	3	Analysis	Equipment check after update process	2	1	Acceptable	Air operations, Maintenance	Functional check
		Engine or system feilure	Battery low, , to low temperature, starter generator broken, ignition failure, plug striken fuel installation, hang start, hot start, contamination or water in fuel	Loss of engine power, engine damage /systems by exceed of limitation (temperature, torque etc), - mission impossible	Monitoring of battery charging. Airworthiness check ccording to HFM	2	3	Analysis	Frequent fuel filter change, battery charging and parameers monitoring – at low temperature battery deinstallation, storage, installation according to manufacturer recommendation, analysis of maintenance programmeOT, monitoring trend of fault	2	2	Acceptable	Maintenance	Fault reporting

		Airfame icing, or winshield, pitot, air inlet icing	Lack of protection against meteorological conditions	Delay or flight cancellation	Handling and protection of the helicopter on lanfield	2	4	Analysis	Monitoriong of icing conditions, preflight check, ussage additional equipment as covers, parking in hangar if possible	1	2	Acceptable	Air operations, Maintenance	Monitoriong of icing conditions, preflight check, ussage additional equipment as covers, parking in hangar if possible
		Helicopter not prepared to mission	Missing or inaccurate preflight inspection	Impossibility of safty take off and scanty quantity of fuel and oil	Strictly observance of check list	3	3	Analysis	Possesion of updated check lists according to Aircraft Flight Manual	3	1	Acceptable	Air operations, Maintenance	Temporary inspestion of check list on borad, thir status, update and completness
		Non wilful interference of observer in flight controls	Not removed flight controls	Loss of helicopter control, possibility to strike to obstacle or ground, property damage, injury or loss of life – materialize durring flight mission	Remove of flight controls elements	4	2	Analysis	Equipment preflight inspection	1	1	Acceptable	Maintenance	Observance of standard opration procedures
		Wrong installation of gimbal system	Time preasure, rutine,	Injury or damage due to loss of gimbal durring flight	Gimbal installation acording to STC and manual	3	2	Analysis	Mounting check as part of preflight inspection	2	1	Acceptable	Maintenance	Personel curant training to maintain and increace qualification level
2.	Repositioning of aircraft to operational airfield	Roll over after landing	Soft groundwork – winter Snow cover irregularity of surface	Main rotor damage, helicopter fire, death, people injure, helicoper structural damage, property blastment	Verification and acceptance of landing place according to information from customer or employer – durring mission planning	4	4	Non-acceptable	Inspecion of landing place from air, landing place check by crew member or veryfication landing place by employe inspection	4	2	Analysis	Air operations	Abiding of standard oprational procedures, transfer of new information by radio, SMS, abiding ruleas concern to landing in unknown terrain
		Severe deterioration of helth with influence for mission and safety of flight	Food toxicosis sickness, alkohol	Stop of mission, preventive, anawares offence, possibility of helicopter damage, posibility of third party injury or property damage	Monitoring of crew feeling	3	3	Analysis	Increaing of crew awarness about results and consequences, selfontrol, alkohol tests, sending for medical check	3	2	Analysis	Air operations	Human Factor and CRM training. Periodic analysis of incidents

(continued)

Table 20.2 (continued)

Faze of action						Rezults				Results after bringing int practice new procedures				
Lp. (Faze of action)		Risk description	Reason	Results	Existing checking procedures	Level of severity	Probability of appearеance	Risk tolerance level	Future procedures reducig risk	Level of severity	Probability of appearеance	Risk tolerance level	To whom or what risk appertain	Monitoring or verifcation of procedures redusig risk
2.	Repositioning of aircraft to operational airfield	Loss of orientation	Insufficiently flight preparation by crew (navigation planning according to meteorological conditio)	Transgression of airspace, entry to bad weather area (fogg, storm, CB etc), exceed of flight time, preventive landing, stop mission, start search and rescue procedure, aggravate of company reputation	Monitoring of mission proces follow standard operational prcedures and operational flight plan	2	2	Acceptable	Ussage of eletronic source of information durring flight	2	1	Acceptable	Air operations, Maintenance	Abiding of standard oprational procedures, periodic database update check
		Icing, heavy rain, snow stor mor hail	Non wilful flight in area of bad weather or CB activity	Loss of visibility, necessity of preventive landing, loss of lift or engine power as resultts of icing, emergency landing	Obserwacja pogody i unikanie niebezpiecznych zjawisk meteorologicznych	3	3	Analysis	Accurate weather analysis : TAF, METAR, AIRMET, SIGNIFICAT, SIGMET, GAMET, ussage de-icing installation monitorinf of ATC frequecy and requst additional weather information. Terrain observation to select lanfield for preventive or emergency landing	2	2	Acceptable	Air operations	Periodic crew training, periodic profficiency check

		Reduced engine power	High temperature, desity altitude	Limitation of MTOW, decraesiong of flight speed, increasing of fuel consumption, possibility of alternate landing place	Obserwacja pogody i unikanie niebezpiecznych zjawisk meteorologicznych, przestrzeganie zaleceń instrukcji użytkowania w locie dt. Ograniczeń temperatury	2	4	Analysis	Recalculation of fuel consumption, selection of alternate landing place, reducing of load, monitoring of weather forecast, avoid time of the day with high temperature	2	2	Acceptable	Air operations	Analysis of past missions, abiding of standard oprational procedures
		Heavy turbulence, wind shear,	Flight into CB or tunderstorm area, mountain interrain, wake turbulence, convective air current	Exceed of limitations, damage of airframe structure, loss of speed – loss of thrust, lift, stall, strike to obstacle or ground, discomfort	Weeather observation and avoidens of hazarus phenomenon Compiliance with helicopter flight manual and time separation	3	4	Analysis	Terrain analysis and airflow, monitoring ATC frequency, continiouse weather observation and analysis.	3	3	Analysis	Air operations	Identification of meteorological phenomenon risk Crew training. Revision of standard operational procedures
3.	Crew and aircraft preparation to mission	Inroad of third parties in area of helicopter operation	Lack of situational awarness or training szkolenia; wrong protected terain	Injury, death, helicopter damage	Training of customer crew concern to safety procedures in flight and on the ground	4	3	Analysis	Periodic briefing and training. Increasing of crew awarness	3	2	Analysis	Air operations,	Verification of crew knowlage, analysis and update of training materials
		Helicopter not prepared to flight	Missing or inaccurate preflight inspection	Impossibility of safty take off and scanty quantity of fuel and oil	Strictly observance of Check list	3	3	Analysis	Updated set of check list on board	3	1	Acceptable	Air operations, Maintenance	Periodic verification of checklists on board. Update check
4.	Durring operation	Loss of orientation	Insufficiently flight preparation by crew (navigation planning according to meteorological condition) Durring pipe control mission low altitude, frequent direction changes; necessity of terrain and obstaclesobservation	Transgression of airspace, entry to bad weather area (fogg, storm, CB etc), exceed of flight time, preventive landing, stop mission, start search and rescue procedure, aggravate of company reputation	Patrol flight with maximum accuracy and specific requirements; Monitoring of operational flight plan. Using route data transfered to GPS	2	2		Ussage of electronic devices durring fligt	2	1		Air operations, Maintenance	Monitoring of mission procedures

(continued)

Table 20.2 (continued)

Faze of action						Rezults				Results after bringing int practice new procedures				
Lp. (Faze of action)		Risk description	Reason	Results	Existing checking procedures	Level of severity	Probability of appearance	Risk tolerance level	Future procedures reducig risk	Level of severity	Probability of appearance	Risk tolerance level	To whom or what risk appertain	Monitoring or verifcation of procedures redusig risk
4.	Durring operation	Landing out of the airfield	On request of customer (gas leak transgression right of way), sickness of observer, preventive or emergency landing	Possibility of helicopter damage, venture of crew or third party	Patrol flight follow right of way usually free of obstacles (30-50m); maintain speed of helicopter out of H/V diagram (55-60kts)-; Over power line alittude 50-80m guarantee possibility of autorotation	3	3	Analysis	Accurate analysis of route and terrain. Initial selection of landing places. Record of GPS coordinates selected findings.	2	2	Acceptable	Air operations	Monitoring and analysis of past missions and record of landing places
		Severe turbulence, wind shear, intense air traffic	Flight into CB or tunderstorm area, mountain interrain, wake turbulence, convective air current	Exceed of limitations, damage of airframe structure, loss of speed – loss of thrust, lift, stall , strike to obstacle or ground , discomfort	Monitoring of ATC frequency and warning information	3	4	Analysis	Terrain analysis and airflow, monitoring ATC frequency, continiouse weather observation and analysis.	2	3	Analysis	Air operations	Raporting to ATC – mission information
		Severe deterioration of helth with influence for mission and safety of flight	Food toxicosis sickness, alkohol	Stop of mission, preventive, anawares offence, possibility of helicopter damage, posibility of third party injury or property damage	Monitoring of crew feeling	3	3	Analysis	Increaing of crew awarness about results and consequences, selfontrol, alkohol tests, sending for medical check	3	2	Analysis	Air operations	Human Factor and CRM training. Periodic analysis of incidents

		Icing, heavy rain, snow storm, hail	Non wilful flight in area of bad weather or CB activity	Loss of visibility, necessity of preventive landing, loss of lift or engine power as resultts of icing, emergency landing	Weeather observation and avoidens of hazarus phenomenon	3	3	Analysis	Accurate weather analysis : TAF, METAR, AIRMET, SIGNIFICAT, SIGMET, GAMET, ussage de-icing installation monitorinf of ATC frequecy and requst additional weather information. Terrain observation to select lanfield for preventive or emergency landing	2	2	Acceptable	Air operations	Imminience – periodic crew training, knowladge check.
		Roll over or collision after landing	Soft groundwork – winter snow cover irregularity of surface, other obstacles	Main rotor damage, helicopter fire, death, people injure, helicoper structural damage, property blastment	Immediat selection of emergency landingfield durring flight mission	4	3	Analysis	Preflight route analysis to select ladfield places	3	2	Analysis	Air operations	Monitoring of flight parameters weather conditions, terrain.
		Helicopter not prepared to flight	Missing or inaccurate preflight inspection	Impossibility of safty take off and scanty quantity of fuel and oil	Strictly observance of Check list	3	3	Analysis	Updated set of check list on board	3	1	Acceptable	Air operations, Maintenance	Temporary inspestion of check list on borad, thir status, update and completness
		Roll over after landing	Soft groundwork – winter Snow cover irregularity of surface	Main rotor damage, helicopter fire, death, people injure, helicoper structural damage, property blastment	Verification and acceptance of landing place according to information from customer or employer– durring mission planning	4	4	Non-acceptable	Inspecion of landing place from air, landing place check by crew member or veryfication landing place by employe inspection	4	2	Analysis	Air operations	Abiding of standard oprational procedures, transfer of new information by radio, SMS, abiding ruleas concern to landing in unknown terrain

(continued)

Table 20.2 (continued)

Faze of action						Rezults				Results after bringing int practice new procedures				
Lp. (Faze of action)		Risk description	Reason	Results	Existing checking procedures	Level of severity	Probability of appearance	Risk tolerance level	Future procedures reducig risk	Level of severity	Probability of appearance	Risk tolerance level	To whom or what risk appertain	Monitoring or verifcation of procedures redusig risk
5.	Return to airbase	Icing, heavy rain, snow storm, hail	Non wilful flight in area of bad weather or CB activity	Loss of visibility, necessity of preventive landing, loss of lift or engine power as resultts of icing, emergency landing	Weeather observation and avoidens of hazarus phenomenon	3	3	**Analysis**	Accurate weather analysis : TAF, METAR, AIRMET, SIGNIFICAT, SIGMET, GAMET, ussage de-icing installation monitorinf of ATC frequecy and requst additional weather information. Terrain observation to select lanfield for preventive or emergency landing	2	2	**Acceptable**	Air operations	Imminience – periodic crew training, knowladge check.
		Severe deterioration of helth with influence for mission and safety of flight	Food toxicosis sickness, alkohol	Stop of mission, preventive, anawares offence, possibility of helicopter damage , posibility of third party injury or property damage	Monitoring of crew feeling	3	3	**Analysis**	Increaing of crew awarness about results and consequences, selfontrol, alkohol tests, sending for medical check	3	2	**Analysis**	Air operations	Human Factor and CRM training. Periodic analysis of incidents
		Severe turbulence, wind shear, intense air traffic	Flight into CB or tunderstorm area, mountain terrain, wake turbulence, convective air current	Exceed of limitations, damage of airframe structure, loss of speed – loss of thrust, lift, stall, strike to obstacle or ground, discomfort	Monitoring of ATC frequency and warning information	3	4	**Analysis**	Terrain analysis and airflow, monitoring ATC frequency, continiouse weather observation and analysis	2	3	**Analysis**	Air operations	Raporting to ATC – mission information

Table 20.3 Risk assessment matrix

Risk probability	*Risk severity*				
	Disasterous 5	*Hazardous 4*	*Seriouse 3*	*Minor 2*	*Non-essential 1*
Frequent 5	*Unacceptable*	*Unacceptable*	*Unacceptable*	*Analysis*	*Analysis*
Ocasionally 4	*Unacceptable*	*Unacceptable*	*Analysis*	*Analysis*	*Analysis*
Unimportant 3	*Unacceptable*	*Analysis*	*Analysis*	*Analysis*	*Acceptable*
Improbable 2	*Analysis*	*Analysis*	*Analysis*	*Acceptable*	*Acceptable*
Extremely fictitiously	*Analysis*	*Acceptable*	*Acceptable*	*Acceptable*	*Acceptable*

20.6 Summary

The basis of proper risk management in transport is awareness of hazard sources found in transport organisations and their surrounding and, thus, awareness of the hazards themselves. Their proper assessment allows to conduct safe carriages (with no unacceptable risk). The safety management system in civil aviation consists of four areas and is divided into two mutually complementary subsystems – the sta te safety management system, the basic document of which is the state safety programme (SSP) in civil aviation, and safety management systems (SMSs) in service providers. The common feature of the Chicago and Union conventions, binding both the aviation authorities as well as organisations in the aviation industry, is definition of safety management through proper safety policy, active risk management, continuous monitoring and improvement of obtained results as well as communication and promotion of safety. (Analisys of proces entry of samety management in civil aviation of Poland 2015).

Although the European Parliament and Council as well as the European Commission have been implementing ICAO SARPs in the Union law successively since 2008 in relation to safety management in civil aviation, there still remains a broad gap within the scope of ICAO principles and requirements that are not regulated by the Union law. The implementation time limits of a great majority of the Union regulations (EC) have already lapsed and they are currently legally binding. Air operators are obliged to prepare hazard identification, analysis and valuation as part of safety management system (SMS).

The legal requirements and good practices applied in aviation within the scope of hazard source detection increase safety level of air operations. Valuation of imminence analysis is a continuous process for each air operator.

Further Suggested Readings

Acceptable Means of Compiliance (AMC) and Guidance Material (GM) of Part–ORA
Acceptable Means of Compliance (AMC) and Guidance Material (GM) of Part–ARA
Analisys of proces entry of samety management in civil aviation of Poland, Raport of Polish Aviation Club, Warszawa 2015

Chicago ICAO convention 7 December 1944 r. Konwencja chicagowska (Dz U z 1959 r., nr 35, poz. 212, z późn. zm.)
Declaration of Polish Government of 20 August 2003 r. in relations to Chicago convention Chicago 7 December 1944 r. Dz U 2003, nr 146, poz. 1413
Guide to the Project Management Body of Knowledge (PMBOK Guide) – Fifth Edition, PMI 2013
ICAO Doc. 9859 Safety Management Manual (SMM)
Kadziński A (2014) Chapter 3, Manual of essential aspects of safety and hygiene work pod red. L. Lewicki i J. Wrzesińska, Wyd. Wyższej Szkoły Logistyki w Poznaniu, Poznań
Krystek R (ed) (2009) Integrated safety system of transport, vol II. Conditions of integration development of safety transport systems, WKŁ, Warszawa
Pritchard C (2001) Risk management in projects. Wig-Press, Warszawa
Projects in Controlled Environments PRINCE2, UK Government Crown 2009

Chapter 21
Resilience Needs in NATO Partner Countries, Global and African Future Earth

Ahmed A. Hady

Abstract Research now demonstrates that the continued functioning of the Earth System, as it has supported the well-being of the human civilization in recent centuries, is at risk. Resilience is thus needed in NATO Partner Countries on national, regional, and global dimensions.

Global Future Earth as an international scientific community, under the umbrella of the International Council of Science Union (ICSU), is responsible to help modify the Earth System to avoid future risks. The purpose of the Africa Future Earth Committee (AFEC) is to be an effective advocate for Future Earth (FE) in Africa, as well as for African interests in the global Future Earth platform. Resilience needs in Future Earth community is very important to avoid the present and future disaster risks, especially in Africa with poor facilities of infrastructure.

21.1 Resilience Needs

21.1.1 Resilience Needs – National Dimensions

The Egyptian population is growing quickly, now approaching 100 million individuals. The effects of an increasing population present a threat to the current infrastructure. Resilience is needed for critical infrastructure safeguarding, especially to areas with an international dimension like the Suez Canal. There is currently a special Early Warning Group working to manage and facilitate the solution before, during and after any disaster risk, with the help of the Egyptian military. The High Dam in Aswan is one of the most important infrastructures, and as its safeguarding is so important for Egypt, there is a special research institute on site working to improve its situation and study the expected disasters (Egyptian Environmental Affair Agency (EEAA) 2009).

A.A. Hady (✉)
Department of Astronomy & Space and Meteorology, Faculty of Science,
Cairo University, Giza, Egypt
e-mail: aahady@sci.cu.edu.eg

© Springer Science+Business Media B.V. 2017

I. Linkov, J.M. Palma-Oliveira (eds.), *Resilience and Risk*, NATO Science for Peace and Security Series C: Environmental Security,
DOI 10.1007/978-94-024-1123-2_21

21.1.2 Resilience Needs in NATO Partner Countries – Regional Dimensions

The needs to implement resilience in NATO Partner Countries include the following items of resilience:

- A resilient system for exchanging information between Partner Countries in the field of terrorism and sabotage.
- Common strategies for NATO Partner Countries in regard to critical infrastructure.
- Initiatc a Technical Support Working Group.
- Establish an early warning unit.
- A method for exchanging experience to reduce the risk of disasters in critical infrastructure due to inexperience and the misuse during operating.
- Establish a scientific system for predicting the risks to critical infrastructure.
- Work to avoid and reduce risks of natural disasters to critical infrastructure, by providing enough information about natural disasters, and encouraging cooperation with Partner Countries to facilitate a fast transition during disasters.
- Reduce the misuse or the excessive use of critical infrastructure's abilities.

21.1.3 Resilience Needs in NATO Partner Countries – Global Dimensions

Partner countries need to reduce industrial pollution, especially those due to car emissions, which raise a huge risk to critical infrastructure. Controlling population explosion is also of the utmost importance. Improving the standard of living around the world is a further step in the same direction.

Life on earth is a monumental and essential task that must be preserved and developed constantly. Preserving the earth improve the quality of human life is Partner Countries main target.

21.2 Global Future Earth (FE)

"Planet under Pressure" Conference (London, March 2012) produced the "Planet Declaration" in response to current research showing that continued functioning of the Earth System is at risk. Current state is doing damage to the socio-economic trends which comprise: world population, urban population, large dams, foreign direct investment, primary energy use, fertilizers consumption, water use, paper production, transportation, telecommunications and international tourism (http://www.futureearth.org).

Earth System Trends: Carbon dioxide – nitrous oxide – methane – coastal nitrogen – surface temperature – marine fish capture – ocean acidification – shrimp and fish aquaculture – tropical forest loss – domesticated land – terrestrial degradation.

Future Earth is a research platform for the anthropogenic sources and research for global sustainability (http://www.futureearth.org). About 50,000 sustainability researchers, from more than 30 countries, are working together on finding solutions to the planet's most pressing challenges. Resilience needs in Future Earth researches is very important to avoid the present and future disaster risks.

21.2.1 The Future Earth Challenges Are

(a) To unite around a common research agenda for global sustainability science; (b) to engage societies in new ways; (c) and to encourage, catalyze and synthesize high quality research to support transformation.

Future Earth addresses these challenges by: (a) Building global communities of practice around key themes in sustainability; (a) Promoting research that informs solutions to real problems around the world; (c) And bringing together researchers, policy experts, businesses, and leaders in civil society and more.

21.2.2 Future Earth Networks Are

Global in scope but designed to inspire transformations at the local level; responsive to the needs of societies around the world; and co-designed and co-produced with the people who will use the results of our research.

- **Natural Assets:** Manages natural assets to preserve human well-being and biodiversity.
- **Oceans:** Addresses the most pressing challenges to ocean sustainability through solutions-oriented research.
- **Water-Energy-Food:** Explores the interactions between water, energy and food, and how these relationships are shaped by environmental and social changes.
- **Finance & Economics:** Supports strategies for linking economic prosperity with social justice and a healthy planet.
- **Health:** Promotes research for a better understanding of the relationships between changing environments and human health.
- **Cities:** Contributes to the transition toward sustainable urban futures where cities are more livable, equitable and resilient.

21.2.3 Sustainable Development Goals

Most important is to promote high-quality scientific research as a tool and approach for achieving the Sustainable Development Goals (SDGs). To co-organized workshops held by SDGs in 2015, 2016 established an organizing committee for 2017 Conference on the Anthropogenic SDGs, in partnership with GEO and financial support from AGPP.

21.2.4 Importance of Future Earth

Future Earth is part of an international community committed to transformation according to a coordinated research agenda: (a) organizing International conferences to meet and share ideas (physical and virtual): (b) to initiate the Intellectual frameworks for the co-design solutions-based research: (c) route to engage with international policy processes: (d) and international support for media, communications, capacity building, and young scientist career development (Future Earth Booklet 2016).

21.3 African Future Earth Committee (AFEC)

The African Future Earth Committee (AFEC) was established in 2015 to be an effective advocate and advisor for Global Future Earth community in Africa. The purpose of this committee to be encourage the research for global sustainability in Africa and scientific advisor for African Union (AU), and as an effective advisor for African interests in the global Future Earth platform (http://www.icsu.org/icsu-africa/about-icsu-roa/about-us/african-future-earth-committee).

The AFEC has two main responsibilities. Firstly, they are responsible for the formulation of a regional strategy for short -, medium -, and long – term deliverables on activities in various domains including capacity building, research, practice, infrastructure, and others. Secondly, they are charged with the development of a comprehensive concept note for Future Earth in Africa that will be submitted to the African Union and other inter-governmental bodies to get buy-in for the Future Earth framework at national, sub-regional and continental levels.

AFEC will initiate National committee allover African country to be the representative of FE and AFEC locally. The resilience needs in African future researches will help the continent in infrastructure safeguarding, modify its abilities and create suggestions of future modifications.

21.3.1 AFEC's Roles and Responsibilities

Resilience will be integrated with African Future Earth responsibilities for the suggested future research in African counties, and lead new ideas for dramatic changes according the following lines:

- Raising awareness of Future Earth agendas, activities and opportunities in African science, policy and practice bodies, at national, regional and continental levels;
- Keeping up to date with Future Earth science and engagement agendas, activities, programmes and other relevant information;
- Consulting with relevant African science, policy, practice bodies on African interests and priorities for the global Future Earth, and ensuring these priorities within the global Future Earth agenda and activities;
- Work with the global Future Earth Platform to oversee the establishment of the African Future Earth Center(s) that will act as secretariat and manage African Future Earth activities;
- Work with the African Future Earth Centre to plan and implement African Future Earth activities;
- Formulate a regional strategy for short-, medium-, and long-term deliverables on activities in various domains including capacity building, research, practice, infrastructure, and others;
- Develop a comprehensive concept note for Future Earth in Africa that will be submitted to the African Union as well as to other intergovernmental bodies.

21.3.2 The Tasks of AFEC

21.3.2.1 The AFEC's Agreement

AFEC's strong agreement that an African science agenda must be predicated on African science and developmental contexts include: (a) inter-phasing in the African 50-Year Strategic Science Agenda (Agenda-2063), with the Global Future Earth programme; (b) obtaining funding support for public engagement and co-creation of an agenda, sensitization and mobilization for Future Earth programmes in Africa; (c) and integrating with FE activates in Africa e.g. updates on the developments/activities of Future Earth global and regional level at offices initiated recently in Africa.

21.3.2.2 Developing FE in Africa (2016–2025)

The risks in Africa need a lot of efforts and modifications of the present situation in the following fields:

- Education and Health are a key priority in Africa and the AFEC was glad to note that this theme features in the Future Earth global plan;
- Updates on the developments/activities of Future Earth global and regional level at offices were initiated recently in Africa (Pretoria, Kigali, and Bibliotheca Alexandrina as a coordinator office);
- Improve the visibility of Future Earth in Africa according to the recent situation. There are weakness and inability to services provided to the African people, especially in education, health, transportation, electricity, paved roads, clean water and other unavailable services.
- Developing a process to articulate key science and other challenges that are of prime interest to Africa to promote an understanding of the African Worldview and in the context of African Development Priorities (current and future).

Sustainability includes the reference to the Sustainable Development Goals and for resilience needs in infrastructure and its development Africa.

The following initial themes have been suggested by AFEC according to (SDGs) roles:

- Natural resource use.
- Understanding the 'Anthropology' of African Peoples in transitions.
- Monitoring and evaluation.
- Well-being and life.
- Technology – new and emerging.
- Africa-driven solutions to infectious orphan diseases.
- Investigating and upgrading traditional solutions.
- Regenerative agriculture based on nutrient-dense, African Indigenous crops. With the capacity to restore the soil.
- Population growth.

Acknowledgements I would like to thank Professor Dr. Paul Shrivastava, the Executive Director of Future Earth, and his team for the valuable discussions during the preparation of this paper, and for supplying the necessary articles and scientific materials. Special thanks are due to Prof. Dr. Gamil Soliman, Cairo University, for revising the language of the manuscript.

Further Suggested Readings

Egyptian Environmental Affair Agency (EEAA) (2009) "Egypt second national declarations", personal communication

Future Earth Booklet (2016), private communication

http://www.igbp.net/publications/stateoftheplanetdeclaration.4.6b007aff13cb59eff6411baa.html

http://www.futureearth.org

http://www.icsu.org/icsu-africa/about-icsu-roa/about-us/african-future-earth-committee

Chapter 22
Contribution to Enhancement of Critical Infrastructure Resilience in Serbia

Branislav Todorovic, Darko Trifunovic, Katarina Jonev, and Marina Filipovic

Abstract This chapter provides an overview of the current situation with critical infrastructure (CI) resilience in the Republic of Serbia, with an emphasis on the possibilities for utilizing the social behaviour in improving CI resilience through the cyber component. It argues that the response to an incident and its impact, as the critical phase in estimating, defining and improving resilience, could play a critical role in enhancing and providing guidance for the CI security in Serbia. To provide a necessary platform, CI resilience should be a main pillar of the new contemporary Law about Critical Infrastructure. On the other hand, the cyber segment is the fastest growing component of CI, making it extremely important for resilience-based CI safeguarding. The pace of technological improvements in the cyber area is not adequately followed by most of its users and their behaviour, making the human factor a constantly growing risk that often renders even the most advanced technological protection measures obsolete. Therefore the strengthening of the social component, followed by appropriate legislation, could provide the required edge in improving the overall resilience of critical infrastructures as the ultimate target.

Keywords Critical infrastructure • Cyber security • Cyber attack • Emergency situation • Guidance • Legislation • Resilience enhancement • Social behaviour • Social engineering • Serbia • Strategy

B. Todorovic (✉)
Faculty of Mechanical Engineering, University of Belgrade, Belgrade, Serbia
e-mail: bt.emit.group@gmail.com

D. Trifunovic • K. Jonev • M. Filipovic
Faculty of Security Studies, University of Belgrade, Belgrade, Serbia
e-mail: galileja@yahoo.com; jonev.katarina@gmail.com; fmarina@fb.bg.ac.rs

© Springer Science+Business Media B.V. 2017

I. Linkov, J.M. Palma-Oliveira (eds.), *Resilience and Risk*, NATO Science for Peace and Security Series C: Environmental Security,
DOI 10.1007/978-94-024-1123-2_22

22.1 Introduction

Resilience ability targets and defines the enhancement of the system's inherent capacity to respond throughout the process of inevitable change – both long and short duration (Linkov et al. 2014), thus invoking a fundamentally temporal perspective (Fig. 22.1). The effectiveness of a resilient infrastructure or enterprise depends on its "*ability to anticipate, absorb, adapt to, and rapidly recover from a potentially disruptive event, whether naturally occurring or human caused*" (NIAC 2009).

It can be noted that the response stage highly influences the resilience ability, requiring special attention in planning and capacity building. Response capabilities are a function of immediate and ongoing activities, tasks, programs, and systems that have been undertaken or developed to respond and adapt to the adverse effects of an event. However, response time directly corresponds to event duration, magnitude and propagation speed; starting from extremely short, but strong incidents (e.g. earthquakes or flash floods) to slowly spreading attacks/incidents, including aftershocks and cascading effects. The near-zero response time, caused by the abruptness of the incident in case of an earthquake, is shown in Fig. 22.2.

Quality of infrastructure in this case is generally expressed as Q(t), a function of time. Specifically, CI performance can range from 0 to 100%, where 100% means no degradation in quality and 0% means total loss of operation. If an earthquake or

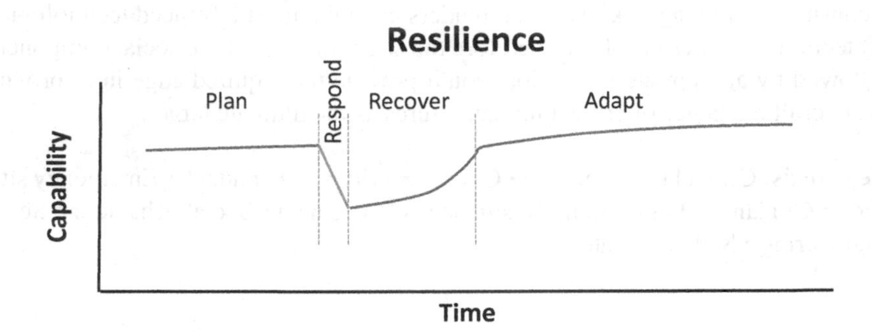

Fig. 22.1 Conceptual model of the stages of resilience as a function of time

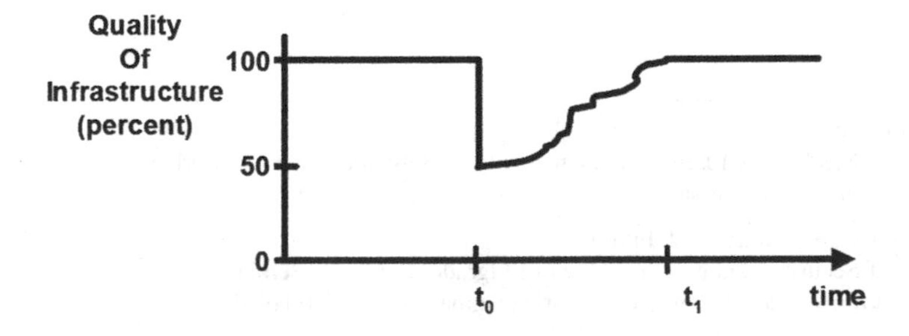

Fig. 22.2 Schematic representation of seismic resilience concept (Bruneau et al. 2003)

other incident occurs at time t0, it could cause sufficient damage to the infrastructure such that the quality measure, Q(t), is immediately reduced (from 100% to 50%, as an example, in Fig. 22.2). Restoration of the infrastructure is expected to occur over time, as indicated in that figure, until time t1 when it is completely repaired and functional (indicated by a quality of 100%) (Bruneau and Reinhorn 2006). Arguable are three aspects: (a) is the loss of CI operation directly proportional to physical damage, (b) what is the critical level of CI operational downgrade that causes social issues besides the financial loss, (c) can the resilience be used effectively to reduce the initial damage to CI, thus directly improving the recovery stage? In particular, aspect (c) is considered very important and potentially beneficial in the case of the Republic of Serbia.

Over the years, there have been numerous attempts to estimate and quantify resilience of critical infrastructures (CI), including the response stage. Despite significant advances in the field, a lot more work is needed to ensure satisfactory levels of CI resilience modeling that could be universally applied.

22.2 Cyber Component

Enhancement of critical infrastructure (CI) resilience is the best safeguarding policy at present. It is complementary with risk & hazard assessment & management and critical infrastructure protection (CIP) activities. For many reasons, including faster technological development than other fields, information & communication technologies (ICT) and consequently cyber security, have become one of the crucial segments of CI resilience management.

Cyber attacks have become a reality and a source of national fear: dangerous programs can secretly be executed on computer systems and send out confidential data straight to terrorists. As computer viruses and worms become "smarter" and better every day, cyber attacks on government and private industry pose an increasing threat to national security. The fact is that cyber attacks represent a new threat to the state and its security. Cyber attackers with different profiles have repeatedly demonstrated their capacity to jeopardize the functioning of the state and national infrastructure, and thus endanger security.

Surely, cyber attacks take different forms. Some of them are acts of cyber crime, stealing sensitive information, defacing websites, creating malicious codes or using some other hacking techniques such as viruses, worms, or Trojan horses, to get access to systems. National infrastructure functioning relies in large part on SCADA controlling systems. It has certain vulnerabilities, and unauthorized control could have far-reaching effects. Attacks can arise from states, groups, individuals, organizations. Nobody can predict the precise timing when the next major cyber attack might happen.

Within the context of CI resilience, the cyber part can be generally divided into three interrelated segments that require attention: (a) internal ICT structure (including Cloud services), (b) integrated Web parts (public and private) and connections with Internet and (c) users of cyber services. The third segment represents the human factor in cyber risk & hazard assessment and is probably the most difficult to handle within CIP efforts and ultimately in building and enhancing resilience.

22.3 Incident – Response Relation and Quantification

Quality of critical infrastructure Q(t) shows the intensity of operational degradation and recovery over time from occurrence until containment of the incident and further during restoration until full repair. Example of time-intensity relation is given in Fig. 22.3, with coloured areas corresponding to response and recovery stages. In the presented case the CI quality Q1 is reduced to 40% before the incident is fully contained and begins recovery.

When a system is better prepared to react to an incident, as part of enhancement efforts, one can see a notable difference in CI behavior. Faster response and incident containment ensures a smaller percentage of infrastructural operational degradation, thus providing a better starting point for the recovery stage. The overall resilience of the system allows for faster and easier restoration of CI operational capability (Fig. 22.4).

Regarding the recovery stage, the corresponding area is often used as an indication of the amount of efforts and resources needed to restore the CI operation. In case of response, the relation is far less straightforward.

Frameworks for the quantitative definition of CI resilience are based on analytical, empirical and combined methodologies. A good example of comprehensive assessment of critical infrastructure systems, targeting support to decision-making

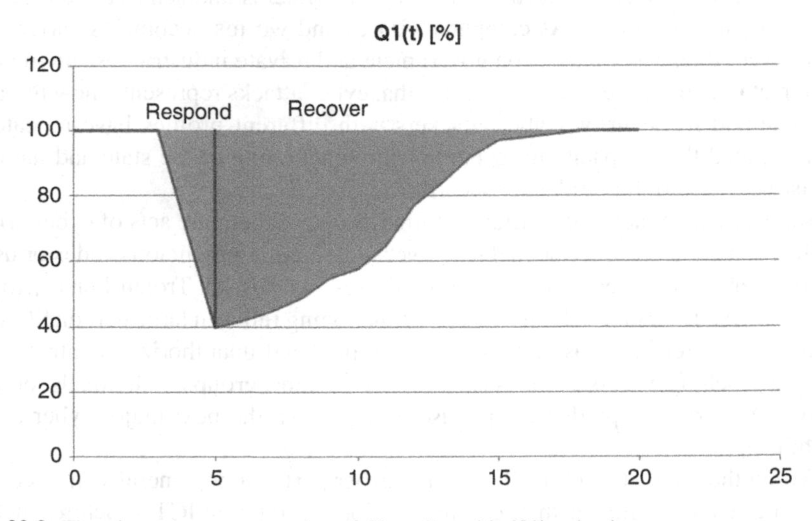

Fig. 22.3 Time-intensity representation of CI quality with 60% reduction

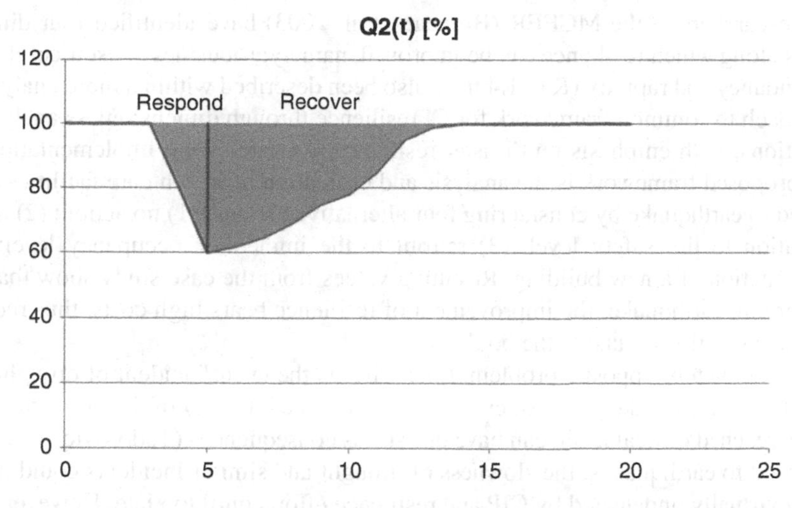

Fig. 22.4 Time-intensity representation of CI quality with 40% reduction

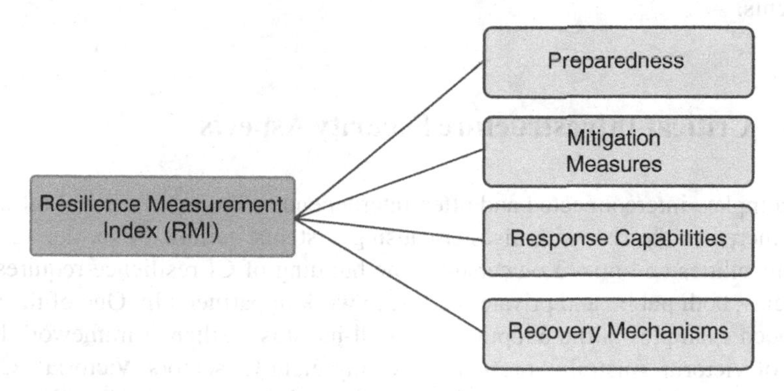

Fig. 22.5 Level 1 Components of the RMI (Petit et al. 2013)

for risk management, disaster response, and business continuity by exploiting empirical information, is the Resilience Measurement Index (RMI) developed by Argonne and DHS in partnership. RMI characterizes the resilience of CI by capturing the fundamental aspects with respect to all hazards and assists in estimating the CI capabilities in terms of resilience (Petit et al. 2013). RMI is the result of long-term efforts and covers all stages of resilience by organizing the collected/evaluated information into four groups called Level 1 components (Fig. 22.5).

The preparedness level is further subdivided into Awareness and Planning, where the latter contains the key components of Business Continuity, Emergency Operation/Emergency Acton and Cyber Plans. For each type of plan, the RMI addresses its characteristics (e.g. level of development and approval), the type of exercises and training defined in the plan, and its content. However, the question of what could be done to improve actual CI responses to unforeseen incidents, remains.

Researchers at the MCEER (Bruneau et al. 2003) have identified four dimensions along which resilience can be improved, namely: robustness, resourcefulness, redundancy and rapidity (R4). R4 have also been described within a more analytical approach to common framework for CI resilience through dimensionless analytical functions, with emphasis on disaster resilience. An interesting implementation of the proposed framework is the analysis and evaluation of health care facilities subjected to earthquake by considering four alternative actions: (1) no action; (2) rehabilitation to life safety level; (3) retrofit to the immediate occupancy level; (4) construction of a new building. Resulting values from the case study show that, in relation to earthquake, the improvement of resilience bears high costs, thus requiring new solutions 'outside the box'.

A completely opposite problem, time-wise, is the event/incident of drought. Its development can take years, or even decades, but the intensity of the problem, once it has reached critical level, can have disastrous consequences (Todorovic 2012). As opposed to earthquakes, the slowness of drought and similar incidents could make them virtually undetected by CIP and resilience efforts until too late. However, due to the limited size of the chapter, further analysis will be restricted to short-term incidents.

22.4 Critical Infrastructure Security Aspects

The complex, interconnected and often interdependent nature of critical infrastructures increases the risk of a disaster-causing systemic failure. Also, due to social responsibilities and mixed ownerships, the building of CI resilience requires governments, both public and private sectors, to work in partnership. One of the recent advanced examples is the adoption of an all-hazards resilience framework by the State of Victoria, Australia, presently covering eight CI sectors. Victoria's Critical Infrastructure Model comprises four levels ('vital,' 'major,' 'significant,' and 'local') with the first three forming the Register, where 'criticality' is defined as a measure of the consequences associated with the loss or degradation of the infrastructure or the service it provides (The State of Victoria – EMV 2015).

'Vital' CI are assessed through a custom-made methodology and participate in the resilience improvement cycle (RIC). RIC contains four stages:

- Risk management planning and documentation
- Exercising
- Validation (through audit)
- Positive assurance

The majority of Victoria's critical infrastructure assets are owned and/or operated by private entities that have strong incentives for risk management. Government works in partnership with these entities to increase the resilience of CI for the wider Victorian community. Key mechanisms for developing these partnerships are Sector Resilience Networks (SRNs) and Sector Resilience Plans (SRPs). Participation in these is mandatory for owners and/or operators of 'vital' infrastructure, and encouraged for others. Operation of CIM is governed by regulations and guidelines.

The example of Victoria State emphasizes the need for a general and integrated approach for building CI resilience. It should be underlined that it is difficult and costly process and any improvement in the presently used methodologies and actions could mean a significant advance.

The complexity of coordinated resilience issues in Europe can easily be illustrated by the current situation in the EU water sector (Table 22.1). Water utility companies as CI units in EU have different types of ownership, priorities, financial capacity and levels of technical development, among other differences.

The Republic of Serbia also has a similar situation, and the Public water utility company (PWUC) in Užice is a good example. It is a medium sized city, among the 15 largest ones in Serbia, and its operation is burdened with a number of every-day operational issues, living little space for resilience building. Some basic facts regarding PWUC Užice:

- Supplies more than 60,000 customers with potable water (15,000 household connections and 1000 companies).
- Water supply network length of around 360 km, 29 reservoirs and 27 pressure zones.
- Due to high pressures, pipe bursts and large water losses are common occurrences.

Table 22.1 Water utility companies (WUC) statistics for some EU countries (Web link 1)

Country	Public/private WUC	No. of providers	Water supply responsibility
Denmark	Service provision only by public and cooperative providers	2,740 (2001)	Local governments
France	High degree of private sector participation using concession and lease contracts	12,400 (2008)	National & local (municipalities > 10.000 inhabitants)
Germany	Only 3.5% entirely privately owned	1,266 larger ones (2005)	Municipalities, regulated by the states
Italy	Public, private or mixed	91 regional utilities; 3,161 providers	National and regional governments
Netherlands	WUC publicly owned, contracting services to the private sector	10 regional WUC	Number of institutions at different levels
Spain	Municipalities 54%, private 33% or mixed	More than 8,000 in municipalities	National & basin agencies
UK	England & Wales – private (23), Scotland-public (1) and Northern Ireland – public (1)	25	Three regulators, one each for England/Wales, Scotland and Northern Ireland

- To reduce water losses and improve efficiency, IPA funds were used to establish the first District Metered Area (DMA) zone capable of operating under pressure control

It is very important to notice that Serbian utility companies, as well as other critical infrastructures, show significant problems as economic operators in covering their every-day operations, leaving little or no space for investing in improvements, including resilience. An additional complication is ownership, since many companies have been privatized during the few last decades, causing issues in regard to responsibilities in a similar manner as previously discussed on the EU level.

22.5 Overview of Latest EC Related Activities in Cyber Domain

The Directive on security of network and information systems (the NIS Directive) was adopted by the European Parliament on 6 July 2016 (Web link 2), entering into force in August 2016. Member States will have 21 months to transpose the Directive into their national laws and six more months to identify operators of essential services. The NIS Directive provides legal measures to boost the overall level of cyber security. Building on those, the Commission will propose how to enhance cross-border cooperation in case of a major cyber-incident, contributing to overall European efforts in enhancing cyber system resilience. Given the speed with which the cyber security landscape is evolving, the Commission will also bring forward its evaluation of the European Union Agency for Network and Information Security (ENISA) (Web link 3), which provides recommendations on cyber security, supports policy development and its implementation, and collaborates with operational teams throughout Europe.

On July 2015 the Commission has launched a new public-private partnership (PPP) on cyber security, which is expected to trigger €1.8 billion of investment by 2020, as a part of a series of new initiatives to better equip Europe against cyber-attacks and to strengthen the competitiveness of its cyber security sector.

Though cooperating on many issues and maintaining the physical single market, Europeans often face barriers when using online tools and services. That has led to the idea of creating a Digital Single Market, where the free movement of goods, persons, services and capital is ensured. In a Digital Single Market there are fewer barriers and more opportunities: it is a seamless area where people and business can trade, innovate and interact legally, safely, securely and at an affordable cost, making their lives easier. It can create opportunities for new start-ups and allow existing companies to grow and profit within a market of over 500 million people (Web link 4). Two of the main objectives are expected to significantly influence the security and resilience of CI cyber domains in relation to social behavior: rapidly concluding negotiations on common EU data protection rules and boosting digital skills and learning.

22.6 Key Aspects of the Cyber Problem

Advanced persistent threats and targeted attacks have proven their ability to penetrate standard security defences and remain undetected for months while siphoning valuable data or carrying out destructive actions. And the companies one relies on most are some of the most likely targets – financial institutions, healthcare organizations, major retailers and others. PC World reported an 81% increase in 2011 of advanced, targeted computer hacking attacks, and according to Verizon's 2012 research findings, there were a staggering 855 cyber security incidents and 174 million compromised records. According to the 2012 Ponemon study on the costs of cybercrime in the US for 56 large organizations, there are 1.8 successful attacks per organization per week, with the median cybercrime cost of $8.9M per organization (Web link 5). These cyber attacks are (Fig. 22.6):

- Social – Targeting and attacking specific people with social engineering and advanced malware
- Sophisticated – Exploiting vulnerabilities, using backdoor controls, stealing and using valid credentials

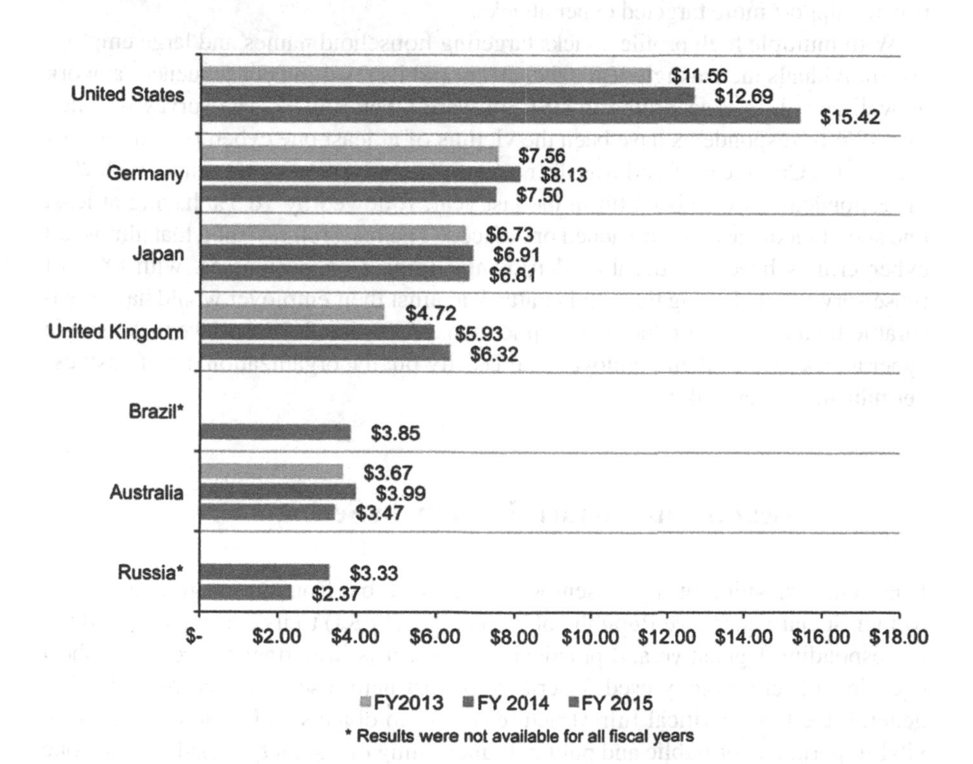

Fig. 22.6 Total cost of cyber crime in seven countries. Cost expressed in US dollars (000,000), n = 252 separate companies (Web link 6)

- Stealthy – Executed in a series of low profile moves that are undetectable to standard security or buried among thousands of other event logs collected every day

Working from home or by remote access creates a specific type of risk to CI cyber security. Cyber attacks in that field are mainly based on tricking employees into breaching security protocol or giving away information, by exploiting weaknesses in people rather than technology. Such cyber criminal tactics are called social engineering. In relation to CI, employees at all levels are targeted, from workers to management. Techniques used involve approach through e-mail (e.g. phishing scams), various communication applications for laptops and smart phones (including social spam), fake buttons on Web pages for malware download, redirection from trusted websites to infected ones, etc.; but also telephone calls or direct observations on the Web. To enhance their attacks, cyber criminals might exploit personal information collected from shared pictures and video clips, or other information gathered via social networking sites. Another common technique is to impersonate some authority to put pressure on clerks or helpdesk staff. The usual goal is to trick users into activating hidden, harmful software. The ultimate target is to obtain unauthorised physical access and, depending on the breach level, to gather information to support more targeted cyber attacks.

With multiple high profile attacks targeting household names and large employers, individuals increasingly fear cyber crime and its resulting consequences at work as well as at home, according to GFI Software (Web link 7). The survey revealed that 47% of respondents have been the victims of at least one cybercrime in the last year alone. Credit card fraud was the most prevalent form of cyber crime, with 20% of respondents having been hit in the last year, followed by 16.5% having at least one social media account breached or defaced. The research revealed that almost all cyber crimes have a noticeable, detrimental impact on businesses, with 88% of those surveyed believing that a cyber attack against their employer would have measurable financial and productivity implications. A further 3.5% believe that a single cyber attack against their employer could easily put the organization out of business permanently (Web link 8).

22.7 Critical Infrastructure Security in Serbia

Due to the transition of the present Republic of Serbia from an integral part of the former Social Federative Republic of Yugoslavia (SFRJ) to its current state and the corresponding legislative and practical changes, it is important to provide a short overview of terminology used for critical infrastructure security and resilience. In general, the term 'Critical Infrastructure' refers to objects and systems that are of vital importance for public and political functioning of a society. Besides those, one should also include systems that are crucial for national security. One good definition coming from the Unites States of America states 'Critical infrastructure and key resources (CIKR) refer to a broad array of assets which are essential to the everyday functionality of social, economic, political and cultural systems in the United States. The interruption of CIKR poses significant threats to the continuity of these systems and can result in property damage, human casualties and significant economic losses.' (Murray and Grubesic 2012). The report of the US Presidential Committee for Critical Infrastructure Protection provides a list of eight most important infrastructures in the state:

The transportation infrastructure moves goods and people within and beyond our borders, and makes it possible for the United States to play a leading role in the global economy. *The oil and gas production and storage infrastructure* fuels transportation services, manufacturing operations, and home utilities. *The water supply infrastructure* assures a steady flow of water for agriculture, industry (including various manufacturing processes, power generation, and cooling), business, firefighting, and our homes. *The emergency services infrastructure in communities* across the country responds to our urgent police, fire, and medical needs, saving lives and preserving property. *The government services infrastructure* consists of federal, state, and local agencies that provide essential services to the public, promoting the general welfare. *The banking and finance infrastructure* manages trillions of dollars, from deposit of individual paychecks to the transfer of huge amounts in support of major global enterprises. *The electrical power infrastructure* consists of generation, transmission, and distribution systems that are essential to all other infrastructures and every aspect of our economy. Without electricity, factories would cease production, televisions would fade to black, and radios would fall silent (even a battery-powered receiver depends on an electric-powered transmitter). Street intersections would suddenly be dangerous. Peoples' homes and businesses would go dark. Computers and telecommunications would no longer operate. *The telecommunications infrastructure* has been revolutionized by advances in information technology in the past two decades to form an information and communications infrastructure, consisting of the Public Telecommunications Network (PTN), the Internet, and the many millions of computers in home, commercial, academic, and government use. Taking advantage of the speed, efficiency and effectiveness of computers and digital communications, all the critical infrastructures are increasingly connected to networks, particularly to the Internet. Thus, they are connected to one another. Networking enables the electronic transfer of funds, the distribution of electrical power, and the control of gas and oil pipeline systems. Networking is essential to a service economy as well as to competitive manufacturing and efficient delivery of raw materials and finished goods. The information and communications infrastructure is basic to responsive emergency services. It is the backbone of the military command and control system. And it is becoming the core of the educational system (PCCIP 1997).

Within the EU, the term European Critical Infrastructure (ECI) refers to CI located on a territory of one member country, the disruption or destruction of which would cause significant consequences to at least two member countries. The level of disruption and its importance for the operation of CI should be estimated on the bases of interdependency criterion. That also includes influence and effects resulting from intersectoral dependency from other types of infrastructure (2008/114/EC).

Within the Republic of Serbia, a certain confusion and chaotic situation can be noted in relation to critical infrastructure security, protection and resilience topics. It continues to exist even after the proclamations of the *Law about emergency situations*, published in the Government Gazette of the Republic of Serbia No. 111/2009 (Zakon o vanrednim situacijama; Službeni glasnik Republike Srbije, broj 111/2009) and the *National strategy for protection and rescue in emergency situations*, published in the Government Gazette of the Republic of Serbia No. 86/2011 (Nacionalna strategija zaštite i spasavanja u vanrednim sitaucijama; Službeni glasnik Republike Srbije, broj 86/2011). It is debatable whether there is a notion in Serbia about CI protection, since legislators haven't clearly defined that area, i.e. terms, scope and targets of CI protection and resilience. Within the described conditions and in order to cover the legislative gaps, the Government of the Republic of Serbia has defined a *Regulation regarding the content and methodology for the development of plans for protection and rescue in emergency situations*, based on article 45, paragraph 4 of the Law about emergency situations. This Regulation officially introduces the term, 'critical infrastructure' in Serbia for the first time. However, it remains unclear which infrastructure the term applies to.

In 1992, during the existence of the SRJ, the Government of the Republic of Serbia released the *Regulation about objects and regions of special importance for the defense of the Republic of Serbia*, published in the Government Gazette of the RS, No. 18/92 (Uredba o objektima i reonima od posebnog značaja za odbranu Republike Srbije; Službeni glasnik RS, br. 18/92). According to that Regulation, objects of special importance for the defense of the Republic of Serbia are considered those for which it can be estimated that their eventual damage, or even the disclosure of their type, purpose or location if considered confidential, might cause serious consequences for the defense and safety of the Republic of Serbia. More specifically, it includes objects and regions in the area of transportation, telecommunication, energy, water and industry, that, if endangered, would jeopardize the proper functioning of the country and the society. Also during the existence of the SRJ, the Federal Government proclaimed the *Decision on defining large technical systems of interest for the defense of the country*, based on article 36, paragraph 3 of the Defense law, published in the Government Gazette of the SRJ, No. 43/94 and 28/96. That Decision defines large technical systems of interest for the defense of the country, providing guidance for selection, construction and development; but also the important technical means required for the proper operation of those systems in the area of communications, informatics, air and railway transportation, electric power supply, water supply and others, with guidance for purchase.

A document that more specifically handles the critical infrastructures in the Republic of Serbia is the *Guidance for methodology for performing the evaluations of vulnerability and plans for protection and rescue in emergency situations*, published in the Government Gazette of the Republic of Serbia, No. 96/2012 (Uputstvo o metodologiji izrade procene ugroženosti i planova zaštite i spasavanja u vanrednim situacijama; Službeni glasnik RS 96/2012). *Guidance for evaluations* is the document which identifies dangers, sources and types of vulnerability, possible effects and consequences, estimates vulnerabilities and risks and provides overview of sources, means and preventive measures required to respond to threats caused by natural disasters and accidents, protection and rescue of human lives and health, domestic animals, material possessions, cultural heritage and environment. Vulnerability and risk assessment of legal entities and commercial companies covers infrastructure and areas in their possession (owned or rented), including the surrounding zones/areas under influence, that can be threatened by natural disasters and accidents in facilities used for their main activities, or otherwise generate negative impact for the immediate surroundings and the broader society in case of operational disruptions or significant disturbances in providing critical services. Assessment and evaluation of vulnerability type and intensity, as well as potential negative effects of natural disasters and accidents, requires firstly to indentify and specify the critical infrastructure systems, particularly in the areas of (GUIDANCE 96/2012):

1. Production and distribution of electricity: hydroelectric power, thermal power, alternative energy, transmission lines, and substations;
2. Supply of energy (energy distribution networks): storage of gas, oil products and other energy sources, with the main oil and gas pipelines and local gas networks;
3. Supply of water: the water distribution system and treatment plants, potable and water sources (e.g. wells); potential polluters of surface and groundwater;
4. Supply of food for the population (production, common storage and distribution): production facilities and capacities, food production facilities, storage rooms for food products, objects and means of distribution, arable areas, fruit plantations, facilities for animal breeding and meat processing;
5. Health care: health security and health-care locations (health centers and facilities, including capacity and technical equipment);
6. Other material and cultural goods: objects of national importance (cultural and historical monuments, museums, legacies, etc.), facilities for cultural events, churches and places of worship;
7. Protected natural resources and the environment: National parks, wildlife reserves;
8. Telecom: transmission paths (underground cables, overhead lines, wireless links), antenna masts, antenna base stations for mobile telephony, telephone exchanges, portable transmission equipment – radio and TV stations (technical equipment for the transmission and broadcasting of audio-visual signals);
9. Traffic: traffic and road network, rail network, inland waterway roads, bridges, viaducts and tunnels;
10. Production of hazardous substances (facilities for production, storage and transport of hazardous substances).

Unfortunately, the published Strategy for the national security in the Republic of Serbia does not recognize the term critical infrastructure. Only in Section II can one find some reference to the elements of critical infrastructure. Section II elaborates on the sensitivity of certain segments of infrastructure, particularly those related to production and transportation of fuels and energy, as well as the possibility of endangering information and telecommunication systems by high-tech criminals. Perhaps that was one of the reasons to emphasize the following from the document "Strategy for development of the informational society in the Republic of Serbia until 2020": "*It is important to develop and improve the protection of critical infrastructure systems from attackers using information technologies, which besides information and telecommunication systems should include other infrastructures that are controlled and regulated through information and telecommunication channels, e.g. electric power system. In relation to the previous it is important to additionally provide the criteria for identification of critical infrastructures from the point of information security, criteria for recognition and characterization of attacks to such infrastructure that are using high-tech information technologies and its relation to other more common types of attacks, as well as the requirements for protection in that field.*" (STRATEGY 51/2010).

22.8 Cyber Segment in Critical Infrastructure Protection & Resilience in Serbia

In developed countries, the operation and functioning of critical national infrastructure relies on computers and ICT technologies and may therefore be an easy target. If we take into account that the national infrastructure includes a number of systems relying on high-tech technology and support, inter alia: energy systems, nuclear power plants, public health, emergency services, government, dams, electricity and water supplies, transport traffic, telecommunications networks, it can be clearly concluded that potential attack on these systems could have enormous consequences to the country and mostly to the civilians (Ophandt 2010). What worries security experts most is the question of whether terrorists may be developing methods and strategies to conduct large-scale cyber attacks with deadly intent and destruction of national vital infrastructure (Collin 1997).

Unsatisfactory protection could allow cyber terrorists to attack and penetrate networks, increasing risk and vulnerability especially on critical infrastructure (HSPD 2003). Cyber attackers could potentially destroy or cause difficulties in operations of the critical infrastructure in one country.

Bearing in mind that this is a global, but also national security issue, many countries have adopted relevant legal frameworks as well as national strategies for cyber security. Also, countries develop bilateral and multilaterally cooperation in this field. A large number of countries have already established the operational mechanisms that enable them to react to cyber incidents. These mechanisms include cooperation between representatives of the state authorities on one side with private sector, academia and the civil society.

Like many other countries in the Balkans, Serbia is lagging behind in these fields.

Operators of electronic communications networks have an obligation to protect their ICT resources, but these measures certainly are not sufficient to ensure complete safety of a country's critical infrastructure from cyber attacks. On the other hand, given that a good part of the critical infrastructure is in the hands of the private (corporate) ownership and management, and the state alone cannot provide enough safety, it is necessary to establish a special form of cooperation between the state and the private sector. In previous sections we have presented an overview of recent cyber oriented EU programs and activities that stress the need to strengthen security of ICT infrastructure and adopt the strategy for a secure information society. Following that, the EU announced a revision of the regulatory framework for electronic communications and services aimed at strengthening the security and integrity of communications networks.

At this point in Serbia it is difficult to recognize similar institutionalized activity. This does not mean that by ignoring this issue Serbia will be spared from a large cyber attack. On the contrary. Serbia has been integrated into the global network as any other state and thus their porous, virtually non-existent digital border currently represents an easy target. Bearing in mind that cyber space does not recognize borders as in the physical world, every country is a potential victim, including the Republic of Serbia.

According to estimates from 2013, a comprehensive cyber attack on Serbia, disabling the key segments of society, such as the state administration, telecommunications and the financial sector, could incur damages exceeding 10 million EUR per each day of such an attack, with significantly higher losses if the attack was to last for several days (Radunović 2013).

With the increase of e-services of the state administration and integrated databases on citizens, the linking of critical infrastructure and industry, and an integrated financial sector, the risks from cyber attack are even higher (CEAS 2016).

22.9 Cyber Activities Related to Social Behaviour in Serbia

The Sector for Analytics, Telecommunication and Information Technology Ministry of Internal Affairs of the Republic of Serbia is responsible for planning, monitoring and the implementation of measures in the field of information security, or the protection of ICT systems and network infrastructure of the Ministry of Internal affairs (Web link 9). The Department of Information Security is responsible for the implementation of work towards the protection of ICT systems, the implementation of measures to protect information security, organization and implementation of crypto, and maintaining confidentiality, integrity, and availability of the Ministry's ICT system.

The Center for responding to attacks on information systems (CERT), participates in the drafting of laws and bylaws and other regulations in the field of information security. Also, it is responsible for the supervision and control of ICT systems, identification and protection of the critical information infrastructure of the Ministry of Internal affairs, analysis and compliance of implemented measures to protect information security in accordance with international standards in the field of information security, periodic risk analysis, and continuous monitoring of the ICT system and the other networks with which the ICT system of Ministry is connected. Since July 2016. CERT is on the list of CERT teams – TI (Trusted introducer) and ENISA (The European Union Agency for Network and Information Security).

Directive on measures to ensure the highest level of network security and information systems (NIS Directive) provides that EU member states are obliged to identify critical information infrastructure. According to the information security, the concept of critical information infrastructure is not mentioned, but the law provides for the ICT systems of particular importance that perform activities of public interest, many of which are critical infrastructure, such as, for example, ICT systems that are used in performing activities in the fields of energy, transport, production and trade of arming and military equipment, utilities, ICT systems in the healthcare and financial institutions. These entities will have the obligation to protect its ICT systems in appropriate ways, and to report incidents to competent bodies, which are to be achieved to raise the level of preparedness of operators and the protection of ICT systems in the Republic of Serbia.

In early 2016, Serbia adopted the Law on Information Security (Web link 10) through which the fundamental legal framework in this area was established, in addition to the existing legislative framework through which the provisions of the Council of Europe Budapest Convention on Cybercrime are implemented. The adoption of the Law on Information Security was additionally envisaged within Serbia's process of accession negotiations with the EU, in the National Program for the Adoption of the EU Acquis (NPAA) for 2014–2018, as well as in the Strategy for the Development of Information Society in the Republic of Serbia by 2020 (Strategy 2010). On the basis of the Law on Information Security, Serbia brings stricter regulations on the ICT system protection measures of particular importance for ensuring the prevention and minimization of damage caused by the occurrence of the incident.

On a proposal from the Ministry for Telecommunications, in March 2016 the Serbian government has formed a technical working group – body for coordination of information security. The aim is to exchange knowledge, experiences and information, but also to create normative documents and link the relevant actors from the public and private sectors, academia and civil sector.

On the basis thereof, the Act on the safety of ICT systems more closely defines the dimensions of information security, a particular way, principles, so that procedures will achieve and maintain an adequate level of information security for ICT systems, as well as the authority and responsibilities of users of ICT resources.

The objectives of the Act of security are, among others:

- Prevention and mitigation of security incidents.
- Raising awareness about the risks and safeguards information security when using ICT resources system.
- Contribution to improving security and control of the implementation of measures to protect ICT systems in accordance with international standards.
- Ensuring confidentiality, authentication, integrity and availability of data.

Despite activities on the governmental level, there is still an open question regarding the level of risks to CI from the every-day behaviour of their personnel. Some of the issues are:

- Use of personal devices (e.g. mobile phones, tablets, laptops etc.) at work and connecting to the Internet through company's network
- Access to social and other personal applications through company's equipment
- Accidental reviling of potentially damaging and/or confidential information related to CI operation through social media; e.g. working shifts, sick leaves or replacement of personal, increased or reduced workloads, celebrations or other special events, etc.
- Lack of regulations and habits related to use of official and/or personal devices in case of incident, that could cause ICT lagging, overload or shutdown, leading to longer duration of response stage and downgrade of CI resilience. Besides panic, even the common urge of untrained personnel to check on family members or other related persons in case of major incident, if not performed in an organised manner, could have serious repercussions.
- Lack of specific predefined activities, as part of resilience response procedures, that would guide the behaviour of responsible personnel towards physical segments of CI in case of ICT failure as part of cyber attack. Newer generations create particular vulnerability since they rely almost exclusively on mobile and Web communications.

Finally, planning and policy creation in Serbia has to take into consideration and be adapted to the difference in mentality between Balkan and leading EU countries. Though the CI personnel in Serbia are at a high level regarding knowledge and professionalism, different social behaviour habits might create unexpected issues and open vulnerabilities in case of simple copy-paste application of EU policies and measures in Serbia.

22.10 Resilience Enhancement

As presented in the previous sections, the Republic of Serbia has to put a lot of effort in building the resilience of critical infrastructures, in particular as part of the accession process to the European Union in order to achieve the same level as other members. On the other hand, the EU still has a lot of diversity and compatibility issues to overcome in coordinating CI protection within all member countries, with CI resilience lagging even further. In the situation where the EU has placed significant efforts in CI protection in recent years, but so far has developed only the coherent policies regarding transportation and energy, it is questionable which strategy and path the Republic of Serbia should follow!

Perhaps the advisable strategy would be not to follow the existing EU procedure step-by-step, but to take a shortcut through combining the latest experience in CI resilience building in the world with existing EU mid-term plans and come up with a custom plan that could elevate the Republic of Serbia directly to 2020 or similar future targets. Sometimes the fact that one system is considerably behind others in a certain area might not be a disadvantage, but an opportunity if handled properly. Countries that are already half way or more in adopting certain procedures could have difficulties in readapting existing legislations and methodologies. New CI resilience development strategies in the Republic of Serbia would give a chance for a success story not only in the security segment, but also as an economic development component by creating new markets and businesses. Advanced resilience development strategy examples from the world, like the presented all-hazards resilience framework developed and applied by the State of Victoria, Australia, could provide the answer.

Comparison of the present CI security situation in the Republic of Serbia against the desired CI resilience level in the EU and the world, in combination with the assessment of efforts needed, indicates that the fastest and most economic way to bring a large number of CI as close to such a level as possible is to significantly improve the response stage. Some work in that area has been done within the infrastructure protection activities, therefore it is necessary to improve in the next period: (a) definition, selection and prioritization of CI in the Republic of Serbia, (b) resilience procedures and methodologies for the response stage, building on existing vulnerability & risk assessment and protection plans. Basic elements that should be covered by (a) have been analysed in Sect. 22.7. Successful achievement of goals under (a) and (b), and in particular for the former as presented in Figs. 22.3 and 22.4, would both create the foundation for full-scale CI resilience development and reduce efforts and costs related to the recovery stage, thus elevating the critical infrastructure situation in the Republic of Serbia to a whole new level.

As indicated in the chapter, incident – response quantification and development of practical measures for the improvement of resilience is an ongoing process. Therefore, the best approach for the Republic of Serbia is yet to be determined. Still, as a first step, one could designate the introduction of adaptable emergency confine and shutdown (AECS) procedures as the most prominent. In the situation where control systems and operators are one of the weak links of the system's resilience and the very first to face the incident/attack, priority should be given to preservation of physical elements of the CI by isolating and/or disconnecting ICT and cyber segments. In cases of very quickly evolving events (e.g. earthquakes) AECS should be handled by specialised software (e.g. Decision Support System – DSS or Artificial Intelligence – AI) or other predetermined cyber based procedures (e.g. computer controlled dampers), with the capability for smart reaction in case of unexpected incidents (i.e. not covered by risk assessment or protection planning). For slow propagating incidents/attacks or ICT/cyber failures the control over CI system should be given to operators and the unexpected should be covered by predefined sets of instructions.

New legislation for CI resilience in the Republic of Serbia should also introduce the specific methodology for assessment of quality of critical infrastructure Q(t), enabling the quantification of resilience and evaluation of plans and measures for improvement. Q(t) would also supply concrete figures to resilience development strategies and scenarios, providing the authorities with means to objectively decide which path is optimal for Serbia and its CI.

22.11 Conclusions

From the provided arguments, it can be concluded that it is necessary for Serbia to perform a detailed analysis of international experience in the area of critical infrastructure protection and resilience and adopt most prominent solution(s), but also to consider utilizing some segments from the discarded National defence and self-protection system from the era of SFRJ ('Sistem Opstenarodne odbrane i samozastite'), in particular those segments that refer to the protection of important infrastructure systems. Furthermore, it might be useful to consider the possibility of returning the jurisdiction of the Sector for emergency situations and related issues from the Ministry of Interior to the Serbian Army. The main point would be that the Army has the required resources to react properly in high-scale emergency situations, including helicopters and other means of transportation, the Army Corps of Engineers unit with special purpose vehicles and systems (e.g. pontoon bridges, heavy construction equipment, etc.), as well as objects for provisional accommodation and care of a large number of people. Therefore it is crucial to prepare and publish the contemporary Law about the Critical Infrastructure, which would also include the resilience approach and quantification, and perform upgrades in the Law about emergency situations that would involve the Serbian Army to a greater extent and in coordination with the Ministry of Interior.

Regarding the cyber domain, larger organisations usually have the resources to protect themselves technically, yet they still routinely fall prey to the type of low-tech cyber attacks aimed at their employees. By improving employee awareness and introducing corresponding simple technical measures, organisations can reinforce protection and enhance CI resilience against social engineering and the risk of a cyber attack and its potential impact on business, customers and data. Such a process could be performed independently from the overall CI resilience approach on the country level. In recent years, Serbia has shown significant advancement in joining the EU towards improving the security situation in the cyber sector. However, the social behaviour of employees within the critical infrastructures in Serbia still presents a weak link in building resilience and needs wide, coordinated action at all levels, starting from the level of authorities and involved responsible institutions, and down to the level of CI units management.

Further Suggested Readings

Bruneau M, Reinhorn A (2006) Overview of the resilience concept, proceedings of the 8th U.S. National Conference on Earthquake Engineering, Paper No. 2040, April 18-22, 2006, San Francisco, California, USA

Bruneau M, Chang S, Eguchi R, Lee G, O'Rourke T, Reinhorn A, Shinozuka M, Tierney K, Wallace W, von Winterfelt D (2003) A framework to quantitatively assess and enhance the seismic resilience of communities. EERI Spectra J 19(4):733–752

CEAS (2016) Guide through information security in the Republic of Serbia, Centre for Euro-Atlantic Studies – CEAS OSCE Mission to Serbia, pp. 19

Cimellaro GP, Reinhorn AM, Bruneau M (2010) Framework for analytical quantification of disaster resilience. Eng Struct 32(2010):3639–3649

Collin B (1997) The Future of Cyber Terrorism: Where the Physical and Virtual Worlds Converge. Paper presented at the 11th Annual International Symposium on Criminal Justice Issues, Chicago, September 23–26, 1997

Council Directive 2008/114/EC of 8 December 2008 on the identification and designation of European critical infrastructures and the assessment of the need to improve their protection, Official Journal of the European Union L, pp 345–375, 23.12.2008

Guidance for methodology for performing the evaluations of vulnerability and plans for protection and rescue in emergency situations, Ministry of Interior, Belgrade, Serbia, No. 96/2012. http://www.vatroival.com/doc/propisi/OZR/04-Uputstvo%20o%20metodologiji%20za%20izradu%20procene%20ugrozenosti%20i%20planova%20zastite%20i%20spasavanja%20u%20vanrednim%20situacijama.pdf

HSPD 2003 7 – Homeland Security Presidential Directive 7. Office of the White House, Press Secretary. http://www.whitehouse.gov

Linkov I, Bridges T, Creutzig F, Decker J, Fox-Lent C, Kröger W et al (2014) Changing the resilience paradigm. Nat Clim Chang 4(6):407–409

Murray AT, Grubesic TH (2012) Critical infrastructure protection: the vulnerability conundrum. Telematics Inform 29(2012):56–65

NIAC (2009) Critical Infrastructure resilience, final report and recommendations. U.S. Department of Homeland Security, Washington, DC

Ophandt Jonathan A (2010) Cyber warfare and the crime of aggression: the need for individual accountability on tomorrow's battlefield. Duke Law Technol Rev, Page 7

Petit FD, Bassett GW, Buehring WA, Whitfield RG et al (2013) Resilience measurement index: an indicator of critical infrastructure Resilience; Argonne National Laboratory Report

President's Commission on Critical Infrastructure Protection (PCCIP), The Report: Critical Foundations Protecting America's Infrastructures, 1997. https://fas.org/sgp/library/pccip.pdf

Radunović V (2013) DDoS – available weapon of mass disruption. Proceedings of the 21st Telecommunications Forum (TELFOR)

Strategy for development of the informational society in the Republic of Serbia until 2020, Government Gazette of the Republic of Serbia No. 51/2010, Chapter 6.2. http://www.paragraf.rs/propisi/strategija_razvoja_informacionog_drustva_u_republici_srbiji.htm

Strategy for the Development of Information Society in the Republic of Serbia by 2020. Official Gazette of the RS, No. 51/2010

The State of Victoria (Emergency Management Victoria – EMV) (2015) Critical infrastructure resilience strategy, Victorian Government document. ISBN 978-0-9943637-2-5

Todorovic B (2012) Drought: Creeping natural hazard – Blue-Green dream of water, PLANETA Magazine for science, research and discoveries 53, Sept.-Oct. 2012 (pp. 60-61)

Web link 1: https://en.wikipedia.org/wiki/List_of_water_supply_and_sanitation_by_country

Web link 2: http://eur-lex.europa.eu/legal-content/EN/TXT/?uri=uriserv:OJ.L_.2016.194.01.0001.01.ENG&toc=OJ:L:2016:194:TOC

Web link 3: https://www.enisa.europa.eu/

Web link 4: http://ec.europa.eu/priorities/digital-single-market_en

Web link 5: http://cloudsecurity.trendmicro.com/us/technology-innovation/cyber-security/
Web link 6: https://heimdalsecurity.com/blog/wp-content/uploads/total-cost-of-cyber-crime-in-2015-in-7-countries.png)
Web link 7: https://www.helpnetsecurity.com/tag/gfi_software/
Web link 8: https://www.helpnetsecurity.com/2015/02/26/the-business-and-social-impacts-of-cyber-security-issues/
Web link 9: http://arhiva.mup.gov.rs/cms_lat/sadrzaj.nsf/uprava-za-IT.h
Web link 10: http://mtt.gov.rs/download/1(2)/Zakon%20o%20informacionoj%20bezbednosti.pdf

Chapter 23
Risk and Resiliency Assessment of Urban Groundwater Supply Sources (Ponds) by Structured Elicitation of Experts Knowledge

Z. Srdjevic, B. Srdjevic, and M. Rajic

Abstract Management/operation failures of urban water supply infrastructure, especially pressurized sub system for distribution of drinking water from groundwater sources (ponds), may have severe social and economic consequences if risk and resilience scenarios are not properly considered during planning, design and/or operation phases. Physical and sanitary protection of ponds is permanent requirement for ensuring proper functioning of ponds and connected distribution network in the city. Because emergency situations may arise in case of natural or other disturbances or disasters, it is of particular importance to identify in advance key risk factors that can cause failures, and also to take into account what should be recovery time once the system is out of order for certain period(s) of time. Being aware that use of expert knowledge in assessing possible risk and resilience scenarios is essential and highly recommended in case of urban water supply of the City of Novi Sad in Serbia, this paper demonstrates how the method for structured elicitation, developed by Smith et al. (Heliyon 1(2015):e00043, 2015), can be used to evaluate important risk and resilience factors within the group decision making process. Namely, there is a plan for enlarging the capacity of one of the three existing ponds within the city area, and we are proposing an application of structured elicitation procedure to properly consider possible risks in operating this important critical part of urban infrastructure. We show that method is sufficiently intuitive in capturing experts' uncertainty, efficient in generating agreement and convenient for communicating the results to the decision-makers. Simulation of decision-making process with three participating experts is aimed to convince city managers and other responsible authorities that recommended use of the method is not only scientifically justified, but also easy and efficient to implement in practice.

Z. Srdjevic (✉) • B. Srdjevic • M. Rajic
Faculty of Agriculture, Department of Water Management, University of Novi Sad, Trg D. Obradovica 8, 21000 Novi Sad, Serbia
e-mail: srdjevicz@polj.uns.ac.rs; bojans@polj.uns.ac.rs; milica@polj.uns.ac.rs

© Springer Science+Business Media B.V. 2017

I. Linkov, J.M. Palma-Oliveira (eds.), *Resilience and Risk*, NATO Science for Peace and Security Series C: Environmental Security,
DOI 10.1007/978-94-024-1123-2_23

23.1 Introduction

US National Infrastructure Advisory Council (National Infrastructure Advisory Council 2010) defines resilience as the 'Ability of a system, community or society exposed to hazards to resist, absorb, accommodate and recover from the effects of a hazard in a timely and efficient manner'. However, resilience related terminology and practice used to manage it differ within different sectors. When it comes to water management, Butler et al. (2014) define resilience as 'the degree to which the system minimizes level of service failure magnitude and duration over its design life when subject to exceptional conditions'.

In order to be able to minimize the failure, one must identify what kind of failures are possible, what can cause failures (risk factors), and what is the likelihood of failure to happened under those risk factors. Six most important water system failures identified by National Drinking Water Advisory Council (NDWAC – National Drinking Water Advisory Council 2005) are:

- Loss of pressurized water for a significant part of the system.
- Long-term loss of water supply, treatment, or distribution.
- Catastrophic release or theft of on-site chemicals affecting public health.
- Adverse impacts to public health or confidence resulting from a contamination threat or incident.
- Long-term loss of wastewater treatment or collection capacity.
- Use of the collection system as a means of attack on other key resources or targets.

It is obvious that consequences of water system failures can be severe for the society and economics, so identification of risk factors and assessment of likelihood of failures of critical water infrastructure under those factors are particularly important.

On the other hand, Water Framework Directive (European Commission 2000) requires public participation (PP) in decisions related to water management, particularly in (1) information supply (requirement), (2) consultation (requirement), and (3) active involvement (encouraged). The WFD does not propose how to introduce PP into practice, and it is evident lack of official guidance documentation (Challies et al. 2016; Ijjas and Botond 2004) and coordination between different frameworks that can result in unrelated participatory procedures (Albrecht 2016). Also, Benson et al. (2014) studied practice of PP between 2006 and 2011 in England and Wales, and show that participation is highly variable in the planning process. Slavíková and Jílková (2011) evaluated the implementation of public participation principle in Czech Republic and found that performance of PP was rather poor and lacking in continuity.

In common terminology, a 'public' means professionals (experts, scientist, officials, practitioners), general public (citizens) or organizations (NGOs, civil society organizations). In problems related to policy making and risk assessment, especially under uncertainty, most common approach is to use the experts' knowledge and not wider public. Guidance on expert knowledge elicitation methods and procedures are given, for example, in European Food Safety Authority (EFSA 2014), United Nations Intergovernmental Panel on Climate Change (Mastrandrea et al. 2010), US Environmental Protection Agency (US Environmental Protection Agency 2011).

Elicitation can be used ‘to estimate unknown quantities, to characterize risk pathways, and to quantify uncertainty’ (Butler et al. 2015). If it is of interest to quantify uncertainty, formal expert elicitation, also known as structured expert judgment (SEJ), should be used. Recent application of SEJ in water management include estimating how long an old earth dam can withstand a leak (Aspinall 2010), assessing nutrient flows in septic tanks (Montangero and Belevi 2007), watershed condition assessment (Gordon and Gallo 2011), assessing the effectiveness of storm water management structures (Koch et al. 2015), developing surface water quality standards in China (Su et al. 2017), etc.

Smith et al. (2015) developed a new SEJ approach that “combines expert calibration and fuzzy based mathematics to capture and aggregate subjective expert estimates of the likelihood that a set of direct risk factors will cause management failure”. Authors improved (Metcalf and Wallace 2013) elicitation method by incorporating ellipse based interval agreement approach and mathematical analysis of expert’s responses. We demonstrate here applicability of Smith et al. (2015) approach in critical water infrastructure management under uncertainty by using as a case study a major groundwater pond which supplies city of Novi Sad in Serbia. The problem is to identify the most important risk factors in managing (operating) wells within the pond, respecting in particular reliability of their proper operation and resilience – recovery time – in cases of malfunctioning and/or failure. Being aware that the urban water distribution system in Novi Sad is a part of more complex critical infrastructure, presented methodological framework respects multiple dimensions of the system’s security and includes considerations of its functioning from various perspectives such as political, temporal, threat, and economic as suggested in Ezell et al. (2000).

23.2 Some Notions on Risks and Resilience of Urban Water Distribution Systems

Disruption of services in urban water distribution system almost always results in disruptions in one or more other infrastructures, usually triggering serious cross sectoral cascading. Failures or malfunctioning of water system in some locations in the city, even for short periods of time, requires ad hoc actions to enable fast recover and return into stable system state. The risks of failure and required time for recovery are greater for aged infrastructures, or if they are already stressed by any reason. For instance, groundwater pond can be stressed if longer over-pumping is applied to fulfill demand levels that exceed capacities of wells as they were designed for.

The essential aspect of a resilient system is that it has an adequate capability to avert adverse consequences under disturbances, and a capacity of self-organization and adaptation. Being resilient means that system can display a greater capacity to provide wanted services. Together with reliability and vulnerability, resilience is regarded as an important property of any water distribution network as critical urban infrastructure. Increase of resilience of such a network assumes increase of resilience of primary sources of water, such as wells and local facilities within protected city area where groundwater pond is located.

Although reliability (or its opposite – risk) and resilience are not necessarily correlated, in most cases improving one property means improvement of the other. Therefore, at various stages of planning, design and management process, these two properties are analyzed at the same time, along with other constructive and non-constructive properties such as disposition of objects, investments, scheduling of implementation etc. Once the water distribution system is in place and is continuously functioning, an important permanent task to be achieved is to establish both qualitative and quantitative risk and resilience assessment framework to fully consider: (a) risk and resilience against crossing a performance threshold, and (b) resilience for response and recovery after a disturbance. Both types of risk and resilience can be modified by self-organization, redundancy, human actions and governance, etc. Therefore, they are dependent to some extent upon the system's adaptive capacity supported by either endogenous or exogenous forces, or both (Wang and Blackmore 2010).

23.3 Method for Structured Elicitation of Experts' Knowledge Under Uncertainty

Formal, or structured, expert elicitation usually has seven stages, as defined by Knol et al. (2010) (Fig. 23.1).

Based on it, Smith et al. (2015) developed an approach which will be briefly described in this section and demonstrated on the case study example presented in the in next section.

By assumption, initial steps of the procedure from Fig. 23.1 can be considered as already completed: problem is defined, experts are selected, and elicitation protocol is already designed. The remaining methodological issues of the approach are related to preparation of the elicitation session, elicitation of expert judgments, and possible aggregation and reporting. We named the remaining steps as: Calibrating and weighting the experts; Elicitation of judgments; Aggregation of experts' likelihoods; and Identification of key risk factors. The last step is on purpose adjusted to the problem of identifying the key risk factors that could cause management/operation failure of critical water infrastructure.

All calculations required in those four steps are performed within the excel file provided as Supplement 2 of the article (Smith et al. 2015). Details on mathematical

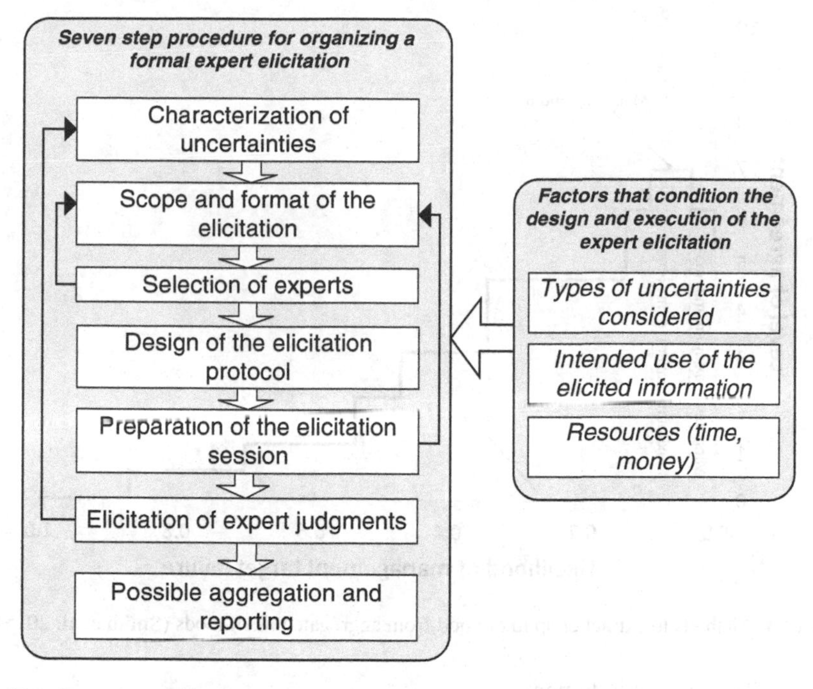

Fig. 23.1 Procedure of formal expert elicitation (Knol et al. 2010)

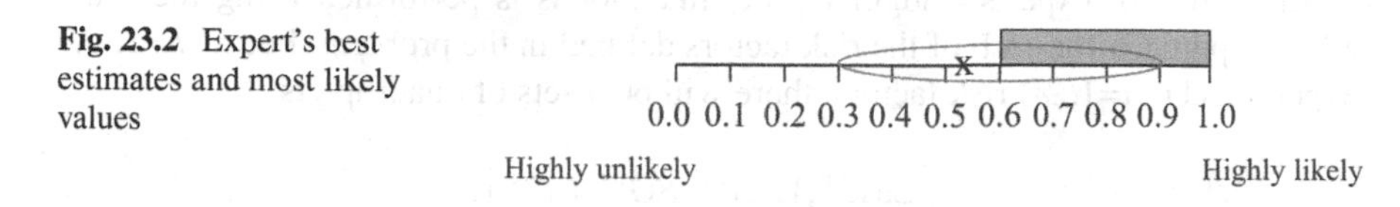

Fig. 23.2 Expert's best estimates and most likely values

background of the approach can be found in Metcalf and Wallace (2013) and Wagner et al. (2014).

Step 1. Calibrating and weighting the experts

Calibration of experts is recommended in order to manage their over-confidence. This procedure is based on series of seeding or calibrating questions relevant to the problem in hand that experts must answer, but on which the expert most probably do not know exact answer.

Each expert draws ellipse with endings representing lowest and highest likelihood (or estimates) of event, marks (x on Fig. 23.2) what he/she believes is true answer and expresses his/hers level of confidence that the correct answer is within the ellipse. An illustration is given on Fig. 23.2 with gray rectangle representing most likely values.

Proportion of the ellipse that overlaps with most likely values and expressed confidence level for each question are used to adjust initial values to desired 80% confidence interval. Adjusted proportions were averaged and used to calculate the weight of the expert.

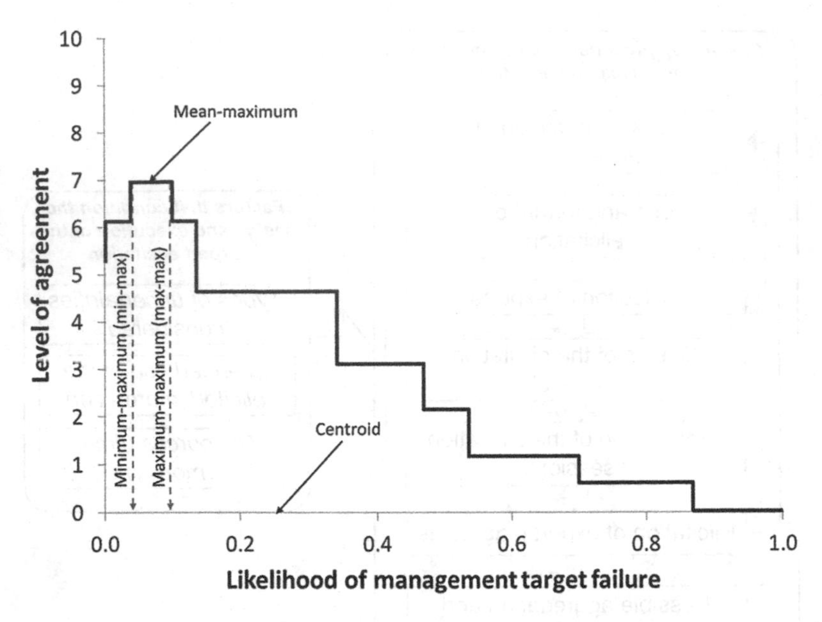

Fig. 23.3 Methods to extract crisp likelihood from aggregated likelihoods (Smith et al. 2015)

Step 2. Elicitation of judgments

Elicitation of experts' judgments i.e. likelihoods is performed using the same ellipse approach for each of the risk factors defined in the problem. For K (k=1,..K) experts and n (i=1,..n) risk factors, there will be n sets of quadruplets

$$\left[\text{Lest}^{i,k}, \text{Hest}^{i,k}, BG^{i,k}, CI^{i,k}\right]$$

at the end of elicitation process. Note that for ith factor and kth expert, $\text{Lest}^{i,k}$ represent lowest estimate, $\text{Hest}^{i,k}$ highest estimate, $\text{BG}^{i,k}$ best guess and $\text{CI}^{i,k}$ confidence level.

Initial likelihood interval [$\text{Lest}^{i,k}$, $\text{Hest}^{i,k}$] is adjusted to desired 80% confidence interval to [$\text{Lest\_calibr}^{i,k}$, $\text{Hest\_calibr}^{i,k}$].

Step 3. Aggregation of experts' likelihoods

Aggregation of experts likelihoods for ith risk factor is performed by multiplying weight of the kth expert with CI adjusted scores [$\text{Lest\_calibr}^{i,k}$, $\text{Hest\_calibr}^{i,k}$]. Result of this step is calculated level of agreement of experts and aggregated likelihoods over the interval [0,1].

As the estimated crisp likelihood that ith factor will cause the management/operation failure, min-max score approach is adopted (Fig. 23.3).

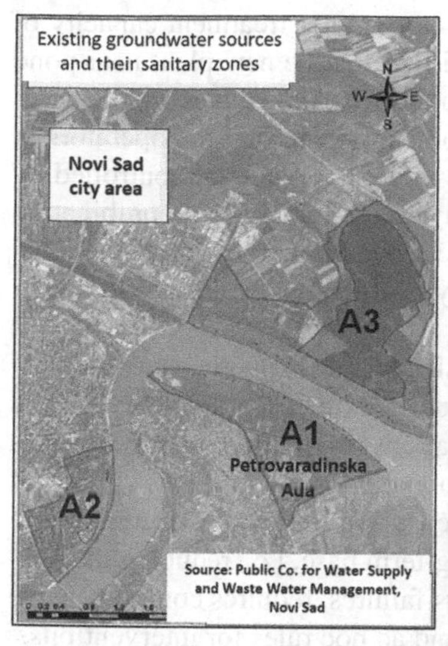

Fig. 23.4 Existing groundwater sources in Novi Sad with sanitary protection zones and Renney well at Perovaradinska Ada groundwater pond (A1)

<u>*Step 4.*</u> Identification of key risk factors

Key risk factors can be identified after setting the thresholds for desired level of agreement of experts and critical crisp likelihood that risk factor will cause management/operation failure.

23.4 Case Study – Risk and Resilience Assessment of Fresh Water Supplier to the City of Novi Sad

A majority of Serbian cities above 30,000 population is facing the problem of assuring required quantity and quality of water for urban supplies. In the City of Novi Sad (second largest city in Serbia) there are three major and two secondary groundwater ponds for supplying customers with fresh water. Major ponds are Petrovaradinska Ada (A1), Strand (A2), and Ratno Ostrvo (A3), all located near the shoreline of the Danube River (Fig. 23.4). Because Danube passes almost through the center of the city, ponds are in fact within the core city area and by location and function represent critical urban infrastructure. All ponds are in full 24-hour operation; their exploitation is supported only for short time periods on an intervening base by two other smaller ponds, also within the city area. Total length of city water distribution network is around 1100 km.

There is a central water treatment plant close to A2 pond Strand, and a new one is under construction close to A1 pond (Pertrovarainska Ada). At the moment, central plant receives water from all three ponds and provides complete treatment process involving aeration, filtering (sand and active coal), ozone treatment, chlorinating and addition of fluoride for dental health. Continuous water treatment capacity of central plant is 1350 l/s with maximum of 1500 l/s, while the new plant near pond A1 will have capacity of 200 l/s. The treatment plant Strand has gauges which monitor inflows and outflows, and all necessary equipment for alerting the operators in case of pumps' failures. Production levels of the treatment plant are controlled by computer-based software, developed during many years, mainly as a combination of water quality standards, network structure, local operational and organizational conditions, and operators' real-life experience relying on judge the required production by observing the multiple- tanks level.

Earlier studies clearly indicated Petrovaradinska Ada (A1) pond as most important water supplier to Novi Sad e.g., (Srdjevic and Srdjevic 2011). Recently, Djuric et al. (2012) conducted hydro geological investigations and hydrodynamic analyses and suggested expansion of this groundwater source by ten tube wells. Being of such importance for water supply of Novi Sad, ensuring the proper operation and management of this groundwater source on a long-term base, i.e., reducing the possibility of its malfunctioning or even more serious failures, requires comprehensive risk and resilience assessment to set prevention and ad hoc rules for interventions.

To reduce the likelihood of operation and management failure in the future (after expansion of the source and for the period of 20 years), three categories of most common water system threats, as defined by van Leuven (2011), are used in the analysis: natural disasters, human caused and workforce/infrastructure. List of risk factors from van Leuven (2011) is modified and expanded to fit the problem in hand (Table 23.1).

Step 1. Calibrating and weighting the experts: Questions used to calibrate experts in this illustrative example were as follows:

- What is the likelihood that half of the population of Novi Sad city would be affected in case of Petrovaradinska Ada operation/management failure?
- What is the likelihood of operation/management failure during the summer season?
- How likely is that the sediments in the area of groundwater ponds in Novi Sad will contain petroleum products above permissible level?

Three experts answered to the calibrating questions by estimating lowest and highest likelihood, best guess and confidence level. According to the answers and required level of confidence of 80%, calculated weights of experts were 0,67, 0,89 and 0,59.

Step 2. Elicitation of judgments: Using the same answering methodology (likelihood ellipses), for each of the 13 risk factors given in Table 23.2 experts provided estimates of lowest and highest likelihood that factor could cause management/operation failure after expansion of the source and for the period of 20 years.

At the end of Step 2, 13 tables (for each risk factor) similar to Table 23.2 were formed.

Step 3. Aggregation of experts' likelihoods: After all 13 risk factors were assessed by experts, aggregation was performed using the calculated calibrated likelihoods

Table 23.1 Set of factors that could cause management/operation failure

Category	Risk factor	Type of failure/implications
Natural disasters	Flood	Above ground elements
	Earthquake	Above/below ground elements
	Drought	Reliability of supply
	Winter storm	Above ground elements
	Seepage stability	Below ground elements
Human caused	Vandal	Above ground elements
	Contamination of Danube River	Above ground elements
	Contamination from petroleum oils	Reliability of supply
	Contamination from septic tanks	Reliability of supply
Workforce/infrastructure	Failure	Above/below ground elements
	Hazardous material release	Reliability of supply
	Overall technical maturity	Above/below ground elements
	Workforce competence and operability	Above/below ground elements

Table 23.2 Experts likelihoods of failure for the risk factor Contamination from septic tanks

Expert	Lowest likelihood	Highest likelihood	Best guess	Confidence	Calibrated lowest likelihood	Calibrated highest likelihood
1	0,700	0,900	0,90	75	0,687	0,900
2	0,800	1,000	0,90	70	0,786	1,000
3	0,700	0,900	0,80	90	0,711	0,889

and weights of experts. Figure 23.5 shows aggregated values for the risk factor Contamination from septic tanks.

Aggregation shows that the level of agreement of experts is 2,15, and that they believe that the likelihood that septic tanks will cause management/operation failure is 80% (recall application of min-max approach to define crisp value). Note that, in this example, theoretical maximum of level of agreement of 3 could be reached only in case that all experts have weight 1. Level of agreement of experts and aggregated values for all 13 risk factors are presented in Table 23.3.

Step 4. Identifying the key risk factors: In order to identify key risk factors that could cause management/operation failure of groundwater pond Petrovaradinska Ada, thresholds for required level of agreement and critical likelihood should be agreed with experts. Here, those values are set to 1,5 (in voting theory, this would be considered as majority) and 70% (usually considered as likely to highly likely event).

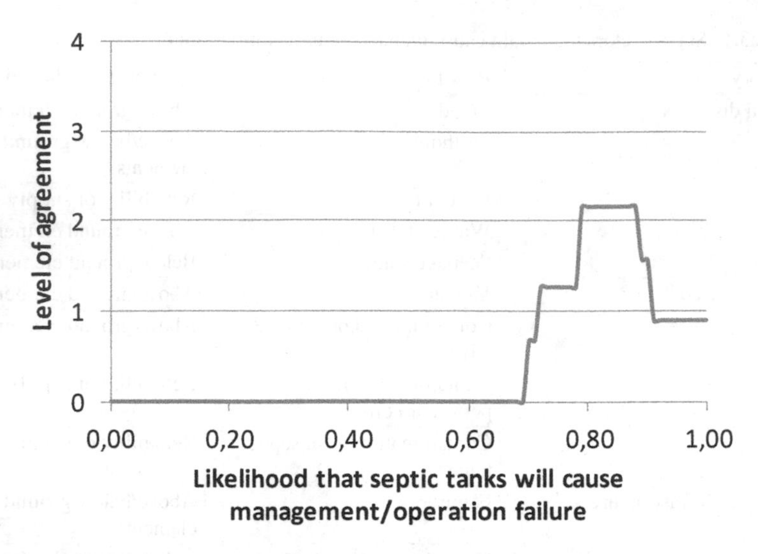

Fig. 23.5 Agreement level of experts and likelihoods of failure due to the Contamination from septic tanks

Table 23.3 Aggregation of experts' likelihoods and agreement level for 13 risk factors

Risk factor	Level of agreement	Min-max likelihood (%)
Flood	1,26	41
Earthquake	1,48	33
Drought	1,56	71
Winter storm	2,15	55
Seepage stability	1,56	78
Vandal	2,15	26
Contamination of Danube River	2,15	40
Contamination from petroleum oils	2,15	71
Contamination from septic tanks	2,15	80
Failure	2,15	80
Hazardous material release	1,48	41
Overall technical maturity	2,15	81
Workforce competence and operability	2,15	71

Data presented in Table 23.3 show that there are seven risk factors (shaded values) that fulfill such requirements: (1) Drought; (2) Seepage stability; (3) Contamination from petroleum oils; (4) Contamination from septic tanks; (5) Failure; (6) Overall technical maturity; and (7) Workforce competence and operability.

23.5 Conclusion

The major source of drinking water for City of Novi Sad customers comes from the aquifer, a large underground water system connected with the Danube river and several smaller aquifers used also for supplying few neighboring municipalities. There are three major ponds with numerous wells at both sides of the Danube and all are considered as critical urban infrastructure. The main source is Petrovaradinska Ada and this paper mainly focuses on risk and resilience factors that are important to respect in ensuring proper functioning of the whole urban water supply system in City of Novi Sad.

Similar to some other urban infrastructure (e.g., electric distribution network, ground and underground transport system), pressurized water distribution networks provide fundamental commodity to humans and are considered as critical infrastructure (van Leuven 2011). Due to effects that have on public health and safety and economics, the assets that ensure functioning of the water systems are vital and should have special attention of managers, operators, as well as local and national policy makers. In order to help making decisions and ensure functioning of the systems for a long period, it is essential to know what are the key threats that can compromise functioning of the system and make system vulnerable, risky and/or no-recoverable.

We presented here a case study example from Serbia and showed how the method of structured judgment elicitation of experts may be used to identify the key risk and resilience factors that could cause management/operation failure of major city groundwater source of fresh water for the City of Novi Sad, capital of Vojvodina Province. The SEJ is a new approach (Smith et al. 2015) and its application within this case study framework provided promising initial results, primarily regarding how intuition 'works' in capturing experts' uncertainty, efficiency in generating agreement, and, in the way, how we can efficiently enable communicating of the results to the end decision-makers.

Our experience showed that, before applying the SEJ, effort should be concentrated on broader planning approach, clear problem definition, selection of experts and their training. These actions are necessary to undertake prior to SEJ implementation in order to ensure meaningful and reliable results in real life application.

Further Suggested Readings

Albrecht J (2016) Legal framework and criteria for effectively coordinating public participation under the Floods Directive and Water Framework Directive: European requirements and German transposition. Environ Sci Pol 55(2):368–375

Aspinall W (2010) A route to more tractable expert advice. Nature 463:294–295

Benson D, Fritsch O, Cook H, Schmid M (2014) Evaluating participation in WFD river basin management in England and Wales: processes, communities, outputs and outcomes. Land Use Policy 38:213–222

Butler D, Farmani R, Fu G, Ward S, Diao K, Astaraie-Imani M (2014) A new approach to urban water management: safe and SuRe. 16th conference on water distribution system analysis, WDSA 2014. Elsevier, Bari, Italy

Butler A, Thomas MK, Pintar KDM (2015) Systematic review of expert elicitation methods as a tool for source attribution of enteric illness. Foodborne Pathog Dis 12(5):367–382

Challies E, Newig J, Thaler T, Kochskämper E, Levin-Keitel M (2016) Participatory and collaborative governance for sustainable flood risk management: an emerging research agenda. Environ Sci Pol 55(2):275–280

Djuric D, Lukic V, Soro A (2012) Hydrodynamic analyses of increasing Petrovaradinska Ada groundwater source capacity in Novi Sad. Vodoprivreda 44:265–272. (in Serbian)

EFSA (2014) Guidance on expert knowledge elicitation in food and feed safety risk assessment. EFSA J 12(6):3734. doi: 10.2903/j.efsa.2014.3734

EPA (US Environmental Protection Agency) (2011) Expert elicitation task force white paper. Available at www.epa.gov/stpc/pdfs/eewhite-paper-final.pdf

European Commission (2000) Directive 2000/60/EC of the European Parliament and of the Council of 23 October 2000 establishing a framework for Community action in the field of water policy [Water Framework Directive]. Off J Euro Communities L 327:1–72

Ezell BC, Farr JV, Wiese I (2000) Infrastructure risk analysis of municipal water distribution system. ASCE J Inf Syst 6:118–122

Gordon SN, Gallo K (2011) Structuring expert input for a knowledge based approach to watershed condition assessment for the Northwest Forest Plan, USA. Environ Monit Assess 172:643–661

Ijjas I, Botond K (2004) Participation and Social Learning in the Implementation of the WFD in Agricultural Water Management: Stakeholder Workshops Report, AWP5 Report of the Harmonicop Project. http://www.harmonicop.info/_files/_down/Hungarian%20National%20SH%20Workshop%20Report.pdf

Knol AB, Slottje P, van der Sluijs JP, Lebret E (2010) The use of expert elicitation in environmental health impact assessment: a seven-step procedure. Environ Health 26:9–19. doi:10.1186/1476-069X-9-19

Koch BJ, Febria CM, Cooke RM, Hosen JD, Baker ME et al (2015) Suburban watershed nitrogen retention: estimating the effectiveness of storm water management structures. Elementa Science of Anthropocene 3: 000063. doi: 10.12952/journal.elementa.000063

Mastrandrea MD, Field CB, Stocker TF, Edenhofer O, Ebi KL et al (2010) Guidance note for lead authors of the IPCC fifth assessment report on consistent treatment of uncertainties. Intergovernmental Panel on Climate Change (IPCC). Available at http://www.ipcc.ch

Metcalf SJ, Wallace K (2013) Ranking biodiversity risk factors in expert groups – treating linguistic uncertainty and documenting epistemic uncertainty. Biol Conserv 162:1–8

Montangero A, Belevi H (2007) Assessing nutrient flows in septic tanks by eliciting expert judgement: a promising method in the context of developing countries. Water Res 41:1052–1064

National Infrastructure Advisory Council (2010) A framework for establishing critical infrastructure resilience goals. Final Report and Recommendations by the Council. http://www.dhs.gov/xlibrary/assets/niac/niac-a-framework-for-establishing-critical-infrastructure-resiliencegoals-2010-10-19.pdf

NDWAC – National Drinking Water Advisory Council (2005) Water Security Group Findings, May 18, 2005, p vii

Slavíková L, Jílková J (2011) Implementing the public participation principle into water management in the Czech Republic: a critical analysis. Reg Stud 45(4):545–557

Smith M, Wallace K, Lewis L, Wagner C (2015) A structured elicitation method to identify key direct risk factors for the management of natural resources. Heliyon 1(2015):e00043

Srdjevic Z, Srdjevic B (2011) Evaluating groundwater ponds as major suppliers to water distribution system of Novi Sad city (Serbia). World Water Week 2011, Abstract vol: 200–201, Stockholm, Sweden

Su J, Ji D, Lin M, Chen Y, Sun Y, Huo S, Zhu J, Xi B (2017) Developing surface water quality standards in China. Resour Conserv Recycl 117(Part B):294–303 van Leuven LJ (2011) Water/wastewater infrastructure security: threats and vulnerabilities. In: Clark RM et al (eds.) Handbook of water and wastewater systems protection, protecting critical infrastructure, doi: 10.1007/978-1-4614-0189-6_2

Wagner C, Miller S, Garibaldi JM, Anderson DT, Havens TC (2014) From interval-valued data to general type-2 fuzzy sets. IEEE Trans Fuzzy Syst 23(2):248–269

Wang CH, Blackmore J (2010) Risk in urban water systems: A demonstration using measures and assessment of rainwater tank use in households. eWater Technical Report, Canberra

Chapter 24
Simulating Reservoir System Operation Under Given Scenarios to Determine Operating Policy with the 'Good' Resilience

B. Srdjevic, Z. Srdjevic, and B. Todorovic

Abstract This chapter provides findings of authors in real-life engineering-style performed assessment of the resilience of complex multipurpose water systems with surface reservoirs. The chapter identifies main steps in modeling the problem in view of water users and operators needs arising in both planning and management phases of the system development and operation. A case study example from Serbia is provided to illustrate authors' approach in creating required input for running computerized river basin simulation models to determine satisfactory operation of reservoirs measured by achieving a 'good resilience' at given demand point within the system.

24.1 Introduction

The systems analysis methods and tools such as mathematical modeling, simulation, and optimization have been widely applied to solving problems in managing water resources for over five decades and obviously, they remain just as relevant today as hitherto. The problems related to operation of large scale water systems seem to have changed radically because of undesired climate changes and in the same time growing water demands subjected to conflicts of water users from different societal sectors. Context in which systems analysis might be applied assumes understanding the challenges related to anticipation of future requirements in front of water system, such as: emergence of stakeholders' participation, respect of environmental ethics, conducting life-cycle analysis, quantifying sustainability, taking

B. Srdjevic (✉) • Z. Srdjevic
Faculty of Agriculture, Department of Water Management, University of Novi Sad, Trg D. Obradovica 8, 21000 Novi Sad, Serbia
e-mail: bojans@polj.uns.ac.rs; srdjevicz@polj.uns.ac.rs

B. Todorovic
Faculty of Mechanical Engineering, University of Belgrade, Kraljice Marije 16, 11120 Belgrade, Serbia
e-mail: bt.emit.group@gmail.com

© Springer Science+Business Media B.V. 2017

I. Linkov, J.M. Palma-Oliveira (eds.), *Resilience and Risk*, NATO Science for Peace and Security Series C: Environmental Security,
DOI 10.1007/978-94-024-1123-2_24

care of industrial impacts on ecology, design for engineering resilience, evaluation and mitigation of risks, estimating vulnerability of technical parts of a system, etc. Although the application of systems analysis experiences permanent innovation, we have to acknowledge that we are still not able to encode all our currently available hypothetical knowledge into a model. Even when there is an obvious progress, this is not verifiable in the conventional, rigorous sense. For instance, in spite of apparently powerful mathematical formulations of the optimization problem, heuristics, metaheuristics and intuition are called upon to reach sufficiently good solutions, by expectation reasonably close to where the optimum is thought to lie.

Any mathematical modeling approach is restricted when describing the real problem. Uncertainty in input data, limitations in the mathematical description of the complex real-world physical phenomena, together with other factors affecting the overall decision-making process (like purely qualitative factors) makes their application, though essential, only part of the process. In modeling system operation strategies to enable simulation of water resources systems and evaluation of consequences of applied strategies, measuring system's resilience is one of very challenging tasks in both planning and implementing phase. Which modeling approach to choose depends greatly on the particular results expected from the analysis. Specific issues to consider are:

- Objectives of the analysis and rate of aggregation;
- Data (output) required to evaluate the strategies and resulted resilience;
- Time, data, money and computational facilities available for the analysis; and
- Modeler's knowledge and skill.

In this paper, we put a focus and discuss several important points related to planning the operation of water resources system with multipurpose surface reservoirs as main regulators of water regime in the river basin (catchment). A case study example from Serbia is used to demonstrate how systems analysis, supported by powerful river basin simulation computer models, can efficiently enable recognition of desired and un-desired system operation and in turn provide information on how much system, sub system or any other demand point in the system is resilient, i.e. capable to recover from undesired status.

24.2 Climate Change and Hydrologic Inputs to Reservoirs

The global temperature rise since the mid-past century has led to the global warming and today it is an important issue that many researchers have recognized as the climatic change which needs special attention in their case study assessments and evaluations of effective actions. Both regional and local scales of the effects of climate variation, together with hydrological uncertainty, are considered as a framework for analyzing human living and determining implications for water resource system management. For example, it is well known that the hydrological uncertainty of river catchment is beyond the certain level of expectation in both quantity and time scale which consequently makes a water resource management tough task to the proper operation. If the multi-purpose water resource system with surface reservoirs is to be properly managed over long periods of uncertain hydrological conditions, the final result could be the more or less mismanagement output with serious economic, political, environmental, and especially social consequences.

The severe drought events and flood damage occur in many local areas, while the increasing tendency of water requirement is likely to be a response to the economic growth, in many cases closely connected with the rising population. To enable a successful and sustainable water management, systems analysis must respect both availability of water and requirements of water users on a long-term base and suggest operators how to take into account limitations and set efficient operating policies according to each or most important local demands.

As far as reservoir system operation is concerned, it is mostly performed under uncertainty of hydrologic conditions and various encompassing factors. Systems approach must enable measurement of the performance of reservoirs by using the information of uncertainty expressed in terms of probability of failure (e.g., being in undesired status over time, so-called risk operation) or of probability of success which is commonly called the reliability (Elshorbagy 2006; Srdjevic and Srdjevic 2016a, b). Reliability is also considered as the complement of probability of failure, or risk.

Performance failure of the reservoir relates to its inability to perform in desired way within the period of interest. Reliability and risk are typical performance indices in evaluation of long-term reservoir behavior, likewise resiliency and vulnerability as two also very important concepts introduced in early 1980s. In (1995), Srdjevic and Obradovic applied the reliability-risk concept in evaluating the control strategies of multi-reservoir water resources system. As reported in Rittima and Vudhivanich (2006), Tsheko (2003) calculated reliability and vulnerability of rainfall data to define the severity and frequency periods of droughts and floods in Botswana. There are also reports from many other countries where assessments of water resources management strategies are conducted using reliability, vulnerability, and resiliency indices accompanied by the simulation models, sometimes all integrated in decision support systems (DSS). In some cases, for instance (Srdjevic and Srdjevic 2016a, b), strategies are evaluated within multi-criteria analysis frameworks supported by ideal-point, utility or outranking methods from the set of decision-making multi-criteria optimization methods.

Apart from evaluating the reservoir performance via many performance indicators such as reliability (risk), resilience, vulnerability, dispersion of reservoir levels from a rule curve, safe water (firm yield) or shortage index (McMahon et al. 2006; Rittima and Vudhivanich 2006; Srdjevic et al. 2004; Srdjevic and Srdjevic 2016a), it is correctly elaborated in many articles that main issue in systems approach is how to perform modeling by engaging both the art and the science and 'apply a limited and imperfect understanding of the "real" world' (Schaake 2002). In (Elshorbagy 2006) it is correctly said that 'such an understanding requires knowledge of the physics of hydrologic processes at different spatial and temporal scales, and information on soils, vegetation, topography, and water and energy forcing variables.'

This paper presents specific modeling approach in assessing the reservoir performance from the resilience point of view. It is rather practical than theoretical approach; for theory, reader may consult multiple sources, e.g., (Hashimoto 1980; Hashimoto et al. 1980; Loucks 1997; Loucks and van Beek 2005; Moy et al. 1986; Sandoval-Solis et al. 2011; Schaake 2002; Srdjevic and Srdjevic 2016a; Srdjević and Obradović 1991, 1995).

24.3 Case Study Example

24.3.1 Background Information

The authors of this paper participated in many studies in Serbia related to river basin planning and management, starting from mid-70-ties of the twentieth century until recently. For instance, the seven-reservoir system, located in central Serbia, is simulated with generic models SIMYLD-II, SIM IV, HEC 3 and HEC 5, delivered under UNDP project from two US sources: Texas Water Development Board (the first two models) and Hydrologic Engineering Center, USCE (the last two models), respectively. Many scientific and professional studies have been completed at that time for the Morava river basin, but also for Mirna river basin in Croatia and elsewhere in former Yugoslavia. Reported applications of aforementioned and many other computerized models, all written in Fortran programming language and installed at the IBM and UNIVAC mainframes, are without exception aimed at determining the best strategy for long-term development of water resources sector in parts of Yugoslavia.

24.3.2 System and Creation of Hydrologic Input to Computer Models

To illustrate an approach in computing reliability and resilience of water resources system, the system in Morava river basin in Serbia with two reservoirs and two diversion structures (Fig. 24.1) is simulated over period of 20 years with typical rule curves at reservoirs (Fig. 24.2). Hydrologic input to the system is represented as incremental monthly net-inflow into reservoirs. It is generated as partially dependent of downstream monthly demands at three diversion structures estimated for the planning horizon. In other words, by assumption, inflow into a single reservoir depends on its natural inflow reduced for immediate downstream demand and proportionalized joint (with the other reservoir) more downstream demand. A proportion ratio is based on relative sizes of local and total catchment areas of two reservoirs. In addition, an assumption is adopted that from larger catchment area more inflow will occur and releases from the reservoir could also be larger. Modeling approach to generate inflows into reservoirs is given by relations shown in Fig. 24.1.

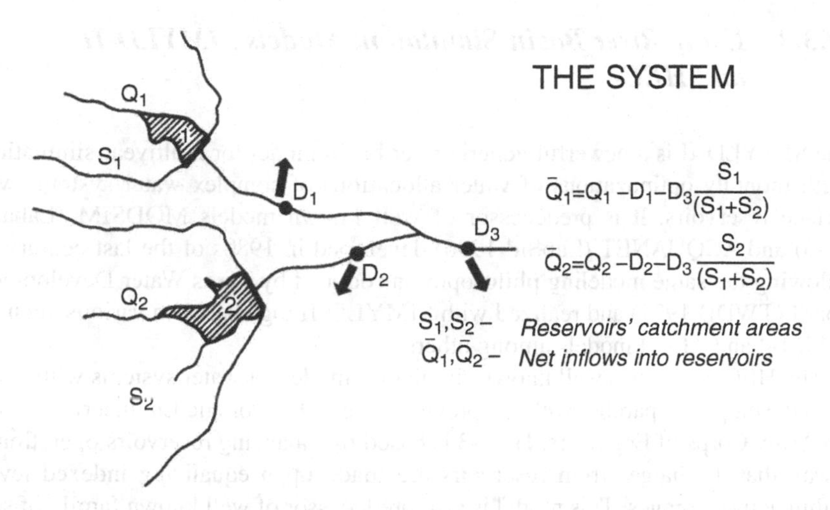

Fig. 24.1 Two-reservoir multipurpose water resources system

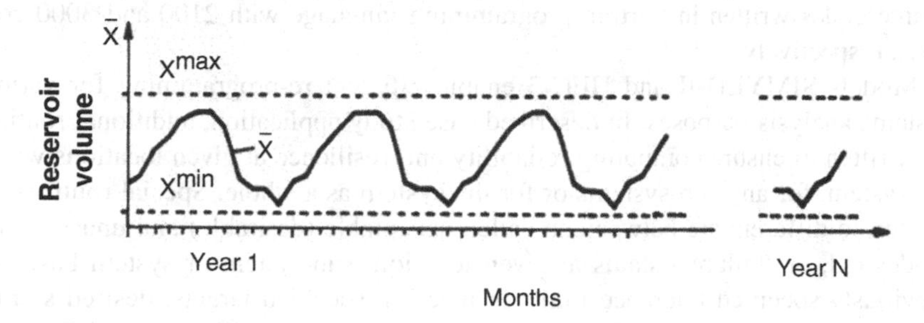

Fig. 24.2 Typical rule curve for surface reservoir

Diversion structures are aimed at supplying irrigation (diversion D_1), municipal (D_2) and combined municipal and industrial users (D_3). With properly defined priorities of demands and by setting operating rules for two reservoirs (with also defined priorities), a complete input data set is created for running two simulation models: (1) network model SIMYLD-II (IMP 1977; TWDB 1972) and (2) HEC-3 (HEC 1972). Worth to mention is that prescribed operating rules at reservoirs 1 and 2 were different because sizes and roles of reservoirs are also different. However, at any single reservoir rules are considered stationary, in a sense that in each year they are the same. In other words, rule for given reservoir does not change during multi-year period because 'deterministic assumption' should be violated, i.e. it is not reliable to predict reservoir inflows for more than a year ahead.

24.3.3 Using River Basin Simulation Models SIMYLD-II and HEC-3

The SIMYLD-II is a powerful generic river basin model for multiyear simulations (with monthly optimizations of water allocation) of complex water systems with surface reservoirs. It is predecessor of well known models MODSIM (Labadie 1986) and ACQUANET (LabSid 1996) developed in 1980s of the last century by following the same modeling philosophy introduced by Texas Water Development Board (TWDB 1971) and realized with SIMYLD-II together with various versions of SIM-4 and ΛL-III models among others.

The HEC-3 model is well known simulation model for water systems with reservoirs developed in parallel with the previous one in Hydrologic Engineering Center US Army Corps of Engineers. HEC-3 is based on balancing reservoirs operation in a way that discharges from reservoirs are made upon equalizing indexed levels within active storages. This model is also predecessor of well known family of several HEC-5 models with the same basic purpose but with some additions (for instance, computing flood damages). Both models SIMYLD-II and HEC-3 are open source codes written in Fortran programming language with 2100 and 3000 code lines, respectively.

Models SIMYLD-II and HEC-3 enable efficient re-programming for various systems analysis purposes. In described case study application, additional routines are written to ensure computing reliability and resilience at given locations within the system, for any sub-systems or for the system as a whole. Special routines are written to differentiate between so called acceptable (desirable) and unacceptable (undesired, or 'failure') status at given location, sub-system or system based on previously specified tolerance limits in meeting specified targets: desired storage levels at reservoirs and supplying demands at diversions. An illustration at Fig. 24.3 is given for a demand point D_2, where tolerant limit of 10% is defined for deficits that may occur at that point due to applied control (operating) policy for the system.

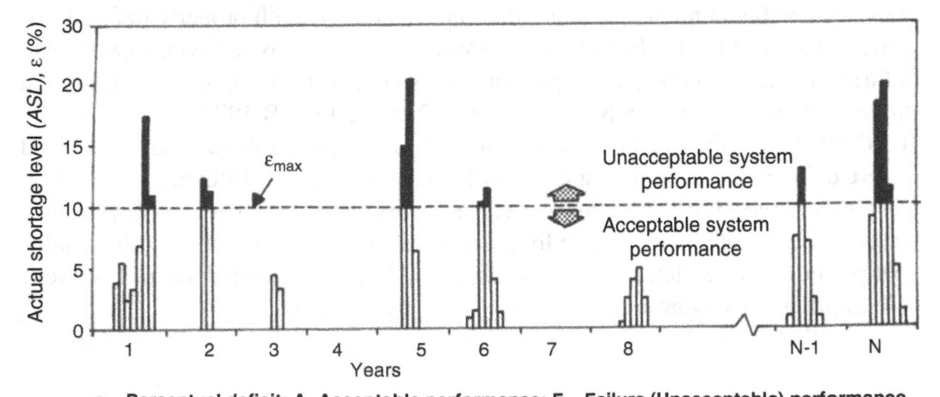

Fig. 24.3 Discriminating A/F system behavior for demand point D_2

Based on the criterion illustrated by discriminating dash line, computer in each model counts acceptable (*A*) and failure (*F*) states and 'recognizes' changes in the system performance at a given point from *A* (acceptable) to *F* (failure), count consecutive months of being in *A* or *F* state, and at the end computes various performance indicators (for points of interest), including reliability and resilience.

'State' variables can be defined in many ways, but they always may be considered as performance indicators, i.e. a consequence of applied control (operating) policy. In this case study example, the operating policy is represented by: (1) rule curves at reservoirs (for SIMYLD-II) and indexed zones within active storages of reservoirs (HEC-3), and (2) by overall priority scheme for water allocation which includes priorities among demand points (D_1, D_2, D_3) and reservoirs 1 and 2. Operating strategy for the system is that priority scheme determines whether and how much water in each month ($i = 1,\ldots,12$) and in each year ($j = 1,\ldots, N$) will be delivered at diversions or kept in the reservoirs while surpluses will be discharged for downstream users outside the system.

24.3.4 Reliability and Resilience

Reliability and resilience of the reservoir, sub-system, or system can be defined in different ways (see for instance Srdjevic and Srdjevic 2016a). Here we adopt definition based on works of Hashimoto et al. (Hashimoto 1980; Hashimoto et al. 1980): It is the indicator of how fast reservoir recovers from undesired (failure) into desired (satisfactory) status. For tolerant shortage specified in advance, reliability may be defined as the probability (or frequency) of satisfactory system performance for control point *k*:

$$\alpha(k) = \frac{\sum_{i=1}^{12}\sum_{j=1}^{N} Z_{i,j}(k)}{12N} \tag{24.1}$$

where Z is a discrete zero-one variable obtaining value 1 if system performance for point *k* was acceptable; otherwise, its value is 0 (Cf. this for months with black parts of histogram above dash line in Fig. 24.3).

A new zero-one variable, *W*, is used to identify changes of system performance at a given point from *A* (acceptable) to *F* (failure):

$$\begin{aligned} W_{i,j}(k) &= 1 \quad \text{if } x_{i,j}(k)\,\varepsilon A \textit{ and } x_{i+1,j}(k)\,\varepsilon F \\ W_{i,j}(k) &= 0 \quad \text{otherwise}^{*} \end{aligned} \tag{24.2}$$

(*The other cases are: $x_{i,j}(k)\,\varepsilon A$ and $x_{i+1,j}(k)\,\varepsilon A$; $x_{i,j}(k)\,\varepsilon F$ and $x_{i+1,j}(k)\,\varepsilon A$; $x_{i,j}(k)\,\varepsilon F$ and $x_{i+1,j}(k)\,\varepsilon F$)

For sufficiently long period of N years, an average of variable W is equal to the probability that a system's performance for point k was in acceptable status (A) in a given month and in failure status (F) in the next month, i.e.:

$$\rho(k) = \mathrm{P}\{x_{i,j}(k)\varepsilon A, x_{i+1,j}(k)\varepsilon F\} = \lim_{N\to\infty}\frac{1}{12N}\sum_{i=1}^{12}\sum_{j=1}^{N}W_{i,j}(k). \tag{24.3}$$

Resilience for point k is now:

$$\gamma(k) = \frac{\mathrm{P}\{x_{i,j}(k)\varepsilon A, x_{i+1,j}\varepsilon F\}}{\mathrm{P}\{x_{i,j}(k)\varepsilon \mathrm{F}\}} = \frac{\rho(k)}{1-\alpha(k)} = \frac{\sum_{i=1}^{12}\sum_{j=1}^{N}W_{i,j}(k)}{12N - \sum_{i=1}^{12}\sum_{j=1}^{N}Z_{i,j}(k)}. \tag{24.4}$$

Differences in resilience indicator of system performance at two different locations is illustrated in Fig. 24.4. Note that reliability and resilience are sometimes, i.e. not necessarily, highly correlated. From the diagram (b) at Fig. 24.4 it also appears that although the diversion D_2 is highly resilient, it might be very vulnerable because of frequent and high deficits close to 20%, twice higher than tolerant.

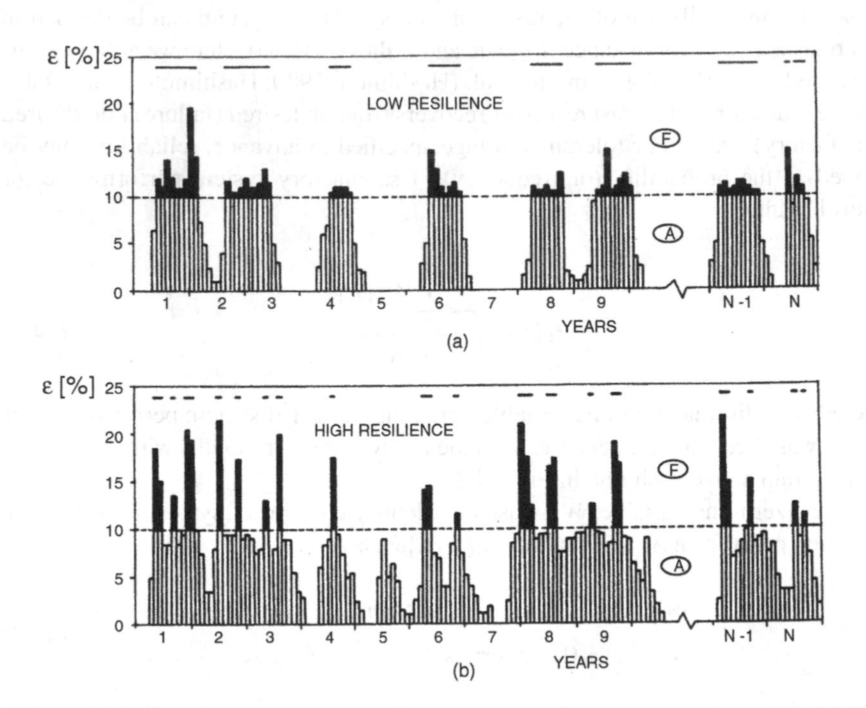

Fig. 24.4 Low and high resilience

Table 24.1 Performance of the system measured through the effects at the diversion D_2

	Calendar months											
Year	1	2	3	4	5	6	7	8	9	10	11	12
1	A	A	F	F	F	A	F	A	A	F	A	A
2	A	A	A	A	F	A	F	F	A	A	A	A
3	F	A	F	F	F	A	A	A	A	A	A	A
4	A	F	F	F	F	A	A	A	A	A	A	A
5	A	A	A	A	A	F	F	F	F	A	A	A
6	A	A	F	A	A	A	F	A	A	F	A	A
7	F	A	A	F	F	A	F	A	A	F	A	F
8	A	A	A	F	F	A	F	F	A	F	A	A
9	F	A	A	F	F	A	A	A	A	F	A	A
10	A	A	F	F	A	A	A	A	F	F	A	A
11	A	A	F	F	A	A	A	F	F	F	F	A
12	A	F	F	F	F	A	A	A	F	F	F	A
13	A	F	A	F	F	F	A	A	A	F	A	A
14	A	F	F	A	F	F	A	F	A	F	F	F
15	F	A	A	F	F	A	F	A	A	F	A	A
16	A	F	F	A	A	F	A	A	A	A	A	A
17	F	F	A	F	F	F	A	F	A	A	A	A
18	A	A	F	F	A	A	A	A	A	F	A	A
19	F	F	A	F	F	A	F	A	A	A	A	A
20	F	A	F	F	A	A	A	A	A	A	A	A

24.3.5 Measuring Resilience for Given Point in a System

The diversion D_2 is selected to illustrate our approach in measuring resilience by evidencing realized supplies at the diversion by comparisons on month-by-month basis with specified requirements at this point. For specified 10% tolerant deficit, the performance of a system for that diversion is determined as shown in Table 24.1. Out of 240 months (for 20 years period of simulation), the system performed in a desired way (less than 10% deficits at control point D_2) during 149 months which gives reliability indicator for this diversion of 0.62 (Table 24.2) (see Eq. (24.1)).

$$\sum_{i=1}^{240} Z_i = 149 \quad \sum_{i=1}^{240} W_i = 51$$

According to relations (24.1) and (24.4) the values of reliability and resilience at point D_2 are:

$$\alpha = \frac{\sum_{i=1}^{240} Z_i}{240} = \frac{149}{240} = 0.62 \qquad \gamma = \frac{\sum_{i=1}^{240} W_i}{240 - \sum_{i=1}^{240} Z_i} = \frac{51}{240 - 149} = 0.56$$

Table 24.2 Zero-one variables for computing resilience at diversion D_2

	Zero-one variables Z/W												Number of 1′	
Year/Month	1	2	3	4	5	6	7	8	9	10	11	12	Z	W
1	1/0	1/1	0/0	0/0	0/0	1/1	0/0	1/0	1/1	0/0	1/0	1/0	8	3
2	1/0	1/0	1/0	1/1	0/0	1/1	0/0	0/0	1/0	1/0	1/0	1/1	9	3
3	0/0	1/1	0/0	0/0	0/0	1/0	1/0	1/0	1/0	1/0	1/0	1/0	8	1
4	1/1	0/0	0/0	0/0	0/0	1/0	1/0	1/0	1/0	1/0	1/0	1/0	8	1
5	1/0	1/0	1/0	1/0	1/1	0/0	0/0	0/0	0/0	1/0	1/0	1/0	8	1
6	1/0	1/1	0/0	1/0	1/0	1/1	0/0	1/0	1/1	0/0	1/0	1/1	9	4
7	0/0	1/0	1/1	0/0	0/0	1/1	0/0	1/0	1/1	0/0	1/0	0/1	7	4
8	1/0	1/0	1/1	0/0	0/0	1/1	0/0	0/0	1/1	0/0	1/0	1/1	7	4
9	0/0	1/0	1/1	0/0	0/0	1/0	1/0	1/0	1/1	0/0	1/0	1/0	8	2
10	1/0	1/1	0/0	0/0	1/0	1/0	1/0	1/1	0/0	0/0	1/0	1/0	8	2
11	1/0	1/1	0/0	0/0	1/0	1/0	1/1	0/0	0/0	0/0	0/0	1/0	6	2
12	1/0	0/0	0/0	0/0	0/0	1/0	1/0	1/1	0/0	0/0	0/0	1/0	5	2
13	1/1	0/0	1/1	0/0	0/0	0/0	1/0	1/0	1/1	0/0	1/0	1/0	7	3
14	1/1	0/0	0/0	1/1	0/0	0/0	1/1	0/0	1/1	0/0	0/0	0/0	4	4
15	0/0	1/0	1/1	0/0	0/0	1/1	0/0	1/0	1/1	0/0	1/0	1/0	7	3
16	1/1	0/0	0/0	1/0	1/1	0/0	1/0	1/0	1/0	1/0	1/0	1/1	9	3
17	0/0	0/0	1/1	0/0	0/0	0/0	1/1	0/0	1/0	1/0	1/0	1/0	6	2
18	1/0	1/1	0/0	0/0	1/0	1/0	1/0	1/0	1/1	0/0	1/0	1/1	9	3
19	0/0	0/0	1/1	0/0	0/0	1/1	0/0	1/0	1/0	1/0	1/0	1/1	7	3
20	0/0	1/1	0/0	0/0	1/0	1/0	1/0	1/0	1/0	1/0	1/0	1/0	9	1
TOTAL													149	51

24.4 Operating Reservoirs to Obtain 'Good Resilience'

The resilience of 0.56 indicates that the overall system performance was not that satisfactory, and that operating strategy for reservoirs or allocation priority scheme could be modified accordingly. To assess different strategies on how to ensure higher resilience in water supply at diversion D_2, more simulations are performed with generated hydrologic input and with alternative rule curves/indexed zones as shown in Fig. 24.5. Simulation of system operation is performed with models SIMYLD-II and HEC-3 and after multiple variation of operational strategies at both reservoirs, eventually the final set of rules and zoning is obtained (Fig. 24.6) that ensure high resilience (0.82) at selected location D_2.

In described case study example, different rule curves are used in SIMYLD-II for dry, average and wet seasons identified as moving 24-month averages of total net inflow into the system, Fig. 24.6. Zoning adjustments used in HEC-3 are aligned with the priority schemes for water allocation (same as for the SIMYLD model!). Both models are used within similar 'running framework', with the same net inflows to the system, as generated as a multi-year stochastic process represented by the sets of sequences of local monthly inflows into the reservoirs minus monthly demands at related downstream diversions.

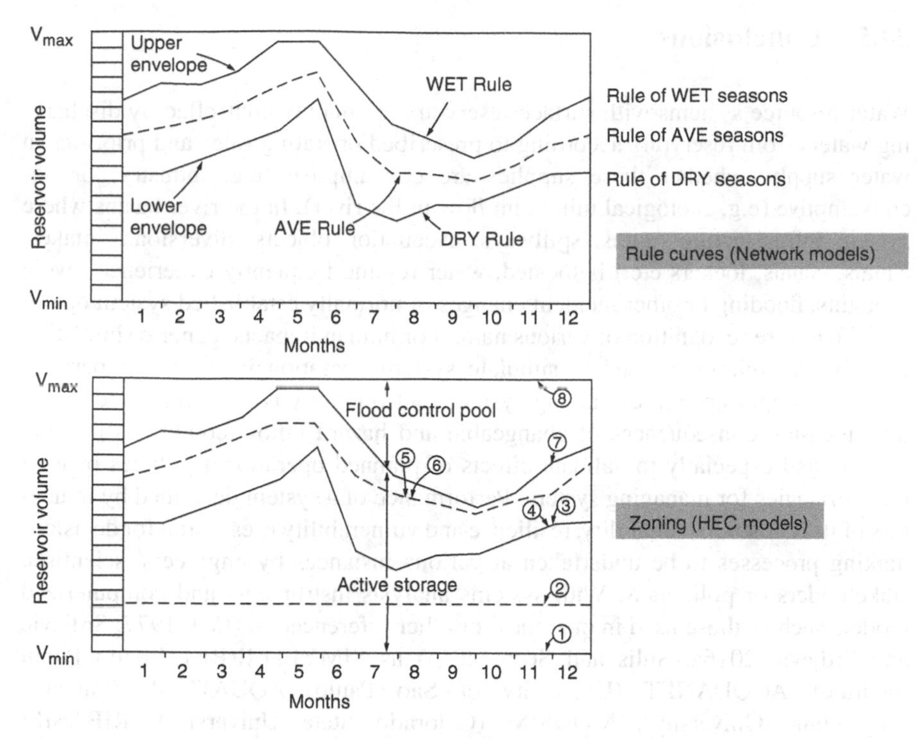

Fig. 24.5 Rule curves and simulated reservoir levels

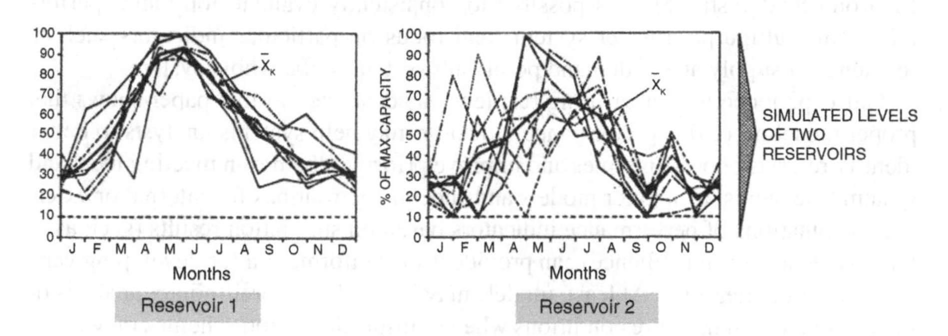

Fig. 24.6 Simulated levels and final rule curves for reservoirs ensuring the highest resilience at the diversion D_2

With representative hydrologic input to the system, manipulation of operating strategy led to determination of multiyear reliability and resilience at demand point D_2, but also at the other points in the system. Resilience at other locations varied between 0.40 (reservoir 2) and 0.75 (reservoir 1). Lowest resilience of reservoir 2 is obtained due to its forced deliveries for diversion D_2 immediately downstream, according to high priority given to this diversion.

24.5 Conclusions

Water resource systems with surface reservoirs are mainly controlled by discharging waters from reservoirs according to prescribed operating rules and priorities in water supply, whether these supplies are consumptive (e.g. industry) or no-consumptive (e.g. ecological minimum flow in the river). In the river basins where system infrastructure (dams, spillways, evacuation objects, diversions, intakes, canals, canals, lockers etc.) is located, water regime frequently experience severe droughts, flooding or other incident changes of normally established system operation. To ensure recognition of various natural or human impacts, generic simulation models are commonly used to simulate system operation in multiyear periods. Particular importance in evaluating system controllability is to identify and somehow measure consequences of changeable and hazard inflow conditions into the system, and especially to validate effects of planned operational policies or long term strategies for managing system. Performance of a system described by indicators of its robustness, reliability, resilience and vulnerability is essential for decision-making processes to be undertaken at various instances by engineers, scientists, stakeholders or politicians. With systems analysis instruments and computerized models such as those used in our study, or other referenced in (IMP 1977; Srdjevic and Srdjevic 2016a; Sulis and Sechi 2013) as SIMYLD-II(P) (Mihailo Pupin Institute), ACQUANET (University of Sao Paulo) AQUATOOL (Valencia Polytechnic University), MODSIM (Colorado State University), RIBASIM (DELTARES), WARGI-SIM (University of Cagliari) and WEAP (Stockholm Environmental Institute), it is possible to consistently evaluate long-term performance of multipurpose water systems and focus on particular indicators such as resilience of supply at any demand point, sub system or the whole system.

Our experience with measuring resilience as described in this paper shows that proper modeling of this performance indicator may help systems analysts to better identify reservoir operating rules and improve priority schemes in meeting local and system's demands. Computer models and specialized routines for internal or external computations of performance indicators based on simulation results (such as in this case indicator of resilience) can provide useful information for developing various mitigation measures. At least, models may be used for a preliminary analysis of system's potential in future conditions when multiple allocation schemes may occur as a request, or different sets of operating rules at reservoirs may be required to control the system with new users, changed types and volumes of demanded waters etc. As stated in (Sulis and Sechi 2013), system analysts at least must respect the fact that 'each model has its own characteristics and uses different approaches to define resources releases from reservoirs and allocation to demand centers' (Schaake 2002, p. 214).

Moreover, the understanding between different technical and social disciplines and interest groups involved in the resilience studies is a prerequisite for a minimum guarantee of success of the work carried out by analysts. Case study example briefly described in this paper, is one more indication of how much such understanding is important. The results obtained in the study of described system's performance are used later for making real-life decisions about the developments in water sector in the Morava river basin in Serbia.

Acknowledgements The authors would like to thank the Ministry of Education and Science of Serbia for funding this study. The study was accomplished as part of the project number 174003 (2011–2014): Theory and application of Analytic Hierarchy Process (AHP) in multi criteria decision making under conditions of risk and uncertainty (individual and group context). Grants received from the Secretariat of Science and Technological Development of Autonomous Province of Vojvodina are also acknowledged.

Further Suggested Readings

Elshorbagy A (2006) Multicriterion decision analysis approach to assess the utility of watershed modeling for management decisions. Water Resour Res 42:W09407. doi:10.1029/2005WR004264

Hashimoto T (1980) Robustness, reliability, resilience and vulnerability criteria for water resources planning. Ph.D. dissertation, Cornel University, Ithaca, NY

Hashimoto T, Stedinger JR, Loucks DP (1980) Reliability, resiliency and vulnerability criteria for water resource system performance evaluation. Water Resour Res 18(1):14–20

HEC (1972) HEC-3 – Reservoir system operation. Hydrologic Engineering Center, U.S. Army Corps of Engineers, Davis, USA

IMP (1977) SIMYLD-II(P) – Program za planiranje sistema akumulacija, Mihajlo Pupin Institute, Beograd (in Serbian)

Labadie JW (1986) MODSIM – river basin network model for water rights planning. Documentation and users Manual, Colorado State University, USA

LabSid (1996) AcquaNet – Modelo integrado para análise de sistemas complexos em recursos hídricos, Universidade de Sao Paulo, Brasil

Loucks DP (1997) Quantifying trends in system sustainability. Hydrol Sci J 42(4):513–530

Loucks DP, van Beek E (2005) Water resources systems planning and management. UNESCO, Paris

McMahon TA, Adeloye AJ, Sen-Lin Z (2006) Understanding performance measures of reservoirs. J Hydrol (Amsterdam) 324:359–382

Moy WS, Cohon JL, Revelle CS (1986) A programming model for analysis of reliability, resilience and vulnerability of a water supply reservoir. Water Resour Res 22(4):489–498

Rittima A, Vudhivanich V (2006) Reliability based multireservoir system operation for Mae Klong river basin. Kasetsart J (Nat Sci) 40:809–823

Sandoval-Solis S, McKinney DC, Loucks DP (2011) Sustainability index for water resources planning and management. J Water Resour Plan Manag 2011:381–390

Schaake JC (2002) Introduction in calibration of watershed models. Water Scientific Applications 6 (ed. Duan Q et al.: 1–7), AGU, Washington DC

Srdjević B, Obradović D (1991) Identification of the reservoir rule curves by use of network models. In: Tsakiris D (ed) Proceedings of the European conference on advances in water resources technology, Athens, Greece, Balkema, Roterdam, pp 483–493

Srdjevic B, Obradovic D (1995) Reliability-risk concept in evaluating control strategies for multireservoir water resources system. 7th IFAC's symposium on large scale systems: theory and applications, London, UK, pp 609–613

Srdjevic B, Srdjevic Z (2016a) Water resources systems analysis with extensions in water management. Faculty of Agriculture Novi Sad, p 321 (in Serbian)

Srdjevic B, Srdjevic Z (2016b) Multicriteria analysis of the water resources system performance. J Water Resour Vodoprivreda, In Press (in Serbian)

Srdjevic B, Medeiros YDP, Faria AS (2004) An objective multicriteria evaluation of water management scenarios. Water Resour Manag 18:35–54

Sulis A, Sechi GM (2013) Comparison of generic simulation models for water resource systems. Environ Model Softw 40:214–225

Tsheko R (2003) Rainfall reliability, drought and flood vulnerability in Botswana. Water SA 29(4):389–392

TWDB (1971) Stochastic optimization and simulation techniques for management of regional water resources systems. Texas Water Development Board, Austin

TWDB (1972) SIMYLD-II – river basin simulation model. Texas Water Development Board, Austin

Printed in the United States
By Bookmasters